SYMBOLS USED IN THIS BOOK

Symbol	Meaning	Page
=	read *is equal to*	4
≠	read *is not equal to*	4
≐	read *is approximately equal to*	46
<	read *is less than*	3
≮	read *is not less than*	4
>	read *is greater than*	3
≯	read *is not greater than*	4
$\sqrt{}$	read *square root*	86
$\sqrt[3]{}$	read *cube root*	87
( )	Parentheses	17
	→ exponent	
$3^4 = 81$	→ 4th power of 3	42
	→ base	
%	read *percent*	228
$3 \times 5 = 3(5) = (3)5$ $= (3)(5) = 3 \cdot 5$	The product of 3 and 5	32
π	pi (≐ 3.14)	315
∧	Caret	181
{ }	Braces. Used to show a set	412
{1, 2, 3, ... }	The set of natural (counting) numbers	2
{0, 1, 2, ... }	The set of whole numbers	3
{0, 1, 2, 3, 4, 5, 6, 7, 8, 9}	The set of digits	4
$4 \in N$	4 is an element of set N	412
$9 \notin A$	9 is not an element of set A	412
$n(A)$	The cardinal number of set A	413
{ } = ∅ = φ	The empty set	413
U	The universal set	414
$A \subseteq B$	Set A is a subset of set B	416
$P \subset Q$	Set P is a proper subset of set Q	416
$R \not\subseteq T$	Set R is not a subset of set T	416
$A \cup B$	The union of set A and set B	418
$A \cap B$	The intersection of set A and set B	419

COMMON ENGLISH MEASURES

```
      1 foot (ft) = 12 inches (in)
      1 yard (yd) = 3 feet (ft)
      1 mile (mi) = 5,280 feet (ft)
   1 minute (min) = 60 seconds (sec)
      1 hour (hr) = 60 minutes (min)
       1 day (da) = 24 hours (hr)
      1 week (wk) = 7 days (da)
      1 year (yr) = 52 weeks (wk) = 365 days (da)
    1 quart (qt) = 2 pints (pt)
  1 gallon (gal) = 4 quarts (qt) = 231 cubic
                       inches (cu in)
    1 pound (lb) = 16 ounces (oz)
           1 ton = 2,000 pounds (lb)
```

COMMON HOUSEHOLD MEASURES

1 tablespoon (tbsp) = 3 teaspoons (tsp)

1 pint (pt) = 2 cups = 16 ounces (oz)

1 cup = 8 ounces (oz) = 16 tablespoons (tbsp)

COMMON METRIC MEASURES

Basic Units

Meter measures length.

Liter measures volume.

Gram measures weight.

Prefixes

Kilo means 1,000.

Milli means $\frac{1}{1,000}$.

Centi means $\frac{1}{100}$.

1 kilometer (km) = 1,000 meters (m)

1 meter (m) = 100 centimeters (cm)

1 meter (m) = 1,000 millimeters (mm)

1 centimeter (cm) = 10 millimeters (mm)

1 kilogram (kg) = 1,000 grams (g)

1 gram (g) = 1,000 milligrams (mg)

1 liter (ℓ) = 1,000 milliliters (ml)

1 cubic centimeter (cc) = 1 milliliter (ml)

1 liter (ℓ) = 1,000 cubic centimeters (cc)

1 cubic centimeter (cc) of water weighs 1 gram (g) at 4°C

USEFUL ENGLISH—METRIC CONVERSIONS

1 inch (in) = 2.54 centimeters (cm)

39.4 inches (in) = 1 meter (m)

0.621 miles (mi) = 1 kilometer (km)

1 mile (mi) = 1.61 kilometers (km)

1 pound (lb) = 454 grams (g)

2.20 pounds (lb) = 1 kilogram (kg)

1.06 quarts (qt) = 1 liter (ℓ)

2.47 acres = 1 hectare (ha)

61.0 cubic inches (cu in) = 1 liter (ℓ)

BASIC METRIC UNITS

Basic Unit	Quantity Measured
meter (m)	length
liter (ℓ)	volume (capacity)
gram (g)	weight
Centigrade degree (°C)	temperature
hectare (ha)	area
second (s)	time
radian (rad)	angle

METRIC PREFIXES

Prefix		Multiply By
tera (T)	$= 10^{12}$	$= 1,000,000,000,000$
giga (G)	$= 10^{9}$	$= 1,000,000,000$
mega (M)	$= 10^{6}$	$= 1,000,000$
kilo (k)	$= 10^{3}$	$= 1,000$
hecto (h)	$= 10^{2}$	$= 100$
deka (da)	$= 10^{1}$	$= 10$
Basic unit	$= 10^{0}$	$= 1$
deci (d)	$= 10^{-1}$	$= .1$
centi (c)	$= 10^{-2}$	$= .01$
milli (m)	$= 10^{-3}$	$= .001$
micro (μ)	$= 10^{-6}$	$= .000\ 001$
nano (n)	$= 10^{-9}$	$= .000\ 000\ 001$
pico (p)	$= 10^{-12}$	$= .000\ 000\ 000\ 001$

UNCOMMON MEASURES

1 nautical mile = 6,080 feet = 10 cables

1 fathom (f or fm) = 6 feet

1 cable = 100 fathoms

1 rod (rd) = $16\frac{1}{2}$ feet

1 mile = 320 rods

1 acre = 43,560 square feet
= 160 square rods

1 chain (ch) = 66 feet

1 furlong (fur) = $\frac{1}{8}$ mile = 40 rods = 660 feet

1 stone = 14 pounds

1 ounce = 28.35 grams

1 carat (car) = 200 milligrams

1 ounce = 16 drams (dr)

1 peck (pk) = 2 gallons

1 bushel (bu) = 8 gallons = 4 pecks

ESSENTIAL ARITHMETIC

SECOND EDITION

C. L. Johnston

Alden T. Willis
East Los Angeles College

Wadsworth Publishing Company, Inc.
Belmont, California

MATHEMATICS EDITOR: Richard Jones

DESIGNER: Nancy Benedict

BOOK PRODUCTION: Greg Hubit Bookworks

TYPING AND LINE DRAWINGS: Susan Rogin

Printed in the United States of America

2 3 4 5 6 7 8 9 10——81 80 79 78 77

Library of Congress Cataloging in Publication Data

Johnston, Carol Lee.
    Essential arithmetic.

    Includes index.
    1. Arithmetic--1961-   I. Willis, Alden T., joint
author. II. Title.
QA107.J64 1977   513'.142   76-51248
ISBN 0-534-00513-6

Preface

This second edition reflects many helpful comments from users of the first edition as well as our own classroom experience in teaching from the book. Following are the major changes in the second edition.

1. A second set of exercises (Set II) has been added to each exercise set in the first edition. These were added because many instructors wanted some exercises that had no answers in the text. The answers for these additional exercises are in the instructor's manual. The level of difficulty of this second set of exercises is the same as that of the original set in every case. Each first set of exercises (Set I) is essentially the same as the corresponding exercise set in the original edition.
2. The metric system, which was only a part of a chapter in the original edition, has been expanded to a full chapter in this second edition.
3. A section on the accuracy of calculations has been added.
4. The unit fraction method of changing units has been used throughout Chapters 7 and 8 on the English and metric systems of measurement. This method eliminates the confusion students often have in deciding whether to multiply or divide when changing units.
5. There are many changes throughout the text that have been made to clarify and simplify concepts, explanations and examples in order to make the book easier for students to read and understand. Examples of such changes are the sections on rounding off numbers.

Chapters 1 through 8 of this book include topics in arithmetic "essential" to daily living and constitute a minimum course in arithmetic. Chapters 11 and 13 include additional topics helpful for those students continuing on to algebra. Chapters 7, 8, 9, 11, and 12 contain additional materials helpful for students going on to courses in science. Chapter 10 gives the student a better understanding and appreciation of our own number system.

Some of the major features of this book are:

1. The contents are arranged in small sections, each with its own examples and exercises.
2. After each topic explanation, many completely worked examples are given before the exercises.
3. There are over 4,000 exercises in this book.

   Set I Exercises: The complete solutions for all odd-numbered Set I exercises are included in the back of this text together with the answers for all even-numbered Set I exercises. In most cases the even-numbered exercises provide practice on problems analogous to the odd-numbered exercises. The student can use the solutions for the odd-numbered Set I exercises as a study aid in doing the even Set I exercises as well as all Set II exercises.

Set II Exercises: The level of difficulty of Set II exercises is the same as that of Set I. Answers to all Set II exercises are included in the instructor's manual. No answers for Set II exercises are given in the text.

4. There is a diagnostic test following each chapter. Complete solutions to all problems on these diagnostic tests, together with section references, appear in the answer section.

5. An instructor's manual contains four different tests for each chapter that may be easily removed and duplicated for class use. These tests are prepared with adequate space for students to work the problems. Answer keys for these tests are provided in the manual.

6. A comprehensive treatment of the metric system is included.

7. For quick reference, English and metric tables, together with conversions, and a list of symbols appear in the front of the book.

8. The percent proportion and aids in identifying the numbers in percent problems eliminate many of the difficulties students have with percent problems.

9. Throughout the book interesting information and everyday problems are included in the exercises.

This textbook can be used in three types of instructional programs:

1. The conventional lecture course. This book is particularly easy to fit into a program of regular assignments because it is divided into many small, self-contained units. Examinations that can be given for each chapter are provided in the instructor's manual. Following each chapter in the textbook is a diagnostic test which students may use for review and diagnostic purposes. This textbook has been class-tested by the authors and many other instructors in the conventional lecture course program.

2. The learning laboratory class. Because of the format of explanation, example, and exercise carried on in each section of the book, together with the diagnostic tests and solutions for each chapter, we have found that a wide degree of latitude in the pace at which a student progresses is possible when using this book in the learning laboratory class.

3. Self-study. This textbook lends itself to self-study because (a) each new topic is short enough to be mastered before continuing, and (b) almost 800 examples are solved and over 1,400 complete solutions are given for the odd-numbered Set I exercises, together with all the answers for the even-numbered Set I exercises. In using this book for self-study, a student could begin by taking the diagnostic test for Chapter 1. When correcting this test himself, he would be directed to specific sections of the book that explain the particular problems he did incorrectly. The student can continue in this manner at his own pace through the book.

We wish to thank our many friends for their valuable suggestions. In particular, we are deeply grateful to Merrill F. Hale, Roosevelt High School, Los Angeles; Heschel Shapiro, Los Angeles Trade Technical College; Edward Rosenberg and Merwin L. Waite, East Los Angeles College; Robert T. Stephens, Moorpark College, California; E. B. Hoff, Glendale Community College, Arizona; and James N. Tipton, Mesa Community College, Arizona, for their thorough reading and useful criticism of the original manuscript. Of special help with this second edition have been Bryn Gary and Clifton Gary, Oscar Rose College, Midwest City, Oklahoma; George L. Holloway, Los Angeles Valley College, California; Samuel Waltmire, Goddard Junior High School, Glendora, California; Jeannie Lazaris and Tom Drouet, East Los Angeles College, California; Susan A. Barker, Atlantic Community College, Mays Landing, New Jersey; and Bill Orr, San Bernardino Valley College, California.

This book is dedicated to our students, who inspired us to do our best to produce a book worthy of their time.

Contents

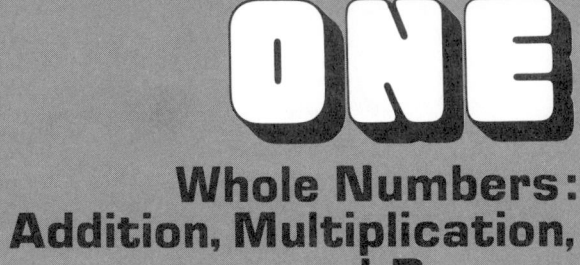

ONE

Whole Numbers: Addition, Multiplication, and Powers

101 Basic Definitions

In this section we introduce some of the names and definitions of numbers and number relations.

<u>NATURAL NUMBERS</u>. The set of numbers

> 1, 2, 3, 4, 5, 6, 7, 8, 9, 10, 11, 12, and so on,

are called the *natural numbers* (or *counting numbers*). These were probably the first numbers invented by man and used to count his possessions, such as sheep, goats, etc.

<u>NUMERALS</u>. Symbols representing numbers are called *numerals*.

<u>Example 1</u>. Examples of symbols used as numerals:

    (a) 1, 5, 30

    (b) four, seven, ten

    (c) I, VI, X

A number is an idea or thought. It is something in our minds. A *numeral* is the symbol we write on paper to show what number we have in mind. From this point on we will not distinguish between a number and its symbol.

<u>NUMBER LINE</u>. Natural numbers can be represented by numbered points equally spaced along a straight line (Figure 101A). Such a line is called a *number line*.

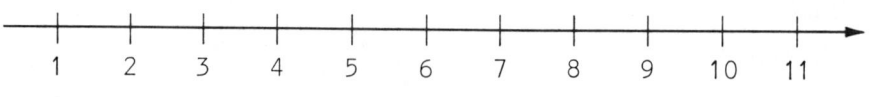

Figure 101A

The arrowhead shows the direction in which numbers get larger. Numbers that follow one another (without interruption) are called *consecutive numbers*. Later we will discuss other kinds of numbers, such as fractions, which can also be placed on the number line.

The smallest natural number is 1. The largest natural number can never be found because no matter how far we count there are always larger natural numbers. Since it is impossible to write all of the natural numbers, it is customary to represent them as follows:

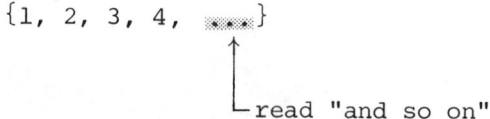

The three dots to the right of the number 4 indicate that the remaining numbers are found by counting in the same way we have begun: Namely, to add 1 to the preceding number to find the next number. We call the set of natural numbers N. So that

$$N = \{1, 2, 3, 4, \ldots\}$$

WHOLE NUMBERS. When 0 is included with the natural numbers, we have the set of numbers known as *whole numbers* (Figure 101B).

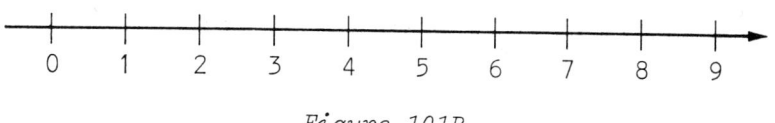

Figure 101B

We call the set of whole numbers W. So that

$$W = \{0, 1, 2, 3, \ldots\}$$

INEQUALITY SYMBOLS. The symbols $<$ and $>$ are called *inequality symbols*. Let X be any number on the number line. Then numbers to the right of X on the number line are said to be *greater than* X, written "$> X$." Numbers to the left of X on the number line are said to be *less than* X, written "$< X$."

Example 2

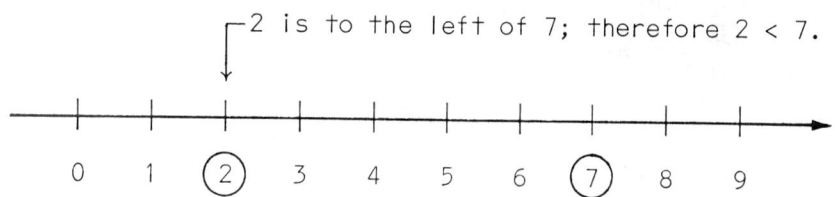

Example 3

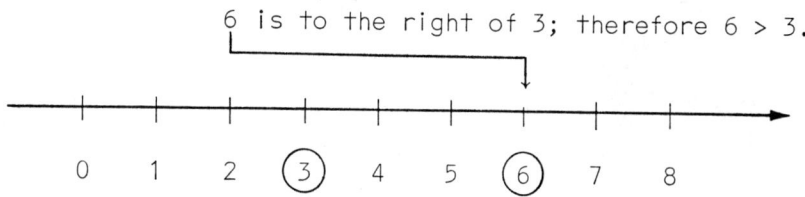

An easy way to remember the meaning of the symbol is to notice that the wide part of the symbol is next to the larger number.

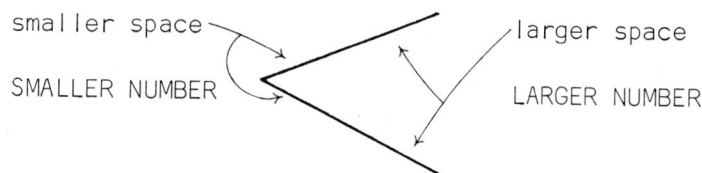

Some people like to think of the symbols $>$ and $<$ as arrowheads that point toward the smaller number.

Example 4

    (a) $7 > 6$ is read "7 is greater than 6."

    (b) $7 > 1$ is read "7 is greater than 1."

(c) 5 < 10 is read "5 is less than 10."

(d) 3 < 4 < 5 is read "3 is less than 4 and 4 is less than 5."

Note that 7 > 6 and 6 < 7 give the same information even though they are read differently.

Another inequality symbol is ≠. A slash line drawn through a symbol puts a "not" in the meaning of the symbol.

Example 5. Examples showing the use of the slash line:

(a) = is read "is equal to";
    ≠ is read "is *not* equal to."

(b) < is read "is less than";
    ≮ is read "is *not* less than."

(c) > is read "is greater than";
    ≯ is read "is *not* greater than."

(d) 4 ≠ 5 is read "4 is not equal to 5."

(e) 3 ≮ 2 is read "3 is not less than 2."

(f) 5 ≯ 6 is read "5 is not greater than 6."

DIGITS. In our number system a digit is any one of the first ten whole numbers {0, 1, 2, 3, 4, 5, 6, 7, 8, 9}. They are shown on the number line in Figure 101B.

Any number can be written by using some or all of these ten digits. For this reason, the digits are sometimes called the building blocks of our number system.

The word digit comes from the Latin word for finger: *digitus*.

Numbers are often referred to as *one-digit* numbers, *two-digit* numbers, *three-digit* numbers, and so on.

Example 6

(a) 35 is a two-digit number.

(b) 7 is a one-digit number.

(c) 275 is a three-digit number.

(d) 100 is a three-digit number.

(e) The first digit of 785 is 7.

(f) The second digit of 785 is 8.

(g) The third digit of 785 is 5.

We show the natural numbers, whole numbers, and digits on the number line in Figure 101C.

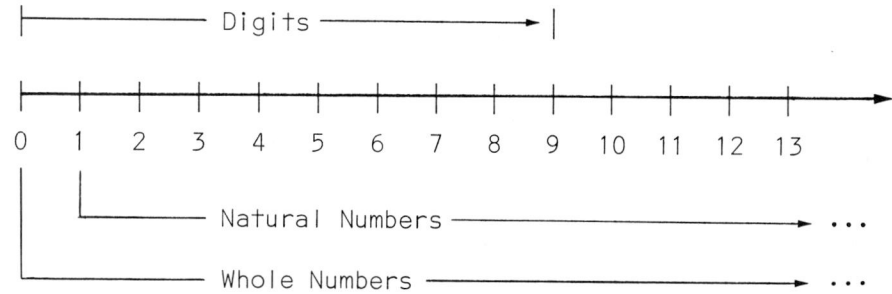

Figure 101C

EXERCISES 101, SET I.

1. What is the second digit of the number 159?
2. What is the smallest natural number?
3. What is the smallest digit?
4. What is the smallest whole number?
5. What is the fourth digit of the number 1,975?
6. What is the smallest two-digit natural number?
7. What is the smallest three-digit whole number?
8. What is the largest one-digit number?
9. What is the largest two-digit number?
10. Write two different symbols that tell the number of days in a week.
11. Is 12 a digit?
12. Is 12 a natural number?
13. Is 12 a whole number?
14. What is the largest natural number?
15. What is the largest digit?
16. Write all the whole numbers < 4.
17. Write all the digits > 5.
18. Write the consecutive natural numbers > 14 and < 17.
19. Write in consecutive order all the digits that are < 5.

In the remaining exercises, determine which of the two symbols > or < should be used to make each statement true.

20. 8 _?_ 7.   21. 0 _?_ 1.
22. 1 _?_ 0.   23. 5 _?_ 8.
24. The weight of a man is _?_ the weight of a baby.
25. The distance to the moon is _?_ the distance to a star.

EXERCISES 101, SET II.

1. What is the second digit in the number 467?
2. What is the largest digit?
3. What is the fourth digit of 3,852?
4. What is the smallest two-digit whole number?
5. What is the largest three-digit number?
6. Is 10 a digit?
7. Is 10 a whole number?
8. What is the largest whole number?
9. Write all the digits greater than 7.
10. Write in consecutive order all the digits that are less than 4.

In the remaining exercises, determine which of the two symbols > or < should be used to make each statement true.

11. 2 _?_ 0.   12. 4 _?_ 6.
13. 0 _?_ 3.
14. The distance from New York City to Alaska is _?_ the distance from New York City to Boston.

Reading and Writing Whole Numbers

We assume that you know how to read one-, two-, and three-digit numbers.

Example 1

    (a) 7 is read "seven."

    (b) 35 is read "thirty-five."

    (c) 256 is read "two hundred fifty-six."

PLACE-VALUE. Our number system is a *place-value* system. This means that the value of each digit in a written number is determined by its place in the number.

Example 2. In this example we show a "1" in three different places.

    1
    └── this means 1 unit

    10
    └── this means 1 ten = 10 units

    100
    └── this means 1 hundred = 10 tens = 100 units

Example 3

    5
    └── this means 5 units

    50
    └── this means 5 tens = 50 units

    500
    └── this means 5 hundreds = 50 tens = 500 units

Notice in the above examples that when a digit is moved one place to the left it becomes 10 times larger.

Example 4

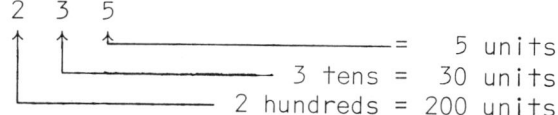

```
2   3   5
↑   ↑   ↑_____ =   5 units
↑   ↑_____ 3 tens =  30 units
↑_____ 2 hundreds = 200 units
```

A more complete discussion of the idea of place-value and our number system is given in Sections 402, 1001, and 1201.

EXERCISES 102A, SET I

1. In the number 576
   (a) The 6 represents how many units?
   (b) The 7 represents how many units?
   (c) The 5 represents how many units?
2. In the number 904
   (a) The 4 represents how many units?
   (b) The 0 represents how many units?
   (c) The 9 represents how many units?
3. In the number 348
   (a) The 4 represents how many tens?
   (b) The 4 represents how many units?
   (c) The 3 represents how many hundreds?
   (d) The 3 represents how many tens?
4. In the number 862
   (a) The 6 represents how many tens?
   (b) The 6 represents how many units?
   (c) The 8 represents how many hundreds?
   (d) The 8 represents how many tens?

EXERCISES 102A, SET II

1. In the number 237
   (a) The 7 represents how many units?
   (b) The 3 represents how many units?
   (c) The 2 represents how many units?
2. In the number 456
   (a) The 5 represents how many tens?
   (b) The 5 represents how many units?
   (c) The 4 represents how many hundreds?
   (d) The 4 represents how many tens?

SEPARATING THE DIGITS INTO GROUPS. Commas are used to separate large numbers into smaller groups of three digits.* In placing the commas, we count from the right as shown in Figure 102A. Note that the first group on the left may have one, two, or three digits. Each group of three has a name: the first group on the right is the units group; the second group from the right is the thousands group; the third group from the right is the millions group; and so on (Figure 102A).

---

*Not all countries use this method of grouping into threes. The English practice, for example, is to mark off into groups of six digits. In England a billion is equal to a million million, whereas in America a billion is equal to a thousand million.

Trillions			Billions			Millions			Thousands			Units (Ones)		
Hundreds	Tens	Ones	Hundreds	Tens	Ones	Hundreds	Tens	Ones	Hundreds	Tens	Ones	Hundreds	Tens	Ones
1	5	, 2	8	6	, 3	4	6	, 7	8	5	, 1	0	5	

| trillion | billion | million | thousand | (units) (ones) |

Figure 102A

To read a number, we read the number formed by the digits in the left group, then say the name of that group. Then read the number formed by the digits in the next group and say the name of that group. We continue in this way until all groups in the number have been read. The name of the units group is usually omitted when reading the number. Study Figure 102A.

In writing numbers in words, commas are used in the same places they are used when the number is written in digits. For example, we read in words the number shown in Figure 102A as follows: Fifteen trillion, two hundred eighty-six billion, three hundred forty-six million, seven hundred eighty-five thousand, one hundred five. The following examples demonstrate the reading and writing of numbers:

Example 1

    (a) 25 is read "twenty-five."

    (b) 237 is read "two hundred thirty-seven."

    (c) 237,000 is read "two hundred thirty-seven thousand."

Notice in examples (d), (e), and (f) that when a group is made up of all zeros it is not read.

    (d) 237,000,000 is read "two hundred thirty-seven million."

    (e) 27,000,005 is read "twenty-seven million, five."

    (f) 2,000,015,145 is read "two billion, fifteen thousand, one hundred forty-five."

Notice hyphens are used in writing two-digit numbers such as 25 (twenty-five), 37 (thirty-seven), 45 (forty-five), etc.

The word "and" is not used in writing whole numbers.

NUMBERS LARGER THAN BILLIONS. Numbers larger than billions have names, but these names are seldom used. For those students who want to know the names of larger numbers, we include Figure 102B. You will not be expected to know the names of numbers larger than trillions.

12 , 315 , 218 , 557 , 314 , 708 , 515 , 900 , 000 , 304 , 017 , 708

Decillion Nonillion Octillion Septillion Sextillion Quintillion Quadrillion Trillion Billion Million Thousand Units

Figure 102B

Large numbers such as billions and trillions do not have much meaning for most of us. For example, one billion miles is farther than 40,000 times the distance around the earth. We often see large numbers written in newspapers and magazines. For example, the Gross National Product of the United States is over one trillion dollars. The distance light travels in one year is almost six trillion miles.

EXERCISES 102B, SET I. In Exercises 1-11, write in words each of the numbers in the statements.

1. The diameter of the earth at the equator is 7,926 miles.
2. The age of an Olduvai Gorge "human" skull is estimated to be 1,890,000 years.
3. The circumference of the earth at the equator is 24,902 miles.
4. The total number of votes cast for President of the United States in 1972 was 76,116,204.
5. The populations of some of the largest countries of the world are:
   (a) China: 750,000,000
   (b) India: 547,000,000
   (c) Russia: 241,748,000
   (d) U.S.A.: 204,765,770
6. The U.S. Gross National Product for 1972 was $1,155,155,000,000.
7. The total area of the earth is 196,940,000 square miles, of which 57,506,000 square miles is land and 139,434,000 square miles is water.
8. In 1970 there were 2,050,000 people injured in automobile accidents in the United States. During that same period, 54,800 people died as a result of automobile accidents.
9. One light-year is the distance light travels in one year. This is about 5,879,200,000,000 miles.
10. The number of registered nurses in the United States in 1971 was 723,000.
11. Sirius, the brightest star in our winter sky, is 50,561,000,000,000 miles from the earth.
12. Mark off with commas, then write out in words the following natural numbers:
    (a) 710405
    (b) 655430186
    (c) 700005009
    (d) 1002003004005
    (e) 10020030040050
    (f) 100200300400500

13. Use digits to write each of the following numbers:
    (a) Eight million, eight thousand, eight hundred eight
    (b) Seven million, seven
    (c) Ten trillion, ten thousand, ten
    (d) One hundred seven billion, thirty-five million, seventy-five

14. Arrange the following numbers in order of size in a vertical column with the smallest number at the top:

    10,004;   34,186,075;   300,156;   75;   1,005

15. Use digits to write each of the following numbers and arrange them in a vertical column in order of size with the smallest number at the top: Seven hundred twenty-one million, forty-nine thousand, eight; fifty-six; five trillion, two hundred thirty-five million, seven hundred ninety-six; three thousand, eighty.

EXERCISES 102B, SET II. In Exercises 1-6, write in words each of the numbers in the statements.

1. The area of Lake Michigan (U.S.) is 22,400 square miles.
2. The height of Angel Falls, Venezuela, is 3,212 feet.
3. The area of some of the largest continents of the earth are:
   (a) Asia: 16,988,000 square miles
   (b) North America: 9,390,000 square miles
   (c) Antarctica: 5,500,000 square miles
4. The population of Asia in 1975 was estimated to be 2,316,312,000.
5. Alpha Centauri, our nearest star, is 25,276,000,000,000 miles from the earth.
6. Mark off with commas, then write out in words the following natural numbers:
   (a) 56714
   (b) 7006345
   (c) 5004003002106
7. Arrange the following numbers in a vertical column with the smallest number at the top:

    47,890;   45,284,035;   200,184;   41;   2,846

8. Use digits to write each of the following numbers and arrange them in a vertical column in order of size with the smallest number at the top: Five hundred forty-seven million, twenty-five thousand, nine; thirty-seven; eight trillion, three hundred eighty-eight million, two hundred fifty-five; six thousand, seventy.

Addition of Whole Numbers

In Section 101 we introduced the whole numbers

$$W = \{0, 1, 2, 3, 4, \ldots\}$$

When whole numbers are added, the numbers being added are called the *addends*, and the answer is called the *sum*.

Example 1

$$
\begin{array}{rl}
2 & \textit{addend} \\
+\ 3 & \textit{addend} \\
\hline
5 & \textit{sum}
\end{array}
$$

Addition is actually repeated counting. For example, suppose you bought a $2 tie and a $3 shirt. You could count out two $1 bills for the tie, then three more $1 bills for the shirt, making a total of $5.

$$\$2 + \$3 = \$5$$

Or you could have counted out three $1 bills for the shirt and then two more $1 bills for the tie, making a total of $5.

$$\$3 + \$2 = \$5$$

That is,

$$\$2 + \$3 = \$3 + \$2 = \$5$$

Example 2

(a) $4 + 5 = 5 + 4 = 9$

(b) $1 + 3 = 3 + 1 = 4$

(c) $10 + 2 = 2 + 10 = 12$

These examples suggest that reversing the order of two whole numbers in an addition problem does not change their sum. This important property of whole numbers is called the *Commutative Property of Addition*.

Commutative Property of Addition

If a and b represent any whole numbers, then

$$a + b = b + a$$

ADDITION ON THE NUMBER LINE. To add 2 and 3 on the number line, start at the 0 mark and move two units to the right. Then from this point move three more units to the right. This makes a total movement of five units to the right. Therefore, $2 + 3 = 5$. See Figure 103A.

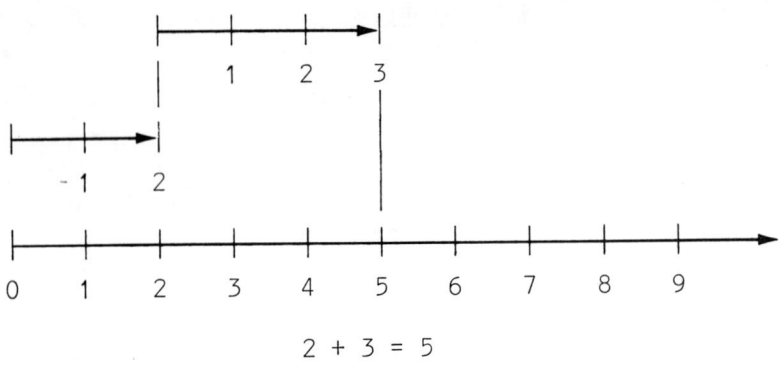

2 + 3 = 5

Figure 103A

In Figure 103B we use the number line to show that
2 + 3 = 3 + 2.

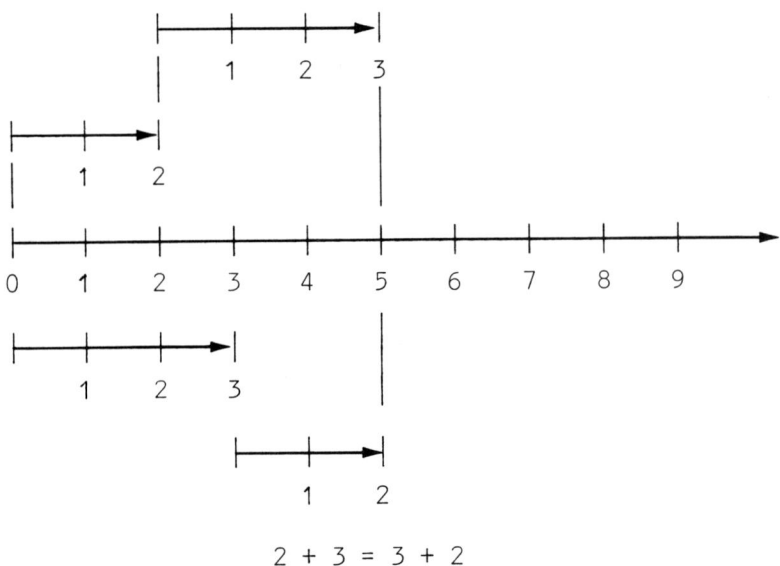

2 + 3 = 3 + 2

Figure 103B

Table of Basic Addition Facts

+	0	1	2	3	4	5	6	7	⑧	9
0	0	1	2	3	4	5	6	7	8	9
1	1	2	3	4	5	6	7	8	9	10
2	2	3	4	5	6	7	8	9	10	11
3	3	4	5	6	7	8	9	10	11	12
4	4	5	6	7	8	9	10	11	12	13
5	5	6	7	8	9	10	11	12	13	14
6	6	7	8	9	10	11	12	13	14	15
⑦	7	8	9	10	11	12	13	14	⑮	16
8	8	9	10	11	12	13	14	15	16	17
9	9	10	11	12	13	14	15	16	17	18

To show how to use the Table of Addition Facts, we give an example. To find the sum 7 + 8 = 15, move right from 7 of the left column until you are directly below 8 of the top row. See the circled numbers in the table.

EXERCISES 103A, SET I. Find the following sums.

1. 6 + 7	2. 8 + 6	3. 9 + 5
4. 6 + 9	5. 9 + 8	6. 7 + 9
7. 8 + 7	8. 8 + 5	9. 9 + 4
10. 7 + 5	11. 9 + 7	12. 8 + 9

EXERCISES 103A, SET II. Find the following sums.

1. 5 + 8	2. 8 + 6	3. 7 + 9
4. 8 + 9	5. 9 + 6	6. 8 + 7

Adding zero to a number gives the identical number for the sum. For this reason, zero is called the *additive identity*. This property of zero can easily be verified by referring to the Table of Basic Addition Facts.

BASIC ADDITION COMBINATIONS. The following addition combinations include every possible pair of digits that can be added together. If a student practices these daily, his or her addition skills will improve. A convenient way of practicing is to place a paper over the answers, then write or say the answers as quickly as possible. For variety in your practice, work from left to right across the page, then from right to left. Also add down one time, then up the next time.

DRILL IN ADDING ONE-DIGIT NUMBERS

1. $\begin{array}{r}1\\0\\\hline1\end{array}$ $\begin{array}{r}0\\2\\\hline2\end{array}$ $\begin{array}{r}0\\4\\\hline4\end{array}$ $\begin{array}{r}3\\0\\\hline3\end{array}$ $\begin{array}{r}5\\0\\\hline5\end{array}$ $\begin{array}{r}7\\0\\\hline7\end{array}$ $\begin{array}{r}0\\6\\\hline6\end{array}$ $\begin{array}{r}9\\0\\\hline9\end{array}$ $\begin{array}{r}0\\8\\\hline8\end{array}$

2. $\begin{array}{r}1\\1\\\hline2\end{array}$ $\begin{array}{r}1\\3\\\hline4\end{array}$ $\begin{array}{r}2\\1\\\hline3\end{array}$ $\begin{array}{r}4\\1\\\hline5\end{array}$ $\begin{array}{r}6\\1\\\hline7\end{array}$ $\begin{array}{r}1\\5\\\hline6\end{array}$ $\begin{array}{r}1\\7\\\hline8\end{array}$ $\begin{array}{r}1\\9\\\hline10\end{array}$ $\begin{array}{r}8\\1\\\hline9\end{array}$

3. $\begin{array}{r}2\\2\\\hline4\end{array}$ $\begin{array}{r}4\\2\\\hline6\end{array}$ $\begin{array}{r}2\\3\\\hline5\end{array}$ $\begin{array}{r}7\\2\\\hline9\end{array}$ $\begin{array}{r}2\\5\\\hline7\end{array}$ $\begin{array}{r}8\\2\\\hline10\end{array}$ $\begin{array}{r}2\\6\\\hline8\end{array}$ $\begin{array}{r}9\\2\\\hline11\end{array}$

4. $\begin{array}{r}3\\3\\\hline6\end{array}$ $\begin{array}{r}3\\4\\\hline7\end{array}$ $\begin{array}{r}6\\3\\\hline9\end{array}$ $\begin{array}{r}3\\5\\\hline8\end{array}$ $\begin{array}{r}7\\3\\\hline10\end{array}$ $\begin{array}{r}3\\9\\\hline12\end{array}$ $\begin{array}{r}8\\3\\\hline11\end{array}$

5. $\begin{array}{r}4\\4\\\hline8\end{array}$ $\begin{array}{r}6\\4\\\hline10\end{array}$ $\begin{array}{r}4\\5\\\hline9\end{array}$ $\begin{array}{r}7\\4\\\hline11\end{array}$ $\begin{array}{r}4\\9\\\hline13\end{array}$ $\begin{array}{r}8\\4\\\hline12\end{array}$

6. $\begin{array}{r}5\\5\\\hline10\end{array}$ $\begin{array}{r}7\\5\\\hline12\end{array}$ $\begin{array}{r}5\\6\\\hline11\end{array}$ $\begin{array}{r}5\\9\\\hline14\end{array}$ $\begin{array}{r}8\\5\\\hline13\end{array}$

7. $\begin{array}{r}6\\6\\\hline12\end{array}$ $\begin{array}{r}8\\6\\\hline14\end{array}$ $\begin{array}{r}6\\7\\\hline13\end{array}$ $\begin{array}{r}6\\9\\\hline15\end{array}$

8. $\begin{array}{r}7\\7\\\hline14\end{array}$ $\begin{array}{r}9\\7\\\hline16\end{array}$ $\begin{array}{r}7\\8\\\hline15\end{array}$

9. $\begin{array}{r}8\\8\\\hline16\end{array}$ $\begin{array}{r}9\\8\\\hline17\end{array}$

10. $\begin{array}{r}9\\9\\\hline18\end{array}$

ADDING A ONE-DIGIT NUMBER TO A TWO-DIGIT NUMBER. We show the method by examples.

Example 3. Find the sum: 63 + 4.

Write as shown: $\begin{array}{r}63\\+\ 4\\\hline67\end{array}$

tens units

```
63
+ 4
67
```
└─── 3 units + 4 units = 7 units ──→

```
         63
        + 4
          7
         →60
          67
```

```
63
+ 4
67
```
└─── 6 tens = 60 units ──┘

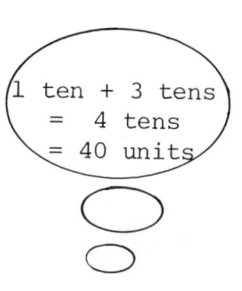

1 ten + 3 tens
= 4 tens
= 40 units

Example 4. Find the sum: 35 + 7.

Write as shown:

```
  1
 35
+ 7
 42
```

Explanation:

tens units

```
 1 ◄──── This small number representing
35        1 ten is called a carry number.
+ 7
42
```
└─── 5 units + 7 units = 12 units
                      = 1 ten + 2 units

```
 1
35
+ 7
42
```
└─── 1 ten + 3 tens = 4 tens = 40 units

In additions of this type, it is not necessary to write down the "1" that was carried. It can be done mentally.

<u>DRILL IN ADDING ONE- AND TWO- DIGIT NUMBERS.</u> In Drill Problems 1-10 place a sheet of unused scratch paper over the answers, then write the sums on the scratch paper as fast as possible. After doing one line, uncover the given answers and correct your answers. Repeat this process until you can work all the additions rapidly without hesitation. In your thinking, begin with the two-digit number. For example, in working the exercises in (1) below we say, "11 and 2 = 13; 12 and 2 = 14; 14 and 2 = 16," and so on.

1.	11	2	14	2	16	2	17	2	19
	2	12	2	13	2	15	2	18	2
	13	14	16	15	18	17	19	20	21

2.	12	3	13	3	17	3	19	3	20
	3	14	3	15	3	16	3	18	4
	15	17	16	18	20	19	22	21	24

3.	13	4	16	4	17	4	18	20	31
	4	14	4	15	4	19	4	5	5
	17	18	20	19	21	23	22	25	36

4.	14	5	17	5	19	5	42	53	66
	5	15	5	16	5	18	5	5	5
	19	20	22	21	24	23	47	58	71

5.	15	6	19	6	16	7	19	7	17
	6	17	6	18	7	17	7	18	8
	21	23	25	24	23	24	26	25	25

6.	75	86	95	20	31	42	53	66	75
	5	5	5	6	6	6	6	6	6
	80	91	100	26	37	48	59	72	81

7.	22	33	44	55	68	77	84	94	66
	6	6	6	6	6	6	6	6	6
	28	39	50	61	74	83	90	100	72

8.	23	82	45	56	69	78	85	96	81
	7	7	7	7	7	7	7	7	7
	30	89	52	63	76	85	92	103	88

9.	26	35	46	57	65	92	83	79	71
	8	8	8	8	8	8	8	8	8
	34	43	54	65	73	100	91	87	79

10.	25	36	43	54	67	76	87	99	89
	9	9	9	9	9	9	9	9	9
	34	45	52	63	76	85	96	108	98

In working the preceding problems, you may have noticed the
following fact: The tens digit remains the same unless the sum
of the units digits is 10 or more. In this case, the tens digit
is increased by one. Knowing this fact makes it unnecessary to
write down the carry number.

Example 5

	(a) 23	(b) 23	(c) 54	(d) 54
	5	8	5	8
	28	31	59	62

EXERCISES 103B, SET I. Find the following sums.

1. 12	2. 24	3. 35	4. 46	5. 78
3	6	4	7	2

6. 95	7. 21	8. 13	9. 83	10. 92
8	7	4	5	7

11. 27	12. 54	13. 87	14. 74	15. 94
5	4	6	9	5

16. 68	17. 75	18. 86	19. 49	20. 36
5	7	8	4	5

21. 88	22. 55	23. 59	24. 87
9	6	4	7

Find the following sums.

1. 13 4	2. 27 3	3. 65 4	4. 94 7	5. 85 3
6. 72 5	7. 78 6	8. 49 8	9. 57 7	10. 94 9
11. 88 9	12. 79 8			

Adding More Than Two Numbers

In adding three numbers, we can do either of the following:

1. Find the sum of the first two numbers, then add that sum to the third number.
2. Find the sum of the last two numbers, then add that sum to the first number.

We use parentheses ( ) to show which two numbers are being added first. Other grouping symbols can be used such as brackets [ ] or braces { }. We only use parentheses ( ) for grouping symbols in this book. When grouping symbols are used, the operations within those grouping symbols must be done first.

Example 1. Add: 2 + 3 + 4.

(a)   (2 + 3) + 4   The parentheses mean that the
  =   5   + 4   2 and 3 are to be added first.
  =   9

(b)   2 + (3 + 4)   Here, the parentheses mean that
  = 2 +   7   the 3 and 4 are to be added
  =   9   first.

Therefore, 2 + 3 + 4 = (2 + 3) + 4 = 2 + (3 + 4)
                            5   + 4 = 2 +   7
                                 9   =   9

In this example, the sum of three numbers was unchanged no matter how we grouped (or associated) the numbers. It is assumed that this property of addition is true when any three whole numbers are added. This is called the *Associative Property of Addition*.

Associative Property of Addition

If a, b, and c represent any whole numbers, then

$$a + b + c = (a + b) + c = a + (b + c).$$

EXERCISES 104A, SET I. Test the Associative Property of Addition by finding each of the following sums in two different ways.

1. $3 + 5 + 4$	2. $7 + 1 + 6$	3. $8 + 2 + 5$
4. $10 + 2 + 4$	5. $9 + 8 + 7$	6. $6 + 5 + 3$

7. 5	8. 8	9. 6
6	9	9
7	7	5

10. 8	11. 10	12. 12
8	4	6
9	6	4

EXERCISES 104A, SET II. Test the Associative Property of Addition by finding each of the following sums in two different ways.

1. $2 + 4 + 7$	2. $8 + 3 + 6$	3. $6 + 7 + 8$

4. 7	5. 4	6. 11
8	5	3
6	7	4

When more than two numbers are to be added, we usually add the first two numbers, then add their sum to the next number. We continue in this way until the final sum is found.

Example 2

$$
\begin{array}{l}
\left. \begin{array}{l} 3 \\ 2 \end{array} \right\} = 5 \\
5 \\
4 \\
\hline 14
\end{array}
\quad
\left. \begin{array}{l} = 5 \\ \\ = 10 \end{array} \right\} = 14
$$

Example 3

$$
4 + 3 + 5 + 7 = 19
$$

$$
7
$$

$$
12
$$

$$
19
$$

EXERCISES 104B, SET I. Find the following sums:

1. 3	2. 5	3. 7	4. 8
4	2	4	4
2	6	6	5
7	3	5	7
1	4	3	3

5. 9	6. 8	7. 5	8. 5
3	9	9	7
4	6	8	9
7	5	7	8
6	4	6	6

9. $7 + 5 + 6 + 3 + 2 + 8 + 4$

10. 6 + 4 + 7 + 8 + 9 + 5 + 5

11. 3 + 9 + 7 + 6 + 5 + 2 + 8

12. 8 + 2 + 4 + 7 + 9 + 5 + 6

13. 10 + 8 + 12 + 25 + 13 + 41

14. 11 + 7 + 15 + 20 + 30 + 40

EXERCISES 104B, SET II. Find the following sums:

1. 4	2. 6	3. 9	4. 8
1	2	5	7
5	5	4	9
3	7	3	6
2	3	8	5

5. 3 + 4 + 5 + 6 + 2 + 3 + 6

6. 2 + 9 + 7 + 5 + 8 + 4 + 6

7. 10 + 9 + 13 + 24 + 32 + 48

8. 12 + 23 + 35 + 46 + 53 + 87

Adding Numbers Having More Than Two Digits

In adding numbers, arrange the numbers in vertical columns with all the units digits in the same vertical line, all the tens digits in the same vertical line, all the hundreds digits in the same vertical line, and so on. Then add the digits in each vertical line. Writing the numbers clearly and keeping the columns straight will help reduce the number of addition errors.

Example 1. Find the sum: 217 + 372.

Write as shown:
```
  217
+ 372
```

Explanation:

```
hundreds
  tens
  units
  217
+ 372
  589
```
↑——7 units + 2 units = 9 units ————

```
  217
+ 372
    9
```

(Explanation continued on next page)

```
  217
+ 372
  589
   ↑
   └──1 ten + 7 tens = 8 tens = 80 units ──────────→
  217
+ 372
  589
   ↑
   └──2 hundreds + 3 hundreds = 5 hundreds
                                = 500 units ──────
```

```
  217
+ 372
    9
   80
  500
  589
```

Example 2. Find the sum: 587 + 265.

Write as shown: 587
                 + 265
                   852

Explanation:

```
hundreds
  tens
  units
 1 1 ←─────────────────────────────────
  587
+ 265
  852
   ↑
   └──7 units + 5 units = 12 units = 1 ten + 2 units
 1 1
  587
+ 265
  852
   ↑
   └──1 ten + 8 tens + 6 tens = 15 tens
                             = 1 hundred + 5 tens
 1 1
  587
+ 265
  852
   ↑
   └──1 hundred + 5 hundreds + 2 hundreds = 8 hundreds
```

Example 3. Add: 85 + 354 + 6 + 5,871.

```
  1 21 ←──── These small numbers are called
    85        carry numbers.
   354
     6
+ 5,871
  6,316
```

EXERCISES 105, SET I. Find the indicated sums.

1. (a)   (b)   (c)   (d)   (e)   (f)   (g)   (h)   (i)   (j)
   10   32   54   60   56   78   85   55   79   98
   15   25   74   37   29   87   58   97   92   77

2.

(a)	(b)	(c)	(d)	(e)	(f)	(g)	(h)	(i)	(j)
17	29	68	59	90	61	46	7	57	98
34	8	23	87	18	19	95	19	78	89
58	15	6	98	46	74	83	80	33	67

3.

(a)	(b)	(c)	(d)	(e)	(f)	(g)	(h)
102	243	354	678	987	555	817	987
213	746	83	54	715	894	209	807

4.

(a)	(b)	(c)	(d)	(e)	(f)	(g)	(h)
308	755	821	908	562	299	506	274
25	9	655	90	267	37	111	666
691	307	777	808	38	451	394	25

5.

(a)	(b)	(c)	(d)	(e)	(f)	(g)
5,840	75	61,304	8	85	1,009	716
218	3,594	7,105	78,481	319	54,311	75
21	315	70,009	4,154	4,667	31,555	48,315
3,009	2,171	801	51,816	8	9,999	8,888

6. Arrange the following numbers in a vertical column and add.

(a) 75,386; 77; 105,706,035; 880,755,009; 28,388,406

(b) 275; 80; 9; 786,410,075; 3,000,000; 259,715,306

7. Use digits to write the following numbers, then arrange them in a vertical column and find their sum. Thirty thousand, six; seventy-five million, one hundred; two billion, five hundred; fifty million, one hundred thousand, ten.

8. Find the sum of the digits.

9. Find the sum of the whole numbers greater than 875 and less than 887.

10. A certain auditorium has three sections. There are 1,032 seats in the center section and 584 seats in each of the side sections. How many people can be seated in the auditorium?

11. The greatest known depth of our oceans is 36,198 feet, and the highest point on the earth is Mt. Everest, 29,028 feet. What is the vertical distance from the lowest point to the highest point on the earth?

12. The populations of some of the nations of the world are: China, over 750,000,000; India, 439,235,082; Russia, 241,748,000; United States, 204,765,770. Find the combined population of China, India, Russia, and the United States.

13. John has $75, Jim has $18, and Marie has $12 more than John and Jim together. Find the total amount of money the three have together.

14. Mary has $34, Jane has $15, and Helen has $27 more than Mary and Jane together. Find the total amount of money the three girls have together.

15. For the numbers below, first find the sum of the numbers in each vertical column, then add these sums across the bottom to find the number for the box at the lower right. Second, add the numbers in the horizontal rows to find the sum of

each row, then add these sums to find the number for the box at the lower right. If your work is correct, this box number will be the same for both horizontal and vertical additions.

```
27 + 15 + 10 + 36 + 71 + 44 =
18 + 30 + 26 + 82 + 28 + 37 =
 9 + 58 + 69 + 77 + 39 +  8 = _____

   +    +    +    +    +   = ┌─────────────────┐
                            │                 │
                            │                 │
                            └─────────────────┘
```

16. Work this exercise in the way explained in Exercise 15.

```
125 + 396 + 704 + 599 + 627 =
288 + 409 + 346 + 655 + 893 =
777 + 644 + 598 + 356 + 289 = _____

   +    +    +    +    =  ┌─────────────────┐
                         │                 │
                         │                 │
                         └─────────────────┘
```

17. The figure shown is called a rectangle. The *perimeter* of a geometric figure is the sum of the lengths of all its sides. The word *perimeter* means "the measure around a figure." What is the perimeter of the rectangle shown?

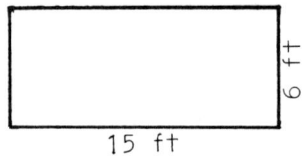

15 ft

18. The figure shown is called a triangle. What is the perimeter of the triangle shown? (For definition of *perimeter*, see problem 17, above).

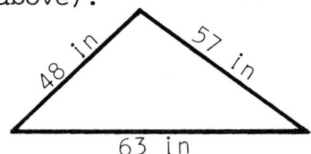

19. What is the perimeter of the figure shown?

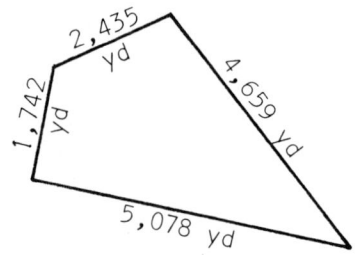

22   Chapter 1: Whole Numbers

20. Find the distance around the building shown in the figure.

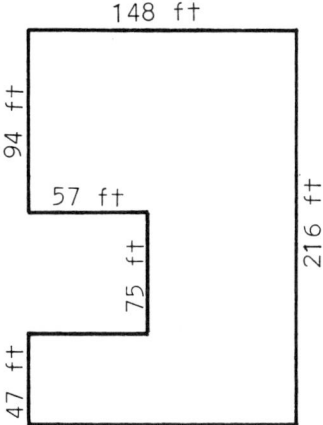

148 ft

94 ft

57 ft

216 ft

75 ft

47 ft

EXERCISES 105, SET II. In Exercises 1-3, find the sums.

1. (a)   (b)   (c)   (d)   (e)   (f)   (g)   (h)   (i)

   10   23   34   70   65   87   39   47   76
   14   35   67   49   38   56   72   89   68

2. (a)   (b)   (c)   (d)   (e)   (f)   (g)

   304   48   406   379   48   496   788
     9   350   780   99   356   984   13
    67   7   95   106   709   39   906

3. (a)   (b)   (c)   (d)   (e)   (f)

   2,370   38   72,035   39,604   74   43,909
     315   4,618   9   7,081   386   7,682
      34   275   3,586   352   14,705   38
   6,009   3,281   784   20,799   3,969   64,366

4. Arrange the following numbers in a vertical column and add.

   359;   40,286;   17;   284,000,189;   70,096,347

5. Use digits to write the following numbers, then arrange them in a vertical column and find their sum. Ten thousand, forty-seven; twenty-three million, five thousand; four billion, six million, seventy-three thousand, forty-two; two hundred million, one hundred fifty-six thousand; five hundred six.

6. Find the sum of the whole numbers greater than 286 and less than 295.

7. The greatest known depth of the Atlantic Ocean is 28,374 feet, and the highest point in Europe is 18,510 feet. What is the vertical distance between these two points?

8. Merv has $85; Irv has $92; and Jan has $17 more than Merv and Irv together. Find the total amount of money the three of them have together.

9. For the numbers below, first find the sum of the numbers in each vertical column, then add these sums across the

bottom to find the number for the box at the lower right. Second, add the numbers in the horizontal rows to find the sum of each row, then add these sums to find the number for the box at the lower right. If your work is correct, this box number will be the same for both horizontal and vertical additions.

```
 34  +  51  +  37  +  48  + 65  =
281  +  49  +  18  +   6  + 78  =
  7  + 356  + 295  + 359  + 39  =  _____

      +      +      +      +     =  ┌─────────────┐
                                   │             │
                                   │             │
                                   └─────────────┘
```

10. The figure shown is a square. The perimeter of a geometric figure is the sum of the lengths of all its sides. The word *perimeter* means "the measure around a figure." What is the perimeter of the square shown?

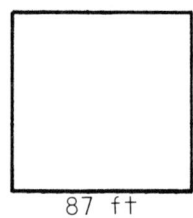

87 ft

Casting out Nines

We can check our work in addition, multiplication, division, etc., by a method called *casting out nines*. We will explain how to use this method for checking calculations. The explanation of why this method works depends upon mathematics not covered in this book.

Associated with each number is a number which we will call its *9-rem*.* We explain how to find the 9-rem of a number with examples.

Example	Number	Adding Digits and Explanation	9-rems
1	10	Add the digits of 10: 1 + 0 = 1	1
2	20	Add the digits of 20: 2 + 0 = 2	2
3	8	The only digit is 8. The 9-rem of numbers less than 9 is the same as the number itself.	8

*"9-rem" is a word coined by the authors of this book. "Rem" is the first three letters of the word "remainder." The 9-rem is what is left after all the nines have been cast out of a number.

Example	Number	Adding Digits and Explanation	9-rems
4	57	5 + 7 = 12 1 + 2 = 3	3
5	575	5 + 7 + 5 12 3 + 5 = 8	8

Finding the 9-rem can be simplified by first casting out nines. Examples 6-9 show this method.

6	93	9 + 3 Take out the 9 before adding; then 3 remains.	3
7	275	② + ⑦ + 5 ⑨ Take out this 9; then the sum is 5.	5
8	41,592	④ + 1 + ⑤ + ⑨ + 2 ⑨ ⑨ After taking out these nines, we have 1 + 2 = 3 remaining.	3
9	1,253	① + 2 + ⑤ + ③ ⑨	2

EXERCISES 106, SET I. Find the 9-rem for each of the following numbers.

1. 11	2. 15	3. 27	4. 81
5. 47	6. 66	7. 293	8. 596
9. 945	10. 819	11. 4,026	12. 7,305
13. 2,837	14. 4,691	15. 70,358	16. 45,209
17. 16,487	18. 83,542	19. 209,736	20. 430,865

EXERCISES 106, SET II. Find the 9-rem for each of the following numbers.

1. 12	2. 14	3. 63	4. 903
5. 357	6. 8,452	7. 1,947	8. 48,906
9. 17,824		10. 34,659	

Checking Addition by Casting out Nines

We use examples to show how to check addition by casting out nines.

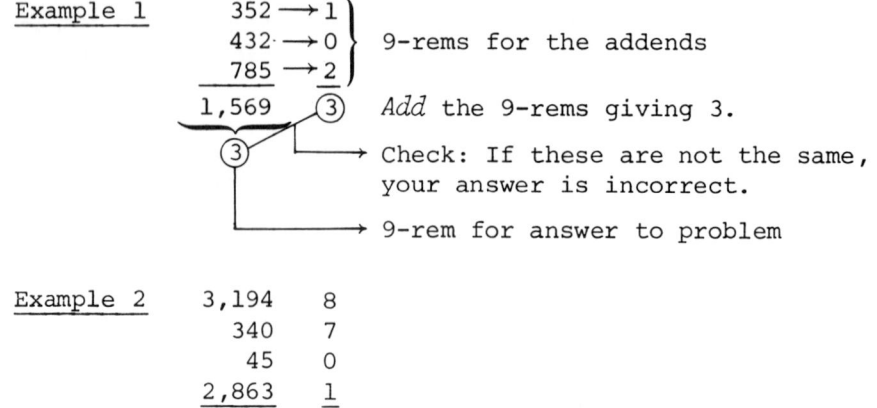

Example 1

$$
\begin{array}{r}
352 \rightarrow 1 \\
432 \rightarrow 0 \\
785 \rightarrow 2 \\
\hline
1,569 \quad ③ \\
③
\end{array}
$$

9-rems for the addends

Add the 9-rems giving 3.

Check: If these are not the same, your answer is incorrect.

9-rem for answer to problem

Example 2

$$
\begin{array}{rr}
3,194 & 8 \\
340 & 7 \\
45 & 0 \\
2,863 & \underline{1} \\
\hline
6,442 & ⑦ \\
⑦
\end{array}
$$

→ Check.

Example 3

$$
\begin{array}{rr}
7,354 & 1 \\
6,135 & 6 \\
2,864 & 2 \\
9,581 & 5 \\
\hline
25,934 & ⑤ \\
⑤
\end{array}
$$

→ Check.

The check by casting out nines can tell us only that the answer is incorrect. When the 9-rems are not the same, the answer is definitely incorrect. However, when the 9-rems are the same, the answer may still be incorrect. For example: If two digits are written in the wrong order, such as 57 instead of 75, the 9-rems would be the same even though the answer is incorrect. Even with this drawback, we feel that the check by casting out nines is helpful.

EXERCISES 107, SET I. Add, then check by casting out nines.

1. 74	2. 29	3. 85	4. 18
38	67	55	36
46	18	94	27
5. 726	6. 3,896	7. 6,123	8. 2,781
49	37	365	351
657	1,582	7,216	9,459
9. 7,856	10. 3,496	11. 7,238	12. 90,748
2,094	5,577	6,272	81,927
5,436	2,793	7,854	31,576
309	8,257	4,259	319

EXERCISES 107, SET II. Add, then check by casting out nines.

1. 44
   82
   93

2. 87
   25
   34

3. 486
    57
   782

4. 7,834
     296
   1,584

5. 6,294
   5,742
      37
   8,673

6. 35,481
   94,567
   13,488
    5,792

Multiplication of Whole Numbers

To find the cost of six 5¢ pencils, we could add as follows:

    5¢ addend
    5¢ addend
    5¢ addend
    5¢ addend
    5¢ addend
 + 5¢ addend
   30¢ sum

A shorter method for finding the cost is to *multiply*.

$$6 \times 5¢ = 30¢$$

Multiplication is a short method for finding the sum of two or more equal numbers. Thus multiplication replaces the repeated addition of the same addend.

TERMS USED IN MULTIPLICATION

      5 *multiplicand*
   × 6 *multiplier*
     30 *product*

$$6 \quad \times \quad 5 \quad = \quad 30$$
factor   *factor*   *product*

The expression 6 × 5 is read "six times five." In the expression 6 × 5 = 30, the numbers 6 and 5 are said to be factors of 30. That is, the multiplier and multiplicand are factors of the product.

Example 1 $3 \quad \times \quad 4 \quad = \quad 12$

         factor   *factor*   *product*

Therefore, 3 and 4 are factors of 12.

Example 2 $2 \quad \times \quad 6 \quad = \quad 12$

         factor   *factor*   *product*

Therefore, 2 and 6 are factors of 12.

Example 3. If 2 is one of the factors of 6, what is the other factor of 6?

Solution: $2 \times ? = 6$

         Since 2 × 3 = 6, 3 is the other factor of 6.

Example 4. If 3 is the multiplicand for a product of 15, what is the multiplier?

Solution:
$$\begin{array}{r} 3 \quad \text{multiplicand} \\ \times\ ? \quad \text{multiplier} \\ \hline 15 \quad \text{product} \end{array}$$

Since 5 × 3 = 15, 5 is the multiplier.

Table of Basic Multiplication Facts

×	0	1	2	3	4	5	6	7	⑧	9
0	0	0	0	0	0	0	0	0	0	0
1	0	1	2	3	4	5	6	7	8	9
2	0	2	4	6	8	10	12	14	16	18
3	0	3	6	9	12	15	18	21	24	27
4	0	4	8	12	16	20	24	28	32	36
5	0	5	10	15	20	25	30	35	40	45
6	0	6	12	18	24	30	36	42	48	54
⑦	0	7	14	21	28	35	42	49	㊶	63
8	0	8	16	24	32	40	48	56	64	72
9	0	9	18	27	36	45	54	63	72	81

To show how to use the Table of Multiplication Facts, we give an example. To find the product, 7 × 8 = 56, move right from 7 of the left column until you are directly below 8 of the top row. See the circled numbers in the table. Therefore, 7 and 8 are factors of 56.

EXERCISES 108, SET I. In Exercises 1-15, use the Table of Multiplication Facts, when needed, to find the products.

1. 6 × 7   2. 8 × 6   3. 9 × 5   4. 6 × 9   5. 9 × 8

6. 7 × 9   7. 8 × 7   8. 8 × 5   9. 9 × 4   10. 7 × 5

11. 9 × 7   12. 8 × 9   13. 8 × 8   14. 7 × 4   15. 6 × 5

16. If 7 is one of the factors of 42, what is the other factor?
17. If 8 is one of the factors of 72, what is the other factor?
18. If 6 is the multiplier for the product 54, what is the multiplicand?

EXERCISES 108, SET II. In Exercises 1-8, use the Table of Multiplication Facts, when needed, to find the products.

1. 8 × 7   2. 6 × 9   3. 7 × 9   4. 9 × 8   5. 5 × 8

6. 9 × 4   7. 9 × 9   8. 8 × 6

9. If 6 is one of the factors of 48, what is the other factor?
10. If 7 is the multiplicand for the product 35, what is the multiplier?

Multiplying any number by one gives the identical number for the product. For this reason, one is called the *multiplicative identity*. This property of one can easily be verified by referring to the Table of Basic Multiplication Facts.

BASIC MULTIPLICATION COMBINATIONS. The following multiplication combinations include every possible pair of digits that can be multiplied together. If a student practices these daily, his multiplication skills will improve. A convenient way of practicing is to place a paper over the answers, then write or say the answers as quickly as possible. To get variety in your practice, work from left to right across the page, then from right to left.

1.	2	4	2	5	2	6	7	2
	2	2	3	2	8	2	2	9
	4	8	6	10	16	12	14	18

2.	3	5	4	3	8	3	9	8
	3	3	3	6	3	7	3	2
	9	15	12	18	24	21	27	16

3.	4	6	4	5	4	8	3	3
	4	4	7	4	8	4	8	9
	16	24	28	20	32	32	24	27

4.	5	7	5	6	5	9	4	7
	5	5	9	5	8	4	8	4
	25	35	45	30	40	36	32	28

5.	6	8	6	9	5	8	9	5
	6	6	7	6	6	5	5	7
	36	48	42	54	30	40	45	35

6.	7	9	8	7	6	6	5	4
	7	7	7	6	8	9	7	5
	49	63	56	42	48	54	35	20

7.	8	9	7	7	4	3	3	8
	8	8	9	8	6	8	9	9
	64	72	63	56	24	24	27	72

8.	9	8	8	7	7	6	6	7
	9	9	8	9	8	9	8	6
	81	72	64	63	56	54	48	42

Multiples

The *multiples* of a number are the products that result when the number is multiplied by the counting numbers.

Example 1

(a)
$$1 \times 2 = 2$$
$$2 \times 2 = 4$$
$$3 \times 2 = 6$$
$$4 \times 2 = 8$$
$$5 \times 2 = 10$$
} Multiples of 2

(b)
$$1 \times 3 = 3$$
$$2 \times 3 = 6$$
$$3 \times 3 = 9$$
$$4 \times 3 = 12$$
$$5 \times 3 = 15$$
} Multiples of 3

(c)
$$1 \times 4 = 4$$
$$2 \times 4 = 8$$
$$3 \times 4 = 12$$
$$4 \times 4 = 16$$
$$5 \times 4 = 20$$
} Multiples of 4

(d)
$$1 \times 5 = 5$$
$$2 \times 5 = 10$$
$$3 \times 5 = 15$$
$$4 \times 5 = 20$$
$$5 \times 5 = 25$$
} Multiples of 5

The Table of Multiplication Facts shows that the numbers in each column are multiples of the number at the top of the column. Also, the numbers in each row are multiples of the number at the left of that row.

EXERCISES 109, SET I

1. Find three multiples of 6.
2. 6 is a multiple of what numbers?
3. 8 is a multiple of what numbers?
4. Find all the multiples of 9 that are less than 50.
5. 45 is a multiple of what numbers?

EXERCISES 109, SET II

1. Find three multiples of 7.
2. 30 is a multiple of what numbers?
3. Find all the multiples of 8 that are less than 20.

Commutative and Associative Properties of Multiplication

Example 1. The Commutative Property of Multiplication.

(a) $7 \times 8 = 8 \times 7 = 56$

(b) $4 \times 5 = 5 \times 4 = 20$

In these examples, when we reversed the order of the factors, the product remained the same. This property of whole numbers is called the *Commutative Property of Multiplication* and is easily verified by referring to the Table of Multiplication Facts in Section 108.

```
┌─────────────────────────────────────────────────────┐
│                                                       │
│   Commutative Property of Multiplication              │
│                                                       │
│   If $a$ and $b$ are whole numbers, then              │
│                                                       │
│                $a \times b = b \times a$              │
│                                                       │
└─────────────────────────────────────────────────────┘
```

Example 2. The Associative Property of Multiplication.

    (a) $(3 \times 4) \times 2 = 3 \times (4 \times 2)$

           $12 \quad \times 2 = 3 \times \quad 8$

              $24 \quad = \quad 24$

    (b) $(5 \times 2) \times 3 = 5 \times (2 \times 3)$

           $10 \quad \times 3 = 5 \times \quad 6$

              $30 \quad = \quad 30$

In these examples, the product of three numbers was unchanged no matter how we grouped the numbers. It is assumed that this property of multiplication is true when any three numbers are multiplied. This property of whole numbers is called the *Associative Property of Multiplication*.

```
┌─────────────────────────────────────────────────────┐
│                                                       │
│   Associative Property of Multiplication              │
│                                                       │
│   If $a$, $b$, and $c$ are whole numbers, then        │
│                                                       │
│         $(a \times b) \times c = a \times (b \times c)$ │
│                                                       │
└─────────────────────────────────────────────────────┘
```

The *Commutative* and *Associative* Properties hold true for both addition and multiplication.

Multiplication Using Zeros

Since multiplication is a method for finding the sum of two or more equal numbers, multiplying a number by zero gives a product of zero. See Example 1.

Example 1

    (a) $3 \times 0 = 0 + 0 + 0 = 0$

    (b) $4 \times 0 = 0 + 0 + 0 + 0 = 0$

Because of the Commutative Property of Multiplication, it follows that

                  $3 \times 0 = 0 \times 3 = 0$

and $4 \times 0 = 0 \times 4 = 0$

$$0 \times \text{any number} = 0$$

$$\text{Any number} \times 0 = 0$$

EXERCISES 111, SET I. In Exercises 1-4, find the indicated product.

1. 0×6 2. 9×0 3. 0×8 4. 5×0

In Exercises 5-8, write the number that should replace each question mark.

5. $3 \times 5 = 5 \times ?$ 6. $0 \times 6 = ? \times 0$

7. $? \times 4 = 4 \times 0$ 8. $7 \times ? = 8 \times 7$

EXERCISES 111, SET II

1. Find the indicated products.

   (a) 0×9 (b) 7×0

2. Write the number that should replace each question mark.

   (a) $4 \times 9 = 9 \times ?$ (b) $? \times 8 = 3 \times 8$

Symbols Used in Multiplication

Multiplication may be shown in several different ways.

Example 1

   (a) $3 \times 2 = 6$

   (b) $3 \cdot 2 = 6$ The multiplication dot "·" is written a little higher than the decimal point.

   (c) $3(2) = 6$ The symbols ( ) are called parentheses.

   (d) $(3)(2) = 6$ In this kind of multiplication, the double parentheses is not necessary.

   (e) $(3)2 = 6$

   (f) $ab = a \times b$ When two expressions are written next to each other in this way, it is understood that they are to be multiplied.

EXERCISES 112, SET I. Write each of the following multiplications in four different ways.

1. 9×8 2. 5×6 3. 3×8 4. 7×4

EXERCISES 112, SET II. Write each of the following multiplications in four different ways.

1. 4×5 2. 6×3 3. 5×8 4. 7×9

Multiplying a One-Digit Number by a Larger Number

Example 1. Find 2 × 43. Write as shown:
$$\begin{array}{r} 43 \\ \times\ 2 \\ \hline 86 \end{array}$$

Explanation:

$$\begin{array}{r} \text{tens units} \\ 43 \\ \times\ 2 \\ \hline 6 \end{array} \leftarrow 2 \text{ units} \times 3 \text{ units} = 6 \text{ units} \longrightarrow \begin{array}{r} 43 \\ \times\ 2 \\ \hline 6 \end{array}$$

$$\begin{array}{r} 43 \\ \times\ 2 \\ \hline 86 \end{array} \leftarrow 2 \text{ units} \times 4 \text{ tens} = 8 \text{ tens} = 80 \text{ units} \longrightarrow \begin{array}{r} 80 \\ \hline 86 \end{array}$$

Example 2. Find 4 × 36. Write as shown:
$$\begin{array}{r} {\scriptstyle 2} \\ 36 \\ \times\ 4 \\ \hline 144 \end{array}$$

Explanation:

$$\begin{array}{r} \text{hundreds tens units} \\ 2 \leftarrow carry\ number \\ 36 \\ \times\ 4 \\ \hline 4 \end{array} \leftarrow 4 \text{ units} \times 6 \text{ units} = 24 \text{ units}$$
$$= 2 \text{ tens} + 4 \text{ units}$$

$$\begin{array}{r} 2 \\ 36 \\ \times\ 4 \\ \hline 144 \end{array}$$

$$\left\{ \begin{array}{l} 4 \text{ units} \times 3 \text{ tens} = 12 \text{ tens;} \\ \qquad\quad \overbrace{}^{carry} \\ 12 \text{ tens} + 2 \text{ tens} = 14 \text{ tens} \\ \qquad\qquad = 1 \text{ hundred} + 4 \text{ tens} \end{array} \right.$$

Example 3. Find 6 × 2,347. Write as shown:
$$\begin{array}{r} {\scriptstyle 2\ 24} \leftarrow carry \\ 2{,}347 \qquad numbers \\ \times\ 6 \\ \hline 14{,}082 \end{array}$$

EXERCISES 113, SET I. Find the following products.

	(a)	(b)	(c)	(d)	(e)	(f)	(g)	(h)
1.	34	74	56	83	48	29	68	95
	2	3	4	6	5	4	7	8
2.	52	78	63	39	46	87	93	26
	3	5	4	6	8	2	7	9

3.

(a)	(b)	(c)	(d)	(e)	(f)	(g)	(h)
135	283	506	209	310	400	687	898
3	4	7	6	8	9	5	8

4.

(a)	(b)	(c)	(d)	(e)	(f)	(g)	(h)
234	417	625	359	678	504	989	600
2	3	8	4	7	6	5	9

5.

(a)	(b)	(c)	(d)	(e)	(f)
2,453	6,987	1,069	6,499	7,088	8,697
3	5	4	8	7	9

6.

(a)	(b)	(c)	(d)	(e)	(f)
23,156	56,041	60,786	70,054	20,009	68,978
3	5	6	7	8	9

7.

(a)	(b)	(c)	(d)	(e)	(f)
700	30,080	12,500	526,000	25,000	70,900
6	7	8	8	4	9

EXERCISES 113, SET II. Find the following products.

1.

(a)	(b)	(c)	(d)	(e)	(f)	(g)	(h)
43	52	76	38	29	86	79	98
6	3	4	6	5	8	7	9

2.

(a)	(b)	(c)	(d)	(e)	(f)	(g)	(h)
245	327	506	490	788	497	800	989
3	4	9	5	6	8	7	9

3.

(a)	(b)	(c)	(d)	(e)	(f)
4,523	4,607	5,794	6,589	7,749	8,096
4	6	7	6	8	9

4.

(a)	(b)	(c)	(d)	(e)	(f)
40,052	50,607	69,897	250,000	80,908	90,087
5	6	7	8	9	8

Problems with Multipliers Having More Than One Digit

Example 1. Find 526 × 34.

Write as shown:

```
   526    multiplicand
 × 34     multiplier
  2104    first partial product
  1578    second partial product
 17884    product (17,884)
```

Explanation:

       tens units

```
    526
  × 3 4
   2104
  1578
  17884
```

```
    526
  × 3 4
   2104
  1578
  17884
```

Method:

1. Multiply 526 by the units digit (4). This gives the first partial product (2104 *units*). For this reason, the 4 in 2104 is placed in the *units* column.
2. Multiply 526 by the tens digit (3). This gives the second partial product (1578 *tens*). For this reason, the 8 in 1578 is placed in the *tens* column.
3. Add all partial products. This gives the final product (17,884).

Example 2. Find 4,385 × 729.

Write as shown:

```
      4385
    × 729
     39465
     8770
    30695
   3196665     product = 3,196,665
```

Explanation: Notice that the right digit of each partial product is directly below the digit of the multiplier that was used to obtain it.

units	tens	hundreds
4385	4385	4385
× 729	× 729	× 729
39465	39465	39465
8770	8770	8770
30695	30695	30695
3196665	3196665	3196665
units	tens	hundreds

EXERCISES 114, SET I. Find the following products.

1.

(a)	(b)	(c)	(d)	(e)	(f)
35	28	56	71	78	95
21	32	43	28	36	42

2.

(a)	(b)	(c)	(d)	(e)	(f)
215	324	555	623	438	897
34	52	47	48	57	98

3.

(a)	(b)	(c)	(d)	(e)	(f)
2,314	4,536	5,076	5,070	7,836	9,267
52	34	45	64	58	49

4.
(a)	(b)	(c)	(d)	(e)
90,236	81,604	10,428	43,871	70,065
54	78	65	68	98

5.
(a)	(b)	(c)	(d)	(e)
235	426	7,023	8,107	23,016
415	351	542	623	524

6.
(a)	(b)	(c)	(d)
4,372	8,092	70,245	97,864
5,168	526	3,619	8,594

EXERCISES 114, SET II. Find the following products.

1.
(a)	(b)	(c)	(d)	(e)	(f)
43	65	87	423	879	698
31	46	29	42	67	89

2.
(a)	(b)	(c)	(d)	(e)
3,452	5,080	8,006	20,284	50,056
27	45	85	56	89

3.
(a)	(b)	(c)	(d)	(e)
453	8,206	33,018	50,384	89,765
284	537	637	794	6,789

Multiplying Numbers Having Zeros in the Multiplier

Example 1. Find 305 × 1,734. Write as shown:

```
  1734
× 305
  8670
  0000
 5202
528870
```
Since the multiplication by zero results in a row of zeros, we shorten the writing by omitting this row of zeros. →

```
  1734
× 305
  8670
 5202
528870 = 528,870
```

Example 2. Find 4,006 × 2,314. Write as shown:

```
   2314
× 4006
  13884
 9256
9269884 = 9,269,884
```

Explanation:

```
   2314
× 4006
  13884
 9256
9269884
```

Notice that the right digit of each partial product is directly below the digit of the multiplier that was used to obtain it.

EXERCISES 115, SET I. Find the following products.

1. 356	2. 1,786	3. 705	4. 3,804	5. 9,067
204	302	206	706	504

6. 30,508	7. 95,046	8. 60,058	9. 7,802	10. 750,009
707	3,007	9,005	5,008	30,007

EXERCISES 115, SET II. Find the following products.

1. 450	2. 805	3. 8,076	4. 85,007	5. 650,008
205	406	405	6,008	40,007

Multiplying Numbers Having Zeros at the End

Example 1. Find 18,000 × 2,341. Rather than write the problem in the form

$$\begin{array}{r} 2341 \\ \times\ 18000 \end{array}$$

it is better to move the multiplier to the right as shown here.

$$\begin{array}{r} 2341 \\ \times\ 18000 \end{array}$$

Draw a vertical line to separate the zeros from the 18. Multiply by 18, then add the zeros later as shown below.

```
    2341
  × 18|000
  ─────────
  18728|
   2341|
  ─────────
  42138|000   = 42,138,000
```

Explanation:

2,341 × 18,000

= 2,341 × (18 × 1,000)

= (2,341 × 18) × 1,000

= 42,138 × 1,000

= 42,138,000

Example 2. Find 5,581 × 7,200.

Solution:
```
      5581
    × 72|00
    ────────
    11162|
    39067|
    ────────
   401832|00    = 40,183,200
```

Example 3. Find 175,000 × 3,200.

Solution:
```
      175|000     3 zeros ⎫ to the right
    ×  32|00      2 zeros ⎬ of the line
    ─────────     ─────────
       350|       5 zeros
       525|
    ─────────
      5600|00000    5 zeros to the right
  = 560,000,000              of the line
```

EXERCISES 116, SET I. Find the following products.

1. 2,500 × 376   2. 12,000 × 507   3. 9,200 × 3,154

4. 300 × 7,855   5. 500 × 3,751   6. 2,000 × 799

7. 6,600 × 449   8. 5,000 × 7,008   9. 9,500 × 7,893

10. 7,500 × 3,500   11. 8,960 × 5,600   12. 38,000 × 7,800

13. A ream of paper contains 500 sheets. A school uses 1,427 reams in a year. How many sheets of paper were used in that time?

14. A town has 7,500 inhabitants who pay property tax. If the average tax paid is $504, what is the town's income from property taxes?

15. If a man's car cost him 8¢ per mile to operate, how much does 16,000 miles of driving cost him?

EXERCISES 116, SET II. Find the following products.

1. 3,500 × 284   2. 8,300 × 2,147   3. 600 × 4,823

4. 7,000 × 6,008   5. 7,500 × 6,398   6. 6,870 × 4,600

7. A ream of paper contains 500 sheets. An office uses 2,500 reams in a year. How many sheets of paper were used in that time?

8. If a man's car cost him 9¢ per mile to operate, how much does 40,000 miles of driving cost him?

Checking Multiplication by Casting out Nines

To Check Multiplication by Casting Out Nines

1. Find the 9-rem for each of the numbers being multiplied.
2. Find the 9-rem for the product of the numbers found in (1).
3. Find the 9-rem for your answer to the problem.
4. The 9-rems found in (2) and (3) must be the same.

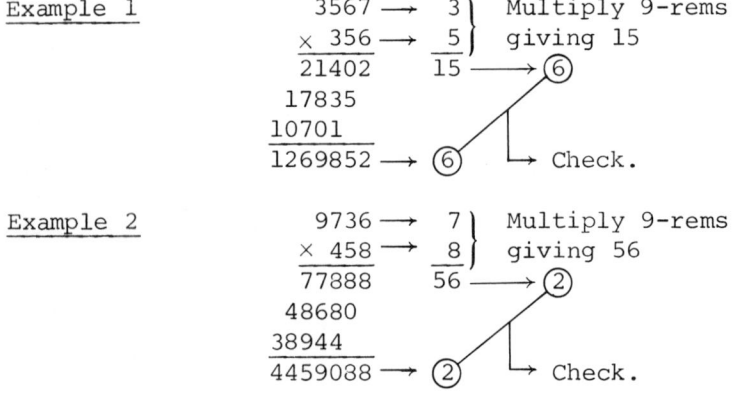

Note that when numbers are multiplied, their 9-rems are multiplied. When numbers are added, their 9-rems are added.

EXERCISES 117, SET I. Find the following products and check by casting out nines.

1. 785	2. 594	3. 834	4. 3,137	5. 5,048
37	35	68	34	56

6. 41,308	7. 18,075	8. 7,508	9. 89,765	10. 70,058
75	284	935	789	1,896

EXERCISES 117, SET II. Find the following products and check by casting out nines.

1. 578	2. 759	3. 4,053	4. 81,057
49	76	65	482

5. 94,856	6. 90,007
2,574	2,005

Area of a Rectangle

A rectangle is shown in Figure 118. The *area* of a rectangle is the space inside the lines. To measure the length of a line, we see how many times a unit of length divides into it.

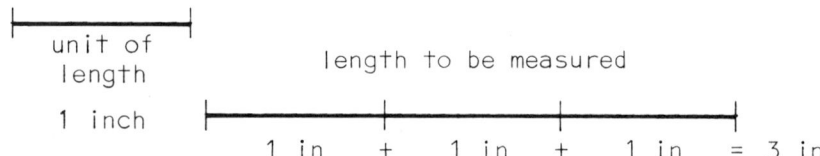

Since the unit of length (inch) fits three times into the length to be measured, we say the length is 3 inches.

To measure the area of the rectangle, we see how many times a unit of area fits into it.

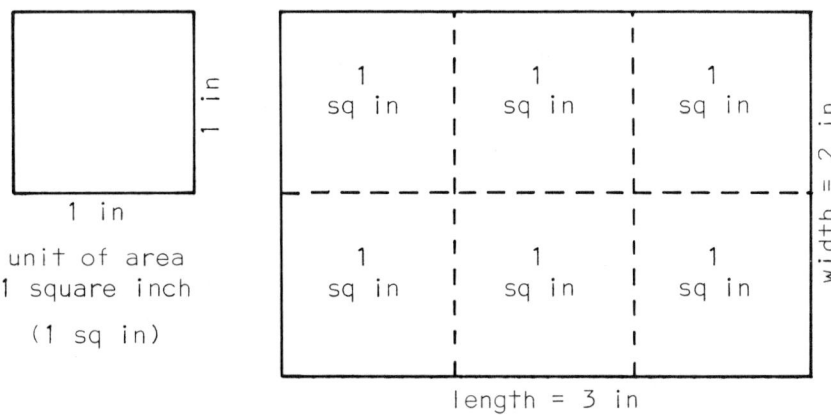

Figure 118

Since the unit of area (1 sq in) fits into the space six times, we say that the area of the rectangle is 6 sq in. We can find this area in another way. Each vertical strip has 2 sq in. There are three such vertical strips, so that the area = 2 sq in + 2 sq in + 2 sq in = 3 × 2 sq in = 6 sq in. (Fig. 118)

```
3 × 2 = 6
│     │     └──── area is 6 sq in
│     └──────────── width is 2 in
└────────────────── length is 3 in
```

So that,   3 in × 2 in = 6 sq in

```
┌─────────────────────────────────────────────┐
│                                               │
│     Area of rectangle = Length × Width        │
│                                               │
│              A = L × W                         │
│                                               │
│                                               │
└─────────────────────────────────────────────┘
```

In a particular problem the length and width of a rectangle are measured in the same length units. The area is measured in square units.

Example 1

Length × Width =   Area
(a) 8 in × 3 in = 24 sq in
(b) 4 ft × 2 ft = 8 sq ft
(c) 5 yd × 4 yd = 20 sq yd

The area shown in Figure 118 can also be found by seeing that each horizontal strip has 3 square inches. There are two such horizontal strips, so that the area equals 3 sq in + 3 sq in = 2 × 3 sq in = 6 sq in. Therefore,

$$A = W \times L$$

Since the area of the rectangle must be the same, no matter how we find it,

$$A = L \times W = W \times L$$

This is another verification of the Commutative Property of Multiplication.

We will learn much more about the application of arithmetic to other familiar geometric figures in Chapter 8.

1. Find the area of the
   rectangle shown below.

2. Find the number of square
   feet in the rectangle below.

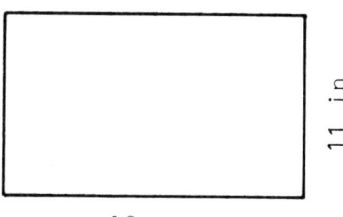

18 in

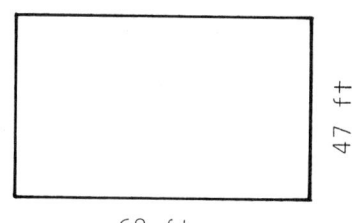

68 ft

2. A room is 5 yards wide and 7 yards long.
   (a) Find the floor area of the room.
   (b) Find the perimeter of the room.
4. A room is 8 yards wide and 9 yards long.
   (a) Find the floor area of the room.
   (b) Find the perimeter of the room.
5. A rectangular window is 7 feet wide and 4 feet high. How
   many square feet of glass are needed for this window?
6. What does it cost to replace a large bathroom mirror that
   measures 7 feet by 3 feet if glass costs $2 per square
   foot?
7. Jan needs drapes that are 3 yards high and 5 yards long to
   cover a picture window in her home. If she pays $7 a square
   yard for the material she likes, what will the drapes cost?
8. The floor of a small balcony is 6 feet wide and 9 feet
   long. How many square feet of material are needed to cover
   this floor?
9. An open box is 8 inches high, 9 inches long, and 6 inches
   wide. Find the total area of the four sides of the box.
   Hint: Each side of the box is a rectangle with an area you
   know how to find.
10. A rectangular room is 6 yards long and 5 yards wide.
    (a) Find the floor area of this room.
    (b) Find the cost to carpet this room if the carpet costs
        $8 per square yard to install.

1. Find the area of
   the rectangle shown.

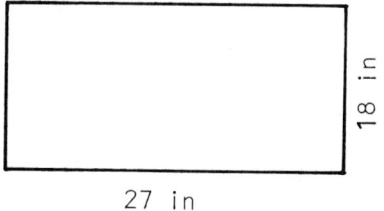

18 in

27 in

2. A room is 4 yards wide and 6 yards long.
   (a) Find the floor area of the room.
   (b) Find the perimeter of the room.
3. A rectangular window is 8 feet wide and 5 feet high. How
   many square feet of glass are needed for this window?
4. The floor of a balcony is 8 feet wide and 18 feet long.
   How many square feet of material are needed to cover this
   floor?

5. An open box is 4 inches high, 5 inches wide, and 7 inches long. Find the total area of the four sides and the bottom of the box. Hint: Each side and the bottom is a rectangle.

 **Powers of Whole Numbers**

Now that we have learned to multiply whole numbers, it is possible to consider products in which the same number is repeated as a factor. For example:

$$3 \cdot 3 \cdot 3 \cdot 3 = 3^4 = 81$$

In the symbol 3^4, the 3 is called the *base*. The 4 is called the *exponent* and is written above and to the right of the base 3. The entire symbol 3^4 is called the *fourth power of three* and is commonly read "three to the fourth power." See Figure 119.

$$3^4 = 81$$

exponent — base — fourth power of 3

Figure 119

Note the importance of the position of the 4.

$$3^4 = 3 \cdot 3 \cdot 3 \cdot 3 = 81$$

the raised 4 indicates repeated *multiplication*

$$3 \cdot 4 = 3 + 3 + 3 + 3 = 12$$

this 4 indicates repeated *addition*

Example 1

(a) $2^3 = 2 \cdot 2 \cdot 2 = 8$ (b) $4^2 = 4 \cdot 4 = 16$

(c) $1^4 = 1 \cdot 1 \cdot 1 \cdot 1 = 1$ (d) $25^2 = 25 \cdot 25 = 625$

(e) $36^3 = 36 \cdot 36 \cdot 36 = 46,656$

ZERO AS A BASE. When 0 is raised to a power other than 0, we get 0 (Example 2).

Example 2

(a) $0^2 = 0 \cdot 0 = 0$ (b) $0^5 = 0 \cdot 0 \cdot 0 \cdot 0 \cdot 0 = 0$

ZERO AS AN EXPONENT. When any whole number (other than 0) is raised to the 0 power, we get 1 (Example 3).

Example 3

(a) $2^0 = 1$ (b) $5^0 = 1$ (c) $10^0 = 1$

ZERO AS BOTH EXPONENT AND BASE. The symbol 0^0 is not defined or used in this book.

EXERCISES 119, SET I. Find the value of each of the following expressions.

1. 3^3 2. 2^4 3. 5^2 4. 6^3 5. 7^2

6. 3^4 7. 0^3 8. 10^0 9. 10^1 10. 10^2

11. 10^3 12. 10^4 13. 10^5 14. 10^6 15. 5^0

16. 2^5 17. 2^6 18. 2^9 19. 2^8 20. 25^2

21. 40^3 22. 0^4 23. 12^3 24. 15^2 25. 1^5

EXERCISES 119, SET II. Find the value of each of the following expressions.

1. 3^2 2. 4^3 3. 5^4 4. 0^5 5. 10^0

6. 10^1 7. 10^4 8. 2^7 9. 3^5 10. 5^3

11. 15^2 12. 4^4 13. 1^4 14. 11^3

Powers of Ten

Powers in which the base is 10 have many important uses in mathematics and science. We show some powers of 10.

	Power of 10	Name
(a)	$10^0 = 1$	= 1 unit
(b)	$10^1 = 10$	= 1 ten
(c)	$10^2 = 100$	= 1 hundred
(d)	$10^3 = 1{,}000$	= 1 thousand
(e)	$10^4 = 10{,}000$	= 1 ten-thousand
(f)	$10^5 = 100{,}000$	= 1 hundred-thousand
(g)	$10^6 = 1{,}000{,}000$	= 1 million
	etc.	

In Example (d), 10^3 has a value of 1 followed by three zeros = 1,000 and is read "one thousand."

In Example (f), 10^5 has a value of 1 followed by five zeros = 100,000 and is read "one hundred-thousand."

Notice also that the successive names of the powers of 10 correspond exactly to the names of the places when we read or write a number. See Figure 120.

Power of 10	10^{14}	10^{13}	10^{12}	10^{11}	10^{10}	10^{9}	10^{8}	10^{7}	10^{6}	10^{5}	10^{4}	10^{3}	10^{2}	10^{1}	10^{0}
Place Name	hundred trillion	ten trillion	one trillion	hundred billion	ten billion	one billion	hundred million	ten million	one million	hundred thousand	ten thousand	one thousand	hundred (unit)	ten (unit)	one (unit)

Figure 120

MULTIPLYING A WHOLE NUMBER BY A POWER OF TEN. Suppose you make $5 a week for mowing the taking care of your neighbor's lawn. As you know, in 10 weeks you would make a total of $50. In 100 weeks you would make a total of $500. That is,

$$10 \times 5 = 50$$
$$100 \times 5 = 500$$
$$1,000 \times 5 = 5,000$$

Because $10 = 10^1$, $100 = 10^2$, $1,000 = 10^3$, and so on,

$$10^1 \times 5 = 50$$
$$10^2 \times 5 = 500$$
$$10^3 \times 5 = 5,000,\text{ and so on.}$$

> When a whole number is multiplied by a power of 10, follow the number by as many zeros as the exponent of 10.

Example 1

(a) $12 \times 10^3 = 12,000$ or $12 \times 1,000 = 12,000$

(b) $275 \times 10^4 = 2,750,000$ or $275 \times 10,000 = 2,750,000$

(c) $4,806 \times 10^2 = 480,600$ or $4,806 \times 100 = 480,600$

(d) $93 \times 10^5 = 9,300,000$ or $93 \times 100,000 = 9,300,000$

(e) Because of the Commutative Property of Multiplication,

$$100 \times 4 = 4 \times 100 = 400$$

(f) $1,000 \times 6 = 6,000$

(g) $10^3 \times 5 = 5,000$

(h) $10^0 \times 7 = 1 \times 7 = 7$

EXERCISES 120, SET I. In the following exercises, write answers without using written calculations.

1. 10×2 2. 2×10 3. 10×35

4. 35×100 5. $1,000 \times 27$ 6. $5 \times 10,000$

7. $10^2 \times 4$ 8. $10^3 \times 7$ 9. 100×10

10. 100×100 11. $1,000 \times 1,000$ 12. $100 \times 1,000$

13. $10^5 \times 7$ 14. $10^0 \times 8$ 15. 8×10^2

16. 5×10^3 17. 7×10 18. 9×10^0

19. $1,000 \times 84$ 20. $75 \times 10,000$

EXERCISES 120, SET II. In the following exercises, write answers without using written calculations.

1. 10×3 2. 10×46 3. 100×48

4. $10^3 \times 4$ 5. 10×100 6. $10^4 \times 8$

7. $100 \times 1,000$ 8. 7×10^2 9. $15 \times 10,000$

10. $88 \times 1,000$

 # Order of Operations

When multiplication and addition are written in the same expression, the multiplication is done first.

Example 1. Find the value of $2 + 5 \times 3$.

Solution: $2 + 15 = 17$

When powers and multiplication are written in the same expression, the powers are worked first.

Example 2. Find the value of $4(2^3)$.

Solution: $4(8) = 32$

When powers, multiplication, and addition are written in the same expression:

1. Raising to powers is done first.
2. Multiplication is done next.
3. Addition is done last.

Example 3. Find the value of $5 + 4(5^2) + 7 \cdot 8$.

Solution: $5 + 4(5^2) + 7 \cdot 8$

$= 5 + 4(25) + 56$ powers first

$= 5 + 100 + 56$ multiplication second

$= 161$ addition last

Example 4. Find the value of $6 \times 100 + 3(2^4) + 6 \cdot 9 + 7 + 3^3$.

Solution: $6 \times 100 + 3(2^4) + 6 \cdot 9 + 7 + 3^3$

$= 6 \times 100 + 3(16) + 6 \cdot 9 + 7 + 27$ powers first

$= 600 + 48 + 54 + 7 + 27$ multiplication second

$= 736$ addition last

EXERCISES 121, SET I. Find the value of each expression, following the proper order of operations.

1. $7 \times 5 + 45$

2. $10 + 8 \cdot 9$

3. $20 + 2^3$

4. $3^2 + 67$

5. $8 \cdot 7 + 9 \cdot 6 + 45$

6. $25 + 6 \times 7 + 13$

7. $(10^2)10 + 100$

8. $1,000 + (10^3)$

9. $2 \cdot 5^2 + 3 \cdot 2^2 + 4$

10. $2 \cdot 3 + 3 + (2^2)(3^2)$

11. $6^0 + 10^0 + 10 + 10^4$

12. $5^0 + 0^2 + 10^5$

EXERCISES 121, SET II. Find the value of each expression, following the proper order of operations.

1. $25 + 4 \times 9$

2. $7 \cdot 8 + 59$

3. $87 + 6 \times 5 + 3 \times 8$

4. $15 + 3^2$

5. $100(10) + 10^2$

6. $5 \cdot 2^3 + 3 \cdot 4$

7. $5^0 + 0^3 + 10^4$

Rounding off Whole Numbers

Numbers are often expressed to the nearest million, to the nearest thousand, or to the nearest hundred, etc. When they say it is 93,000,000 miles from the earth to the sun, it is understood that 93,000,000 has been rounded off to the nearest million. The distance around the earth at the equator is 24,902 miles, but in speaking of this distance it is more common to say 25,000 miles. We say that 24,902 miles has been "rounded off to the nearest thousand" miles. When a whole number is rounded off, we must say what place it has been rounded off to.

The symbol $\doteq$, read "is approximately equal to," is often used when numbers have been rounded off.

Example 1.

(a) 9**3**,004,000 ≐ 93,000,000 rounded off to the nearest *million*.

(b) 2**4**,902 ≐ 25,000 rounded off to the nearest *thousand*.

(c) **4**95 ≐ 500 rounded off to the nearest *hundred*.

(d) **5**10 ≐ 500 rounded off to the nearest *hundred*.

IDENTIFYING THE PLACE WE ARE ROUNDING OFF TO

Example 2. When rounding off 243,280 to the nearest thousand, the place we are rounding off to is shown shaded.

$$243\underline{3}280 \quad \text{————thousands place}$$

Example 3. When rounding off 5,624 to the nearest ten, the place we are rounding off to is shown shaded.

$$56\underline{2}4 \quad \text{———— tens place}$$

ROUNDING OFF WHOLE NUMBERS

CASE I. *When the first digit to the right of the place to which we are rounding off is less than 5* (0, 1, 2, 3, or 4),

1. The digit in the place we are rounding off to remains unchanged.
2. *All digits to the right* of the place we are rounding off to *are replaced by zeros.*
3. *All digits to the left* of the place we are rounding off to *remain unchanged.*

Example 4. Round off 243,280 to the nearest thousand.

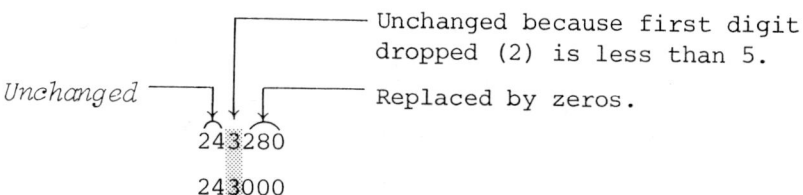

Therefore, 243,280 ≐ 243,000 rounded off to thousands.

Example 5. Round off 5,624 to the nearest ten.

```
                    ┌──────────── Unchanged because the first digit
                    │             dropped (4) is less than 5.
                    ↓
              5624

              5620
```

Therefore, 5,624 ≐ 5,620 rounded off to tens.

CASE II. *When the first digit to the right of the place we are rounding off to is greater than 5 (6, 7, 8, or 9),*

1. The digit in the place we are rounding off to is increased by one.
2. *All digits to the right* of the place we are rounding off to *are replaced by zeros.*
3. *All digits to the left* of the place we are rounding off to *remain unchanged.* (Special case: Example 8)

Example 6. Round off 90,671 to the nearest hundred.

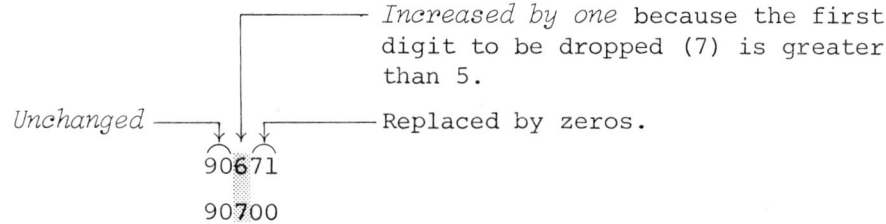

```
                        ┌──────── Increased by one because the first
                        │         digit to be dropped (7) is greater
                        │         than 5.
                        ↓
   Unchanged ──────┐  ┌───────── Replaced by zeros.
                   ↓ ↓↓
              90671

              90700
```

Therefore, 90,671 ≐ 90,700 rounded off to hundreds.

Example 7. Round off 53,817,000 to the nearest million.

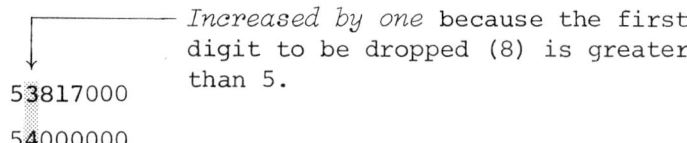

```
                        ┌──────── Increased by one because the first
                        │         digit to be dropped (8) is greater
                        ↓         than 5.
           53817000

           54000000
```

Therefore, 53,817,000 ≐ 54,000,000 rounded off to millions.

Example 8. Round off 31,972 to the nearest thousand.

```
                    ┌──────────── When this is 9 and must be in-
                    │             creased by 1, the 9 is replaced
                    ↓             by 0, and the first digit to its
           31972                  left is increased by 1.

           32000
```

Therefore, 31,972 ≐ 32,000 to the nearest thousand.

EXERCISES 122A, SET I. Round off the following numbers to the indicated place.

1.	4,728	Nearest hundred
2.	256,491	Nearest thousand
3.	926	Nearest ten
4.	28,619,000	Nearest million
5.	63,195	Nearest thousand
6.	28,232	Nearest hundred

7. 798	Nearest ten
8. 629,653	Nearest thousand
9. 52,461,000	Nearest million
10. 853	Nearest ten
11. 3,472	Nearest hundred
12. 78,415	Nearest ten-thousand

EXERCISES 122A, SET II. Round off the following numbers to the indicated place.

1. 3,446	Nearest hundred
2. 385,516	Nearest thousand
3. 54,500	Nearest thousand
4. 42,610,000	Nearest million
5. 365	Nearest ten
6. 26,417	Nearest ten-thousand
7. 8,451	Nearest thousand

CASE III. *When the first digit to the right of the place to which we are rounding off is 5,*

> A. *and there are some nonzero digits to the right of 5,* increase the preceding digit by one. (See Example 9.)
> B. *or 5 followed only by zeros,*
>
> > 1. If the digit in the place we are rounding off to is *even*, it remains unchanged. (See Example 10.)
> > 2. If the digit in the place we are rounding off to is *odd*, increase it by one. (See Example 11.)

Example 9. Round 422,501 off to the nearest thousand.

422501 — Increase 2 by one.

423000

Since the first digit to be dropped is 5 and there are some nonzero digits to the right of it, increase the preceding digit by one.

Therefore, 422,501 ≐ 423,000 rounded off to thousands.

Example 10. Round 76,450 off to the nearest hundred.

76450 — When this is *even*, it remains unchanged.

76400

Since the first digit to be dropped is 5 followed only by zeros, and the digit in the place we are rounding off to is *even*, it remains unchanged.

Therefore, 76,450 ≐ 76,400 rounded off to hundreds.

Example 11. Round 8,350 off to the nearest hundred.

Increase 3 by one.

8350

8400

Since the first digit to be dropped is 5 followed only by zeros, and the digit in the place we are rounding off to is *odd*, increase it by one.

Therefore, 8,350 $\doteq$ 8,400 to the nearest hundred.

Example 12. Round 999,507 to the nearest thousand.

999,507

1,000,000

Since the first digit to be dropped is 5 and there are some nonzero digits to the right of it, increase the preceding digit by one. This changes the preceding 999 to 1,000.

Therefore, 999,507 $\doteq$ 1,000,000 rounded off to thousands.

EXERCISES 122B, SET I. Round off the following numbers to the indicated place.

1.	753	Nearest hundred
2.	485	Nearest ten
3.	19,500,000	Nearest million
4.	26,500	Nearest thousand
5.	995	Nearest ten
6.	1,850	Nearest hundred
7.	39,500	Nearest thousand
8.	86,500,001	Nearest million
9.	265	Nearest ten
10.	63,501	Nearest thousand

EXERCISES 122B, SET II. Round off the following numbers to the indicated place.

1.	1,351	Nearest hundred
2.	745	Nearest ten
3.	82,500	Nearest thousand
4.	36,500,001	Nearest million
5.	199,500	Nearest thousand

The three cases of rounding off are summarized as follows.

To Round Off a Whole Number

CASE I. When the first digit to be dropped is less than 5,

```
       ↓————————Unchanged
    347216
    347000
```

Therefore, 347,216 ≐ 347,000 rounded off to thousands.

CASE II. When the first digit to be dropped is greater than 5,

```
       ↓————————Increased by 1
    9482
    9500
```

Therefore, 9,482 ≐ 9,500 rounded off to hundreds.

CASE III. When the first digit to be dropped is 5,

A. and there are some nonzero digits to the right of 5,

```
       ↓————————Increased by 1
    26501
    27000
```

Therefore, 26,501 ≐ 27,000 rounded off to thousands;

B. or 5 followed only by zeros, and

1. digit in place we are rounding off to is *even*,

```
       ↓————————Even unchanged
    745
    740
```

Therefore, 745 ≐ 740;

2. digit in place we are rounding off to is *odd*,

```
       ↓————————Odd is increased by 1
    91500
    92000
```

Therefore, 91,500 ≐ 92,000.

When the first digit dropped is 5, there are other methods of rounding off which are sometimes used. We have described the method commonly used in mathematics.

The following exercise set gives practice in all three cases of rounding off.

EXERCISES 122C, SET I. Round off the following numbers as indicated.

1. Depth of the deepest well:
   30,050 feet . . . . Nearest thousand
2. Height of Mt. Whitney:
   14,495 feet . . . . Nearest thousand
3. The Dead Sea is 1,299 feet
   below sea level . . . . Nearest hundred
4. The lowest point in Death
   Valley is 282 feet below
   sea level . . . . Nearest hundred
5. In 1970 the population of the
   United States was 204,765,770 . . Nearest million
6. In 1970 the population of the
   Soviet Union was 241,748,000. . . Nearest million
7. The speed of light is said to
   be 186,380 miles per second . . . Nearest thousand
8. The area of Africa is
   11,506,000 square miles . . . . . Nearest million
9. The area of North America is
   9,390,000 square miles. . . . . . Nearest hundred-thousand
10. The length of the Nile River
    is 4,145 miles. . . . . Nearest hundred
11. The length of the Mississippi
    River is 2,348 miles. . . . . . . Nearest hundred
12. The speed of a point on the
    earth's equator due to the
    rotation of the earth is
    1,040 miles per hour. . . . . . . Nearest hundred
13. The velocity of Los Angeles due
    to the rotation of the earth
    is 859 miles per hour . . . . . . Nearest ten
14. The depth of Lake Victoria is
    265 feet. . . . . Nearest ten
15. $245,429,444. . . . . Nearest ten-thousand
16. $245,429,444. . . . . Nearest hundred-million

EXERCISES 122C, SET II. Round off the following numbers as indicated.

1. Depth of the Caspian Sea:
   3,360 feet. . . . . Nearest hundred
2. Height of Mt. McKinley:
   20,320 feet . . . . Nearest thousand
3. The population of North
   America: 342,700,000. . . . . . . Nearest million
4. The speed New York City is
   moving east due solely to the
   rotation of the earth about its
   axis: 788 miles per hour. . . . . Nearest ten

5. The population of South
   America: 219,000,000. . . . . . . Nearest hundred-million
6. The area of Greenland:
   840,000 square miles. . . . . . . Nearest hundred-thousand
7. The earth's deepest gorge (Hell's
   Canyon, Idaho): 7,900 feet. . . . Nearest thousand
8. $245,000. . . . . Nearest ten-thousand

Chapter Summary

Numbers. *Natural numbers* are the numbers that start with 1 and
continue on forever. When 0 is added to the natural numbers,
we have the *whole numbers*. The *digits* are the first ten whole
numbers. The digits are the building blocks used in writing
numbers, no matter how large or small. We show the natural
numbers, whole numbers, and digits on the number line in Figure 123.

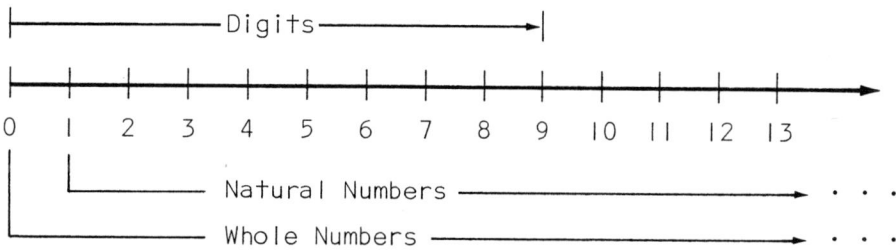

Figure 123

To Read Whole Numbers (Sec. 102)

   1. Start at the right end of the number and separate the
      digits into groups of three by means of commas.
   2. Start at the left end of the number and read the first
      group and follow it by that group's name. Continue in
      this way until all groups have been read.

To Write Whole Numbers in Words. (Sec. 102) Write the number
in the same way it is read, putting commas in the same places
they appear when the number is written in digits.

Powers of Whole Numbers. (Sec. 119) In the expression $2^3 = 8$,
2 is the *base*, 3 is the *exponent*, and 8 is the *power*.

Repeated Operations. (Secs. 103, 119) *Addition* is repeated
counting; *multiplication* is repeated addition; and *raising a
number to a power* is repeated multiplication.

Commutative and Associative Properties (Secs. 103, 104, 110)

   1. The *Commutative Property* says that *changing the order*
      of the numbers in an addition or multiplication problem
      gives the same answer.
   2. The *Associative Property* says that *changing the group-
      ing* of the numbers in an addition or multiplication
      problem gives the same answer.

Examples

Commutative Property of Addition: $4 + 3 = 3 + 4$

Commutative Property of Multiplication: $4 \times 3 = 3 \times 4$

Associative Property of Addition: $2 + (3 + 4) = (2 + 3) + 4$

Associative Property of Multiplication: $2(3 \times 4) = (2 \times 3)4$

REVIEW EXERCISES 123, SET I

1. Write the number 3,075,600,008 in words.
2. Write "five million, seventy-two thousand, six" using digits.
3. Find the sum of $7,825 + 84 + 900 + 45,788 + 9 + 2,000,085$.
4. Find the following sums:

(a) 756
    384

(b) 1,045
      896

(c) 785
    29
    396

(d) 78,619
    788
    6,907

5. Find the following products:

(a) 786
    35

(b) 342
    28

(c) 2,847
    9

(d) 3,967
    867

(e) $704 \times 9,207$ (f) $10^3 \cdot 586$ (g) $3,500 \times 176,000$

6. Find the following products:

(a) 295
    42

(b) 7,814
    8

(c) 219
    35

(d) 7,564
    789

(e) $10^2 \cdot 54$ (f) $2,800 \times 175$ (g) $2,005 \times 78,186$

In Exercises 7-12, find the value of each expression using the correct order of operations.

7. $8 + 7 \cdot 9$ 8. $8(3^4) + 5 \times 10$ 9. $5 \times 0 + 8 \cdot 9 + 0$

10. $2 + 3^2$ 11. $25 \times 4 + 10^2$ 12. $0(10^5) + 10^3$

13. Find the number of square feet in a rectangle that is 3 feet long and 2 feet wide.
14. Fencing costs $7 per foot. What will 256 feet of fencing cost?
15. How many square feet of tile are needed to cover a rectangular floor that is 12 feet long and 9 feet wide?
16. Find the cost to carpet a rectangular room that is 7 yards long and 5 yards wide if carpet costs $6 a square yard to install.
17. Each tanker in an oil fleet can carry 503,024 gallons of oil. If the fleet has 207 tankers, how many gallons of oil can the fleet transport at once?
18. The average page in a particular textbook has about 703 words. If there are 356 pages in the book, how many words would you estimate there are in the book?
19. Each student at a particular college is charged a $6 fee to cover the cost of his student activities. How much is added to the student body fund if there are 7,568 student

fees paid this semester?

20. Two hundred seventy-five families are invited to a neighborhood picnic. The planning committee estimates the average family size to be four. How many people should they plan on providing refreshments for?

21. Enrique has a job paying $168 per week. How much does he earn in one year (52 weeks)?

22. Suppose the gasoline tank of your car holds 22 gallons, and you average 16 miles to the gallon. How far can you drive with a tankful of gasoline?

23. Suppose you were offered a job that pays 2¢ the first day, 4¢ the second day, 8¢ the third day, and continues to increase in this manner for 30 days. Study the following short table to discover how to work the following exercises.

Days worked	Pay per day
1	$2¢ = 2^1¢$
2	$4¢ = 2^2¢$
3	$8¢ = 2^3¢$

(a) How much would you make the tenth day?
(b) What would your total salary for the first ten days be?
(c) How much would you make the fifteenth day?
(d) How much would you make the twentieth day?

If you are ever offered a job of this kind, take it! Your pay on the thirtieth day would be $10,737,418.24, and the total amount you would make in 30 days would be $21,474,836.47. However, don't get excited, because we are certain you will not be offered a 30-day contract of this kind.

24. Brain Teaser. Two fishermen, each weighing 200 pounds, and their two sons, weighing 100 pounds each, cross a river in a small boat which can carry only 200 pounds. How do they all manage to get across?

25. Brain Teaser. Are the operations of putting on your shoes and socks commutative?

REVIEW EXERCISES 123, SET II

1. Write the number 2,000,500,064,006 in words.
2. Find the sum of 387 + 63 + 4,300 + 8 + 5,006,041.
3. Find the following products:

(a) 684
       9

(b) 753
      46

(c) 5,905
      804

(d) 4,908
      205

(e) $10^4 \times 56$

(f) 2,500 × 908,000

In Exercises 4-6, find the value of each expression.

4. $5 \cdot 9 + 7$ 5. $6 \times 0 + 5 \cdot 4 + 3$ 6. $10^3 + 10(10^2)$

7. Find the number of square feet in a rectangle that is 17 feet long and 12 feet wide.
8. Find the cost to carpet a rectangular room that is 6 yards wide and 8 yards long if carpet costs $9 a square yard to install.
9. The average page in a particular book has about 645 words. If there are 427 pages in the book, how many words would you estimate there are in the book?
10. Mary Lou has a job that pays her $4 an hour. How much does she earn in a month in which she works 156 hours?
11. Suppose the gasoline tank of your car holds 24 gallons and you average 15 miles to the gallon. How far can you drive with a tankful of gasoline?
12. Suppose you were offered a job that pays 3¢ the first day, 9¢ the second day, 27¢ the third day, and continues to increase in this manner for 30 days. Study the following short table to discover how to work the following exercises.

Days worked	Pay per day
1	$3¢ = 3^1¢$
2	$9¢ = 3^2¢$
3	$27¢ = 3^3¢$

(a) How much would you make the fifth day?
(b) How much would you make the tenth day?

At this rate you would make over two trillion dollars the thirtieth day.

Chapter 1: Diagnostic Test

Name_____

_____

The purpose of this test is to see how well you understand addition, multiplication, and powers of whole numbers. We recommend that you work this diagnostic test *before* your instructor tests you on this chapter. Allow yourself about an hour to do this test.

Complete solutions for all the problems on this test, together with section references, are given in the Answer Section at the end of the book. You should study the sections referred to for the problems you do incorrectly.

1. Write all the digits greater than 7.

   (1) _____

2. Write all the whole numbers less than 4.

   (2) _____

3. Write the largest two-digit natural number.

   (3) _____

4. Which of the two symbols, > or < , should be used to make each statement true?

   (a) 6 _?_ 14   (4a) _____

   (b) 12 _?_ 0   (4b) _____

5. Light travels about 5,879,200,000,000 miles in 1 year. Write this number in words.

   (5) _____

   _____

6. Use digits to write the number fifty-four billion, seven million, five hundred six thousand, eighty.

   (6) _____

7. Add horizontally. 27 + 35 + 19 + 8 + 63 + 90

   (7) _____

8. Add, then check your answer by casting out nines.

```
7,856    _____
3,942    _____
8,674    _____   } 9-rems
3,597    _____
```

(8) _____

---

9. Arrange the following numbers in a vertical column, then add: 75; 3,086; 70,500,006; 108; 8,009.

(9) _____

---

10. Multiply: (a) 7,546   (b) 3,084
                     89   706

(10a) _____

(10b) _____

11. Multiply: 75,000 × 8,600

(11) _____

12. Multiply, then check your answer by casting out nines.

```
836    _____
 74    _____   } 9-rems

       _____  →  _____
```

_____   (12) _____

---

13. In the following, just look at the problem, then write your answer. Do not use written calculations.

(a) 75 × 100   (13a) _____

(b) 8 × 10³   (13b) _____

(c) 300 × 1,000   (13c) _____

(d) 508 × 10⁴   (13d) _____

(e) 10² × 10³   (13e) _____

---

14. Find the value of 29^2.

(14) _____

---

15. Find the value of each of the following expressions, being careful to use the correct order of operations.

(a) $10 + 8 \cdot 9$ (15a) _____

(b) $2^3 \cdot 3 + 5$ (15b) _____

(c) $5^2 + 6^1 + 10^0 + 10^3$ (15c) _____

---

16. A rectangular play area is shown.

(a) Find the perimeter of this play area.

65 ft

48 ft

(16a) _____

(b) Find the cost of fencing this play area if the fencing costs $4 per foot of length.

(16b) _____

---

17. John has $45. Harry has $23. Bill has $12 more than John and Harry together. Find the total amount of money the three have together.

(17) _____

---

18. Mr. Matucek plans to carpet his living room and a bedroom. The living room is 7 yards by 5 yards. The bedroom is 4 yards by 4 yards.

(a) Find the total area of the two rooms in square yards.

(18a)_____

(b) Find the total cost of carpeting the two rooms if the carpet costs $6 a square yard installed.

(18b)_____

_____

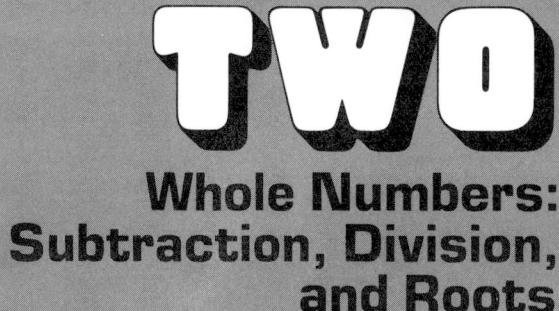

TWO

Whole Numbers: Subtraction, Division, and Roots

In Chapter 1 we showed that addition is repeated counting, multiplication is repeated addition, and raising a number to a power is repeated multiplication. In this chapter we reverse the operations. We will show that subtraction is the inverse of addition, division is the inverse of multiplication, and finding roots is the inverse of raising a number to a power.

Subtraction

Suppose you want to buy a second-hand car that costs $1,000. If you have $200, how much more must you add to your $200 to make the $1,000? That is,

(a) $200 + ? = $1,000

You can see that $800 must be added to $200 to make $1,000. We could write the problem another way:

Subtraction symbol

(b) $1,000 – $200 = ?   (Read "$1,000 *minus* $200 = ?")

You know the answer is $800. We say "$200 has been *subtracted* from $1,000."

In statement (a) we used addition. In statement (b) we used subtraction.

<u>TERMS USED IN SUBTRACTION</u>. The problem just discussed can also be written as follows:

$$\begin{array}{rl} \$1,000 & \textit{minuend} \\ -\ \ 200 & \textit{subtrahend} \\ \hline \$\ \ 800 & \textit{difference} \end{array}$$

Note that 5 is where we begin and 5 is where we end.

Example 1   5 + 4 = 9,   9 – 4 = 5
                    └addition   └subtraction
                      of 4   of 4

In Example 1, a subtraction of 4 "undoes" an addition of 4. For this reason, subtraction is called the *inverse* of addition.

<u>METHODS OF SUBTRACTION</u>. There are several different ways of subtracting whole numbers. We will discuss only two methods:

1. The Austrian Method
2. The Take-away Borrow Method

We think the Austrian Method is the simplest and fastest way to do subtraction. This method of subtraction is taught in many countries, such as Austria, Mexico, parts of Canada, South America, Germany, Israel; it is also used by many people in the United States. By using the Austrian Method (1) we avoid having to learn the Basic Subtraction Facts, and (2) we strengthen our knowledge of the Basic Addition Facts.

Austrian Subtraction Method

The Austrian Method changes a subtraction problem into an addition problem. For example, in solving 8 - 2 we ask ourselves "2 + ? = 8." Then

$$\begin{array}{r} 8 \\ -\ 2 \\ \hline 6 \end{array} \quad \text{because} \quad 2 + 6 = 8$$

Study the following examples to familiarize yourself with this method.

Example 1

(a) $\begin{array}{r} 8 \\ -\ 6 \\ \hline ? \end{array}$ We ask ourselves "6 + ? = 8." $\begin{array}{r} 8 \\ -\ 6 \\ \hline 2 \end{array}$
Since 6 + 2 = 8, then

(b) $\begin{array}{r} 5 \\ -\ 2 \\ \hline ? \end{array}$ We ask ourselves "2 + ? = 5." $\begin{array}{r} 5 \\ -\ 2 \\ \hline 3 \end{array}$
Since 2 + 3 = 5, then

(c) $\begin{array}{r} 7 \\ -\ 3 \\ \hline 4 \end{array}$ because 3 + 4 = 7

(d) 11 - 7 = 4   because 7 + 4 = 11

(e) 14 - 8 = 6   because 8 + 6 = 14

(f) 15 - 0 = 15 because 0 + 15 = 15

(g) $\begin{array}{r} 7 \\ -\ 5 \\ \hline 2 \end{array}$ because 5 + 2 = 7

(h) $\begin{array}{r} 9 \\ -\ 5 \\ \hline 4 \end{array}$ because 5 + 4 = 9

(i) $\begin{array}{r} 4 \\ -\ 0 \\ \hline 4 \end{array}$ because 0 + 4 = 4

(j) $\begin{array}{r} 0 \\ -\ 0 \\ \hline 0 \end{array}$ because 0 + 0 = 0

(k) $\begin{array}{r} 3 \\ -\ 3 \\ \hline 0 \end{array}$ because 3 + 0 = 3

BASIC SUBTRACTION COMBINATIONS. Verify the following differences by thinking in terms of addition. Place a paper over the answers, then write or say the differences as fast as possible. You should practice on these examples until you can give each difference without hesitation.

DRILL IN SUBTRACTING ONE-DIGIT NUMBERS

1.	2 - 1 1	3 - 2 1	4 - 1 3	1 - 1 0	3 - 1 2	4 - 2 2	3 - 0 3	5 - 1 4	6 - 2 4
2.	2 - 2 0	4 - 3 1	5 - 2 3	6 - 3 3	5 - 3 2	6 - 4 2	7 - 1 6	8 - 2 6	7 - 3 4
3.	7 - 2 5	4 - 4 0	6 - 5 1	5 - 4 1	6 - 0 6	7 - 4 3	8 - 3 5	6 - 1 5	6 - 6 0
4.	5 - 5 0	7 - 5 2	8 - 4 4	7 - 6 1	8 - 5 3	7 - 7 0	9 - 2 7	10 - 4 6	9 - 4 5
5.	8 - 6 2	9 - 3 6	10 - 2 8	10 - 6 4	9 - 5 4	10 - 3 7	9 - 6 3	9 - 8 1	8 - 7 1

SUBTRACTION WHEN MINUEND AND SUBTRAHEND HAVE MORE THAN ONE DIGIT

Example 2. Find 578 - 346.

Solution: Always write units digits below units digits, tens digits below tens digits, hundred digits below hundreds digits, etc.

        578   Think: "6 + ? = 8." The answer to this question
      - 346   is 2. Therefore, write 2 in the units column. In
        232   like manner, 4 + 3 = 7 and 3 + 2 = 5.

Example 3. Subtract 2,502 from 78,514.

        78,514   minuend
      - 2,502   subtrahend
        76,012   difference

EXERCISES 202A, SET I. In Exercises 1-4, do the indicated subtractions.

	(a)	(b)	(c)	(d)
1.	7,564 - 2,341	38,921 - 17,211	77,806 - 35,002	91,105 - 1,102
2.	6,278 - 170	5,787 - 782	87,909 - 405	65,678 - 2,356
3.	9,673 - 3,542	2,745 - 324	84,273 - 61,022	14,863 - 3,521

4.
(a)	(b)	(c)	(d)
5,497	62,483	72,156	325,764
− 2,105	− 1,332	− 21,034	− 12,432

5. Subtract 281 from 785.

6. Subtract 10,322 from 18,566.

7. The average distance to the sun is about 92,889,000 miles. The average distance to the moon is about 239,000 miles. How much farther is it to the sun than to the moon?

8. A man gave his son the following riddle. "Take a 5-gallon and a 3-gallon bucket down to the lake, and bring back exactly 4 gallons of water." How would you solve this riddle?

EXERCISES 202A, SET II. In Exercises 1 and 2, do the indicated subtractions.

1.
(a)	(b)	(c)	(d)
8,473	45,869	62,905	86,027
− 5,241	− 22,513	− 42,303	− 3,024

2.
(a)	(b)	(c)	(d)
7,592	3,957	56,208	16,084
− 4,380	− 241	− 16,005	− 5,062

3. Subtract 524 from 876.

4. On a particular day the sun is about 92,786,000 miles from earth, while the moon is only about 245,000 miles. How much farther is it to the sun than to the moon?

CHECKING SUBTRACTION. Subtraction is checked very easily by addition. When subtraction was explained in Section 201, we saw that

$$subtrahend + difference = minuend$$

Example 4

(a)
$$
\begin{array}{r}
6 \\
- 2 \\
\hline
4
\end{array}
\quad
\begin{array}{l}
\text{minuend} \\
\leftarrow \text{subtrahend} \\
\leftarrow \text{difference}
\end{array}
\quad
\begin{array}{r}
2 \\
+ 4 \\
\hline
6 \quad \text{minuend}
\end{array}
$$

This same check can be done as follows:

check:
$$
\begin{array}{r}
6 \\
- 2 \\
+ 4 \\
\hline
6
\end{array}
\quad \text{add}
$$

(b) check:
$$
\begin{array}{r}
426 \\
- 359 \\
+ 67 \\
\hline
426
\end{array}
\quad \text{add}
$$

(c) check:
$$
\begin{array}{r}
3,010 \\
- 1,324 \\
+ 1,686 \\
\hline
3,010
\end{array}
\quad \text{add}
$$

Subtraction can also be checked by casting out nines. However, we think the method just shown is faster and easier.

<u>EXERCISES 202B, SET I</u>. In Exercises 1-10, subtract, then check by adding the difference and subtrahend to get the minuend.

1. 37	2. 48	3. 79	4. 83	5. 7,864
− 24	− 15	− 51	− 30	− 2,033

6. 8,907	7. 2,806	8. 5,568	9. 7,186	10. 98,765
− 702	− 1,502	− 2,563	− 7,136	− 4,565

<u>EXERCISES 202B, SET II</u>. In Exercises 1-5, subtract, then check by adding the difference and subtrahend to get the minuend.

1. 65	2. 97	3. 5,948	4. 8,703	5. 74,096
− 22	− 61	− 3,024	− 5,102	− 2,063

<u>SUBTRACTION WHEN SOME DIGITS OF SUBTRAHEND ARE LARGER THAN SOME DIGITS OF MINUEND</u>. We illustrate by example.

Example 5

$$8^13$$
$$- \; 2 \; 7$$
$$\underline{\quad 1 \quad}$$
$$5 \; 6$$

Since 7 is greater than 3, we can't add anything to 7 and expect to get 3. Since this is the case, we write a small 1 to the left of the 3 (making it 13), and write a small 1 under the 2 (making it 3). Then because 7 + 6 = 13, we write 6 in the answer. And because 3 + 5 = 8, we write 5 in the answer. (1+2)

Example 6

$$9^11$$
$$- \; 3 \; 8$$
$$\underline{\quad 1 \quad}$$
$$5 \; 3$$

Because 8 + 3 = 11, we write 3 in the answer.
Because 4 + 5 = 9, we write 5 in the answer. (3+1)

Example 7

$$6 \; 9^12$$
$$- \; 4 \; 5 \; 6$$
$$\underline{\quad 1 \quad}$$
$$2 \; 3 \; 6$$

Because 6 + 6 = 12, we write 6 in the answer.
Because 6 + 3 = 9, we write 3 in the answer. (5+1)
Because 4 + 2 = 6, we write 2 in the answer.

<u>Explanation</u>: In Example 5, by writing the small 1 to the left of the 3 and making it 13, we have added 10 to the minuend. By writing a 1 under the 2 (in the tens column) we have arranged to take another 10 away from the minuend. By adding one 10 *to* and subtracting one 10 *from* the minuend, we have not changed its value.

Austrian Method of Subtraction

1. If a digit in the subtrahend is greater than the digit appearing above it in the minuend, write a small one in front of that digit in the minuend, and a small one under the next digit to the left in the subtrahend.
2. If the digit in the subtrahend is less than or equal to the digit appearing above it in the minuend, subtract as before.

Example 8

$$4^1 0\ 3$$
$$-\ \ 5\ 1$$
$$\underline{\ _1\qquad\ }$$
$$3\ 5\ 2$$

$1 + 2 = 3$
$5 + 5 = 10$
$1 + 3 = 4$

Example 9

$$7,^1 0\ 9^1 2$$
$$-\ 5,3\ 4\ 8$$
$$\underline{\ _1\ \ \ _1\ \ }$$
$$1,7\ 4\ 4$$

$8 + 4 = 12$
$5 + 4 = 9$
$3 + 7 = 10$
$6 + 1 = 7$

Example 10

$$5^1 2,^1 0\ 9^1 3$$
$$-\ 4,1\ 6^.7$$
$$\underline{\ _1\ _1\ \ \ _1\ }$$
$$4\ 7,9\ 2\ 6$$

$7 + 6 = 13$
$7 + 2 = 9$
$1 + 9 = 10$
$5 + 7 = 12$
$1 + 4 = 5$

Example 11

$$7^1 1,^1 0^1 0^1 3$$
$$-\ \ 9,4\ 5\ 6$$
$$\underline{\ _1\ _1\ \ _1\ _1\ }$$
$$6\ 1,5\ 4\ 7$$

$6 + 7 = 13$
$6 + 4 = 10$
$5 + 5 = 10$
$10 + 1 = 11$
$1 + 6 = 7$

EXERCISES 202C, SET I. Find the following differences.

	(a)	(b)	(c)	(d)	(e)	(f)	(g)
1.	25 − 8	74 − 9	36 − 7	55 − 8	92 − 5	80 − 6	88 − 9
2.	52 − 7	47 − 9	63 − 5	24 − 8	60 − 6	91 − 4	78 − 9
3.	42 − 28	54 − 37	25 − 19	73 − 46	60 − 54	81 − 62	97 − 79
4.	43 − 15	72 − 38	54 − 27	61 − 39	70 − 14	60 − 28	50 − 44

5.

(a)	(b)	(c)	(d)	(e)	(f)	(g)
723	425	684	541	256	300	802
− 218	− 157	− 295	− 207	− 177	− 157	− 509

6.

(a)	(b)	(c)	(d)	(e)	(f)	(g)
315	642	308	790	615	833	427
− 107	− 208	− 79	− 156	− 287	− 295	− 388

7.

(a)	(b)	(c)	(d)	(e)	(f)
2,108	6,914	3,005	1,804	60,701	28,964
− 1,896	− 6,057	− 684	− 308	− 10,808	− 4,972

8.

(a)	(b)	(c)	(d)
7,164	2,107	60,084	693,421
− 586	− 308	− 10,529	− 355,555

(e)	(f)	(g)
173,041	7,060,503	500,014
− 88,350	− 6,829,711	− 34,926

9. If the population of China is 750,000,000 and the population of the United States is 204,776,770, how many more people are there in China than in the United States?

10. If at the beginning of a trip your odometer* reading was 67,856 miles and at the end of the trip it read 71,304 miles, how many miles did you drive?

11. The height of Mt. Everest is twenty-nine thousand, twenty-eight feet. The height of Mt. Whitney is fourteen thousand, four hundred ninety-five feet. How much higher is Mt. Everest than Mt. Whitney?

12. A new car had a list price of $4,120. Mr. Jones bought the car and was given a discount of $455. What did he pay for the car?

13. After trading his car in on a new car, Mr. Perez owed a balance of $2,664. He pays this off in equal monthly payments of $74 per month. Find the unpaid balance after he has made 27 payments.

14. At the beginning of the month Mr. Hanson's checking account balance was $356. He made deposits of $225, $57, and $375. He wrote checks for $56, $135, $157, $38, and $417. What was his balance at the end of the month?

15. Jim has $75 and Joe has $57. Jack has $17 more than Jim, and Mike has $13 less than Jim and Joe together. How much money do all four boys have together?

16. Brain Teaser. Two coins have a value of 60¢. One is not a dime. What are the coins?

17. Brain Teaser. Four sheep ahead of a sheep, four sheep behind a sheep, and a sheep in the middle. How many sheep

---

*The instrument on your car that is commonly called the "speedometer" is both a speedometer and an odometer. The part that tells how many miles the car has been driven is the odometer. The part that tells how fast you are going is the speedometer.

are there?

EXERCISES 202C, SET II. Find the following differences.

1. (a)   (b)   (c)   (d)   (e)   (f)   (g)

(a)	(b)	(c)	(d)	(e)	(f)	(g)
31	65	47	73	82	70	68
− 7	− 8	− 9	− 5	− 6	− 7	− 9

2.

(a)	(b)	(c)	(d)	(e)	(f)	(g)
73	46	67	82	90	62	51
− 38	− 17	− 49	− 56	− 73	− 24	− 18

3.

(a)	(b)	(c)	(d)	(e)	(f)	(g)
647	401	370	752	983	502	800
− 309	− 164	− 283	− 318	− 427	− 236	− 591

4.

(a)	(b)	(c)	(d)	(e)	(f)
7,825	3,209	5,002	2,083	40,902	36,855
− 6,309	− 2,754	− 698	− 406	− 20,903	− 5,827

5.

(a)	(b)	(c)	(d)	(e)
52,703	40,000	275,061	603,405	8,500,906
− 5,047	− 29,056	− 90,909	− 324,197	− 7,820,952

6. If the population of Asia is 2,316,312,000 and the population of North America is 342,700,000, how many more people are there in Asia than in North America?

7. The highest point in North America, Mt. McKinley, is twenty thousand, three hundred twenty feet. The highest point in South America, Mt. Aconcagua, is twenty-two thousand, eight hundred thirty-four feet. How much higher is Mt. Aconcagua than Mt. McKinley?

8. After trading in her old car for a new one, Beth owed a balance of $3,329. She pays this off in equal monthly payments of $89 per month. Find the unpaid balance after she has made 28 payments.

9. At the beginning of the month Mr. Rodarte's checking account balance was $417. He made deposits of $179, $83, and $216. He wrote checks for $68, $24, $162, $46, and $185. What was his balance at the end of the month?

Take-Away Borrow Method

In Section 202 we explained how to subtract using the Austrian Method. We hope students who have tried that method are just as convinced as we are that it is the best method for subtraction. If you like the Austrian Method of subtraction and wish to continue using it, then it is not necessary for you to study this section. At this time, we present the Take-Away Borrow Method that is commonly used. Those students familiar with this method may wish to continue using it.

Example 1. Find
$$83$$
$$-\ 27$$

Solution:

```
  7 13
  8 3
- 2 7
  5 6
```

We borrow 1 ten from 8 tens, leaving 7 tens. The borrowed 1 ten (= 10 units) is added to the 3 units we already have, making 13 units.
Then, 13 - 7 = 6 units,
and 7 - 2 = 5 tens.

Example 2. Find
$$692$$
$$-\ 456$$

Solution:

```
  8 12
  6 9 2
- 4 5 6
  2 3 6
```

We borrow 1 ten from 9 tens, leaving 8 tens. The borrowed 1 ten (= 10 units) is added to the 2 units we already have, making 12 units.
Then, 12 - 6 = 6 units,
and 8 - 5 = 3 tens,
and 6 - 4 = 2 hundreds.

Example 3. Find   52,093
$$-\ 4,167$$

Solution:

```
   4 11
   1   10 8 13
  5 2 , 0 9 3
- 4 , 1 6 7
  4 7 , 9 2 6
```

We borrow 1 ten from 9 tens, leaving 8 tens.
        10 + 3 = 13 units
Then, 13 - 7 = 6 units,
and 8 - 6 = 2 tens.

We borrow 1 thousand from 2 thousands, leaving 1 thousand.
        10 + 0 = 10 hundreds
Then, 10 - 1 = 9 hundreds.

We borrow 1 ten-thousand from 5 ten-thousands, leaving 4 ten-thousands.
        10 + 1 = 11 thousands
Then, 11 - 4 = 7 thousands,
and 4 - 0 = 4 ten-thousands.

EXERCISES 203. Work Exercises 202C, Sets I and II, using the Take-Away Borrow Method.

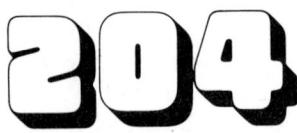

Subtraction Is Not Commutative

In Chapter 1 we found that the operations of addition and multiplication are both commutative. That is,

$$a + b = b + a \text{ and } a \cdot b = b \cdot a$$

That is, interchanging the order of the numbers does not change the sum or product. Can this be done with subtraction? We need only take a single example to see that *subtraction is not commutative.*

Example 1. 7 - 4 ≠ 4 - 7

The symbol ≠ is read "is not equal to." 7 - 4 = 3, whereas 4 - 7 does not represent a whole number.

Subtraction Is Not Associative

In Chapter 1 we found that the operations of addition and multiplication are both associative. That is

$$(a + b) + c = a + (b + c) \text{ and } (a \cdot b) \cdot c = a \cdot (b \cdot c)$$

That is, changing the way the numbers are grouped does not change the sum or product. Is this also true for subtraction? Again we need take only a single example to show that *subtraction is not associative.*

Example 2. (9 - 5) - 2 ≠ 9 - (5 - 2)

because (9 - 5) - 2 = 4 - 2 = 2

whereas 9 - (5 - 2) = 9 - 3 = 6

2 ≠ 6

Summary for Subtraction

Subtraction is the inverse of addition.
Subtraction is not commutative or associative.

PROCEDURE FOR SUBTRACTION. When any digit in the subtrahend is larger than its corresponding digit in the minuend, write a small 1 to the left of that digit in the minuend and add a small 1 to the digit of the bottom number in the next column to the left. To find the digits for the answer, ask yourself, "What must be added to the digit in the subtrahend to give the digit(s) in the minuend?" Remember, the small 1 written under a digit in the subtrahend increases that digit by one.

Division of Whole Numbers

THE MEANING OF DIVISION. The ordinary dictionary meaning of division is "the separating of something into parts." If we have 6 ÷ 2 (read "6 divided by 2"), we can think of this as a separation of 6 into two equal parts, each part being 3 (Figure 207A). We see from the figure that 6 can be divided into two equal parts, each part being 3, so that we say 6 ÷ 2 = 3.

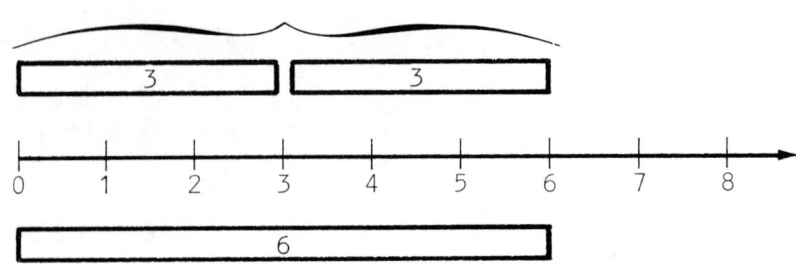

Figure 207A

The following example illustrates another important meaning of division.

Example 1. Suppose a man has $6 in his pocket. A store has a special on men's shirts for $2 each. How many of these shirts can he buy, ignoring sales tax? He has to figure how many times he can take $2 out of his $6. He can do this by seeing how many times he can *subtract* $2 from his $6.

```
He starts with      6                      You can see he
                  - 2  ←——1st shirt        would be able to
He has left         4                      buy three shirts
                  - 2  ←——2nd shirt        because we can
He has left         2                      subtract 2 from 6
                  - 2  ←——3rd shirt        a total of three
He has left         0                      times.
```

We see that in this case *division can be considered as repeated subtraction*.

We have shown that there are two ways of considering a division of whole numbers such as 6 ÷ 2:

1. The separation of 6 into 2 equal parts of 3 each.
2. The repeated subtraction of 2 from 6, 3 times.

Example 1 also shows an important relation between multiplication and division.

Multiplication	*Division* *(Inverse of multiplication)*
When we think	When we think

2 + 2 + 2 = 6,	6 - 2 ←——1st subtraction of 2 4 - 2 ←——2nd subtraction of 2 2 - 2 ←——③rd subtraction of 2 0
we say 3 · 2 = 6.	we say 6 ÷ 2 = ③
Here we have a *repeated addition* of three twos = 6.	Here we have a *repeated subtraction* of three twos from 6.

The division of 6 by 2 caused us to *subtract three twos* from 6, undoing the result of multiplying 3 by 2, which caused us to *add three twos*. For this reason, *division is called the inverse of multiplication*.

Example 2. Examples of Division.

(a) Find 12 ÷ 3.

$$
\begin{array}{r}
12 \\
- \ 3 \\
\hline
9 \\
- \ 3 \\
\hline
6 \\
- \ 3 \\
\hline
3 \\
- \ 3 \\
\hline
0
\end{array}
$$

   ←—— 1st subtraction of 3
   ←—— 2nd subtraction of 3
   ←—— 3rd subtraction of 3
   ←—— (4)th subtraction of 3

Therefore, 12 ÷ 3 = (4). (Note that 4 • 3 = 12.)

(b) Find 14 ÷ 7.

$$
\begin{array}{r}
14 \\
- \ 7 \\
\hline
7 \\
- \ 7 \\
\hline
0
\end{array}
$$

   ←—— 1st subtraction of 7
   ←—— (2)nd subtraction of 7

Therefore, 14 ÷ 7 = (2). (Note that 2 • 7 = 14.)

EXERCISES 207A, SET I. Find the following quotients by repeated subtraction.

1. 8 ÷ 2	2. 6 ÷ 3	3. 10 ÷ 5	4. 10 ÷ 2
5. 16 ÷ 4	6. 18 ÷ 6	7. 20 ÷ 4	8. 21 ÷ 3
9. 35 ÷ 7	10. 32 ÷ 8	11. 45 ÷ 5	12. 48 ÷ 6
13. 36 ÷ 9	14. 33 ÷ 11	15. 48 ÷ 12	16. 60 ÷ 15

17. Tickets to a show cost $3 each. How many tickets can you buy with $15?
18. If the price of beans is $4 a bushel, how many bushels can a farmer buy for $20?

EXERCISES 207A, SET II. Find the following quotients by repeated subtraction.

1. 12 ÷ 3	2. 15 ÷ 5	3. 14 ÷ 7	4. 24 ÷ 4
5. 45 ÷ 9	6. 40 ÷ 8	7. 44 ÷ 11	8. 60 ÷ 12

9. Racquet balls cost $2 a can. How many cans can you buy for $16?

TERMS USED IN DIVISION

The *dividend* is the number being divided.
The *divisor* is the number doing the dividing.
The *quotient* is the answer in a division problem. It is the number the divisor is multiplied by to give the dividend.

Refer to Figure 207B for help in identifying these terms.

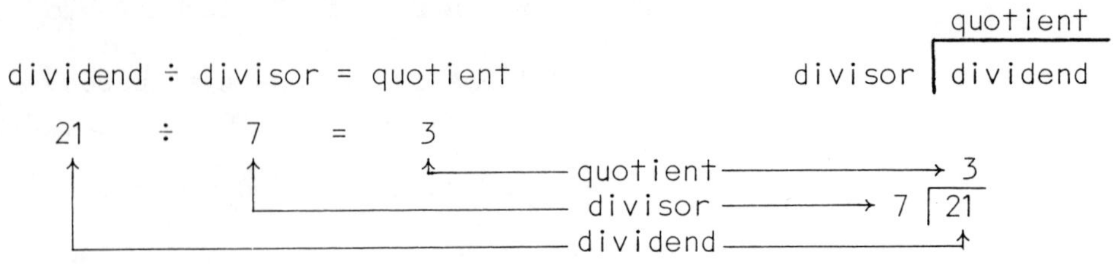

$$\text{dividend} \div \text{divisor} = \text{quotient} \qquad \text{divisor} \overline{\smash{)}\text{dividend}}^{\text{quotient}}$$

$$21 \div 7 = 3$$

quotient ⟶ 3

divisor ⟶ 7)‾21

dividend

Figure 207B

Other names are used for the various numbers in a division problem. Since 21 = 3 • 7, the numbers 3 and 7 are called *divisors* of 27. Also, 21 is said to be a *multiple* of 3 as well as of 7. We say that 7 (and 3 as well) *divides 21 exactly* or that 3 (and 7 as well) *goes into 21 evenly.*

Example 1

   (a) 15 is a multiple of 1, 3, 5, 15; and 1, 3, 5, 15 are divisors of 15 because 15 = 1 • 15 = 3 • 5.

   (b) 12 is a multiple of 1, 2, 3, 4, 6, 12; and 1, 2, 3, 4, 6, 12 are divisors of 12 because 12 = 1 • 12 = 2 • 6 = 3 • 4.

   (c) 34 is a multiple of 1, 2, 17, 34; and 1, 2, 17, 34 are divisors of 34 because 34 = 1 • 34 = 2 • 17.

EXERCISES 207B, SET I. Find all the divisors for each of the following numbers.

1. 18   2. 12   3. 35   4. 48   5. 39

6. 54   7. 7   8. 64   9. 13   10. 81

EXERCISES 207B, SET II. Find all the divisors for each of the following numbers.

1. 14   2. 17   3. 22   4. 16   5. 42

AN IMPORTANT RELATION IN A DIVISION PROBLEM

$$\boxed{\text{divisor} \bullet \text{quotient} = \text{dividend} \qquad \text{divisor} \overline{\smash{)}\text{dividend}}^{\text{quotient}}}$$

Example 2

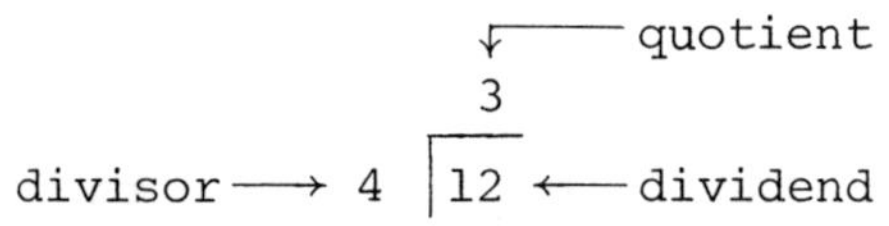

$$\text{divisor} \longrightarrow 4 \overline{\smash{)}12}^{3 \; \text{quotient}} \longleftarrow \text{dividend}$$

$$\text{divisor} \times \text{quotient} = \text{dividend}$$
$$4 \times 3 = 12$$

Division by Zero Not Possible

Let's try to divide some number (not 0) by 0.
For example: $4 \div 0 = 0 \overline{)4}$. Suppose the quotient
is some unknown number we call x. Then

$0 \overline{)\overset{x}{4}}$ means $x \cdot 0 = 4$, which is certainly false.

$x \cdot 0 \neq 4$ because any number multiplied by $0 = 0$.
Therefore, dividing any number (except 0) by 0
is impossible. What about $0 \div 0$? Consider the
following examples:

$0 \overline{)\overset{1}{0}}$ means $1 \cdot 0 = 0$, which is true.

$0 \overline{)\overset{17}{0}}$ means $17 \cdot 0 = 0$, which is true.

$0 \overline{)\overset{156}{0}}$ means $156 \cdot 0 = 0$, which is true.

In other words, $0 \div 0 = 1$, 17, and also 156. In
fact, it can be any number. Therefore, we don't
know what answer to put for $0 \div 0$. For these
reasons, we say *division by 0 is not possible*.

ZERO DIVIDED BY ANY NUMBER OTHER THAN ZERO IS ZERO

Example 3

(a) $\dfrac{0}{2} = 0 \div 2 = 2\overline{)\overset{0}{0}}$ because $2 \cdot 0 = 0$
(divisor $\times$ quotient = dividend)

(b) $\dfrac{0}{5} = 0 \div 5 = 5\overline{)\overset{0}{0}}$ because $5(0) = 0$

THE REMAINDER. In all the division examples done up until
now, the dividend has been an exact multiple of the divisor.
By this we mean that when the divisor was repeatedly subtracted from the dividend, we were eventually left with 0.
The *remainder* is the number left over after the divisor has
been subtracted as many times as possible. For example, find
$34 \div 8$.

$$
\begin{array}{r}
34 \\
- 8 \quad \leftarrow \text{1st subtraction of 8} \\
\hline
26 \\
- 8 \quad \leftarrow \text{2nd subtraction of 8} \\
\hline
18 \\
- 8 \quad \leftarrow \text{3rd subtraction of 8} \\
\hline
10 \\
- 8 \quad \leftarrow \text{4th subtraction of 8} \\
\hline
② \leftarrow \text{the } remainder
\end{array}
$$

8 has been subtracted four times; therefore, the quotient = 4, and we have a remainder of 2.

Example 4

(a) 9 ÷ 2

$$2 \overline{)9} \quad \overset{4 \quad \text{R } 1}{}$$

because 9 = 2 · 4 + 1   2 can be subtracted
        =   8   + 1   4 times, leaving
        =   9   a remainder of 1.

(b) 8 ÷ 3

$$3 \overline{)8} \quad \overset{2 \quad \text{R } 2}{}$$

because 8 = 3 · 2 + 2   3 can be subtracted
        =   6   + 2   2 times, leaving
        =   8   a remainder of 2.

(c) 45 ÷ 6

$$6 \overline{)45} \quad \overset{7 \quad \text{R } 3}{}$$

because 45 = 6 · 7 + 3   6 can be subtracted
        =   42   + 3   7 times, leaving
        =   45   a remainder of 3.

(d) 77 ÷ 9

$$9 \overline{)77} \quad \overset{8 \quad \text{R } 5}{}$$

because 77 = 9 · 8 + 5   9 can be subtracted
        =   72   + 5   8 times, leaving
        =   77   a remainder of 5.

(e) 29 ÷ 3

$$3 \overline{)29} \quad \overset{9 \quad \text{R } 2}{}$$

because 29 = 3 · 9 + 2   3 can be subtracted
        =   27   + 2   9 times, leaving
        =   29   a remainder of 2.

EXERCISES 207C, SET I. In Exercises 1-16, use repeated subtraction to find the quotient and remainder.

1. 7 ÷ 2	2. 8 ÷ 3	3. 9 ÷ 5	4. 11 ÷ 2
5. 15 ÷ 4	6. 17 ÷ 6	7. 12 ÷ 5	8. 23 ÷ 8
9. 15 ÷ 7	10. 37 ÷ 5	11. 33 ÷ 10	12. 42 ÷ 9
13. 49 ÷ 6	14. 52 ÷ 7	15. 64 ÷ 11	16. 57 ÷ 12

17. There are 45 chairs in a classroom. If they are arranged with nine chairs in a row, how many rows can be formed?
18. A student has saved $600. If he spends $120 a month, how many months will it last?
19. A rectangle is 15 feet by 27 feet. What is its area in square yards? (Remember there are 3 feet in 1 yard and 9 square feet in 1 square yard.)
20. Hamburgers cost 35¢ at a neighborhood drive-in. How many can a father buy for his family with $2? How much change does he receive? (Disregard any sales tax.)
21. A family wants to save $5,000 to make a trip to Europe in 4 years. How much must they save each year? If they save $100 each month, will they meet their goal?

EXERCISES 207C, SET II. In Exercises 1-8, use repeated subtraction to find the quotient and remainder.

1. $7 \div 3$ 2. $10 \div 6$ 3. $17 \div 5$ 4. $19 \div 4$
5. $37 \div 8$ 6. $25 \div 9$ 7. $53 \div 10$ 8. $52 \div 11$

9. Maria brought along an extra $90 on her 15-day vacation just for buying souvenirs. What is the average amount she can spend each day for this purpose?
10. A classroom contains 8 rows of 6 chairs each. How many more chairs are needed to seat 53 students?
11. How many 39¢ cans of diet soda can Roger buy for $3? How much change would he receive? (Disregard any sales tax.)

Short Division

We now consider a method of division called *short division*, which we can use <u>when the divisor has one digit</u> and the dividend has any number of digits.

Method of Short Division

1. (a) *If the first digit of the dividend can be divided by the divisor*, do that division. Then write this first digit of the quotient above the *first* digit of the dividend. Write this first remainder to the left of the next digit in the dividend.
   (b) *If the first digit of the dividend cannot be divided by the divisor*, divide the first two digits of the dividend by the divisor. Write this first digit of the quotient above the *second* digit of the dividend. Write this first remainder to the left of the next digit in the dividend.
2. Divide the number formed by the first remainder and the next digit of the dividend by the divisor.
3. Continue this procedure until a digit in the quotient has been placed above the last digit of the dividend. The final remainder is written to the right of the quotient as before.

Example 1. $74 \div 3$

Solution: $3\overline{)7^14}$ $2\ 4$ R 2   Notice how the first remainder, 1, is written to the left of 4, making 14. Then 14 is divided by 3.

Example 2. $34 \div 5$

Solution: $5\overline{)3\ 4}$ 6 R 4   Because 3 cannot be divided by 5, we divide 34 by 5.

If there are more digits in the dividend, this procedure is continued.

Example 3. 733 ÷ 4

Solution:

```
    1 8 3  R 1
  _____
4 | 7³3¹3
```

Example 4. 274 ÷ 6

Solution:

```
    4 5  R 4
  _____
6 | 2 7³4
```

CHECKING DIVISION. In any division problem:

$$\text{dividend} = \text{divisor} \times \text{quotient} + \text{remainder}$$

Example 5

(a)
```
    5 8  R 4
  _____
9 | 5 2⁷6
```

Check: 526 = 9 · 58 + 4
            =   522   + 4
            =   526

A word of caution: Many students leave this final zero off. Remember that there must be a digit in the quotient above each final digit in the dividend.

(b)
```
    1,2 9 0  R 5
  _____
7 | 9²,0⁶3 5
```

Check: 9,035 = 7 · 1,290 + 5
              =   9,030   + 5
              =   9,035

(c)
```
    6,3 3 8  R 5
  _____
8 | 5 0²,7³0⁶9
```

Check: 50,709 = 8 · 6,338 + 5
               =   50,704 + 5
               =   50,709

If you had trouble following the examples just given, look closely at the comments written in explaining the following short division problem. Then take a second look at the three preceding examples.

```
        8,9 7 1  R 6
  _____
7 | 6 2,⁶8⁵0¹3
```

7 · 8 = 56; 62 - 56 = 6
7 · 9 = 63; 68 - 63 = 5
7 · 7 = 49; 50 - 49 = 1
7 · 1 = 7; 13 - 7 = 6

Check: 62,803 = 7 · 8,971 + 6
               =   62,797 + 6
               =   62,803

Work the following problems using short division.

1. 28 ÷ 3   2. 57 ÷ 4   3. 79 ÷ 5   4. 94 ÷ 2

5. 147 ÷ 3   6. 265 ÷ 4   7. 378 ÷ 2   8. 513 ÷ 3

9. 893 ÷ 6   10. 904 ÷ 7   11. 685 ÷ 5   12. 2,124 ÷ 6

13. 1,144 ÷ 4 14. 2,553 ÷ 3   15. 1,512 ÷ 7   16. 1,235 ÷ 8

17. 7,166 ÷ 9 18. 5,254 ÷ 5   19. 20,441 ÷ 6   20. 54,375 ÷ 8

21. At a market, a 7-ounce bottle of shampoo costs 98¢. How much is this per ounce?
22. Seven girls pooled their money for a lottery ticket and won $2,500. When they divide the money in dollars, there is a remainer. They draw straws to see who will get the remainder. What was the total amount received by the winner?
23. Eight boys going on a camping trip are trying to divide 100 small boxes of raisins up evenly among their packs. The leader (one of the eight) will take the remainder as well as his share. How many boxes did the leader have to carry?
24. Form numbers by arranging the digits 1 through 9 in several different orders. Divide each number so obtained by 9. Compare the remainders in each division.

EXERCISES 208, SET II. Work the following problems using short division.

1. 31 ÷ 4   2. 84 ÷ 6   3. 132 ÷ 5   4. 297 ÷ 2

5. 739 ÷ 7   6. 632 ÷ 8   7. 1,206 ÷ 3   8. 1,473 ÷ 9

9. 8,093 ÷ 6   10. 20,804 ÷ 7

11. A nine-man softball team won a $5,000 prize. When they attempt to divide the prize equally (in dollars), there is a remainder. If they give the remainder to the captain, how much does he receive?
12. Select any digit. Divide the sum of the remaining digits by 9. Subtract the remainder found from this division from 9. Compare this result with the digit you originally selected. Does it make any difference what digit you select?

Long Division

Short division involves doing most of the work mentally—only writing down the quotient, remainder, and at most the numbers that are carried. In *long division* the numbers are too large to do the work mentally, so that all steps should be written. The basic process of division is the same in both long and short division. We will use the *trial divisor method* for long division. The *trial divisor* is usually the first digit of the divisor. We use examples to illustrate the method.

Example 1. $725 \div 31$

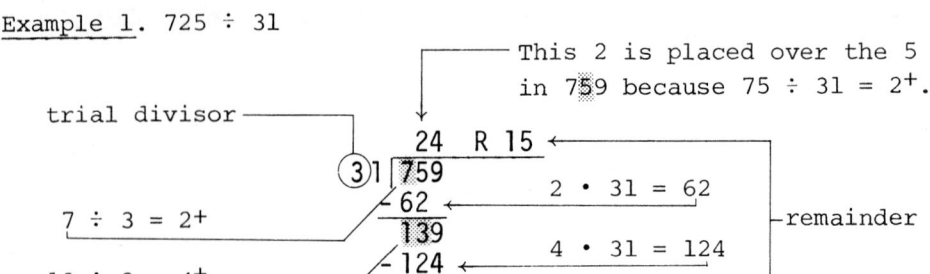

This 2 is placed over the 5 in 7**5**9 because $75 \div 31 = 2^+$.

trial divisor

```
               24   R 15
        ③1 759
7 ÷ 3 = 2⁺   - 62          2 · 31 = 62
              139
13 ÷ 3 = 4⁺  - 124         4 · 31 = 124
               15
```
— remainder

In the following example, instead of using the first digit, 3, of divisor 39 for our trial divisor, we use 4 because 39 is closer to four tens than it is to three tens.

Example 2. $6,842 \div 39$

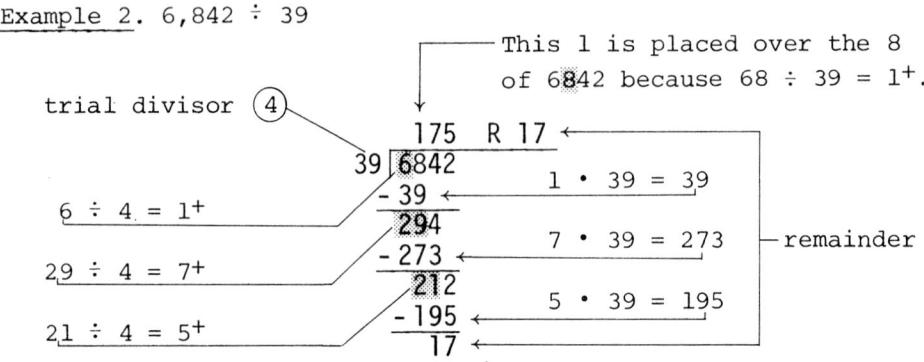

This 1 is placed over the 8 of 6**8**42 because $68 \div 39 = 1^+$.

trial divisor ④

```
              175   R 17
        39 6842
6 ÷ 4 = 1⁺   - 39          1 · 39 = 39
              294
29 ÷ 4 = 7⁺  - 273         7 · 39 = 273
              212
21 ÷ 4 = 5⁺  - 195         5 · 39 = 195
               17
```
— remainder

Example 3

(a) $4,908 \div 67$

This 7 is placed over the 0 in 49**0**8 because $490 \div 67 = 7^+$.

trial divisor ⑦

```
              73   R 17
        67 4908
           - 469
             218
           - 201
              17
```

(b) $52,643 \div 218$

This 2 is placed over the 6 in 52**6**43 because $526 \div 218 = 2^+$.

trial divisor ②

```
              241   R 105
       218 52643
           - 436
             904
           - 872
             323
           - 218
             105
```

Example 4. 15,730 ÷ 26

This 0 is placed over the 3 in 15730 because 13 ÷ 26 = 0+.

trial divisor ③

```
        605
  26 )15730
     -156
       130
      -130
```

EXERCISES 209, SET I. Work the following long division problems.

1. 238 ÷ 14
2. 625 ÷ 25
3. 1,645 ÷ 35
4. 1,428 ÷ 28
5. 745 ÷ 21
6. 847 ÷ 31
7. 519 ÷ 12
8. 275 ÷ 16
9. 629 ÷ 13
10. 885 ÷ 43
11. 1,404 ÷ 52
12. 2,415 ÷ 69
13. 5,230 ÷ 78
14. 8,866 ÷ 31
15. 4,677 ÷ 87
16. 22,815 ÷ 27
17. 80,240 ÷ 85
18. 97,514 ÷ 91
19. 185,503 ÷ 89
20. 65,058 ÷ 185
21. 147,379 ÷ 206
22. 425,743 ÷ 301
23. 565,090 ÷ 715
24. 14,372,693 ÷ 895
25. 21,979,818 ÷ 784
26. 31,932,670 ÷ 921

27. A man can make 38 machine parts on a lathe in 1 hour. How many hours will he need to make 2,356 parts?
28. A certain brand of soap sells for 21¢ a bar.
    (a) How many bars can be bought with $1.50?
    (b) How much change is left?
29. Frank figures a tire on his car lasts for 24,000 miles. How many tires will he need to drive his car 96,000 miles?
30. If it takes 6 minutes to saw a log into three pieces, how long will it take to saw the same log into four pieces?

EXERCISES 209, SET II. Work the following long division problems.

1. 217 ÷ 13
2. 709 ÷ 15
3. 754 ÷ 23
4. 895 ÷ 42
5. 1,574 ÷ 37
6. 2,158 ÷ 59
7. 6,302 ÷ 72
8. 25,093 ÷ 81
9. 732,016 ÷ 28
10. 160,840 ÷ 935
11. 806,074 ÷ 627
12. 45,076,903 ÷ 871
13. 80,700,010 ÷ 805

14. Jerry earns $78 a day. How many days must he work to earn $2,028?
15. Janet figures glass-belt tires last 36,000 miles on her car. How many such tires would she buy in driving the car 80,000 miles? Would she need any more tires to drive 100,000 miles?

Average

If a student gets 70 on one test and 90 on another test, you probably know his average is 80. Why is 80 called his average?

It is because

and

$$70 + 90 = 160$$

$$80 + 80 = 160$$

— same total points

— 80 on *every* test gives the same total points

In other words, the average is the score he would have to make on *every* test in order to get the same total points. To find the average, we take the sum of all the grades (160) and divide it by the number of grades (2).

We can find the average of any kinds of quantities such as grades, weights, speeds, and costs. The average is found by dividing the sum of the quantities by the number of quantities.

$$\text{Average} = \frac{\text{Sum of all the quantities being averaged}}{\text{The number of quantities being averaged}}$$

Example 1

(a) Find the average grade for 70, 86, 90, 66.

$$\text{average} = \frac{70 + 86 + 90 + 66}{4} = \frac{312}{4} = 78$$

(b) Find the average grade for 73, 84, 88, 92, 68.

$$\text{average} = \frac{73 + 84 + 88 + 92 + 68}{5} = \frac{405}{5} = 81$$

(c) Find the average weight of a group of people whose weights are 126, 159, 216.

$$\text{average} = \frac{126 + 159 + 216}{3} = \frac{501}{3} = 167$$

(d) The last six times Bob had his car repaired the bills were $24, $13, $38, $86, $61, $42.

$$\text{average} = \frac{24 + 13 + 38 + 86 + 61 + 42}{6} = \frac{264}{6} = 44$$

EXERCISES 210, SET I. In Exercises 1-10, find the average of each set of numbers.

1. {7, 5}   2. {8, 6}   3. {3, 6, 9}

4. {3, 5, 7}   5. {6, 8, 9, 5}   6. {9, 0, 6, 9}

7. {21, 24, 33}   8. {7, 10, 8, 5, 11, 7}

9. {74, 88, 85, 69}   10. {96, 92, 95, 89, 88}

11. Maria's examination scores during the semester were 75, 83, 74, 86, 95, and 61. What was her average score?

12. Mrs. Lindstrom recorded her weight each Monday morning for 6 weeks. The weights were 155 lb., 150 lb., 149 lb., 148 lb., 150 lb., and 142 lb. What is her average weight?

13. Five basketball players have heights of 76 in., 78 in., 84 in., 72 in., and 75 in. What is the average height of the team?

14. Traveling across the country, a family stayed in motels. The cost per night for the motels was as follows: $18, $21, $19, $24, $15, $20, $23, $14, and $17. Find the average cost per night for motels.

15. Find the average of all the whole numbers from 73 through 79 (inclusive).

16. Cheryl's clothing expenses for 5 successive months were $27, $0, $13, $25, and $20. What is her average monthly clothing expense?

17. The monthly rainfall for a city was: 17 inches for January, 14 inches for February, 19 inches for March, 15 inches for April, 7 inches for May, 2 inches for June, 1 inch for July, 2 inches for August, 4 inches for September, 7 inches for October, 11 inches for November, and 9 inches for December. What is its average monthly rainfall?

18. Two groups of five students took a test. The students in group A made scores of 78, 85, 97, 76, and 84. Those in group B made scores of 95, 87, 78, 80, and 55. Which group made the higher average, and by how much?

EXERCISES 210, SET II. In Exercises 1-5, find the average of each set of numbers.

1. {13, 19}   2. {12, 9, 18}   3. {42, 37, 29, 44}

4. {75, 63, 79, 91, 87}   5. {16, 21, 11, 23, 15, 9, 24}

6. A defensive line's front four weigh 276 lb., 251 lb., 265 lb., and 284 lb. What is their average weight?

7. Find the average of all the even numbers between 75 and 87.

8. A model's weekly cleaning bill ran $16, $27, $9, $13, $4, $18, $12, and $21 during eight consecutive weeks. What was her average weekly cleaning bill?

9. Ron's exam scores were 61, 73, 48, and 81. What grade will he need to get on his fifth test so that his average will be 70 for all five tests?

Checking Division by Casting out Nines

Up until now we have used

$$dividend = divisor \times quotient + remainder$$

to check our division problems. If we find the 9-rem of each of these quantities, the numbers we get are smaller and easier to work with. Consider the following division problem and its check.

$$\text{9-rem} = 1 \text{ (quotient)}$$

$$\text{9-rem} = 5 \text{ (divisor)} \longrightarrow 32 \overline{)2357} \longleftarrow \text{9-rem} = 8 \text{ (dividend)}$$
$$73 \text{ R } 21 \longleftarrow \text{9-rem} = 3 \text{ (remainder)}$$
$$\underline{-224}$$
$$117$$
$$\underline{-96}$$
$$21$$

Check: dividend = divisor × quotient + remainder

8	=	5	×	1	+	3
8	=		5		+	3
8	=	8				

Example 1. Study the checks of the following division problems to be sure you understand the method.

(a)
$$\longrightarrow 8$$
$$53 \text{ R } 348 \longrightarrow 6$$
$$438 \overline{)23562} \longrightarrow 0$$
$$6 \longleftarrow \underline{-2190}$$
$$1662$$
$$\underline{-1314}$$
$$348$$

(b)
$$\longrightarrow 6$$
$$78 \text{ R } 299 \longrightarrow 2$$
$$537 \overline{)42185} \longrightarrow 2$$
$$6 \longleftarrow \underline{-3759}$$
$$4595$$
$$\underline{-4296}$$
$$299$$

This "=" means the 9-rems are equal.

Check: 0 = 8 · 6 + 6
= 48 + 6
= 3 + 6
0 = 0

Check: 2 = 6 · 6 + 2
= 36 + 2
= 0 + 2
2 = 2

EXERCISES 211. Check the first fifteen exercises in Exercises 209, Set I, by casting out nines.

Division Is Not Commutative

When addition and multiplication were defined and explained, we discovered that both these operations are commutative. That is,

$$a + b = b + a \text{ and } a \cdot b = b \cdot a$$

But subtraction is not commutative. In other words,

$$a - b \neq b - a$$

What is the case with division? A single example will show that *division is not commutative.*

$$6 \div 3 \neq 3 \div 6$$

because $6 \div 3 = 2$, whereas $3 \div 6$ does not represent a whole number.

Division Is Not Associative

We found that the operations of addition and multiplication are both associative. That is

$$(a + b) + c = a + (b + c) \text{ and } (a \cdot b) \cdot c = a \cdot (b \cdot c)$$

On the other hand, we found that the operation of subtraction is not associative. That is,

$$(a - b) - c \neq a - (b - c)$$

What is the case with division? A single example will show that *division is not associative*.

$$(16 \div 4) \div 2 \neq 16 \div (4 \div 2)$$

because $\quad (16 \div 4) \div 2 = 4 \div 2 = 2$

whereas $\quad 16 \div (4 \div 2) = 16 \div 2 = 8 \quad \Big\}\, 2 \neq 8$

Summary for Division

Division is the inverse of multiplication. (Sec. 207) Division of whole numbers can be considered either as (1) *the separation of a number into equal parts*, or (2) *repeated subtraction*. The number being divided is called the *dividend*. The number doing the dividing is called the *divisor*. The answer to the division problem is called the *quotient*. The remainder is the number left after the divisor has been subtracted from the dividend as many times as possible.

Division by zero is not possible. (Sec. 207)

Short division is used when we have one-digit divisors. (Sec. 208)

Long division is used when divisors are greater than one digit. (Sec. 209)

The trial divisor is the divisor rounded off to its left digit. (Sec. 209)

Division is not commutative or associative. (Secs. 212, 213)

To check a division problem: (Sec. 211)

$$\text{dividend} = \text{divisor} \times \text{quotient} + \text{remainder}$$

To check by casting out nines, use the 9-rem in place of each above-mentioned quantity.

Roots of Numbers

Finding the root of a number is the inverse of raising a number to a power. In this section we will talk only about roots that can be found by inspection.

SQUARE ROOTS. Study the following examples.

Example 1

$$3^2 = \boxed{3} \times 3 = \boxed{9}$$

→3 is called the *square root* of 9.

9 is called the *square* of 3.

Note the difference between the meaning of *the square root of a number* and *the square of a number*. Example 2 shows this difference.

Example 2. The square root of 9 is 3. $\sqrt{9} = 3$

The square of 9 is 81. $9^2 = 81$

Example 3

$$10^2 = \boxed{10} \times 10 = \boxed{100}$$

→10 is called the square root of 100.

100 is called the square of 10.

Example 4. Find the square root of 25, written $\sqrt{25}$.

$$\sqrt{25} = 5 \text{ because } 5^2 = 25$$

Example 5. Find the square root of 16, written $\sqrt{16}$.

$$\sqrt{16} = 4 \text{ because } 4^2 = 16$$

Example 6

(a) $\sqrt{4}$ = 2 because $2^2 = 4$

(b) $\sqrt{9}$ = 3 because $3^2 = 9$

(c) $\sqrt{36}$ = 6 because $6^2 = 36$

(d) $\sqrt{0}$ = 0 because $0^2 = 0$

(e) $\sqrt{1}$ = 1 because $1^2 = 1$

EXERCISES 215A, SET I

1. Complete the table.

Number	Square of Number		Number	Square of Number
0			9	
1			10	
2			11	
3			12	
4			13	
5			14	
6			15	
7			16	
8				

In Exercises 2-10, find the indicated square roots or powers.

2. $\sqrt{25}$ 3. $\sqrt{81}$ 4. $\sqrt{49}$ 5. $\sqrt{100}$ 6. $\sqrt{144}$

7. $\sqrt{196}$ 8. $\sqrt{256}$ 9. 11^2 10. 8^2

EXERCISES 215A, SET II. Find the following square roots or powers.

1. $\sqrt{64}$ 2. $\sqrt{121}$ 3. $\sqrt{169}$ 4. $\sqrt{225}$ 5. 14^2 6. 9^2

HIGHER ROOTS

Example 7.

2 is called the *cube root* of 8.

In the expression $2^3 = 2 \times 2 \times 2 = 8$

8 is called the *cube* of 2.

Note the difference between the meaning of *the cube root of a number* and *the cube of a number*. Example 8 shows this difference.

Example 8. The cube root of 8 is 2. $\sqrt[3]{8} = 2$

The cube of 8 is 512. $8^3 = 8 \cdot 8 \cdot 8 = 512$

Example 9.

2 is called the *fourth root* of 16.

In the expression $2^4 = 2 \times 2 \times 2 \times 2 = 16$

16 is called the *fourth power* of 2.

SOME SYMBOLS USED TO INDICATE ROOTS

$\sqrt[3]{}$ indicates cube root. Therefore, $\sqrt[3]{8} = 2$, because $2^3 = 8$.

$\sqrt[4]{}$ indicates fourth root. Therefore, $\sqrt[4]{16} = 2$, because $2^4 = 16$.

$\sqrt[5]{}$ indicates fifth root. Therefore, $\sqrt[5]{32} = 2$, because $2^5 = 32$.

Finding the root of a number is the *inverse* of raising that number to a power because it "undoes" raising to that power.

Example 10

Note that 2 is where we begin and 2 is where we end.

$$2^4 = 16 \qquad \sqrt[4]{16} = 2$$

finding the 4th root
raising to the 4th power

EXERCISES 215B, SET I

1. Complete the table.

Number	Square of Number	Cube of Number	Fourth Power of Number
0			
1			
2			
3			
4			
5			
6			
10			

In Exercises 2-10, find each of the roots or powers.

2. $\sqrt[4]{81}$ 3. $\sqrt[3]{27}$ 4. $\sqrt[3]{125}$ 5. $\sqrt[4]{1}$ 6. $\sqrt[3]{0}$

7. $\sqrt[3]{1,000}$ 8. $\sqrt[5]{32}$ 9. 7^4 10. 8^3

EXERCISES 215B, SET II. Find each of the indicated roots or powers.

1. $\sqrt[5]{1}$ 2. $\sqrt[3]{64}$ 3. $\sqrt[4]{16}$ 4. $\sqrt[5]{243}$ 5. 3^5 6. 9^3

Order of Operations

When addition, subtraction, multiplication, division, powers, and roots are written in the same expression, and there are no parentheses to change the normal order, we perform the operations in the following order:

Order of Operations

1st: Powers and roots are done in any order.
2nd: Multiplication and division are done in order from left to right.
3rd: Addition and subtraction are done in order from left to right.

Example 1. Find $4^2 + \sqrt{25} + 6$.

$$16 + 5 + 6 = 27$$

Example 2. Find $16 \div 2 \cdot 4$.

$$8 \cdot 4 = 32$$

A word of caution: Remember multiplication and division are done in the order they are written from left to right.

Example 3. Find $5 \cdot 2^3 \div 2 - 4$.

$$5 \cdot 8 \div 2 - 4$$
$$40 \div 2 - 4$$
$$20 - 4 = 16$$

Example 4. Find $6\sqrt{9} + 14 - 4$.

$$6 \cdot 3 + 14 - 4$$
$$18 + 14 - 4 = 28$$

EXERCISES 216, SET I. In working the following exercises, be sure to perform the operations in the correct order.

1. $10 \div 2 \times 5$
2. $20 \times 15 \div 5$
3. $3(2^4)$
4. $4 \cdot 3 + 15 \div 5$
5. $10 \cdot 15^2 - 4^3$
6. $(785)^3(0) + 1^5$
7. $(10^2)\sqrt{16} \cdot 5$
8. $(5)\sqrt{25} + 4(6) - 6$
9. $2 \cdot 3 + 3^2 - 4 \cdot 2$
10. $100 \div 5^2 \cdot 6 + 8 \cdot 75$
11. $4 + 77 \div 11 \cdot 2 - 5$
12. $10^4(4^2) + 100(15)$
13. $(10^2)\sqrt{4} - 5 + 36$
14. $2 + 3(100) \div 25$
15. $28 \div 4 \cdot 2(6)$

EXERCISES 216, SET II. In working the following exercises, be sure to perform the operations in the correct order.

1. $12 \div 3 \times 2$

2. $6 \cdot 5 + 20 \div 5$

3. $4 \cdot 12^2 - 5^3$

4. $(10^2)\sqrt{25} \cdot 4$

5. $5 \cdot 7 + 2^3 - 6 \cdot 3$

6. $15 + 44 \div 11 \cdot 2 - 14$

7. $(13)^2\sqrt{0} + 18 - 5$

8. $36 \div 9 \cdot 2(7)$

Chapter Summary

Addition is repeated counting; its _inverse is subtraction_. (Sec. 201)

Multiplication is repeated addition; its _inverse is division_. Division is repeated subtraction. (Sec. 207)

Raising a number to a power is repeated multiplication; its _inverse is finding the root of a number_. (Sec. 215)

Terms Used in Subtraction (Sec. 201)

$$
\begin{array}{rl}
12 & \text{minuend} \\
-\ 8 & \text{subtrahend} \\
\hline
4 & \text{difference}
\end{array}
$$

To check a subtraction problem, add the subtrahend to the difference to get the minuend. (Sec. 202)

Subtraction and division are not commutative or associative. (Secs. 204, 205, 212, 213)

Terms Used in Division (Sec. 207)

$$
\begin{array}{r}
4 \quad \text{quotient} \\
\text{divisor} \quad 16\overline{)68} \quad \text{dividend} \\
-\,64 \\
\hline
4 \quad \text{remainder}
\end{array}
$$

To check a division problem, (Sec. 211) remember that the

$$\text{dividend} = \text{divisor} \times \text{quotient} + \text{remainder}$$

$Average = \dfrac{\text{Sum of all the quantities being averaged}}{\text{The number of quantities being averaged}}$

(Sec. 210)

Terms Used in Finding a Root (Sec. 215)

$$\sqrt{16} = 4 \quad \text{square root}$$
$$\sqrt[3]{8} = 2 \quad \text{cube root}$$
$$\sqrt[4]{16} = 2 \quad \text{fourth root}$$

To check a square root, square your answer. (Sec. 215)

$$\sqrt{25} = 5 \quad \text{Check:} \quad 5^2 = 25$$

To check a cube root, cube your answer. (Sec. 215)

$$\sqrt[3]{27} = 3 \quad \text{Check:} \quad 3^3 = 27$$

Order of Operations (Sec. 216)

1. Powers and roots are done in any order.
2. Multiplication and division are done in the order in which they are written, from left to right.
3. Addition and subtraction are worked in order from left to right.

REVIEW EXERCISES 217, SET I. In Exercises 1-10, write the word(s) or number that makes the statement correct.

1. Division is the _____ of multiplication.
2. Division of whole numbers can be considered repeated _____.
3. The number being divided is called the _____.
4. $6 \div 0 =$ _____.
5. Since $5 \cdot 3 = 15$, 5 and 3 are _____ of 15.
6. $0 \div 5 =$ _____.
7. When we have one-digit divisors, we use _____ division.
8. In many cases, the first digit of the divisor is used as the _____.
9. $0 \div 0 =$ _____.
10. Which of the operations—addition, subtraction, multiplication, and division—are neither associative nor commutative? _____
11. Do the following subtractions and check your answers.

    (a)   7,506   (b) 8,247   (c) 31,562   (d) 700,051
       − 2,789   − 358   − 20,099   − 20,893

12. Work the following short division problems and check your answers.

    (a) $3\overline{)7,896}$ (b) $4\overline{)3,416}$ (c) $7\overline{)245,730}$

13. Work the following long division problems and check your answers.

    (a) $756 \div 31$ (b) $198 \div 51$ (c) $704 \div 77$

    (d) $3,156 \div 84$ (e) $60,074 \div 126$ (f) $3,174 \div 708$

14. How many 3-ounce bottles can a druggist fill from a bottle that contains 16 ounces of peroxide?
15. Mr. Jones earns an annual salary of $9,096. What is his monthly salary?
16. Mr. Smith's yearly tax on his home is $576. How much is this a month?
17. After trading his car in for a new car, Joe had a balance of $2,016 due. What equal monthly payments must he make to pay this off in three years?

18. A man owed $2,365 on his car. After making 35 payments of $67 each, how much was left to be paid?

19. Lee makes monthly payments to pay off a $1,350 debt. He makes 23 payments of $58 each and then one final payment to pay off the balance of the debt. How much was the final payment?

20. A man pays off a debt of $3,048 in equal monthly payments over a period of two years. Find the amount he pays each month.

21. The full price including tax and financing on a certain color television set was $456. Mr. White bought the set. He agreed to pay for it in 24 equal monthly payments.
(a) Find the amount of his monthly payment.
(b) After making 17 payments, how much does he still owe?

22. During the semester, Mary took six examinations in arithmetic. The sum of all her examination scores was 498 points. What was her average examination score?

23. During the semester, Jim received the following scores on examinations in arithmetic: 77, 89, 95, 61, 81, 92, and 86. Find his average score.

24. In air, light travels 983,584,800 feet per second and sound travels 1,129 feet per second. The speed of light is how many times the speed of sound?

25. To get a gasoline mileage check, Mr. Perez filled his gasoline tank and wrote down the odometer reading, which was 53,408 miles. The next time he got gasoline his tank took 19 gallons and his odometer reading was 53,731 miles. How many miles did he get per gallon?

26. <u>Brain Teaser</u>: A number of birds are resting on two limbs of a tree. One limb is above the other. A bird on the lower limb says to the birds on the upper limb, "If one of you will come down here, we will have an equal number on each limb." A bird from above replied, "If one of you will come up here we will have twice as many up here as you have down there." How many birds sat on each branch at first?

REVIEW EXERCISES 217, SET II.

1. Evaluate (if possible) each of the following divisions.

   (a) 8 ÷ 0   (b) 0 ÷ 7

2. State which of the following operations are commutative: addition, subtraction, multiplication, division.

3. Which of the following operations is the inverse of addition? Multiplication, division, subtraction.

4. Do the following subtractions and check your answers.

   (a)   9,012   (b)   800,061
        - 8,607   - 50,984

5. Work the following short division problems and check your answers.

   (a) 6⟌3,042   (b) 8⟌328,639

6. Work the following long division problems and check your answers.

   (a) 501 ÷ 62   (b) 5,017 ÷ 78   (c) 30,580 ÷ 609

7. Pam bought a bottle of 50 vitamin C tablets. If she takes three each day, in how many days will she have to buy a new bottle?

8. The property tax on Mr. Itahara's home was $1,272 for the year. How much is this a month?

9. Jan owed $3,076 on her car. After making 29 payments of $87 each, how much was left to be paid?

10. Ruben got grades of 56, 73, 82, 68, 95, 64, 87, and 83 on his astronomy exams. Find his average score.

11. In water, light travels about 737,690,450 feet per second and sound travels 1,435 feet per second. The speed of light in water is how many times the speed of sound?

12. Jim filled his car's gas tank and noted that his car had been driven a total of 48,019 miles. It took 16 gallons to fill his tank the next time he got gas. At that time the car had been driven 48,403 miles.
    (a) How many miles had he driven between fill-ups?
    (b) How many miles did he get per gallon?

Chapter 2: Diagnostic Test

Name_____

_____

The purpose of this test is to see how well you know how to subtract, divide, and find roots of whole numbers. We recommend that you work this diagnostic test *before* your instructor tests you on this chapter. Allow yourself about an hour to do this test.

Complete solutions for all the problems on this test, together with section references, are given in the Answer Section. You should study the sections referred to for the problems you do incorrectly.

---

1. Write the term used for each part in the following subtraction example:

8	(1a)_____
− 2	(1b)_____
6	(1c)_____

---

2. Work the following subtraction problems:

(a) 4,768   (b) 3,564   (c) 50,406
  − 3,205   − 782   − 35,008

(2a)_____

(2b)_____

(2c)_____

---

3. If at the beginning of a trip your odometer reading was 67,856 miles and at the end of the trip it read 71,304 miles, how many miles did you drive?

(3)_____

---

4. Write the term used for each part in the following division example:

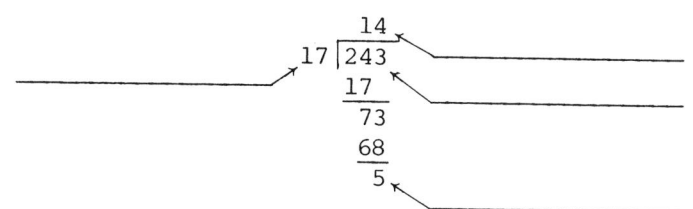

$$17\overline{)243}$$
with quotient 14, 17 subtracted, 73, 68, remainder 5

---

5. In each of the following, if division is possible, write the quotient. If division is not possible, write "not possible."

(a) $\dfrac{0}{5}$ (b) $\dfrac{4}{4}$ (c) $\dfrac{7}{0}$

(5a) _____

(5b) _____

(5c) _____

---

6. Find all the divisors of 12.

(6) _____

---

7. Work the following division problems:

(a) $8\overline{)78,407}$ (b) $65\overline{)2,210}$

(c) $495\overline{)349,596}$

(7a) _____

(7b) _____

(7c) _____

---

8. A man can make 38 machine parts on a lathe in 1 hour. How many hours will he need to make 2,356 parts?

(8) _____

---

9. After trading his car in on a new car, Joe had a balance of $2,016 due. What equal monthly payments must he make to pay this off in 3 years?

(9) _____

10. John's examination scores during the semester were 73, 84, 88, 92, and 68. What was his average score?

(10) _____

11. Find each of the indicated roots.

(a) $\sqrt{25}$ (b) $\sqrt[3]{8}$

(c) $\sqrt{100}$ (d) $\sqrt[4]{16}$

(11a) _____

(11b) _____

(11c) _____

(11d) _____

12. Work the following exercises, being careful to perform the operations in the correct order.

(a) $100 \div 4 \cdot 5 - 3$

(b) $2(5^2) - 3\sqrt{16}$

(c) $2^4 + 10^2 + \sqrt[3]{8}$

(12a) _____

(12b) _____

(12c) _____

13. Susan makes monthly payments to pay off a $1,350 debt. She makes 23 payments of $58 each, and then a final payment to pay off the balance. How much was the final payment?

(13)_____

THREE
Fractions

So far we have been concerned mainly with whole numbers and the operations that can be performed on them. In this chapter, we introduce a new kind of number, *fractions*, and show how to perform the basic operations on fractions.

The Meaning of Fraction

The dictionary meaning of fraction is "a breaking of something," "a piece broken off." It is like division. In division, we broke a whole number into equal parts, where each of the parts was a *whole* number. In fractions, we usually break up a whole number into parts that are *not whole* numbers. Suppose we take a whole rectangle and divide it into two equal parts. (See Figure 301A.)

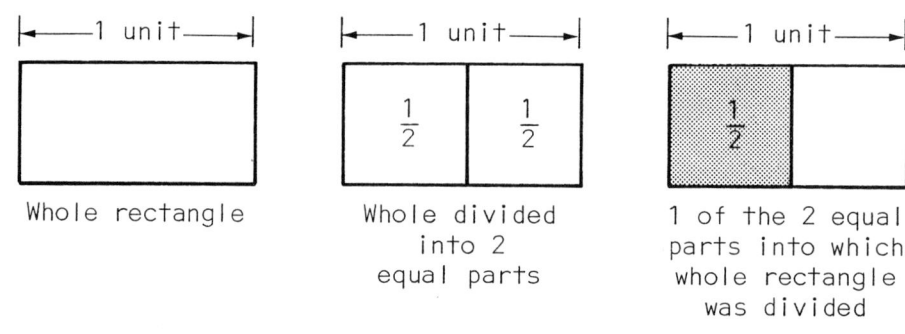

|——1 unit——|	|——1 unit——|	|——1 unit——|

Whole rectangle   Whole divided   1 of the 2 equal
                         into 2   parts into which
                         equal parts   whole rectangle
                                                was divided

Figure 301A

Next we take a whole circle and divide it into three equal parts. (See Figure 301B.)

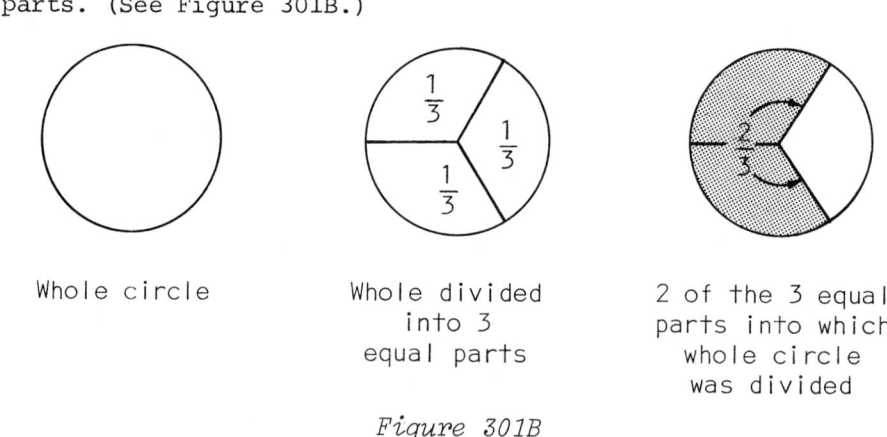

Whole circle   Whole divided   2 of the 3 equal
                      into 3   parts into which
                      equal parts   whole circle
                                             was divided

Figure 301B

DEFINITION OF A FRACTION. A *fraction* is part of a whole. It is written $\frac{a}{b}$. a and b are called the *terms* of the fraction. b is called the *denominator*. It tells how many parts the whole was divided into. a is called the *numerator*. It tells the number of those equal parts used. (See Figure 301C.)

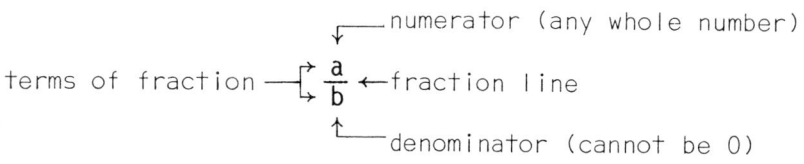

terms of fraction ──{ $\dfrac{a}{b}$ ←fraction line
numerator (any whole number)
denominator (cannot be 0)

Figure 301C

Example 1. Examples of Fractions.

(a) In Figure 301A, we divided the whole rectangle into two equal parts. The expression $\dfrac{1}{2}$, read "one-half," means we have one of the two equal parts of the whole.

(one-half) $\dfrac{1}{2}$ ← ① of the ② equal parts

(b) In Figure 301B, we divided the whole circle into three equal parts. Then $\dfrac{1}{3}$ means we have one of the three equal parts.

(one-third) $\dfrac{1}{3}$ ← ① of the ③ equal parts

(c) Two-thirds, $\dfrac{2}{3}$, means we have two of the three equal parts.

(two-thirds) $\dfrac{2}{3}$ ← ② of the ③ equal parts

(d) Three-thirds, $\dfrac{3}{3}$, means that we have three of the three equal parts, but as the circle has only three equal parts, $\dfrac{3}{3} = 1$ (the whole). In the same way,

(two-halves) $\dfrac{2}{2} = 1$;   (four-fourths) $\dfrac{4}{4} = 1$;

(five-fifths) $\dfrac{5}{5} = 1$.

In general, any number (except zero) divided by itself is one.

PROPER AND IMPROPER FRACTIONS. A *proper fraction* is a fraction whose numerator is less than its denominator. Examples of proper fractions are:

$$\dfrac{1}{2},\ \dfrac{2}{3},\ \dfrac{3}{4},\ \dfrac{4}{5},\ \dfrac{7}{8},\ \dfrac{18}{35}$$

An *improper fraction* is a fraction whose numerator is larger than (or equal to) its denominator. Examples of improper fractions are:

$$\frac{4}{3}, \quad \frac{5}{4}, \quad \frac{7}{2}, \quad \frac{11}{11}, \quad \frac{131}{17}$$

EXERCISES 301, SET I

1. If we divide a whole into eight equal parts and take three of them, what fraction would represent the part taken?
2. If you cut a pie into five equal pieces and serve two pieces, what fractional part of the pie is left?
3. If we divide a class into six equal groups and take five of the groups, what fractional part of the class was *not* taken?
4. In a football game, three of the eleven first-string players were injured. What fractional part of the team's first-string players were injured?

5. 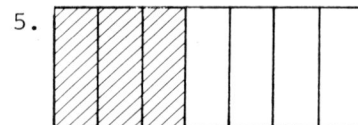 What fraction represents the shaded portion of the rectangle?

6. 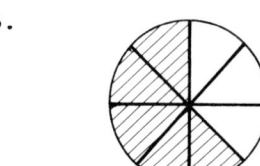 What fraction represents the shaded portion of the circle?

For Exercises 7-12, use the following list:

$$\frac{5}{11}, \quad \frac{28}{13}, \quad \frac{17}{22}, \quad \frac{8}{8}, \quad \frac{1}{2}, \quad \frac{98}{107}, \quad \frac{316}{219}, \quad \frac{1}{31}, \quad \frac{4}{4}$$

7. Which fractions in the list are proper fractions?
8. Which fractions in the list are improper fractions?
9. What is the numerator of the first fraction?
10. What is the denominator of the second fraction?
11. What are the terms of the third fraction?
12. Name any of the fractions that are equal in the above list.

EXERCISES 301, SET II

1. If we divide a whole into five equal parts and take two of them, what fraction would represent the part taken?
2. If we divide a class into eight equal groups and take six of the groups, what fractional part of the class was *not* taken?

3. 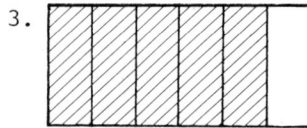 What fraction represents the shaded part of the rectangle?

For Exercises 4-7, use the following list.

$$\frac{3}{7}, \quad \frac{6}{6}, \quad \frac{36}{37}, \quad \frac{1}{3}, \quad \frac{41}{40}, \quad \frac{5}{5}, \quad \frac{21}{17}$$

4. Which fractions in the list are proper fractions?
5. What is the denominator in the first fraction?
6. Name any of the fractions that are equal.
7. Which fractions in the list are improper fractions?

Multiplication of Fractions

When dealing with whole numbers, addition was introduced first, and all other operations on whole numbers were based on an understanding of addition. With fractions, multiplication is introduced first, and all other operations follow.

> To Multiply Two Fractions
>
> $$\frac{\text{product of}}{\text{two fractions}} = \frac{\text{product of their numerators}}{\text{product of their denominators}}$$
>
> In symbols, this is
>
> $$\frac{a}{b} \cdot \frac{c}{d} = \frac{a \cdot c}{b \cdot d}$$
>
> (Definition of the product of two fractions)

Example 1. Product of two fractions.

(a) $\dfrac{3}{5} \cdot \dfrac{2}{7} = \dfrac{3 \cdot 2}{5 \cdot 7} = \dfrac{6}{35}$ (b) $\dfrac{13}{6} \cdot \dfrac{3}{8} = \dfrac{13 \cdot 3}{6 \cdot 8} = \dfrac{39}{48}$

(c) $\dfrac{5}{12} \cdot \dfrac{7}{16} = \dfrac{5 \cdot 7}{12 \cdot 16} = \dfrac{35}{192}$

EXERCISES 302, SET I. Multiply the following fractions.

1. $\dfrac{2}{3} \cdot \dfrac{5}{7}$ 2. $\dfrac{1}{2} \cdot \dfrac{5}{3}$ 3. $\dfrac{3}{4} \cdot \dfrac{7}{8}$ 4. $\dfrac{5}{8} \cdot \dfrac{3}{2}$

5. $\dfrac{4}{5} \cdot \dfrac{3}{5}$ 6. $\dfrac{5}{6} \cdot \dfrac{7}{8}$ 7. $\dfrac{4}{9} \cdot \dfrac{2}{3}$ 8. $\dfrac{7}{3} \cdot \dfrac{5}{9}$

9. $\dfrac{3}{4} \cdot \dfrac{3}{4}$ 10. $\dfrac{6}{7} \cdot \dfrac{6}{7}$ 11. $\dfrac{5}{12} \cdot \dfrac{5}{8}$ 12. $\dfrac{3}{8} \cdot \dfrac{5}{16}$

13. $\dfrac{11}{32} \cdot \dfrac{3}{2}$ 14. $\dfrac{7}{12} \cdot \dfrac{13}{15}$ 15. $\dfrac{7}{8} \cdot \dfrac{11}{13}$ 16. $\dfrac{37}{16} \cdot \dfrac{15}{43}$

17. $\dfrac{24}{23} \cdot \dfrac{6}{17}$ 18. $\dfrac{41}{34} \cdot \dfrac{19}{29}$ 19. $\dfrac{81}{28} \cdot \dfrac{13}{55}$ 20. $\dfrac{117}{84} \cdot \dfrac{163}{289}$

21. What is the area of a rectangle having a length of $\dfrac{11}{15}$ miles and a width of $\dfrac{2}{5}$ miles?

22. Which is larger, a $\frac{1}{2}$-inch square or $\frac{1}{2}$ square inch?

EXERCISES 302, SET II. Multiply the following fractions.

1. $\frac{3}{5} \cdot \frac{4}{7}$ 2. $\frac{2}{3} \cdot \frac{4}{5}$ 3. $\frac{3}{7} \cdot \frac{4}{7}$

4. $\frac{2}{3} \cdot \frac{2}{3}$ 5. $\frac{4}{9} \cdot \frac{4}{5}$ 6. $\frac{5}{8} \cdot \frac{3}{16}$

7. $\frac{6}{7} \cdot \frac{9}{11}$ 8. $\frac{31}{41} \cdot \frac{13}{15}$ 9. $\frac{51}{19} \cdot \frac{11}{9}$

10. Which is larger, a $\frac{1}{4}$-inch square or $\frac{1}{4}$ square inch?

MULTIPLICATIVE IDENTITY. When any whole number is multiplied by 1, the product is the identical whole number we started with. For example:

$$7 \cdot 1 = 7$$
$$25 \cdot 1 = 25$$
$$13 \cdot 1 = 13$$

For this reason, 1 is called the *multiplicative identity* for whole numbers.

Since $\frac{1}{1} = 1$, then $\frac{a}{b} \cdot 1 = \frac{a}{b} \cdot \frac{1}{1} = \frac{a \cdot 1}{b \cdot 1} = \frac{a}{b}$. Because the fraction $\frac{a}{b}$ when multiplied by 1 gives the identical fraction $\frac{a}{b}$, we see that 1 is the multiplicative identity for fractions as well as for whole numbers.

Example 2

(a) $\frac{2}{3} \cdot 1 = \frac{2}{3} \cdot \frac{1}{1} = \frac{2 \cdot 1}{3 \cdot 1} = \frac{2}{3}$.   (b) $\frac{5}{9} \cdot 1 = \frac{5}{9}$

A Fraction Is Equivalent to a Division

Refer to Figure 303A.

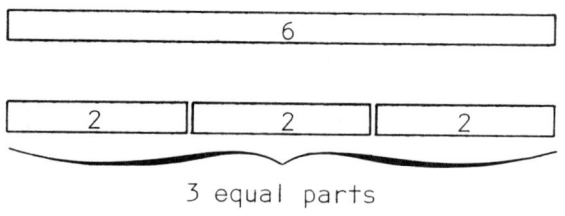

3 equal parts

Figure 303A

Fraction Interpretation. If we split six units up into three equal parts, we can say that each of the pieces is $\frac{1}{3}$ of 6. From the figure, we can see that $\frac{1}{3}$ of 6 = 2.

explained below

$$\frac{1}{3} \text{ of } 6 = \frac{1}{3} \cdot 6 = \frac{1}{3} \cdot \frac{6}{1} = \frac{1 \cdot 6}{3 \cdot 1} = \frac{6}{3} = 2 \qquad \text{(from the figure)}$$

<u>Division Interpretation</u>. We already know $6 \div 3 = 2$. We just showed that $\frac{6}{3} = 2$. Then, since both $\frac{6}{3}$ and $6 \div 3$ represent the same thing (in this case, 2), we can say:

$$\frac{6}{3} = 6 \div 3 = 2$$

If a and b are whole numbers, $(b \neq 0)$, then

$$\frac{a}{b} = a \div b$$

<u>Example 1</u>

(a) $\frac{4}{2} = 4 \div 2 = 2\overline{\smash{)}4} = 2$ (b) $\frac{36}{4} = 36 \div 4 = 4\overline{\smash{)}36} = 9$

(c) $\frac{39}{11} = 39 \div 11 = 11\overline{\smash{)}39}$ Here the answer is not a whole number. Divisions of this type will be discussed in Section 305.

<u>EXERCISES 303A, SET I</u>. Change the fractions to whole numbers.

1. $\frac{8}{2}$ 2. $\frac{9}{3}$ 3. $\frac{24}{6}$ 4. $\frac{42}{7}$ 5. $\frac{35}{35}$

6. $\frac{84}{7}$ 7. $\frac{144}{16}$ 8. $\frac{751}{751}$ 9. $\frac{2,231}{97}$ 10. $\frac{128,934}{551}$

<u>EXERCISES 303A, SET II</u>. Change the fractions to whole numbers.

1. $\frac{6}{3}$ 2. $\frac{16}{4}$ 3. $\frac{18}{18}$ 4. $\frac{108}{12}$ 5. $\frac{1,692}{141}$

<u>WHOLE NUMBERS CAN BE WRITTEN AS FRACTIONS</u>. Since $\frac{8}{1} = 8 \div 1 = 8$, in general, $\frac{a}{1} = a \div 1 = a$. This means that any whole number can be written as a fraction by writing it over 1. That is, the whole number becomes the numerator of a fraction whose denominator is 1.

<u>Example 2</u>

(a) $5 = \frac{5}{1}$ (b) $29 = \frac{29}{1}$ (c) $117 = \frac{117}{1}$

This makes it possible to multiply fractions by whole numbers.

Example 3

(a) $2 \cdot \frac{4}{9} = \frac{2}{1} \cdot \frac{4}{9} = \frac{2 \cdot 4}{1 \cdot 9} = \frac{8}{9}$

(b) $\frac{3}{4} \cdot 17 = \frac{3}{4} \cdot \frac{17}{1} = \frac{3 \cdot 17}{4 \cdot 1} = \frac{51}{4}$

(c) $5 \cdot \frac{13}{9} = \frac{5}{1} \cdot \frac{13}{9} = \frac{5 \cdot 13}{1 \cdot 9} = \frac{65}{9}$

(d) $\frac{3}{5}$ of a class are girls. How many of its 35 members are girls?

$\frac{3}{5} \cdot 35 = \frac{3}{5} \cdot \frac{35}{1} = \frac{3 \cdot 35}{5 \cdot 1} = \frac{105}{5} = 5\overline{\smash)105}^{\,21} = 21$ girls.

EXERCISES 303B, SET I. In Exercises 1-10, find the products.

1. $2 \cdot \frac{1}{3}$ 2. $3 \cdot \frac{2}{7}$ 3. $\frac{1}{5} \cdot 4$ 4. $\frac{1}{8} \cdot 5$ 5. $3 \cdot \frac{5}{2}$

6. $6 \cdot \frac{3}{5}$ 7. $\frac{7}{12} \cdot 7$ 8. $5 \cdot \frac{3}{16}$ 9. $5 \cdot \frac{1}{12}$ 10. $\frac{3}{32} \cdot 9$

11. If $\frac{2}{3}$ of the graduating class of 420 seniors are girls, how many girls are graduating?

12. A rectangular airfield is 2 miles long and $\frac{2}{5}$ mile wide. What is its area?

13. A school athletic field is 120 yards long and 60 yards wide. If the grass on $\frac{2}{3}$ of it had to be reseeded, how many square yards were reseeded?

EXERCISES 303B, SET II. In Exercises 1-5, find the products.

1. $3 \cdot \frac{1}{5}$ 2. $\frac{1}{4} \cdot 5$ 3. $4 \cdot \frac{3}{5}$ 4. $\frac{5}{9} \cdot 5$ 5. $\frac{3}{31} \cdot 8$

6. If $\frac{3}{5}$ of a graduating class of 320 are boys, how many boys are graduating?

7. A rectangular field is 3 miles long and $\frac{3}{5}$ mile wide. What is its area?

Mixed Numbers

Since   5 thirds = 3 thirds + 2 thirds

then $\dfrac{5}{3}$ = $\dfrac{3}{3}$ + $\dfrac{2}{3}$

and $\dfrac{5}{3}$ = 1 + $\dfrac{2}{3}$

so that $\dfrac{5}{3}$ = 1 unit + $\dfrac{2}{3}$ of another unit
(Figure 304).

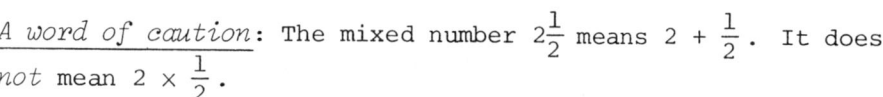

$\dfrac{5}{3}$ = 1 + $\dfrac{2}{3}$ (written $1\dfrac{2}{3}$)

1 unit + $\dfrac{2}{3}$ of unit

Figure 304

In arithmetic, this result is written without the plus sign as $1\dfrac{2}{3}$, and is read "one and two thirds." Such a number is called a *mixed number* because it is made up of both a whole number part (1) and a fraction part ($\dfrac{2}{3}$).

A word of caution: The mixed number $2\dfrac{1}{2}$ means $2 + \dfrac{1}{2}$. It does *not* mean $2 \times \dfrac{1}{2}$.

Example 1. Examples of mixed numbers.

$$2\dfrac{1}{2}, \quad 3\dfrac{5}{8}, \quad 5\dfrac{1}{4}, \quad 12\dfrac{3}{16}$$

Changing an Improper Fraction to a Mixed Number

In Section 303, we showed that a fraction is equivalent to a division. That is, $\dfrac{a}{b} = a \div b = b\overline{)a}$. We know that in a division

$$dividend = divisor \times quotient + remainder \qquad (1)$$

This statement can also be expressed as

$$\frac{dividend}{divisor} = quotient + \frac{remainder}{divisor} \qquad (2)$$

Consider the improper fraction $\frac{11}{8}$. Since $\frac{11}{8} = 11 \div 8 = $

$8\overline{)11}^{\;1\ R\ 3}$, putting this result in formula (2), we have

$$\frac{11}{8} = 1 + \frac{3}{8} = 1\frac{3}{8}$$

So we see that *an improper fraction may be changed to a mixed number by division* if its numerator is greater than its denominator.

To Change an Improper Fraction to a Mixed Number

1. Divide the numerator by the denominator.
2. Write the quotient followed by the fraction: remainder over denominator.

$$quotient\ \frac{remainder}{denominator}$$

Example 1

(a) $\frac{3}{2} = 3 \div 2 = 2\overline{)3}^{\;1\ R\ 1} = 1 + \frac{1}{2} = 1\frac{1}{2}$

(b) $\frac{15}{11} = 15 \div 11 = 11\overline{)15}^{\;1\ R\ 4} = 1\frac{4}{11}$

(c) $\frac{37}{5} = 37 \div 5 = 5\overline{)37}^{\;7\ R\ 2} = 7\frac{2}{5}$

(d) $\frac{51}{19} = 51 \div 19 = 19\overline{)51}^{\;2\ R\ 13} = 2\frac{13}{19}$

EXERCISES 305, SET I. Change the following improper fractions to mixed numbers.

1. $\frac{5}{3}$ 2. $\frac{7}{4}$ 3. $\frac{9}{5}$ 4. $\frac{11}{4}$ 5. $\frac{13}{5}$

6. $\frac{11}{3}$ 7. $\frac{15}{4}$ 8. $\frac{35}{2}$ 9. $\frac{23}{6}$ 10. $\frac{86}{9}$

11. $\frac{16}{13}$ 12. $\frac{32}{23}$ 13. $\frac{20}{7}$ 14. $\frac{56}{15}$ 15. $\frac{207}{19}$

16. $\frac{37}{21}$ 17. $\frac{54}{17}$ 18. $\frac{87}{31}$ 19. $\frac{136}{45}$ 20. $\frac{237}{64}$

Change the following improper fractions to mixed numbers.

1. $\frac{6}{5}$ 2. $\frac{9}{4}$ 3. $\frac{14}{3}$ 4. $\frac{23}{7}$ 5. $\frac{67}{8}$

6. $\frac{32}{29}$ 7. $\frac{58}{19}$ 8. $\frac{127}{31}$ 9. $\frac{135}{43}$ 10. $\frac{241}{81}$

Changing a Mixed Number to an Improper Fraction

Consider the mixed number $2\frac{3}{8}$. This means $2 + \frac{3}{8} = 2$ units + 3 eighths. Each unit is 8 eighths; therefore, 2 units = 2 · 8 eighths = 16 eighths. So that $2\frac{3}{8}$ = 16 eighths + 3 eighths = 19 eighths = $\frac{19}{8}$.

To Change a Mixed Number to an Improper Fraction

1. Multiply the whole number by the denominator of the fraction.
2. Add the numerator of the fraction to the product found in (1).
3. Write that sum found in (2) over the denominator of the fraction.

Example 1

(a) $2\frac{1}{3} = \frac{2 \cdot 3 + 1}{3} = \frac{6 + 1}{3} = \frac{7}{3}$

(b) $7\frac{2}{5} = \frac{7 \cdot 5 + 2}{5} = \frac{35 + 2}{5} = \frac{37}{5}$

(c) $4\frac{7}{13} = \frac{4 \cdot 13 + 7}{13} = \frac{52 + 7}{13} = \frac{59}{13}$

(d) $25\frac{19}{23} = \frac{25 \cdot 23 + 19}{23} = \frac{575 + 19}{23} = \frac{594}{23}$

EXERCISES 306, SET I. Change the following mixed numbers to improper fractions.

1. $1\frac{1}{2}$ 2. $2\frac{3}{5}$ 3. $3\frac{1}{4}$ 4. $2\frac{5}{8}$ 5. $4\frac{5}{6}$

6. $4\frac{1}{2}$ 7. $3\frac{7}{10}$ 8. $3\frac{5}{16}$ 9. $5\frac{7}{12}$ 10. $6\frac{2}{7}$

11. $12\frac{2}{3}$ 12. $3\frac{9}{13}$ 13. $6\frac{3}{4}$ 14. $4\frac{5}{11}$ 15. $3\frac{7}{15}$

16. $1\frac{8}{17}$ 17. $15\frac{23}{44}$ 18. $21\frac{17}{33}$ 19. $2\frac{8}{63}$ 20. $8\frac{23}{117}$

<u>EXERCISES 306, SET II</u>. Change the following mixed numbers to improper fractions.

1. $1\frac{1}{3}$ 2. $2\frac{3}{4}$ 3. $4\frac{3}{5}$ 4. $2\frac{7}{10}$ 5. $4\frac{5}{12}$

6. $12\frac{3}{5}$ 7. $5\frac{2}{15}$ 8. $2\frac{9}{13}$ 9. $20\frac{10}{29}$ 10. $7\frac{25}{113}$

Equivalent Fractions

A fraction is equivalent to a division. *Equivalent fractions* are fractions having the same quotient. They are different ways of writing the same number. A look at Figure 307 will convince you that $\frac{3}{4}$ and $\frac{9}{12}$ are equivalent fractions because they represent the same thing.

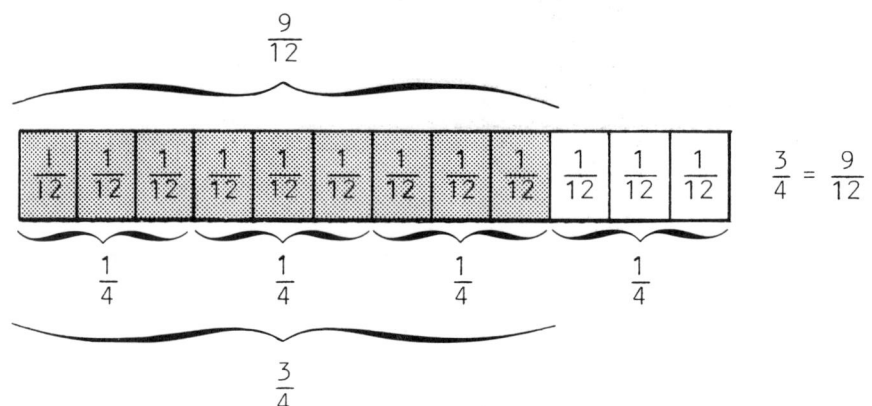

Figure 307

Another way to see that $\frac{3}{4} = \frac{9}{12}$ is

$$\frac{3}{4} = \frac{3}{4} \cdot 1 = \frac{3}{4} \cdot \frac{3}{3} = \frac{3 \cdot 3}{4 \cdot 3} = \frac{9}{12}$$

↑——this is 1

So that $\frac{3}{4}$ and $\frac{9}{12}$ are equivalent fractions (or simply, $\frac{3}{4} = \frac{9}{12}$). In general, $\frac{a}{b} = \frac{a}{b} \cdot 1 = \frac{a}{b} \cdot \frac{c}{c} = \frac{a \cdot c}{b \cdot c}$. So that $\frac{a}{b}$ and $\frac{a \cdot c}{b \cdot c}$ are equivalent fractions. Or simply,

$$\frac{a}{b} = \frac{a \cdot c}{b \cdot c} \qquad \text{where } c \neq 0 \qquad (1)$$

This means: *If the numerator and denominator of a fraction are multiplied by the same number (not zero), an equivalent fraction is formed.* An easy way to check if two fractions are equivalent is to see if their *cross-products* are equal.

$$\frac{a}{b} \bowtie \frac{c}{d} \quad \text{if} \quad a \cdot d = b \cdot c$$

cross ——→   ⌊————————⌋ cross-products

If the product $a \cdot d$ is greater than the product $b \cdot c$, then $\frac{a}{b} > \frac{c}{d}$. If the product $a \cdot d$ is less than the product $b \cdot c$, then $\frac{a}{b} < \frac{c}{d}$. For a more complete explanation of how to compare the size of fractions, see Section 325.

Example 1

(a) Is $\frac{6}{3} = \frac{10}{5}$? Yes, because $6 \cdot 5 = 3 \cdot 10$
$$30 \;=\; 30$$

(b) Is $\frac{18}{34} = \frac{1}{2}$? No, because $18 \cdot 2 \neq 34 \cdot 1$
$$36 \;\neq\; 34$$

Since $18 \cdot 2 > 34 \cdot 1$, then $\frac{18}{34} > \frac{1}{2}$.

(c) Is $\frac{23}{25} = \frac{99}{107}$? No, because $23 \cdot 107 \neq 25 \cdot 99$
$$2{,}461 \;\neq\; 2{,}475$$

Since $23 \cdot 107 < 25 \cdot 99$, then $\frac{23}{25} < \frac{99}{107}$.

(d) Is $\frac{46}{69} = \frac{34}{51}$? Yes, because $46 \cdot 51 = 69 \cdot 34$
$$2{,}346 \;=\; 2{,}346$$

EXERCISES 307, SET I. Find out whether the given fractions are equivalent (have equal values). If the fractions are not equivalent, state which one is larger.

1. $\frac{2}{3}$, $\frac{8}{12}$ 2. $\frac{3}{5}$, $\frac{9}{15}$ 3. $\frac{7}{27}$, $\frac{1}{4}$

4. $\frac{3}{22}$, $\frac{1}{7}$ 5. $\frac{25}{30}$, $\frac{5}{6}$ 6. $\frac{22}{6}$, $\frac{44}{12}$

7. $\frac{30}{35}$, $\frac{24}{28}$ 8. $\frac{24}{21}$, $\frac{72}{54}$ 9. $\frac{8}{12}$, $\frac{10}{18}$

10. $\frac{15}{18}$, $\frac{30}{42}$ 11. $\frac{10}{8}$, $\frac{15}{12}$ 12. $\frac{18}{24}$, $\frac{15}{20}$

13. $\frac{28}{40}$, $\frac{24}{30}$ 14. $\frac{40}{48}$, $\frac{35}{42}$ 15. $\frac{54}{70}$, $\frac{48}{56}$

EXERCISES 307, SET II. Find out whether the given fractions are equivalent (have equal values). If the fractions are not equivalent, state which one is larger.

1. $\dfrac{3}{4}$, $\dfrac{9}{12}$ 2. $\dfrac{3}{5}$, $\dfrac{20}{35}$ 3. $\dfrac{7}{35}$, $\dfrac{1}{5}$

4. $\dfrac{14}{21}$, $\dfrac{18}{27}$ 5. $\dfrac{35}{42}$, $\dfrac{42}{48}$ 6. $\dfrac{26}{39}$, $\dfrac{27}{42}$

7. $\dfrac{44}{77}$, $\dfrac{48}{84}$

Raising a Fraction to Higher Terms

Raising a fraction to higher terms means finding an equivalent fraction having numerator and denominator larger than those of the original fraction. Formula (1) in Section 307 makes it easy to raise a fraction to higher terms.

To raise a fraction to higher terms, multiply both the numerator and the denominator by the same number (greater than one).

$$\frac{a}{b} = \frac{a \cdot c}{b \cdot c} \qquad\qquad (1)$$

Example 1

(a) Raise $\dfrac{3}{4}$ to higher terms.

$$\frac{3}{4} = \frac{3 \cdot 2}{4 \cdot 2} = \frac{6}{8}$$

Both numerator and denominator were multiplied by 2. We took $c = 2$ in Formula (1).

(b) Raise $\dfrac{7}{9}$ to higher terms.

$$\frac{7}{9} = \frac{7 \cdot 5}{9 \cdot 5} = \frac{35}{45}$$

(c) Raise $\dfrac{5}{8}$ to a fraction having a denominator of 24.
In other words, $\dfrac{5}{8} = \dfrac{?}{24}$. Since the denominator, 8, must be multiplied by 3 to give 24, we must multiply the numerator by 3, giving 15.

$$\frac{5}{8} = \frac{5 \cdot 3}{8 \cdot 3} = \frac{15}{24}$$

(d) Raise $\dfrac{11}{15}$ to a fraction having a numerator of 44.
In other words, $\dfrac{11}{15} = \dfrac{44}{?}$. Since the numerator, 11, must be multiplied by 4 to give 44, we must multiply the denominator by 4, giving 60.

$$\frac{11}{15} = \frac{11 \cdot 4}{15 \cdot 4} = \frac{44}{60}$$

(e) Raise $\frac{9}{12}$ to a fraction having a denominator of 60.

In other words, $\frac{9}{12} = \frac{?}{60}$. Since the denominator must be multiplied by 5 to give 60, we must multiply the numerator by 5, giving 45.

$$\frac{9}{12} = \frac{9 \cdot 5}{12 \cdot 5} = \frac{45}{60}$$

EXERCISES 308, SET I. In Exercises 1-20, replace the question mark with a number that makes the fractions equivalent (have equal value).

1. $\frac{1}{2} = \frac{?}{4}$ 2. $\frac{1}{2} = \frac{3}{?}$ 3. $\frac{1}{2} = \frac{?}{10}$ 4. $\frac{1}{3} = \frac{?}{6}$

5. $\frac{2}{3} = \frac{?}{6}$ 6. $\frac{2}{3} = \frac{4}{?}$ 7. $\frac{1}{4} = \frac{3}{?}$ 8. $\frac{3}{4} = \frac{?}{40}$

9. $\frac{2}{5} = \frac{8}{?}$ 10. $\frac{5}{6} = \frac{?}{12}$ 11. $\frac{3}{5} = \frac{?}{15}$ 12. $\frac{4}{7} = \frac{24}{?}$

13. $\frac{9}{13} = \frac{?}{52}$ 14. $\frac{12}{16} = \frac{60}{?}$ 15. $\frac{15}{18} = \frac{?}{36}$ 16. $\frac{5}{9} = \frac{35}{?}$

17. $\frac{8}{11} = \frac{?}{55}$ 18. $\frac{14}{25} = \frac{42}{?}$ 19. $\frac{20}{26} = \frac{?}{52}$ 20. $\frac{6}{15} = \frac{48}{?}$

EXERCISES 308, SET II. In Exercises 1-10, replace the question mark with a number that makes the fractions equivalent (have equal value).

1. $\frac{1}{3} = \frac{?}{6}$ 2. $\frac{1}{2} = \frac{?}{8}$ 3. $\frac{3}{4} = \frac{?}{8}$ 4. $\frac{3}{5} = \frac{9}{?}$

5. $\frac{2}{3} = \frac{?}{30}$ 6. $\frac{3}{7} = \frac{12}{?}$ 7. $\frac{9}{11} = \frac{?}{44}$ 8. $\frac{12}{16} = \frac{48}{?}$

9. $\frac{13}{20} = \frac{?}{60}$ 10. $\frac{8}{12} = \frac{32}{?}$

Reducing a Fraction to Lower Terms

To reduce a fraction to lower terms, we must find an equivalent fraction whose numerator and denominator are smaller than those of the original fraction. For example:

$$\frac{12}{18} = \frac{2 \cdot 6}{3 \cdot 6} = \frac{2}{3} \cdot \frac{6}{6} = \frac{2}{3} \cdot 1 = \frac{2}{3}$$

Here we found a number, 6, which is a factor (or divisor) of both terms of $\frac{12}{18}$. The same problem can be done another way:

$$\frac{12}{18} = \frac{12 \div 6}{18 \div 6} = \frac{2}{3}$$

Here we divided numerator and denominator by the same number, 6. This same fraction, $\frac{12}{18}$, can be reduced to different lower terms.

$$\frac{12}{18} = \frac{12 \div 2}{18 \div 2} = \frac{6}{9}$$

Here we divided numerator and denominator by the same number, 2. $\frac{12}{18}$ has been reduced to $\frac{6}{9}$ as well as to $\frac{2}{3}$, where $\frac{2}{3}$ has terms lower than $\frac{6}{9}$. In Section 311, we will discuss methods of reducing a fraction to its *lowest* terms.

> *To reduce a fraction to lower terms,* divide both the numerator and the denominator by a number (greater than one) that is a divisor of both.
>
> $$\frac{a}{b} = \frac{a \div c}{b \div c}$$

(3)

Example 1

(a) Reduce $\frac{15}{20}$ to lower terms: $\frac{15}{20} = \frac{15 \div 5}{20 \div 5} = \frac{3}{4}$

(b) Reduce $\frac{24}{16}$ to lower terms: $\frac{24}{16} = \frac{24 \div 2}{16 \div 2} = \frac{12}{8}$

(c) Reduce $\frac{36}{18}$ to a fraction having a denominator of 9.
In other words, $\frac{36}{18} = \frac{?}{9}$. Since the denominator, 18, must be divided by 2 to give 9, we must divide the numerator by 2, giving 18.

$$\frac{36}{18} = \frac{36 \div 2}{18 \div 2} = \frac{18}{9}$$

(d) Reduce $\frac{14}{35}$ to a fraction having a numerator of 2.
In other words, $\frac{14}{35} = \frac{2}{?}$. Since the numerator must be divided by 7 to give 2, we must divide the denominator by 7, giving 5.

$$\frac{14}{35} = \frac{14 \div 7}{35 \div 7} = \frac{2}{5}$$

(e) Reduce $\frac{24}{40}$ to a fraction having a denominator of 5.
Since the denominator must be divided by 8 to give 5, we must divide the numerator by 8, giving 3.

$$\frac{24}{40} = \frac{24 \div 8}{40 \div 8} = \frac{3}{5}$$

EXERCISES 309, SET I. Reduce each of the given fractions to lower terms having the given numerator or denominator.

1. $\frac{3}{6} = \frac{?}{2}$ 2. $\frac{2}{4} = \frac{1}{?}$ 3. $\frac{6}{8} = \frac{?}{4}$ 4. $\frac{2}{10} = \frac{1}{?}$

5. $\frac{6}{10} = \frac{?}{5}$ 6. $\frac{9}{12} = \frac{3}{?}$ 7. $\frac{6}{16} = \frac{?}{8}$ 8. $\frac{15}{20} = \frac{3}{?}$

9. $\frac{14}{20} = \frac{?}{10}$ 10. $\frac{10}{16} = \frac{5}{?}$ 11. $\frac{18}{45} = \frac{?}{15}$ 12. $\frac{63}{35} = \frac{9}{?}$

13. $\frac{36}{54} = \frac{?}{9}$ 14. $\frac{6}{14} = \frac{3}{?}$ 15. $\frac{72}{24} = \frac{?}{3}$ 16. $\frac{80}{100} = \frac{20}{?}$

17. $\frac{85}{34} = \frac{?}{2}$ 18. $\frac{27}{78} = \frac{9}{?}$ 19. $\frac{33}{77} = \frac{?}{7}$ 20. $\frac{72}{60} = \frac{6}{?}$

EXERCISES 309, SET II. Reduce each of the given fractions to lower terms having the given numerator or denominator.

1. $\frac{4}{8} = \frac{?}{2}$ 2. $\frac{6}{9} = \frac{?}{3}$ 3. $\frac{9}{12} = \frac{?}{4}$ 4. $\frac{16}{20} = \frac{4}{?}$

5. $\frac{10}{12} = \frac{5}{?}$ 6. $\frac{42}{28} = \frac{3}{?}$ 7. $\frac{6}{15} = \frac{2}{?}$ 8. $\frac{60}{100} = \frac{15}{?}$

9. $\frac{81}{111} = \frac{27}{?}$ 10. $\frac{77}{66} = \frac{7}{?}$

Review

1. A fraction, $\frac{5}{8}$, may be thought of as taking five of eight equal parts a unit is divided into. It may also be thought of as 5 ÷ 8. (Sec. 301)
2. The numerator of a fraction can be any whole number (including 0), but its denominator cannot be 0. (Sec. 301)
3. A *proper fraction* is a fraction in which the numerator is less than the denominator. An *improper fraction* is a fraction in which the numerator is greater than (or equal to) the denominator. (Sec. 301)
4. *To change an improper fraction to a mixed number*, divide the numerator by the denominator. (Sec. 305) Write the quotient, followed by the fraction: remainder over denominator.
5. *To change a mixed number to an improper fraction*: (Sec. 306)

   (1) Multiply the whole number by the denominator of the fraction.
   (2) Add the numerator of the fraction to the product found in (1).
   (3) Write the sum found in (2) over the denominator of the fraction.

6. *To multiply fractions*, multiply their numerators and put

that product over the product of their denominators. (Sec. 302)

7. If a whole number or a fraction is multiplied by 1, its value remains the same.

8. *Two fractions are equivalent* (have equal value) if their cross products are equal. (Sec. 307)

$$\frac{a}{b} = \frac{c}{d} \quad \text{if} \quad a \cdot d = b \cdot c$$

9. *To raise a fraction to higher terms*, multiply both numerator and denominator by the same number (greater than one). (Sec. 308)

10. *To reduce a fraction to lower terms*, divide both numerator and denominator by a number (greater than one) that is a divisor of both. (Sec. 309)

EXERCISES 310, SET I. In Exercises 1-6, change the improper fraction to a mixed number.

1. $\frac{5}{2}$

2. $\frac{7}{4}$

3. $\frac{8}{3}$

4. $\frac{18}{5}$

5. $\frac{25}{16}$

6. $\frac{83}{19}$

In Exercises 7-12, change the mixed number to an improper fraction.

7. $1\frac{2}{3}$

8. $2\frac{1}{4}$

9. $4\frac{1}{2}$

10. $9\frac{3}{5}$

11. $3\frac{7}{12}$

12. $25\frac{3}{16}$

In Exercises 13-18, find the products. If your answer is an improper fraction, change it to a mixed number.

13. $\frac{7}{8} \cdot \frac{3}{4}$

14. $\frac{8}{9} \cdot \frac{2}{3}$

15. $\frac{7}{3} \cdot \frac{5}{13}$

16. $\frac{18}{7} \cdot \frac{4}{5}$

17. $\frac{17}{32} \cdot \frac{21}{13}$

18. $\frac{53}{37} \cdot \frac{46}{83}$

In Exercises 19-32, replace the question mark with a number that makes the fractions equivalent (have equal value).

19. $\frac{7}{9} = \frac{?}{72}$

20. $\frac{24}{16} = \frac{6}{?}$

21. $\frac{3}{5} = \frac{12}{?}$

22. $\frac{18}{27} = \frac{?}{3}$

23. $\frac{2}{7} = \frac{?}{42}$

24. $\frac{21}{36} = \frac{7}{?}$

25. $\frac{16}{27} = \frac{32}{?}$

26. $\frac{33}{15} = \frac{?}{5}$

27. $\frac{4}{3} = \frac{?}{54}$

28. $\frac{45}{80} = \frac{?}{16}$

29. $\frac{7}{9} = \frac{56}{?}$

30. $\frac{56}{40} = \frac{?}{5}$

31. $\frac{15}{11} = \frac{?}{99}$

32. $\frac{35}{84} = \frac{5}{?}$

33. Which of the following fractions are improper?

     (a) $\frac{3}{7}$ (b) $\frac{23}{19}$ (c) $\frac{19}{18}$ (d) $\frac{12}{12}$ (e) $\frac{97}{103}$

In Exercises 34-37, determine which of the fractions are equivalent.

34. $\frac{14}{10}$, $\frac{24}{20}$ 35. $\frac{24}{40}$, $\frac{18}{30}$ 36. $\frac{14}{12}$, $\frac{28}{24}$, $\frac{7}{6}$ 37. $\frac{27}{18}$, $\frac{45}{30}$, $\frac{9}{5}$

38. A rectangle has a length of $\frac{5}{8}$ yard and a width of $\frac{3}{4}$ yard. What is its area in square yards?
39. Jan cut a pie into five equal pieces and served three pieces. What fraction of the pie was left?
40. Bertha used five eggs in making a cake. What fraction of a dozen did she use?
41. John took one bottle of soda from a pack of six bottles. What fraction of the pack did he take?
42. Harry took 17 cards from a deck of 52 cards. What fraction of the deck did he take?
43. If we are in school five days a week, what fraction of the week are we not in school?

EXERCISES 310, SET II. In Exercises 1-3, change the improper fraction to a mixed number.

1. $\frac{7}{3}$ 2. $\frac{17}{6}$ 3. $\frac{40}{17}$

In Exercises 4-6, change the mixed number to an improper fraction.

4. $2\frac{1}{3}$ 5. $6\frac{3}{4}$ 6. $25\frac{5}{8}$

In Exercises 7-9, find the products. If your answer is an improper fraction, change it to a mixed number.

7. $\frac{5}{6} \cdot \frac{5}{3}$ 8. $\frac{8}{3} \cdot \frac{7}{9}$ 9. $\frac{15}{19} \cdot \frac{13}{11}$

In Exercises 10-16, replace the question mark with a number that makes the fractions equivalent (have the same value).

10. $\frac{5}{8} = \frac{?}{48}$ 11. $\frac{3}{7} = \frac{12}{?}$ 12. $\frac{18}{15} = \frac{6}{?}$

13. $\frac{15}{24} = \frac{?}{8}$ 14. $\frac{7}{8} = \frac{?}{40}$ 15. $\frac{9}{7} = \frac{45}{?}$

16. $\frac{12}{20} = \frac{15}{?}$

17. Which of the following fractions are improper fractions?

$$\frac{6}{6}, \ \frac{3}{4}, \ \frac{7}{5}, \ \frac{4}{9}$$

In Exercises 18 and 19, determine which of the fractions are equivalent.

18. $\frac{6}{5}$, $\frac{12}{10}$ 19. $\frac{7}{9}$, $\frac{10}{13}$

20. A rectangle has a length of $\frac{5}{4}$ yards and a width of $\frac{5}{9}$ yards. What is its area in square yards?

21. Win used seven eggs in making a cake. What fraction of a dozen did she use?

Reducing Fractions to Lowest Terms

In Section 309, we learned how to reduce a fraction to *lower* terms.

$$\frac{12}{18} = \frac{12 \div 2}{18 \div 2} = \frac{6}{9} \quad \text{and} \quad \frac{12}{18} = \frac{12 \div 3}{18 \div 3} = \frac{4}{6}$$

Since we have reduced the fraction $\frac{12}{18}$ to two different fractions having *lower* terms, the question arises: "How can we reduce $\frac{12}{18}$ to a fraction having the *lowest* terms?"

CANCELING. In Section 309, we learned that a fraction can be reduced to lower terms by dividing both numerator and denominator by the same number. We wrote

$$\frac{12}{18} = \frac{12 \div 6}{18 \div 6} = \frac{2}{3}$$

This division of numerator and denominator can be written

$$\frac{\overset{2}{\cancel{12}}}{\underset{3}{\cancel{18}}} = \frac{2}{3}$$

and we say "6 has been canceled from both numerator and denominator." Since both the numerator and denominator of the fraction $\frac{2}{3}$ cannot be divided by any number other than 1, we say the fraction is in lowest terms.

Example 1. Examples of reducing fractions to lowest terms.

    (a) Reduce $\frac{6}{8}$ to lowest terms.

$\dfrac{\overset{3}{\cancel{6}}}{\underset{4}{\cancel{8}}} = \dfrac{3}{4}$ Both 6 and 8 were divided by 2.

Another way: $\dfrac{6}{8} = \dfrac{\cancel{2} \cdot 3}{\cancel{2} \cdot 4} = \dfrac{3}{4}$

We canceled like factors (2) from numerator and denominator.

(b) Reduce $\frac{10}{15}$ to lowest terms.

$\frac{\overset{2}{\cancel{10}}}{\underset{3}{\cancel{15}}}$ Both 10 and 15 were divided by 5.

Another way: $\frac{10}{15} = \frac{2 \cdot \cancel{5}}{3 \cdot \cancel{5}} = \frac{2}{3}$

We canceled like factors (5) from numerator and denominator.

(c) Reduce $\frac{30}{42}$ to lowest terms.

$\frac{\overset{5}{\cancel{\overset{15}{\cancel{30}}}}}{\underset{7}{\cancel{\underset{21}{\cancel{42}}}}} = \frac{5}{7}$ We first divide 30 and 42 by 2.
Then we divide 15 and 21 by 3.

When we divide first by one number, then by a second number, we call this *repeated canceling*.

Sometimes the same factor can be divided from both numerator and denominator more than once.

(d) Reduce $\frac{18}{45}$ to lowest terms.

$\frac{\overset{2}{\cancel{\overset{6}{\cancel{18}}}}}{\underset{5}{\cancel{\underset{15}{\cancel{45}}}}} = \frac{2}{5}$ We first divide 18 and 45 by 3.
Then we divide 6 and 15 by 3.

After canceling a factor from numerator and denominator, be sure to check whether the same factor can be canceled again.

EXERCISES 311A, SET I. Reduce to lowest terms by canceling like factors from numerator and denominator.

1. $\frac{6}{9}$ 2. $\frac{15}{20}$ 3. $\frac{12}{16}$ 4. $\frac{14}{21}$ 5. $\frac{30}{40}$

6. $\frac{12}{15}$ 7. $\frac{24}{32}$ 8. $\frac{36}{90}$ 9. $\frac{10}{8}$ 10. $\frac{16}{24}$

11. $\frac{32}{40}$ 12. $\frac{48}{64}$ 13. $\frac{54}{90}$ 14. $\frac{135}{315}$ 15. $\frac{84}{105}$

16. $\frac{42}{84}$ 17. $\frac{33}{55}$ 18. $\frac{44}{66}$ 19. $\frac{210}{270}$ 20. $\frac{3,000}{4,200}$

EXERCISES 311A, SET II. Reduce to lowest terms by canceling like factors from numerator and denominator.

1. $\frac{4}{12}$ 2. $\frac{9}{15}$ 3. $\frac{20}{30}$ 4. $\frac{42}{48}$ 5. $\frac{15}{21}$

6. $\frac{39}{52}$ 7. $\frac{63}{84}$ 8. $\frac{36}{60}$ 9. $\frac{150}{180}$ 10. $\frac{1,400}{1,600}$

Canceling can also be used to shorten multiplication of fractions.

Example 2

(a) $\dfrac{7}{12} \cdot \dfrac{4}{21}$ Solution: $\dfrac{\overset{1}{\cancel{7}}}{\underset{3}{\cancel{12}}} \cdot \dfrac{\overset{1}{\cancel{4}}}{\underset{3}{\cancel{21}}} = \dfrac{1 \cdot 1}{3 \cdot 3} = \dfrac{1}{9}$

(b) $\dfrac{4}{15} \cdot \dfrac{5}{8}$ Solution: $\dfrac{\overset{1}{\cancel{4}}}{\underset{3}{\cancel{15}}} \cdot \dfrac{\overset{1}{\cancel{5}}}{\underset{2}{\cancel{8}}} = \dfrac{1}{6}$

This method can be used when multiplying more than two fractions.

(c) $\dfrac{3}{7} \cdot \dfrac{35}{21} \cdot \dfrac{7}{9}$ Solution: $\dfrac{\overset{1}{\cancel{3}}}{\underset{1}{\cancel{7}}} \cdot \dfrac{\overset{5}{\cancel{35}}}{\underset{3}{\cancel{21}}} \cdot \dfrac{\overset{1}{\cancel{7}}}{\underset{3}{\cancel{9}}} = \dfrac{1 \cdot 5 \cdot 1}{1 \cdot 3 \cdot 3} = \dfrac{5}{9}$

(d) $\dfrac{\overset{1}{\underset{\underset{1}{2}}{\cancel{\cancel{5}}}}}{\underset{2}{\cancel{8}}} \cdot \dfrac{\overset{1}{\cancel{4}}}{\underset{3}{\cancel{15}}} \cdot \dfrac{\overset{1}{\cancel{2}}}{3} = \dfrac{1}{9}$

←————————— *repeated canceling*

EXERCISES 311B, SET I. Find the following products, canceling to simplify your work.

1. $\dfrac{2}{3} \cdot \dfrac{6}{7}$ 2. $\dfrac{2}{5} \cdot \dfrac{5}{8}$ 3. $\dfrac{6}{8} \cdot \dfrac{4}{9}$

4. $\dfrac{7}{8} \cdot \dfrac{12}{7}$ 5. $\dfrac{10}{18} \cdot \dfrac{9}{5}$ 6. $\dfrac{25}{14} \cdot \dfrac{21}{10}$

7. $\dfrac{35}{72} \cdot \dfrac{18}{56}$ 8. $\dfrac{80}{45} \cdot \dfrac{27}{60}$ 9. $\dfrac{22}{75} \cdot \dfrac{20}{44}$

10. $\dfrac{3}{5} \cdot \dfrac{5}{8} \cdot \dfrac{1}{6}$ 11. $\dfrac{5}{7} \cdot \dfrac{3}{10} \cdot \dfrac{14}{15}$ 12. $\dfrac{4}{5} \cdot \dfrac{5}{6} \cdot \dfrac{10}{15}$

13. $\dfrac{16}{49} \cdot \dfrac{14}{8} \cdot \dfrac{9}{12}$ 14. $\dfrac{40}{14} \cdot \dfrac{21}{5} \cdot \dfrac{2}{27}$ 15. $\dfrac{4}{15} \cdot \dfrac{10}{28} \cdot \dfrac{21}{10}$

16. $\dfrac{4}{3} \cdot \dfrac{5}{14} \cdot \dfrac{7}{12} \cdot \dfrac{6}{15}$

EXERCISES 311B, SET II. Find the following products, canceling to simplify your work.

1. $\dfrac{3}{4} \cdot \dfrac{8}{9}$ 2. $\dfrac{2}{5} \cdot \dfrac{5}{4}$ 3. $\dfrac{24}{25} \cdot \dfrac{5}{8}$

4. $\dfrac{60}{45} \cdot \dfrac{15}{80}$ 5. $\dfrac{2}{3} \cdot \dfrac{2}{4} \cdot \dfrac{6}{8}$ 6. $\dfrac{3}{7} \cdot \dfrac{4}{5} \cdot \dfrac{7}{6}$

7. $\dfrac{20}{16} \cdot \dfrac{8}{10} \cdot \dfrac{5}{15}$ 8. $\dfrac{2}{7} \cdot \dfrac{21}{4} \cdot \dfrac{16}{18} \cdot \dfrac{3}{4}$

TESTS FOR DIVISIBILITY. When canceling, use the following rules to tell when a number is divisible by 2, 3, 5, or 10. We have listed only rules that will be most useful.

Divisibility by 2. A number is divisible by 2 if its last digit is 0, 2, 4, 6, or 8.

Examples. 1②; 3 0⓪; 2, 0 3④; 5 7⑧ are divisible by 2.

Divisibility by 3. A number is divisible by 3 if the sum of its digits is divisible by 3.

Examples. 210 is divisible by 3 because 2 + 1 + 0 = 3, which is divisible by 3.

5,162 is not divisible by 3 because 5 + 1 + 6 + 2 = 14, which is not divisible by 3.

Divisibility by 5. A number is divisible by 5 if its last digit is 0 or 5.

Examples. 2 5⓪ and 7 5⑤ are both divisible by 5.

1, 1 2⑦ is not.

Divisibility by 10. A number is divisible by 10 if its last digit is 0.

Examples. 5⓪; 1, 7 2⓪; 3 2⓪ are divisible by 10.

2 5⑥; 6, 4 3⑦; 1 0, 5 9③ are not divisible by 10.

EXERCISES 311C, SET I. Use the rules for divisibility to help you reduce the following fractions to lowest terms.

1. $\dfrac{39}{51}$ 2. $\dfrac{62}{74}$ 3. $\dfrac{35}{55}$ 4. $\dfrac{60}{80}$ 5. $\dfrac{45}{75}$

6. $\dfrac{42}{260}$ 7. $\dfrac{44}{264}$ 8. $\dfrac{85}{215}$ 9. $\dfrac{290}{610}$ 10. $\dfrac{102}{138}$

11. $\dfrac{423}{531}$ 12. $\dfrac{148}{188}$ 13. $\dfrac{175}{325}$ 14. $\dfrac{120}{225}$ 15. $\dfrac{583}{689}$

EXERCISES 311C, SET II. Use the rules for divisibility to help you reduce the following fractions to lowest terms.

1. $\dfrac{42}{111}$ 2. $\dfrac{54}{66}$ 3. $\dfrac{105}{120}$ 4. $\dfrac{45}{72}$ 5. $\dfrac{1,200}{1,500}$

6. $\dfrac{180}{240}$ 7. $\dfrac{160}{360}$

EUCLID'S ALGORITHM. When the numbers in a fraction make it difficult to reduce to lowest terms, a procedure called Euclid's Algorithm can be used. We use Problem 15 of Exercises 311C, Set I, to show the method. In that fraction $\dfrac{583}{689}$, 53 is the largest number that divides both numerator and denominator.* To show how 53 is found, study the following procedure.

---

*The largest number that divides each number in a set of numbers is called their greatest common divisor (GCD).

$$\begin{array}{r} 1 \\ 583\overline{)689} \end{array}$$ ←————————Divide smaller term into larger term.

$$\begin{array}{r} 583 \qquad 5 \\ \overline{106\,)583} \end{array}$$ ←————Divide remainder into previous divisor.

$$\begin{array}{r} 530 \qquad 2 \\ \overline{53\,)106} \end{array}$$ ←————Divide remainder into previous divisor.

Largest
number that
divides 583 ↑ $\dfrac{106}{0}$ ←————Final remainder.
and 689 —┘

$$\begin{array}{r} 11 \\ 53\overline{)583} \\ \underline{53} \\ 53 \\ \underline{53} \\ 0 \end{array} \qquad\qquad \begin{array}{r} 13 \\ 53\overline{)689} \\ \underline{53} \\ 159 \\ \underline{159} \\ 0 \end{array}$$

Therefore, $\dfrac{583}{689} = \dfrac{583 \div 53}{689 \div 53} = \dfrac{11}{13}$

in lowest terms.

Example 3. Use Euclid's Algorithm to reduce $\dfrac{943}{2047}$ to lowest terms.

$$\begin{array}{r} 2 \\ 943\overline{)2047} \end{array}$$ ←————————Divide smaller term into larger term.

$$\begin{array}{r} 1886 \qquad 5 \\ \overline{161\,)943} \end{array}$$ ←————————Divide remainder into previous divisor.

$$\begin{array}{r} 805 \qquad 1 \\ \overline{138\,)161} \end{array}$$ ←————————Divide remainder into previous divisor.

$$\begin{array}{r} 138 \qquad 6 \\ \overline{23\,)138} \end{array}$$ ←————Repeat process.

Largest
number that
divides 943 ↑ $\dfrac{138}{0}$ ←————Final remainder.
and 2,047 —┘

$$\begin{array}{r} 41 \\ 23\overline{)943} \\ \underline{92} \\ 23 \\ \underline{23} \\ 0 \end{array} \qquad\qquad \begin{array}{r} 89 \\ 23\overline{)2047} \\ \underline{184} \\ 207 \\ \underline{207} \\ 0 \end{array}$$

Therefore, $\dfrac{943}{2047} = \dfrac{943 \div 23}{2047 \div 23} = \dfrac{41}{89}$

in lowest terms.

Example 4. Use Euclid's Algorithm to reduce $\dfrac{3951}{2764}$ to lowest terms.

$$\begin{array}{r} 1 \\ 2764\overline{)3951} \end{array}$$ ←————————Divide smaller term into larger term.

$$\begin{array}{r} 2764 \qquad 2 \\ \overline{1187\,)2764} \end{array}$$ ←————Divide remainder into previous divisor.

$$\begin{array}{r} 2374 \qquad 3 \\ \overline{390\,)1187} \end{array}$$ ←————Divide remainder into previous divisor.

$$\begin{array}{r} 1170 \qquad 22 \\ \overline{17\,)390} \end{array}$$ ←————Repeat process.

$$\begin{array}{r} 34 \\ \overline{50} \end{array}$$

$$\begin{array}{r} 34 \qquad 1 \\ \overline{16\,)17} \end{array}$$

$$\begin{array}{r} 16 \qquad 16 \\ \overline{1\,)16} \end{array}$$

Largest number that
divides 2,764 and ↑ $\dfrac{16}{0}$ ← Final remainder.
3,951 —————————————┘

Therefore, $\dfrac{3951}{2764}$ is already reduced to lowest terms.

To Reduce a Fraction to Lowest Terms

1. Divide numerator and denominator by any number that obviously divides both.
2. Use Euclid's Algorithm if you think it may be reduced further.

<u>Example 5.</u> Reduce $\dfrac{1302}{1488}$ to lowest terms.

<u>Solution:</u>

$$\dfrac{\overset{\overset{651}{217}}{1302}}{\underset{\underset{744}{248}}{1488}} = \dfrac{651}{744} = \dfrac{217}{248}$$

$$217\overline{)248}^{1}$$
$$\underline{217}$$
$$31\overline{)217}^{7} \qquad 31\overline{)217}^{7} \qquad 31\overline{)248}^{8}$$
$$\underline{217} \qquad\qquad \underline{217} \qquad\qquad \underline{248}$$
$$0 \qquad\qquad\quad 0 \qquad\qquad\quad 0$$

└── Largest number that divides 217 and 248.

Therefore, $\dfrac{1302}{1488} = \dfrac{217}{248} = \dfrac{7}{8}$, reduced to lowest terms.

Note: A fraction should always be reduced to lowest terms unless otherwise directed.

EXERCISES 311D, SET I. Reduce to lowest terms.

1. $\dfrac{182}{234}$ 2. $\dfrac{363}{429}$ 3. $\dfrac{555}{820}$ 4. $\dfrac{230}{345}$

5. $\dfrac{576}{656}$ 6. $\dfrac{236}{265}$ 7. $\dfrac{188}{235}$ 8. $\dfrac{116}{174}$

9. $\dfrac{354}{885}$ 10. $\dfrac{1,539}{2,394}$ 11. $\dfrac{1,909}{1,577}$ 12. $\dfrac{15,257}{19,019}$

EXERCISES 311D, SET II. Reduce to lowest terms.

1. $\dfrac{205}{246}$ 2. $\dfrac{434}{496}$ 3. $\dfrac{63}{84}$ 4. $\dfrac{165}{198}$

5. $\dfrac{319}{377}$ 6. $\dfrac{247}{285}$

Adding Like Fractions

Like fractions are fractions that have the same denominator.

Example 1

(a) $\frac{2}{3}$, $\frac{5}{3}$, $\frac{1}{3}$ are like fractions.

(b) $\frac{5}{8}$, $\frac{3}{8}$, $\frac{7}{8}$, $\frac{1}{8}$ are like fractions.

Unlike fractions are fractions that have different denominators.

Example 2

(a) $\frac{1}{3}$, $\frac{2}{7}$, $\frac{3}{8}$ are unlike fractions.

(b) $\frac{7}{10}$, $\frac{5}{8}$, $\frac{1}{2}$, $\frac{3}{4}$ are unlike fractions.

ADDING LIKE FRACTIONS

We know that

 1 car   + 3 cars + 7 cars = (1 + 3 + 7) cars   = 11 cars.

Using the same reasoning,

2 thirds + 5 thirds + 1 third = (2 + 5 + 1) thirds = 8 thirds

which can be written

$$\frac{2}{3} \;+\; \frac{5}{3} \;+\; \frac{1}{3} \;=\; \frac{2+5+1}{3} \;=\; \frac{8}{3}$$

> To Add Like Fractions
>
> 1. Add their numerators.
> 2. Write the sum found in (1) over the same de-
>   nominator as that of the like fractions being
>   added.
> 3. Reduce the resulting fraction to lowest terms.
> 4. Change any improper fraction found in (3) to
>   a mixed number.

Example 3

(a) Add: $\frac{1}{9} + \frac{4}{9} + \frac{7}{9} = \frac{1+4+7}{9} = \frac{12}{9} = \frac{\overset{4}{\cancel{12}}}{\underset{3}{\cancel{9}}} = \frac{4}{3} = 1\frac{1}{3}$

(b) Add: $\dfrac{11}{23} + \dfrac{5}{23} = \dfrac{11 + 5}{23} = \dfrac{16}{23}$ (already lowest terms)

(c) Add: $\dfrac{3}{12} + \dfrac{1}{12} + \dfrac{4}{12} + \dfrac{7}{12} = \dfrac{3 + 1 + 4 + 7}{12} = \dfrac{\overset{5}{\cancel{15}}}{\underset{4}{\cancel{12}}} = \dfrac{5}{4} = 1\dfrac{1}{4}$

EXERCISES 312, SET I.   Find the following sums.

1. $\dfrac{1}{6} + \dfrac{3}{6}$

2. $\dfrac{3}{8} + \dfrac{1}{8}$

3. $\dfrac{3}{5} + \dfrac{2}{5}$

4. $\dfrac{3}{4} + \dfrac{1}{4}$

5. $\dfrac{5}{6} + \dfrac{5}{6}$

6. $\dfrac{3}{5} + \dfrac{4}{5}$

7. $\dfrac{1}{2} + \dfrac{3}{2} + \dfrac{5}{2}$

8. $\dfrac{2}{3} + \dfrac{5}{3} + \dfrac{1}{3}$

9. $\dfrac{5}{8} + \dfrac{4}{8} + \dfrac{7}{8}$

10. $\dfrac{1}{6} + \dfrac{5}{6} + \dfrac{3}{6}$

11. $\dfrac{3}{15} + \dfrac{1}{15} + \dfrac{6}{15}$

12. $\dfrac{11}{24} + \dfrac{4}{24}$

13. $\dfrac{1}{12} + \dfrac{5}{12} + \dfrac{2}{12} + \dfrac{3}{12}$

14. $\dfrac{5}{16} + \dfrac{7}{16}$

15. $\dfrac{35}{80} + \dfrac{27}{80}$

16. $\dfrac{29}{45} + \dfrac{16}{45} + \dfrac{3}{45}$

EXERCISES 312, SET II. Find the following sums.

1. $\dfrac{1}{5} + \dfrac{2}{5}$

2. $\dfrac{4}{6} + \dfrac{2}{6}$

3. $\dfrac{2}{7} + \dfrac{5}{7}$

4. $\dfrac{23}{40} + \dfrac{7}{40}$

5. $\dfrac{3}{8} + \dfrac{5}{8} + \dfrac{1}{8}$

6. $\dfrac{11}{16} + \dfrac{3}{16}$

7. $\dfrac{5}{14} + \dfrac{9}{14}$

8. $\dfrac{18}{35} + \dfrac{16}{35} + \dfrac{6}{35}$

Subtracting Like Fractions

We know that

$$5 \text{ planes} - 3 \text{ planes} = 2 \text{ planes}.$$

Using the same reasoning,

$$5 \text{ eighths} - 3 \text{ eighths} = 2 \text{ eighths}$$

which can be written

$$\dfrac{5}{8} - \dfrac{3}{8} = \dfrac{5 - 3}{8} = \dfrac{2}{8} = \dfrac{\overset{1}{\cancel{2}}}{\underset{4}{\cancel{8}}} = \dfrac{1}{4}$$

```
┌─────────────────────────────────────────────────────────────┐
│                                                               │
│   To Subtract Like Fractions                                  │
│                                                               │
│   1. Subtract their numerators.                               │
│   2. Write the difference found in (1) over the               │
│      same denominator as that of the like fractions           │
│      being subtracted.                                        │
│   3. Reduce the resulting fraction to lowest terms.           │
│   4. Change any improper fraction found in (3) to             │
│      a mixed number.                                          │
│                                                               │
└─────────────────────────────────────────────────────────────┘
```

Example 1

(a) Subtract: $\dfrac{25}{8} - \dfrac{11}{8} = \dfrac{25 - 11}{8} = \dfrac{\overset{7}{\cancel{14}}}{\underset{4}{\cancel{8}}} = \dfrac{7}{4} = 1\dfrac{3}{4}$

(b) Subtract: $\dfrac{13}{48} - \dfrac{7}{48} = \dfrac{13 - 7}{48} = \dfrac{\overset{1}{\cancel{6}}}{\underset{8}{\cancel{48}}} = \dfrac{1}{8}$

(c) Subtract: $\dfrac{23}{31} - \dfrac{17}{31} = \dfrac{23 - 17}{31} = \dfrac{6}{31}$ (already lowest terms)

EXERCISES 313, SET I. Subtract the following.

1. $\dfrac{3}{4} - \dfrac{1}{4}$ 2. $\dfrac{5}{6} - \dfrac{1}{6}$ 3. $\dfrac{7}{8} - \dfrac{3}{8}$ 4. $\dfrac{1}{2} - \dfrac{1}{2}$

5. $\dfrac{5}{3} - \dfrac{2}{3}$ 6. $\dfrac{9}{5} - \dfrac{3}{5}$ 7. $\dfrac{7}{10} - \dfrac{5}{10}$ 8. $\dfrac{11}{12} - \dfrac{3}{12}$

9. $\dfrac{6}{7} - \dfrac{6}{7}$ 10. $\dfrac{5}{9} - \dfrac{2}{9}$ 11. $\dfrac{9}{14} - \dfrac{5}{14}$ 12. $\dfrac{27}{35} - \dfrac{6}{35}$

13. $\dfrac{45}{52} - \dfrac{6}{52}$ 14. $\dfrac{70}{81} - \dfrac{19}{81}$ 15. $\dfrac{123}{144} - \dfrac{43}{144}$ 16. $\dfrac{171}{235} - \dfrac{126}{235}$

EXERCISES 313, SET II. Subtract the following.

1. $\dfrac{5}{6} - \dfrac{1}{6}$ 2. $\dfrac{7}{9} - \dfrac{4}{9}$ 3. $\dfrac{9}{10} - \dfrac{4}{10}$ 4. $\dfrac{7}{12} - \dfrac{4}{12}$

5. $\dfrac{7}{8} - \dfrac{3}{8}$ 6. $\dfrac{18}{25} - \dfrac{3}{25}$ 7. $\dfrac{47}{66} - \dfrac{25}{66}$ 8. $\dfrac{197}{228} - \dfrac{86}{228}$

Prime and Composite Numbers

A *prime number* is a natural number greater than 1 that can be divided only by itself and 1. A prime number has *no* factors other than itself and 1.

A *composite number* is a natural number that can be divided by some number other than itself and 1. A composite number *has* factors other than itself and 1.

Example 1

    (a) 9 is composite because 3 • 3 = 9, so that 9 has a
        factor other than itself or 1.

    (b) 17 is prime because 1 and 17 are the only factors
        of 17.

    (c) 45 is composite because it has factors 3, 5, 9, and
        15 other than 1 and 45.

    (d) 31 is prime because 1 and 31 are the only factors of
        31.

EXERCISES 314, SET I. State whether each of the following numbers is prime or composite.

1. 5	2. 8	3. 13	4. 15	5. 12
6. 11	7. 18	8. 19	9. 23	10. 21
11. 55	12. 41	13. 49	14. 31	15. 51
16. 42	17. 111	18. 101		

EXERCISES 314, SET II. State whether each of the following
numbers is prime or composite.

| 1. 7 | 2. 9 | 3. 17 | 4. 20 | 5. 29 |
| 6. 37 | 7. 61 | 8. 77 | 9. 143 | |

Adding Unlike Fractions

To add $\frac{1}{2}$ dollar and $\frac{1}{4}$ dollar (one quarter), we change the $\frac{1}{2}$
dollar to two $\frac{1}{4}$ dollars (two quarters). Then

$$\$\frac{1}{2} + \$\frac{1}{4} = \$\frac{2}{4} + \$\frac{1}{4} = \frac{2 + 1}{4} = \$\frac{3}{4}$$

We converted $\frac{1}{2}$ into $\frac{2}{4}$, which has the same denominator as $\frac{1}{4}$.
In this problem, 4 is called the *lowest common denominator* of
the fractions.

Example 1. Find $\frac{1}{3} + \frac{5}{6}$.

Solution: $\frac{1}{3} = \frac{2}{6}$. Therefore, $\frac{1}{3} + \frac{5}{6} = \frac{2}{6} + \frac{5}{6} = \frac{2 + 5}{6} = \frac{7}{6} = 1\frac{1}{6}$

Here 6 is the lowest common denominator.

Example 2. Find $\frac{1}{2} + \frac{1}{3}$.

Solution: $\frac{1}{2} = \frac{3}{6}$ and $\frac{1}{3} = \frac{2}{6}$.

Therefore, $\dfrac{1}{2} + \dfrac{1}{3} = \dfrac{3}{6} + \dfrac{2}{6} = \dfrac{3 + 2}{6} = \dfrac{5}{6}$

The lowest common denominator is 6.

<u>Example 3</u>. Find $\dfrac{1}{2} + \dfrac{2}{3} + \dfrac{3}{4}$.

<u>Solution</u>: $\dfrac{1}{2} = \dfrac{6}{12}$, $\dfrac{2}{3} = \dfrac{8}{12}$, and $\dfrac{3}{4} = \dfrac{9}{12}$.

Therefore, $\dfrac{1}{2} + \dfrac{2}{3} + \dfrac{3}{4} = \dfrac{6}{12} + \dfrac{8}{12} + \dfrac{9}{12} = \dfrac{6 + 8 + 9}{12} = \dfrac{23}{12} = 1\dfrac{11}{12}$

The lowest common denominator is 12.

<u>Example 4</u>. Find $\dfrac{4}{15} + \dfrac{7}{18} + \dfrac{5}{12}$.

You probably cannot find the lowest common denominator in this case. We now present a method that will make it possible for you to find the lowest common denominator in every case. After you have learned this method, we will solve this example.

<u>LOWEST COMMON DENOMINATOR (LCD)</u>. The *lowest common denominator* is the smallest number that each of the denominators will divide into exactly.

To Find the LCD

1. Write the denominators in a horizontal line (see Example 5).
2. Find a prime number that divides exactly into at least two of the denominators. Write the quotient on the next line.
3. Any denominator not divided exactly by the prime number is brought down to the next line unchanged.
4. Continue this procedure until no prime number divides at least two numbers on the last line.
5. The LCD is the product of all the prime divisors and the numbers in the last line.

Special Case:

When no prime number divides exactly into at least two of the denominators, the LCD is the product of all the denominators.

Example 5

(a) Find the LCD for $\frac{1}{6}$, $\frac{5}{12}$, $\frac{7}{18}$.

$$
\begin{array}{c|ccc}
2 & 6 & 12 & 18 \\
3 & 3 & 6 & 9 \\
& 1 & 2 & 3
\end{array}
$$

— LCD is product of all these numbers.
$LCD = 2 \cdot 3 \cdot 1 \cdot 2 \cdot 3 = 36$

(b) Find the LCD for $\frac{4}{15}$, $\frac{5}{24}$.

$$
\begin{array}{c|cc}
3 & 15 & 24 \\
& 5 & 8
\end{array}
$$

— LCD is product of all these numbers.
$LCD = 3 \cdot 5 \cdot 8 = 120$

(c) Find the LCD for $\frac{3}{8}$, $\frac{7}{10}$, $\frac{2}{5}$.

$$
\begin{array}{c|ccc}
5 & 8 & 10 & 5 \\
2 & 8 & 2 & 1 \\
& 4 & 1 & 1
\end{array}
$$

$LCD = 5 \cdot 2 \cdot 4 \cdot 1 \cdot 1 = 40$

(d) Find the LCD for $\frac{2}{9}$, $\frac{5}{14}$, $\frac{13}{21}$.

$$
\begin{array}{c|ccc}
3 & 9 & 14 & 21 \\
7 & 3 & 14 & 7 \\
& 3 & 2 & 1
\end{array}
$$

$LCD = 3 \cdot 7 \cdot 3 \cdot 2 \cdot 1 = 126$

(e) Find the LCD for $\frac{23}{72}$, $\frac{7}{54}$, $\frac{13}{90}$.

$$
\begin{array}{c|ccc}
3 & 72 & 54 & 90 \\
3 & 24 & 18 & 30 \\
2 & 8 & 6 & 10 \\
& 4 & 3 & 5
\end{array}
$$

$LCD = 3 \cdot 3 \cdot 2 \cdot 4 \cdot 3 \cdot 5 = 1080$

(f) Find the LCD for $\frac{3}{5}$, $\frac{1}{2}$, $\frac{2}{3}$.

$$
\begin{array}{|ccc}
5 & 2 & 3
\end{array}
$$

Since no prime number divides evenly into at least two of the denominators, the LCD is the product of all the denominators. Therefore, *LCD* = 5 • 2 • 3 = *30*.

EXERCISES 315A, SET I. Assume the following sets to be denominators of fractions. In Exercises 1-6, find the LCD by inspection, then check it by the method explained in this section.

1. {2, 3, 4}   2. {4, 5, 10}   3. {2, 8, 4}

4. {3, 6, 9}   5. {3, 5, 15}   6. {7, 2, 14}

In Exercises 7-20, find the LCD by the method described in this section.

7. {14, 10}   8. {16, 12}   9. {7, 5}

10. {4, 9}   11. {6, 8, 9}   12. {4, 15, 18}

13. {40, 15, 25}   14. {3, 7, 5}   15. {4, 5, 21}

16. {4, 6, 9, 12}   17. {6, 13, 26}   18. {45, 63, 98}

19. {66, 33, 132}   20. {24, 40, 48, 56}

EXERCISES 315A, SET II. Assume the following sets to be denominators of fractions. In Exercises 1-3, find the LCD by inspection, then check by the method explained in this section.

1. {3, 4, 6}   2. {5, 10, 15}   3. {2, 8, 3}

In Exercises 4-10, find the LCD by the method described in this section.

4. {12, 10}   5. {6, 9}   6. {4, 8, 9}

7. {4, 5, 7}   8. {3, 6, 9, 15}   9. {4, 8, 12, 16}

10. {6, 9, 8, 12}

To Add Unlike Fractions

1. Find the LCD.
2. Convert all fractions to equivalent fractions having the LCD as denominator.
3. Add the like fractions as before.
4. Reduce resulting fraction to lowest terms.
5. Change any improper fraction found in (4) to a mixed number.

Example 6. Add $\frac{3}{4} + \frac{2}{3} + \frac{7}{12}$.

Solution: $\dfrac{3}{4} = \dfrac{9}{12}$

$\dfrac{2}{3} = \dfrac{8}{12}$ ⎤— Add like fractions.

$\dfrac{7}{12} = \dfrac{7}{12}$

$$\dfrac{24}{12} = \dfrac{\overset{2}{\cancel{24}}}{\underset{1}{\cancel{12}}} = \dfrac{2}{1} = 2$$

2	4	3	12
2	2	3	6
3	1	3	3
	1	1	1

LCD = 2 · 2 · 3 · 1 · 1 · 1 = 12

Example 7. Add $\dfrac{5}{6} + \dfrac{3}{10} + \dfrac{4}{15}$.

Solution: $\dfrac{5}{6} = \dfrac{25}{30}$

$\dfrac{3}{10} = \dfrac{9}{30}$ ⎤— Add like fractions.

$\dfrac{4}{15} = \dfrac{8}{30}$

2	6	10	15
3	3	5	15
5	1	5	5
	1	1	1

LCD = 2 · 3 · 5 · 1 · 1 · 1 = 30

$$\dfrac{42}{30} = \dfrac{\overset{7}{\cancel{42}}}{\underset{5}{\cancel{30}}} = \dfrac{7}{5} = 1\dfrac{2}{5}$$

Example 8. Add $\dfrac{3}{16} + \dfrac{5}{12} + \dfrac{5}{24}$.

2	16	12	24
2	8	6	12
2	4	3	6
3	2	3	3
	2	1	1

Solution: $\dfrac{3}{16} = \dfrac{9}{48}$

$\dfrac{5}{12} = \dfrac{20}{48}$ ⎤— Add like fractions.

$\dfrac{5}{24} = \dfrac{10}{48}$

LCD = 2 · 2 · 2 · 3 · 2 · 1 · 1 = 48

$$\dfrac{39}{48} = \dfrac{\overset{13}{\cancel{39}}}{\underset{16}{\cancel{48}}} = \dfrac{13}{16}$$

Example 9. Add $\dfrac{4}{15} + \dfrac{7}{18} + \dfrac{5}{12}$. This is Example 4 on page 128, which we said would be solved after you learned how to find the LCD.

Solution: $\dfrac{4}{15} = \dfrac{48}{180}$

$\dfrac{7}{18} = \dfrac{70}{180}$

$\dfrac{5}{12} = \dfrac{75}{180}$

$\dfrac{193}{180} = 1\dfrac{13}{180}$

3	15	18	12
2	5	6	4
	5	3	2

LCD = 3 · 2 · 5 · 3 · 2 = 180

EXERCISES 315B, SET I. First reduce fractions to lowest terms, then add. Reduce your answers to lowest terms. Write any improper fraction as a mixed number.

1. $\frac{1}{2} + \frac{3}{4}$ 2. $\frac{1}{3} + \frac{5}{6}$ 3. $\frac{3}{5} + \frac{3}{10}$

4. $\frac{5}{8} + \frac{1}{2}$ 5. $\frac{2}{3} + \frac{1}{4}$ 6. $\frac{2}{5} + \frac{1}{3}$

7. $\frac{3}{4} + \frac{1}{12}$ 8. $\frac{2}{6} + \frac{6}{8}$ 9. $\frac{3}{6} + \frac{5}{15}$

10. $\frac{2}{7} + \frac{3}{14}$ 11. $\frac{5}{16} + \frac{3}{5}$ 12. $\frac{5}{6} + \frac{3}{7}$

13. $\frac{3}{20} + \frac{1}{8}$ 14. $\frac{1}{2} + \frac{2}{3} + \frac{3}{4}$ 15. $\frac{3}{5} + \frac{1}{2} + \frac{3}{10}$

16. $\frac{7}{10} + \frac{3}{5} + \frac{2}{3}$ 17. $\frac{1}{4} + \frac{7}{12} + \frac{5}{8}$ 18. $\frac{5}{6} + \frac{4}{12} + \frac{3}{8}$

19. $\frac{4}{6} + \frac{6}{14} + \frac{2}{3}$ 20. $\frac{6}{9} + \frac{8}{12} + \frac{9}{10}$ 21. $\frac{5}{12} + \frac{6}{16} + \frac{12}{32}$

22. $\frac{1}{3} + \frac{3}{5} + \frac{2}{11}$ 23. $\frac{3}{4} + \frac{2}{14} + \frac{1}{2}$ 24. $\frac{1}{12} + \frac{3}{16} + \frac{4}{10}$

25. $\frac{5}{6} + \frac{3}{8} + \frac{1}{12}$ 26. $\frac{3}{7} + \frac{5}{16} + \frac{5}{8}$ 27. $\frac{2}{28} + \frac{2}{16} + \frac{3}{21}$

28. $\frac{4}{12} + \frac{6}{18} + \frac{5}{25}$ 29. $\frac{13}{28} + \frac{5}{42}$ 30. $\frac{23}{72} + \frac{17}{56} + \frac{8}{63}$

EXERCISES 315B, SET II. First reduce fractions to lowest terms, then add. Reduce your answers to lowest terms. Write any improper fraction as a mixed number.

1. $\frac{2}{3} + \frac{1}{6}$ 2. $\frac{3}{8} + \frac{1}{2}$ 3. $\frac{2}{3} + \frac{1}{4}$

4. $\frac{2}{4} + \frac{3}{12}$ 5. $\frac{2}{6} + \frac{6}{9}$ 6. $\frac{3}{5} + \frac{2}{3}$

7. $\frac{1}{3} + \frac{1}{2} + \frac{3}{4}$ 8. $\frac{2}{5} + \frac{1}{2} + \frac{7}{10}$ 9. $\frac{3}{4} + \frac{1}{3} + \frac{5}{12}$

10. $\frac{4}{6} + \frac{2}{3} + \frac{6}{9}$ 11 $\frac{3}{5} + \frac{2}{3} + \frac{1}{4}$ 12. $\frac{3}{12} + \frac{4}{16} + \frac{8}{32}$

13. $\frac{4}{20} + \frac{2}{3} + \frac{6}{8}$ 14. $\frac{5}{42} + \frac{7}{36}$ 15. $\frac{4}{21} + \frac{3}{14} + \frac{1}{7} + \frac{1}{3}$

 **Subtracting Unlike Fractions**

```
To Subtract Unlike Fractions

1. Find the LCD.
2. Convert to equivalent fractions with LCD as
   denominator.
3. Subtract the like fractions as before.
4. Reduce the difference to lowest terms.
5. Change any improper fraction found in (4)
   to a mixed number.
```

Example 1

(a) Subtract: $\dfrac{4}{5} - \dfrac{3}{10}$.

Solution: $\dfrac{4}{5} = \dfrac{8}{10}$

$\dfrac{3}{10} = \dfrac{3}{10}$ ⎱ Subtract like fractions.

$\dfrac{5}{10} = \dfrac{\cancel{5}^{1}}{\cancel{10}_{2}} = \dfrac{1}{2}$

$$5\,\big|\,\underline{5\quad 10}$$
$$1\quad 2$$
$$\text{LCD} = 5 \cdot 1 \cdot 2 = 10$$

(b) Subtract: $\dfrac{7}{15} - \dfrac{5}{12}$.

Solution: $\dfrac{7}{15} = \dfrac{28}{60}$

$\dfrac{5}{12} = \dfrac{25}{60}$ ⎱ Subtract like fractions.

$\dfrac{3}{60} = \dfrac{\cancel{3}^{1}}{\cancel{60}_{20}} = \dfrac{1}{20}$

$$3\,\big|\,\underline{15\quad 12}$$
$$5\quad 4$$
$$\text{LCD} = 3 \cdot 5 \cdot 4 = 60$$

(c) Subtract: $\dfrac{11}{18} - \dfrac{7}{24}$.

Solution: $\dfrac{11}{18} = \dfrac{44}{72}$

$\dfrac{7}{24} = \dfrac{21}{72}$

$\dfrac{23}{72}$ (already in lowest terms)

$$2\,\big|\,\underline{18\quad 24}$$
$$3\,\big|\,\underline{9\quad 12}$$
$$3\quad 4$$
$$\text{LCD} = 2 \cdot 3 \cdot 3 \cdot 4 = 72$$

EXERCISES 316, SET I. First reduce fractions to lowest terms, then subtract. Reduce your answers to lowest terms.

1. $\dfrac{3}{4} - \dfrac{1}{2}$ 2. $\dfrac{2}{3} - \dfrac{1}{6}$ 3. $\dfrac{5}{8} - \dfrac{2}{4}$ 4. $\dfrac{7}{10} - \dfrac{3}{5}$

5. $\dfrac{6}{12} - \dfrac{1}{3}$ 6. $\dfrac{3}{4} - \dfrac{2}{5}$ 7. $\dfrac{6}{7} - \dfrac{4}{12}$ 8. $\dfrac{10}{16} - \dfrac{5}{12}$

9. $\dfrac{12}{30} - \dfrac{1}{5}$ 10. $\dfrac{12}{16} - \dfrac{5}{12}$ 11. $\dfrac{25}{32} - \dfrac{3}{4}$ 12. $\dfrac{13}{16} - \dfrac{5}{8}$

13. $\dfrac{56}{64} - \dfrac{14}{24}$ 14. $\dfrac{14}{35} - \dfrac{5}{20}$ 15. $\dfrac{11}{18} - \dfrac{4}{15}$ 16. $\dfrac{9}{14} - \dfrac{15}{42}$

17. $\dfrac{5}{12} - \dfrac{2}{15}$ 18. $\dfrac{15}{16} - \dfrac{5}{24}$ 19. $\dfrac{19}{35} - \dfrac{5}{14}$ 20. $\dfrac{17}{40} - \dfrac{3}{16}$

21. $\dfrac{26}{36} - \dfrac{11}{20}$ 22. $\dfrac{61}{84} - \dfrac{35}{72}$

EXERCISES 316, SET II. First reduce fractions to lowest terms, then subtract. Reduce your answers to lowest terms.

1. $\dfrac{5}{6} - \dfrac{1}{3}$ 2. $\dfrac{3}{4} - \dfrac{1}{2}$ 3. $\dfrac{4}{8} - \dfrac{2}{6}$ 4. $\dfrac{12}{16} - \dfrac{1}{4}$

5. $\dfrac{6}{8} - \dfrac{2}{4}$ 6. $\dfrac{9}{12} - \dfrac{4}{6}$ 7. $\dfrac{14}{24} - \dfrac{20}{36}$ 8. $\dfrac{15}{18} - \dfrac{6}{15}$

9. $\dfrac{12}{16} - \dfrac{3}{24}$ 10. $\dfrac{30}{35} - \dfrac{12}{15}$ 11. $\dfrac{16}{24} - \dfrac{14}{56}$

Division of Fractions

> To Divide Fractions
>
> Invert the second fraction and multiply.

Example 1

(a) Find $\dfrac{3}{5} \div \dfrac{3}{7}$.

Solution: $\dfrac{3}{5} \div \dfrac{3}{7} = \dfrac{3}{5} \cdot \dfrac{7}{3} = \dfrac{\overset{1}{\cancel{3}}}{5} \cdot \dfrac{7}{\underset{1}{\cancel{3}}} = \dfrac{7}{5} = 1\dfrac{2}{5}$

(b) $\dfrac{5}{6} \div \dfrac{2}{3} = \dfrac{5}{6} \cdot \dfrac{3}{2} = \dfrac{5}{\underset{2}{\cancel{6}}} \cdot \dfrac{\overset{1}{\cancel{3}}}{2} = \dfrac{5}{4} = 1\dfrac{1}{4}$

(c) $\dfrac{18}{5} \div \dfrac{6}{25} = \dfrac{\overset{3}{\cancel{18}}}{\underset{1}{\cancel{5}}} \cdot \dfrac{\overset{5}{\cancel{25}}}{\underset{1}{\cancel{6}}} = \dfrac{15}{1} = 15$

(d) $\dfrac{34}{63} \div \dfrac{17}{9} = \dfrac{\overset{2}{\cancel{34}}}{\underset{7}{\cancel{63}}} \cdot \dfrac{\overset{1}{\cancel{9}}}{\underset{1}{\cancel{17}}} = \dfrac{2}{7}$ 　　　　(e) $\dfrac{15}{32} \div 2 = \dfrac{15}{32} \cdot \dfrac{1}{2} = \dfrac{15}{64}$

(f) $7 \div \dfrac{3}{8} = \dfrac{7}{1} \cdot \dfrac{8}{3} = \dfrac{56}{3} = 18\dfrac{2}{3}$

This method works because:

$$\dfrac{3}{5} \div \dfrac{4}{7} = \dfrac{\frac{3}{5}}{\frac{4}{7}} = \dfrac{\frac{3}{5}}{\frac{4}{7}} \cdot \left(\dfrac{\frac{7}{4}}{\frac{7}{4}}\right) = \dfrac{\frac{3}{5} \cdot \frac{7}{4}}{\frac{4}{7} \cdot \frac{7}{4}} = \dfrac{\frac{3}{5} \cdot \frac{7}{4}}{1} = \dfrac{3}{5} \cdot \dfrac{7}{4}$$

This is 1

Therefore, $\dfrac{3}{5} \div \dfrac{4}{7} = \dfrac{3}{5} \cdot \dfrac{7}{4}$

EXERCISES 317, SET I. Find the following quotients.

1. $\dfrac{3}{4} \div \dfrac{1}{2}$ 　　2. $\dfrac{2}{5} \div \dfrac{1}{3}$ 　　3. $\dfrac{1}{2} \div \dfrac{2}{3}$ 　　4. $4 \div \dfrac{2}{5}$

5. $\dfrac{1}{2} \div 5$ 　　6. $\dfrac{4}{3} \div \dfrac{8}{6}$ 　　7. $\dfrac{5}{2} \div \dfrac{5}{8}$ 　　8. $\dfrac{7}{8} \div 7$

9. $1 \div \dfrac{1}{2}$ 　　10. $\dfrac{100}{150} \div \dfrac{4}{9}$ 　　11. $\dfrac{3}{5} \div \dfrac{3}{10}$ 　　12. $\dfrac{7}{12} \div \dfrac{1}{3}$

13. $\dfrac{3}{16} \div \dfrac{9}{20}$ 　　14. $\dfrac{13}{28} \div 39$ 　　15. $34 \div \dfrac{17}{56}$ 　　16. $\dfrac{8}{15} \div \dfrac{24}{10}$

17. $\dfrac{35}{16} \div \dfrac{42}{22}$ 　　18. $\dfrac{7}{18} \div \dfrac{21}{15}$ 　　19. $36 \div \dfrac{4}{5}$ 　　20. $\dfrac{4}{9} \div 36$

21. $\dfrac{6}{35} \div \dfrac{8}{15}$ 　　22. $\dfrac{11}{84} \div \dfrac{22}{60}$ 　　23. $\dfrac{14}{24} \div 210$ 　　24. $22 \div \dfrac{11}{5}$

25. $\dfrac{56}{15} \div \dfrac{28}{90}$ 　　26. $\dfrac{44}{80} \div \dfrac{11}{60}$

EXERCISES 317, SET II. Find the following quotients.

1. $\dfrac{2}{3} \div \dfrac{5}{6}$ 　　2. $\dfrac{2}{3} \div \dfrac{1}{6}$ 　　3. $\dfrac{3}{4} \div \dfrac{6}{8}$ 　　4. $\dfrac{6}{7} \div 3$

5. $3 \div \dfrac{1}{3}$ 　　6. $\dfrac{2}{5} \div \dfrac{4}{5}$ 　　7. $\dfrac{11}{23} \div \dfrac{22}{7}$ 　　8. $\dfrac{8}{9} \div \dfrac{24}{27}$

9. $\dfrac{25}{16} \div \dfrac{15}{32}$ 　　10. $\dfrac{5}{9} \div 20$ 　　11. $\dfrac{6}{25} \div \dfrac{12}{35}$ 　　12. $14 \div \dfrac{7}{8}$

318 Complex Fractions

A *simple fraction* has only one fraction line.

Example 1

$$\dfrac{2}{3} , \quad \dfrac{3 + 5}{12} , \quad \dfrac{7 - 4}{6} , \quad \dfrac{18}{5} , \quad \dfrac{2}{7 - 3} , \quad \dfrac{13 + 6}{9 - 2}$$

A *complex fraction* is a fraction having more than one fraction line.

Example 2

$$\frac{\frac{2}{5}}{3} \quad , \quad \frac{8}{\frac{1}{2}} \quad , \quad \frac{\frac{4}{7}}{\frac{5}{9}} \quad , \quad \frac{\frac{3}{2} - \frac{2}{5}}{\frac{5}{6} + \frac{3}{4}}$$

To Simplify Complex Fractions

1. Simplify the numerator and the denominator when possible.
2. Divide the simplified numerator by the simplified denominator.

Example 3

(a) $\dfrac{\frac{2}{3}}{\frac{1}{2}} = \frac{2}{3} \div \frac{1}{2} = \frac{2}{3} \cdot \frac{2}{1} = \frac{4}{3} = 1\frac{1}{3}$

(b) $\dfrac{\frac{5}{12}}{\frac{8}{15}} = \frac{5}{12} \div \frac{8}{15} = \frac{5}{\overset{}{\underset{4}{\cancel{12}}}} \cdot \frac{\overset{5}{\cancel{15}}}{8} = \frac{25}{32}$

(c) $\dfrac{6}{\frac{3}{5}} = 6 \div \frac{3}{5} = \frac{\overset{2}{\cancel{6}}}{1} \cdot \frac{5}{\underset{1}{\cancel{3}}} = 10$

(d) $\dfrac{\frac{4}{5}}{12} = \frac{4}{5} \div \frac{12}{1} = \frac{\overset{1}{\cancel{4}}}{5} \cdot \frac{1}{\underset{3}{\cancel{12}}} = \frac{1}{15}$

(e) $\dfrac{\frac{1}{6} + \frac{2}{3}}{\frac{5}{8} - \frac{1}{4}} = \dfrac{\frac{1+4}{6}}{\frac{5-2}{8}} = \dfrac{\frac{5}{6}}{\frac{3}{8}} = \frac{5}{6} \div \frac{3}{8} = \frac{5}{\underset{3}{\cancel{6}}} \cdot \frac{\overset{4}{\cancel{8}}}{3} = \frac{20}{9} = 2\frac{2}{9}$

(f) $\dfrac{\frac{3}{5} + 2}{2 - \frac{3}{8}} = \dfrac{\frac{3+10}{5}}{\frac{16-3}{8}} = \dfrac{\frac{13}{5}}{\frac{13}{8}} = \frac{\overset{1}{\cancel{13}}}{5} \cdot \frac{8}{\underset{1}{\cancel{13}}} = \frac{8}{5} = 1\frac{3}{5}$

EXERCISES 318, SET I. Simplify these complex fractions.

1. $\dfrac{\frac{3}{4}}{\frac{1}{6}}$

2. $\dfrac{\frac{15}{16}}{\frac{12}{5}}$

3. $\dfrac{\frac{2}{3}}{\frac{1}{2}}$

4. $\dfrac{\frac{3}{4}}{\frac{7}{8}}$

5. $\dfrac{\frac{3}{5}}{\frac{3}{10}}$

6. $\dfrac{\frac{7}{16}}{\frac{7}{24}}$

7. $\dfrac{\frac{3}{8}}{\frac{5}{12}}$

8. $\dfrac{\frac{5}{7}}{\frac{10}{21}}$

9. $\dfrac{6}{\frac{2}{3}}$

10. $\dfrac{\frac{15}{6}}{9}$

11. $\dfrac{14}{\frac{8}{5}}$

12. $\dfrac{\frac{3}{4}}{8}$

13. $\dfrac{\frac{1}{4} + \frac{2}{5}}{\frac{1}{6}}$

14. $\dfrac{\frac{1}{8} + \frac{3}{4}}{\frac{1}{2} - \frac{1}{3}}$

15. $\dfrac{4 + \frac{1}{4}}{2 - \frac{1}{2}}$

16. $\dfrac{\frac{3}{16} + 5}{6 - \frac{7}{8}}$

17. $\dfrac{\frac{11}{4} - \frac{5}{9}}{\frac{7}{18} + \frac{13}{36}}$

18. $\dfrac{\frac{1}{7} + \frac{9}{28}}{\frac{13}{14} - \frac{3}{7}}$

19. $\dfrac{\frac{16}{5} - \frac{7}{15}}{\frac{9}{30} + \frac{3}{10}}$

20. $\dfrac{\frac{13}{18} - \frac{11}{24}}{\frac{5}{12} - \frac{7}{36}}$

EXERCISES 318, SET II. Simplify these complex fractions.

1. $\dfrac{\frac{2}{3}}{\frac{3}{4}}$

2. $\dfrac{\frac{1}{2}}{\frac{2}{3}}$

3. $\dfrac{\frac{7}{8}}{\frac{21}{4}}$

4. $\dfrac{\frac{6}{7}}{\frac{9}{21}}$

5. $\dfrac{\frac{12}{13}}{6}$

6. $\dfrac{\frac{5}{3}}{10}$

7. $\dfrac{\frac{2}{3} + \frac{1}{5}}{\frac{4}{5}}$

8. $\dfrac{\frac{4}{9} + 2}{\frac{3}{4} + 2}$

9. $\dfrac{\frac{1}{6} + \frac{7}{18}}{\frac{11}{12} - \frac{2}{3}}$

10. $\dfrac{\frac{9}{5} - \frac{7}{15}}{\frac{3}{20} + \frac{1}{30}}$

Review

To reduce a fraction to lowest terms: (Sec. 311)

    1. Divide both numerator and denominator by all factors common to both.
    2. Use Euclid's Algorithm if you think it may be reduced further.

To add like fractions: (Sec. 312)

    1. Add their numerators.

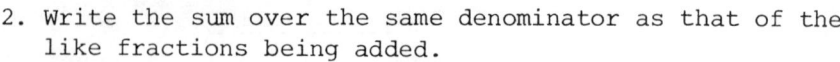

2. Write the sum over the same denominator as that of the like fractions being added.
3. Reduce the resulting fraction to lowest terms.
4. Change any improper fraction found in (3) to a mixed number.

To subtract like fractions: (Sec. 313)

1. Subtract their numerators.
2. Write the difference over the same denominator as that of the like fractions being subtracted.
3. Reduce the resulting fraction to lowest terms.
4. Change any improper fraction found in (3) to a mixed number.

To add unlike fractions: (Sec. 315)

1. Find the LCD.
2. Convert all fractions to equivalent fractions having the LCD as denominator.
3. Add the like fractions as before.
4. Reduce resulting·fraction to lowest terms.

To subtract unlike fractions: (Sec. 316)

1. Find the LCD.
2. Convert to equivalent fractions with LCD as denominator.
3. Subtract the like fractions as before.
4. Reduce the difference to lowest terms.

To divide one fraction by a second fraction, invert the second fraction and multiply. (Sec. 317)

A *simple fraction* has only one fraction line. (Sec. 318)

A *complex fraction* is a fraction having more than one fraction line. (Sec. 318)

To simplify a complex fraction, simplify its numerator and denominator, then divide its numerator by its denominator. (Sec. 318)

A *prime number* is a natural number greater than 1 that can be divided only by itself and 1. A prime number has *no* factors other than itself and 1. (Sec. 314)

A *composite number* is a natural number that can be divided by some number other than itself and 1. A composite number *has* factors other than itself and 1. (Sec. 314)

EXERCISES 319, SET I. In Exercises 1-10, reduce the fractions to lowest terms.

1. $\dfrac{36}{60}$ 2. $\dfrac{105}{75}$ 3. $\dfrac{99}{143}$ 4. $\dfrac{97}{53}$ 5. $\dfrac{208}{286}$

6. $\dfrac{198}{231}$ 7. $\dfrac{210}{252}$ 8. $\dfrac{913}{1,411}$ 9. $\dfrac{2,535}{3,120}$ 10. $\dfrac{6,360}{9,328}$

In Exercises 11-20, perform the indicated operations.

11. $\dfrac{6}{7} + \dfrac{3}{7}$ 12. $5 \div \dfrac{1}{2}$ 13. $\dfrac{5}{8} - \dfrac{3}{8}$ 14. $\dfrac{10}{\dfrac{5}{7}}$

15. $\dfrac{\dfrac{6}{9}}{12}$ 16. $\dfrac{3}{4} + \dfrac{5}{8}$ 17. $\dfrac{3}{5} \div \dfrac{7}{10}$ 18. $\dfrac{\dfrac{5}{12}}{\dfrac{20}{18}}$

19. $\dfrac{2 + \dfrac{2}{3}}{\dfrac{1}{2} - \dfrac{1}{6}}$ 20. $\dfrac{35}{12} \div 14$

21. On an average day, Robert spends $\dfrac{1}{3}$ of his time sleeping, $\dfrac{1}{4}$ of his time working, and $\dfrac{1}{12}$ of his time eating. What part of his day is left for other things?

22. Find the perimeter of a triangle whose sides have lengths of $\dfrac{3}{4}$ inch, $\dfrac{5}{8}$ inch, and $\dfrac{13}{16}$ inch.

23. Mrs. Martinez gave a 5-pound box of candy to her three children and told them to share it equally. How much did each child receive?

24. A patient must take $\dfrac{2}{3}$ grain of vitamin B$_2$ each day. How many $\dfrac{1}{12}$-grain tablets must he take?

EXERCISES 319, SET II. In Exercises 1-5, reduce the fractions to lowest terms.

1. $\dfrac{24}{36}$ 2. $\dfrac{99}{111}$ 3. $\dfrac{225}{275}$ 4. $\dfrac{930}{1,116}$ 5. $\dfrac{1,260}{1,440}$

In Exercises 6-10, perform the indicated operations.

6. $\dfrac{5}{12} + \dfrac{1}{12}$ 7. $6 \div \dfrac{1}{3}$ 8. $\dfrac{\dfrac{2}{3}}{\dfrac{4}{9}}$ 9. $\dfrac{2 - \dfrac{5}{6}}{\dfrac{3}{4} + \dfrac{2}{3}}$

10. $\dfrac{5}{8} \div \dfrac{15}{16}$

11. On an average day, Jim spends $\dfrac{1}{3}$ of his time sleeping, $\dfrac{1}{6}$ of his time in school, $\dfrac{1}{4}$ of his time studying and $\dfrac{1}{12}$ of his time eating. What fractional part of his day is left for other things?

12. Mr. Mock gave a 4-pound box of candy to his six children and told them to share it equally. How much should each child receive?

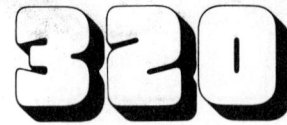

Operations with Mixed Numbers

Mixed numbers were introduced in Section 304. We learned that a mixed number is equivalent to an improper fraction. One way to carry out all the arithmetic operations with mixed numbers is to change them to improper fractions (Section 306), then do the problem with fractions as we have already learned.

ADDITION OF MIXED NUMBERS

Example 1

(a) $2\frac{1}{2} + 4\frac{1}{3} = \frac{5}{2} + \frac{13}{3} = \frac{15}{6} + \frac{26}{6} = \frac{41}{6} = 6\frac{5}{6}$

(b) $3 + 2\frac{1}{4} = \frac{3}{1} + \frac{9}{4} = \frac{12}{4} + \frac{9}{4} = \frac{21}{4} = 5\frac{1}{4}$ (Can you find this answer just by inspection?)

(c) $3\frac{1}{5} + 2 + 1\frac{2}{3} = \frac{16}{5} + \frac{2}{1} + \frac{5}{3} = \frac{48}{15} + \frac{30}{15} + \frac{25}{15} = \frac{103}{15} = 6\frac{13}{15}$

EXERCISES 320A, SET I. Find the following sums.

1. $2\frac{1}{4} + 1\frac{3}{4}$ 2. $3\frac{1}{5} + 1\frac{2}{5}$ 3. $1\frac{1}{3} + 2\frac{1}{6}$ 4. $4\frac{1}{4} + 1\frac{1}{2}$

5. $2\frac{3}{5} + 1\frac{1}{10}$ 6. $1\frac{3}{8} + 2\frac{1}{4}$ 7. $3\frac{1}{3} + 2\frac{1}{2}$ 8. $5\frac{3}{6} + 1\frac{1}{2}$

9. $4\frac{1}{6} + 3\frac{2}{3}$ 10. $2\frac{1}{2} + 3\frac{3}{4}$ 11. $1\frac{5}{8} + 2\frac{1}{2}$ 12. $3\frac{1}{8} + 2\frac{1}{5}$

13. $7\frac{5}{6} + 3\frac{2}{3}$ 14. $8\frac{1}{5} + 2\frac{7}{8}$ 15. $2\frac{5}{8} + 3$ 16. $4 + 2\frac{3}{5}$

17. $1\frac{1}{2} + 2\frac{1}{3} + 3\frac{1}{4}$ 18. $2\frac{1}{3} + 1\frac{3}{4} + 3\frac{5}{6}$

19. $5\frac{1}{4} + 3 + 2\frac{3}{8}$ 20. $6 + 2\frac{3}{5} + 1\frac{7}{10}$

21. Brain Teaser. How can you place three nines so that they will equal 10?

EXERCISES 320A, SET II. Find the following sums.

1. $3\frac{1}{3} + 1\frac{2}{3}$ 2. $3\frac{1}{4} + 2\frac{5}{8}$ 3. $4\frac{3}{5} + 2\frac{3}{10}$ 4. $2\frac{1}{3} + 5\frac{1}{2}$

5. $5\frac{1}{6} + 3\frac{2}{3}$ 6. $1\frac{5}{8} + 3\frac{1}{2}$ 7. $5\frac{1}{3} + 4\frac{5}{6}$ 8. $5\frac{3}{4} + 7$

9. $2\frac{1}{3} + 1\frac{1}{2} + 3\frac{3}{4}$ 10. $5\frac{2}{3} + 4 + 7\frac{1}{5}$

SUBTRACTION OF MIXED NUMBERS

Example 2

(a) $5\frac{3}{8} - 3\frac{1}{4} = \frac{43}{8} - \frac{26}{8} = \frac{17}{8} = 2\frac{1}{8}$

(b) $4 - 1\frac{5}{6} = \frac{4}{1} - \frac{11}{6} = \frac{24}{6} - \frac{11}{6} = \frac{13}{6} = 2\frac{1}{6}$

EXERCISES 320B, SET I. Find the following differences.

1. $3\frac{4}{5} - 1\frac{1}{5}$ 2. $5\frac{2}{3} - 2\frac{1}{3}$ 3. $2\frac{1}{3} - 1\frac{3}{5}$ 4. $2\frac{7}{8} - 1\frac{3}{4}$

5. $5 - 2\frac{3}{8}$ 6. $4 - 1\frac{5}{6}$ 7. $3\frac{3}{4} - 2$ 8. $7\frac{1}{5} - 3$

9. $5\frac{8}{9} - 1\frac{2}{3}$ 10. $6\frac{11}{12} - 2\frac{7}{8}$ 11. $3\frac{3}{4} - 2\frac{1}{3}$ 12. $4\frac{3}{7} - 1\frac{5}{14}$

13. Mr. Segal has $5\frac{3}{4}$ square yards of carpet left after carpeting his living room. How much more will he need to
carpet a bathroom that has a floor area of 7 square yards?

14. Jim wants three boards for shelves that measure $5\frac{3}{4}$ feet,
$2\frac{5}{12}$ feet, and $3\frac{1}{2}$ feet. Can he cut all three shelves from
a 12-foot board, allowing $\frac{1}{4}$ foot for waste?

EXERCISES 320B, SET II. Find the following differences.

1. $5\frac{3}{4} - 2\frac{1}{4}$ 2. $4\frac{1}{3} - 1\frac{2}{3}$ 3. $6 - 2\frac{3}{5}$ 4. $4\frac{1}{6} - 2$

5. $8\frac{5}{12} - 3\frac{1}{4}$ 6. $5\frac{9}{10} - 3\frac{2}{5}$

7. Manuel needs three boards for shelves that measure $2\frac{1}{6}$
feet, $3\frac{1}{3}$ feet, and $4\frac{1}{4}$ feet. Can he cut all three shelves
from a 10-foot board, allowing $\frac{1}{12}$ foot for waste?

MULTIPLICATION OF MIXED NUMBERS

Example 3

(a) $2\frac{11}{12} \cdot 1\frac{3}{5} = \frac{\overset{7}{\cancel{35}}}{\underset{3}{\cancel{12}}} \cdot \frac{\overset{2}{\cancel{8}}}{\underset{1}{\cancel{5}}} = \frac{14}{3} = 4\frac{2}{3}$

(b) $3\frac{1}{8} \cdot 16 = \frac{25}{\underset{1}{\cancel{8}}} \cdot \frac{\overset{2}{\cancel{16}}}{1} = 50$

DIVISION OF MIXED NUMBERS

Example 4

(a) $6\frac{4}{5} \div 1\frac{7}{10} = \frac{34}{5} \div \frac{17}{10} = \frac{\cancel{34}^{2}}{\cancel{5}_{1}} \cdot \frac{\cancel{10}^{2}}{\cancel{17}_{1}} = 4$

(b) $12 \div 2\frac{2}{3} = \frac{12}{1} \div \frac{8}{3} = \frac{\cancel{12}^{3}}{1} \cdot \frac{3}{\cancel{8}_{2}} = \frac{9}{2} = 4\frac{1}{2}$

EXERCISES 320C, SET I. In Exercises 1-20, perform the indicated operations.

1. $1\frac{2}{3} \times 2\frac{1}{2}$ 2. $1\frac{1}{4} \times 2\frac{2}{5}$ 3. $1\frac{3}{7} \div 1\frac{1}{4}$ 4. $1\frac{7}{9} \div 2\frac{2}{3}$

5. $2\frac{2}{3} \times 2\frac{1}{4}$ 6. $2\frac{4}{5} \times 2\frac{1}{7}$ 7. $2\frac{3}{5} \div 1\frac{4}{35}$ 8. $3\frac{2}{3} \div 1\frac{7}{15}$

9. $8 \times 3\frac{3}{4}$ 10. $4\frac{2}{3} \times 6$ 11. $7 \div 4\frac{2}{3}$ 12. $3\frac{4}{5} \div 19$

13. $2\frac{5}{8} \times 4$ 14. $6 \times 2\frac{5}{12}$ 15. $3\frac{1}{3} \div 5$ 16. $11 \div 3\frac{1}{7}$

17. $3\frac{3}{10} \times \frac{6}{11} \times 1\frac{2}{3}$ 18. $1\frac{1}{8} \times \frac{4}{9} \times 1\frac{5}{6}$

19. $3\frac{1}{5} \times 75 \times \frac{7}{10}$ 20. $\frac{7}{8} \times 1\frac{3}{14} \times 64$

21. Each of eight hikers carries a food pack weighing $2\frac{7}{16}$ pounds. How much food are they carrying in all?
22. How many tablets, each containing 3 milligrams of a heart medicine, must be used to make up a $4\frac{1}{2}$-milligram dosage?

EXERCISES 320C, SET II. In Exercises 1-10, perform the indicated operations.

1. $3\frac{3}{4} \times 2\frac{2}{5}$ 2. $2\frac{1}{6} \div 3\frac{1}{4}$ 3. $3\frac{2}{5} \div 2\frac{4}{15}$ 4. $5\frac{3}{4} \times 12$

5. $22 \div 3\frac{2}{3}$ 6. $3\frac{5}{6} \div 46$ 7. $12 \times 3\frac{7}{12}$ 8. $15 \div 2\frac{6}{7}$

9. $1\frac{2}{7} \times 2\frac{1}{3} \times 2\frac{1}{6}$ 10. $2\frac{2}{3} \times 2\frac{1}{4} \times 12$

11. How many tablets, each containing $1\frac{1}{2}$ milligrams of a medicine, must be used to make up a 6-milligram dosage?

Another Way to Add Mixed Numbers

This method is recommended when the whole number parts are large.

<u>Example 1.</u> Find $2\frac{1}{2} + 4\frac{1}{3}$.

<u>Solution:</u> $2\frac{1}{2} = 2\frac{3}{6}$ The fraction parts were changed to equivalant fractions having the LCD for denominator.

$$4\frac{1}{3} = 4\frac{2}{6}$$

$$6\frac{5}{6}$$

<u>Example 2.</u> Find $12\frac{3}{4} + 21\frac{3}{8} + 45\frac{1}{2}$.

<u>Solution:</u> $12\frac{3}{4} = 12\frac{6}{8}$

$$21\frac{3}{8} = 21\frac{3}{8}$$

$$45\frac{1}{2} = 45\frac{4}{8}$$

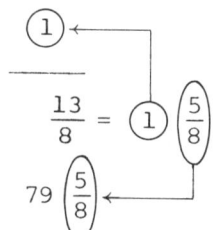

$$\frac{13}{8} = ①\left(\frac{5}{8}\right)$$

$$79\left(\frac{5}{8}\right)$$

<u>EXERCISES 321, SET I.</u> In Exercises 1–18, find the sums.

1. $13\frac{1}{5}$
   $4\frac{2}{3}$

2. $15\frac{5}{8}$
   $23\frac{1}{16}$

3. $12\frac{1}{2}$
   $23\frac{3}{4}$

4. $21\frac{3}{4}$
   $16\frac{4}{5}$

5. $37\frac{5}{6}$
   $44\frac{1}{4}$

6. $56\frac{2}{3}$
   $67\frac{5}{8}$

7. $125\frac{3}{7}$
   208

8. 379
   $417\frac{7}{9}$

9. $72\frac{2}{3}$
   $81\frac{3}{4}$
   $93\frac{1}{2}$

10. $68\frac{2}{9}$
    $97\frac{1}{3}$
    $55\frac{5}{6}$

11. $17\frac{1}{3}$
    $28\frac{2}{5}$
    $15\frac{4}{15}$

12. $32\frac{3}{14}$
    $78\frac{19}{28}$
    $21\frac{5}{7}$

13. $56\frac{3}{4}$
    72
    $48\frac{2}{3}$

14. 59
    $47\frac{7}{10}$
    $93\frac{2}{3}$

15. $107\frac{5}{6}$
    $293\frac{1}{3}$
    $480\frac{7}{9}$

16. $320\frac{5}{8}$
    $209\frac{1}{16}$
    $743\frac{3}{4}$

17. $156\frac{4}{5}$
    93
    $81\frac{7}{15}$
    $204\frac{3}{10}$

18. $148\frac{7}{16}$
    $\frac{19}{32}$
    37
    $8\frac{1}{2}$

19. Find the sum of $7\frac{3}{4}$ ounces, $5\frac{7}{8}$ ounces, $8\frac{5}{16}$ ounces, and $10\frac{2}{5}$ ounces.

20. Tony weighs $145\frac{1}{2}$ pounds. Mike weighs $157\frac{3}{4}$ pounds, and Pat weighs $7\frac{3}{4}$ pounds more than Tony. Find the weight of the three men.

<u>EXERCISES 321, SET II</u>. In Exercises 1-9, find the sums.

1. $12\frac{1}{4}$
   $5\frac{2}{3}$

2. $15\frac{1}{6}$
   $8\frac{3}{4}$

3. $45\frac{3}{5}$
   $28\frac{7}{10}$

4. 287
   $359\frac{5}{6}$

5. $49\frac{5}{9}$
   $53\frac{2}{3}$
   $76\frac{1}{6}$

6. $19\frac{5}{14}$
   $72\frac{3}{7}$
   $85\frac{1}{28}$

7. 84
   $7\frac{3}{10}$
   $156\frac{1}{4}$

8. $506\frac{2}{3}$
   $390\frac{3}{4}$
   $104\frac{1}{2}$

9. $175\frac{3}{8}$
   $\frac{15}{16}$
   $289\frac{5}{6}$

10. Tom weighs $175\frac{3}{4}$ pounds. Merv weighs $168\frac{1}{2}$ pounds, and Pat weighs $5\frac{3}{4}$ pounds more than Merv. Find the weight of the three men.

Another Way to Subtract Mixed Numbers

This method is recommended when the whole number parts are large.

Example 1. Find $5\frac{3}{8} - 3\frac{1}{4}$.

Solution:
$$5\frac{3}{8} = 5\frac{3}{8}$$
$$-3\frac{1}{4} = -3\frac{2}{8}$$
$$\overline{\phantom{-3\frac{1}{4}}} \quad \overline{2\frac{1}{8}}$$

Example 2. Find $12\frac{1}{6} - 4\frac{1}{2}$.

Solution:
We cannot subtract $\frac{3}{6}$ from $\frac{1}{6}$.

$$12\frac{1}{6} = 12\frac{1}{6} = \overset{11+1}{12} + \frac{1}{6} = 11 + \frac{6}{6} + \frac{1}{6} = 11\frac{7}{6} = 11\frac{7}{6}$$
$$-4\frac{1}{2} = -4\frac{3}{6}$$

1 is *"borrowed"* from 12, leaving 11. This 1 is changed to $\frac{6}{6}$. Then

$$12\frac{1}{6} = 11\frac{6}{6} + \frac{1}{6} = 11\frac{7}{6}.$$

$$-4\frac{3}{6}$$
$$\overline{7\frac{4}{6} = 7\frac{2}{3}}$$

We shorten the work by writing the problem as follows:

$$12\frac{1}{6} = 12\overset{11\frac{7}{6}}{\cancel{\frac{1}{6}}}$$
$$-4\frac{1}{2} = -4\frac{3}{6}$$
$$\overline{\phantom{-4\frac{1}{2}}} \quad \overline{7\frac{4}{6} = 7\frac{2}{3}}$$

Example 3. Find $58\frac{2}{7} - 25\frac{1}{3}$.

Solution:
$$58\frac{2}{7} = 58\overset{57\frac{27}{21}}{\cancel{\frac{6}{21}}}$$
$$25\frac{1}{3} = -25\frac{7}{21}$$
$$\overline{\phantom{25\frac{1}{3}}} \quad \overline{32\frac{20}{21}}$$

Example 4. Find $15 - 5\frac{3}{8}$.

Solution:
$$\overset{14\frac{8}{8}}{\cancel{15}}$$
$$-5\frac{3}{8}$$
$$\overline{9\frac{5}{8}}$$

In the last three examples, we saw that if the fraction being subtracted (subtrahend) is larger than the fraction it is being subtracted from (minuend), "1" can be "borrowed" from the whole number part of the minuend to make the subtraction of the fractions possible.

EXERCISES 322, SET I. Find the following differences.

1. $14\frac{3}{4}$
   $- 10\frac{1}{4}$

2. $21\frac{5}{6}$
   $- 18\frac{1}{6}$

3. $17\frac{3}{4}$
   $- 5\frac{1}{8}$

4. $19\frac{3}{5}$
   $- 6\frac{1}{10}$

5. 8
   $- 4\frac{1}{2}$

6. 7
   $- 3\frac{1}{3}$

7. 6
   $- 2\frac{3}{5}$

8. 4
   $- 1\frac{5}{6}$

9. $4\frac{1}{4}$
   $- 1\frac{3}{4}$

10. $5\frac{1}{3}$
    $- 1\frac{2}{3}$

11. $12\frac{2}{5}$
    $- 7\frac{3}{10}$

12. $54\frac{2}{3}$
    $- 39\frac{1}{6}$

13. $3\frac{1}{12}$
    $- 1\frac{1}{6}$

14. $23\frac{5}{8}$
    $- 17\frac{3}{4}$

15. 45
    $- 38\frac{2}{3}$

16. 32
    $- 28\frac{7}{15}$

17. $68\frac{5}{16}$
    $- 53\frac{3}{4}$

18. $107\frac{2}{3}$
    $- 99\frac{1}{6}$

19. $234\frac{5}{14}$
    $- 157\frac{3}{7}$

20. $7,005\frac{2}{5}$
    $- 2,867\frac{2}{3}$

21. When a $1\frac{1}{4}$-pound steak was trimmed of fat, it weighed $\frac{7}{8}$ pound. How much did the fat weigh?

22. Mr. Angelini took $7\frac{1}{2}$ days to paint the interior of his house. A professional painter said he could do it in $2\frac{1}{3}$ days. How much time would be saved having the painter do it?

EXERCISES 322, SET II. Find the following differences.

1. $18\frac{5}{6}$
   $- 12\frac{1}{6}$

2. $28\frac{3}{4}$
   $- 15\frac{1}{2}$

3. 10
   $- 3\frac{1}{4}$

4. 8
   $- 2\frac{3}{5}$

5. $7\frac{1}{4}$
   $- 2\frac{3}{4}$

6. $63\frac{1}{3}$
   $- 19\frac{2}{3}$

7. $35\frac{3}{8}$
   $- 27\frac{3}{4}$

8. 75
   $- 28\frac{11}{12}$

9. $209\frac{5}{16}$
   $- 30\frac{3}{4}$

10. $8,003\frac{1}{4}$
    $- 3,806\frac{2}{3}$

11. When a $1\frac{5}{8}$-pound steak was trimmed of fat it weighed $1\frac{1}{4}$ pounds. How much did the fat weigh?

Commutative and Associative Rules for Fractions

ADDITION OF FRACTIONS IS COMMUTATIVE. If we reverse the order of the fractions in the sum, the sum remains the same.

order reversed
$$\frac{2}{3} + \frac{4}{5} = \frac{10}{15} + \frac{12}{15} = \frac{10 + 12}{15} = \frac{22}{15}$$
$$\frac{4}{5} + \frac{2}{3} = \frac{12}{15} + \frac{10}{15} = \frac{12 + 10}{15} = \frac{22}{15}$$
equal value

Therefore,
$$\frac{2}{3} + \frac{4}{5} = \frac{4}{5} + \frac{2}{3}$$

MULTIPLICATION OF FRACTIONS IS COMMUTATIVE. If we reverse the order of the fractions in a product, the product remains the same.

order reversed
$$\frac{7}{8} \cdot \frac{3}{5} = \frac{7 \cdot 3}{8 \cdot 5} = \frac{21}{40}$$
$$\frac{3}{5} \cdot \frac{7}{8} = \frac{3 \cdot 7}{5 \cdot 8} = \frac{21}{40}$$
equal value

Therefore,
$$\frac{7}{8} \cdot \frac{3}{5} = \frac{3}{5} \cdot \frac{7}{8}$$

SUBTRACTION OF FRACTIONS IS NOT COMMUTATIVE. If we change the order in which fractions are subtracted, the differences are *not* the same.

order reversed
$$\frac{1}{2} - \frac{1}{3} = \frac{3 - 2}{6} = \frac{1}{6}$$
$$\frac{1}{3} - \frac{1}{2} = \frac{(2 - 3)}{6}$$ ← This is not a whole number; therefore, the differences are not the same.

Therefore,
$$\frac{1}{2} - \frac{1}{3} \neq \frac{1}{3} - \frac{1}{2}$$

DIVISION OF FRACTIONS IS NOT COMMUTATIVE. If we change the order in which fractions are divided, the quotients are *not* the same.

order reversed
$$\frac{1}{4} \div \frac{1}{2} = \frac{1}{4} \cdot \frac{2}{1} = \frac{2}{4} = \frac{1}{2}$$
$$\frac{1}{2} \div \frac{1}{4} = \frac{1}{2} \cdot \frac{4}{1} = \frac{4}{2} = 2$$
not the same

Therefore, $$\frac{1}{4} \div \frac{1}{2} \neq \frac{1}{2} \div \frac{1}{4}$$

ADDITION OF FRACTIONS IS ASSOCIATIVE. If we change the way we group when fractions are added, we get the same sum.

different grouping $\left\{\begin{array}{l} \left(\dfrac{2}{3}+\dfrac{1}{2}\right)+\dfrac{4}{5} = \dfrac{4+3}{6}+\dfrac{4}{5} = \dfrac{7}{6}+\dfrac{4}{5} = \dfrac{35+24}{30} = \dfrac{59}{30} \\[2mm] \dfrac{2}{3}+\left(\dfrac{1}{2}+\dfrac{4}{5}\right) = \dfrac{2}{3}+\dfrac{5+8}{10} = \dfrac{2}{3}+\dfrac{13}{10} = \dfrac{20+39}{30} = \dfrac{59}{30} \end{array}\right\}$ same sum

Therefore, $$\left(\frac{2}{3}+\frac{1}{2}\right)+\frac{4}{5} = \frac{2}{3}+\left(\frac{1}{2}+\frac{4}{5}\right)$$

MULTIPLICATION OF FRACTIONS IS ASSOCIATIVE. If we change the way we group when fractions are multiplied, we get the same product.

different grouping $\left\{\begin{array}{l} \left(\dfrac{2}{3} \cdot \dfrac{1}{5}\right) \cdot \dfrac{4}{7} = \dfrac{2 \cdot 1}{3 \cdot 5} \cdot \dfrac{4}{7} = \dfrac{(2 \cdot 1) \cdot 4}{(3 \cdot 5) \cdot 7} = \dfrac{8}{105} \\[2mm] \dfrac{2}{3} \cdot \left(\dfrac{1}{5} \cdot \dfrac{4}{7}\right) = \dfrac{2}{3} \cdot \dfrac{1 \cdot 4}{5 \cdot 7} = \dfrac{2 \cdot (1 \cdot 4)}{3 \cdot (5 \cdot 7)} = \dfrac{8}{105} \end{array}\right\}$ equal value

Therefore, $$\left(\frac{2}{3} \cdot \frac{1}{5}\right) \cdot \frac{4}{7} = \frac{2}{3} \cdot \left(\frac{1}{5} \cdot \frac{4}{7}\right)$$

SUBTRACTION OF FRACTIONS IS NOT ASSOCIATIVE. If we change the way we group when fractions are subtracted, we do *not* get the same difference.

different grouping $\left\{\begin{array}{l} \left(\dfrac{1}{2}-\dfrac{1}{3}\right)-\dfrac{1}{10} = \dfrac{3-2}{6}-\dfrac{1}{10} = \dfrac{1}{6}-\dfrac{1}{10} = \dfrac{5-3}{30} = \dfrac{2}{30} = \dfrac{1}{15} \\[2mm] \dfrac{1}{2}-\left(\dfrac{1}{3}-\dfrac{1}{10}\right) = \dfrac{1}{2}-\dfrac{10-3}{30} = \dfrac{1}{2}-\dfrac{7}{30} = \dfrac{15-7}{30} = \dfrac{8}{30} = \dfrac{4}{15} \end{array}\right\}$ not the same

Therefore, $$\left(\frac{1}{2}-\frac{1}{3}\right)-\frac{1}{10} \neq \frac{1}{2}-\left(\frac{1}{3}-\frac{1}{10}\right)$$

DIVISION OF FRACTIONS IS NOT ASSOCIATIVE. If we change the way we group when fractions are divided, we get *different* quotients.

different grouping $\left\{\begin{array}{l} \left(\dfrac{2}{3} \div \dfrac{1}{5}\right) \div \dfrac{1}{2} = \left(\dfrac{2}{3} \cdot \dfrac{5}{1}\right) \div \dfrac{1}{2} = \dfrac{10}{3} \div \dfrac{1}{2} = \dfrac{10}{3} \cdot \dfrac{2}{1} = \dfrac{20}{3} = 6\dfrac{2}{3} \\[2mm] \dfrac{2}{3} \div \left(\dfrac{1}{5} \div \dfrac{1}{2}\right) = \dfrac{2}{3} \div \left(\dfrac{1}{5} \cdot \dfrac{2}{1}\right) = \dfrac{2}{3} \div \dfrac{2}{5} = \dfrac{2}{3} \cdot \dfrac{5}{2} = \dfrac{5}{3} = 1\dfrac{2}{3} \end{array}\right\}$ not the same

Therefore, $$\left(\frac{2}{3} \div \frac{1}{5}\right) \div \frac{1}{2} \neq \frac{2}{3} \div \left(\frac{1}{5} \div \frac{1}{2}\right)$$

Combined Operations

In evaluating expressions with more than one operation, the following order of operations is used:

> Order of Operations
>
> 1. If there are any parentheses in the expression, that part of the expression within a pair of parentheses is evaluated first, then the entire expression.
> 2. Any evaluation always proceeds in three steps:
>   *First:* Powers and roots are done in any order.
>   *Second:* Multiplication and division are done in order from left to right.
>   *Third:* Addition and subtraction are done in order from left to right.

Example 1

$$2 - \frac{3}{4} + 3\frac{1}{2}$$

$$= \frac{2}{1} - \frac{3}{4} + \frac{7}{2}$$

$$= \frac{8}{4} - \frac{3}{4} + \frac{14}{4}$$

$$= \frac{5}{4} + \frac{14}{4} = \frac{19}{4} = 4\frac{3}{4}$$

Example 2

$$2\frac{4}{5} \cdot \frac{1}{2} \div 1\frac{3}{10}$$

$$= \frac{\overset{7}{\cancel{14}}}{5} \cdot \frac{1}{\underset{1}{\cancel{2}}} \div \frac{13}{10}$$

$$= \frac{7}{5} \div \frac{13}{10}$$

$$= \frac{7}{\underset{1}{\cancel{5}}} \cdot \frac{\overset{2}{\cancel{10}}}{13} = \frac{14}{13} = 1\frac{1}{13}$$

Example 3

$$2\frac{1}{3} + \frac{5}{6} \div 1\frac{3}{4}$$

$$= \frac{7}{3} + \frac{5}{6} \div \frac{7}{4}$$

$$= \frac{7}{3} + \frac{5}{\underset{3}{\cancel{6}}} \cdot \frac{\overset{2}{\cancel{4}}}{7}$$

$$= \frac{7}{3} + \frac{10}{21}$$

$$= \frac{49}{21} + \frac{10}{21} = \frac{59}{21} = 2\frac{17}{21}$$

Example 4

$$8 - \frac{2}{3} \cdot 2\frac{1}{2}$$

$$= \frac{8}{1} - \frac{2}{3} \cdot \frac{5}{2}$$

$$= \frac{8}{1} - \frac{5}{3}$$

$$= \frac{24}{3} - \frac{5}{3} = \frac{19}{3} = 6\frac{1}{3}$$

Example 5   Example 6

$$\left(\frac{3}{4}\right)^2 + 2\frac{4}{5} \cdot 1\frac{1}{4} \qquad\qquad \left(1\frac{3}{8} - \frac{1}{2}\right) \div 1\frac{5}{16}$$

$$= \quad \frac{3}{4} \cdot \frac{3}{4} + \frac{\overset{7}{\cancel{14}}}{\cancel{5}} \cdot \frac{\overset{1}{\cancel{5}}}{\cancel{4}} \qquad = \left(\frac{11}{8} - \frac{1}{2}\right) \div \frac{21}{16}$$

$$= \qquad \frac{9}{16} \quad + \quad \frac{7}{2} \qquad = \left(\frac{11}{8} - \frac{4}{8}\right) \div \frac{21}{16}$$

$$= \qquad \frac{9}{16} + \frac{56}{16} = \frac{65}{16} = 4\frac{1}{16} \qquad = \qquad \frac{7}{8} \div \frac{21}{16}$$

$$= \quad \frac{\cancel{7}^{1}}{\cancel{8}_{1}} \cdot \frac{\cancel{16}^{2}}{\cancel{21}_{3}} = \frac{2}{3}$$

EXERCISES 324, SET I. In working the following exercises, be sure to perform the operations in the correct order.

1. $10 - 3 + 2$ 2. $4 + 10 - 12$

3. $8 \cdot 4 \div 2$ 4. $12 \div 6 \cdot 2$

5. $8 \div 4 \cdot 2$ 6. $12 \cdot 6 \div 2$

7. $5 \cdot 2 + 3 \div 6$ 8. $6 \cdot 4 + 8 \div 12$

9. $6 \cdot \frac{1}{2} - 2 \div 8$ 10. $8 \cdot \frac{3}{4} - 5 \div 15$

11. $(5)^2 + 3 \cdot 1\frac{1}{3}$ 12. $(4)^2 + 8 \cdot 2\frac{1}{4}$

13. $\left(\frac{3}{4}\right)^2 + \frac{1}{4} \cdot 1\frac{3}{4}$ 14. $\left(\frac{4}{5}\right)^2 + \frac{1}{5} \cdot 1\frac{4}{5}$

15. $4 - \frac{2}{3} + 1\frac{1}{2}$ 16. $\frac{4}{5} \cdot 1\frac{2}{3} \div 1\frac{7}{9}$

17. $\frac{3}{4} + 33 \div 4\frac{1}{8}$ 18. $8\frac{1}{4} - 1\frac{2}{3} \cdot 3$

19. $\left(8\frac{1}{4} - 1\frac{2}{3}\right) \cdot 3$ 20. $\left(1\frac{1}{2}\right)^2 + 2\frac{1}{3} \cdot \frac{3}{4}$

21. $2\frac{2}{5} \div \left(3 - \frac{3}{10}\right)$ 22. $2\frac{2}{5} \div 3 - \frac{3}{10}$

23. $2\frac{2}{3} \cdot \left(\frac{3}{7} + 2\right) \div 4\frac{6}{7}$ 24. $1\frac{5}{6} + \frac{2}{3} \cdot 2\frac{1}{2} - 3$

EXERCISES 324, SET II. In working the following exercises, be sure to perform the operations in the correct order.

1. $15 + 3 - 4$ 2. $12 \cdot 6 \div 2$

3. $8 \cdot 2 + 2 \div 4$ 4. $10 \cdot \frac{1}{2} - 2 \div 8$

5. $(6)^2 + 4 \cdot 1\frac{1}{4}$ 6. $\left(\frac{2}{5}\right)^2 + \frac{1}{3} \cdot 1\frac{4}{5}$

7. $3 - \dfrac{3}{4} + 5\dfrac{3}{4}$

8. $\dfrac{5}{6} \cdot 3\dfrac{3}{4} \div \dfrac{5}{8}$

9. $9\dfrac{1}{6} - 1\dfrac{3}{4} \cdot 2$

10. $\left(1\dfrac{1}{3}\right)^2 + 3\dfrac{1}{2} \cdot \dfrac{2}{3}$

11. $4\dfrac{4}{5} \div 8 - \dfrac{3}{10}$

12. $3\dfrac{1}{3}\left(\dfrac{3}{5} + 3\right) \div 2\dfrac{2}{5}$

Comparing the Size of Fractions

Sometimes we need to recognize which of a group of fractions is largest. *We can compare the size of fractions by converting each of the given fractions to an equivalent fraction having the LCD as its denominator.*

Example 1. Arrange the following fractions in order of size—the largest first: $\dfrac{3}{8}$, $\dfrac{1}{3}$, $\dfrac{5}{12}$.

Solution: LCD = 2 • 2 • 3 • 2 • 1 • 1 = 24

$$\dfrac{3}{8} = \dfrac{9}{24}$$

$$\dfrac{1}{3} = \dfrac{8}{24}$$

$$\dfrac{5}{12} = \dfrac{10}{24}$$

```
2 | 8   3   12
2 | 4   3    6
3 | 2   3    3
    2   1    1
```

Arranging the *equivalent* fractions in order of size, we have:

$$\dfrac{10}{24} > \dfrac{9}{24} > \dfrac{8}{24}$$

Therefore, the *original* fractions arranged in order are:

$$\dfrac{5}{12} > \dfrac{3}{8} > \dfrac{1}{3}$$

EXERCISES 325, SET I. In Exercises 1-6, arrange the given fractions in order of size—the largest first.

1. $\dfrac{3}{4}$, $\dfrac{5}{6}$, $\dfrac{2}{3}$

2. $\dfrac{2}{9}$, $\dfrac{5}{12}$, $\dfrac{1}{3}$

3. $\dfrac{5}{14}$, $\dfrac{3}{7}$, $\dfrac{3}{4}$

4. $\dfrac{4}{5}$, $\dfrac{3}{4}$, $\dfrac{7}{10}$

5. $\dfrac{2}{15}$, $\dfrac{1}{6}$, $\dfrac{3}{10}$

6. $\dfrac{5}{8}$, $\dfrac{19}{32}$, $\dfrac{9}{16}$

7. A stock price went from $12\dfrac{5}{8}$ to $12\dfrac{3}{4}$. Did it go up or down? By how much?

8. A stock price went from $16\dfrac{3}{8}$ to $16\dfrac{1}{4}$. Did it go up or down? By how much?

EXERCISES 325, SET II. In Exercises 1-3, arrange the given fractions in order of size—the largest first.

1. $\frac{1}{3}$, $\frac{5}{12}$, $\frac{1}{4}$ 2. $\frac{5}{8}$, $\frac{3}{4}$, $\frac{11}{16}$ 3. $\frac{5}{6}$, $\frac{2}{3}$, $\frac{7}{8}$

4. A stock price went from $6\frac{3}{4}$ to $6\frac{7}{8}$. Did it go up or down? By how much?

Fractions and Whole Numbers Are Part of the Real Number System

In Chapter 1, the number line was introduced.

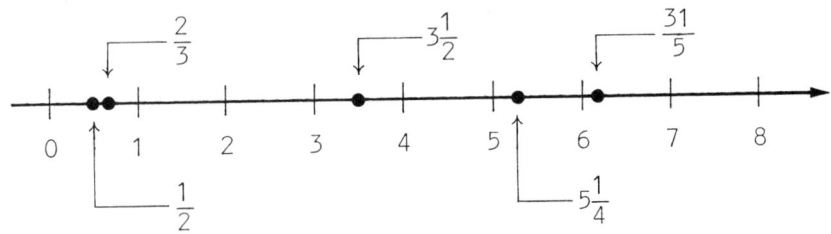

Figure 326

At that time, we explained that any natural number or any whole number can be represented by a point on the number line. All the numbers that can be represented by points on the number line are called *real numbers*. Since any fraction or mixed number can be represented by a point on the number line, fractions and mixed numbers are part of the real number system. The points representing some fractions and mixed numbers are shown in Figure 326.

Chapter Summary

A *proper fraction* is a fraction whose numerator is less than its denominator. (Sec. 301)

An *improper fraction* is a fraction whose numerator is equal to or greater than its denominator. (Sec. 301)

A *mixed number* has a whole number part and a fraction part. (Sec. 304)

A *simple fraction* is a fraction with one fraction line. (Sec. 318)

A *complex fraction* is a fraction with more than one fraction line. (Sec. 318)

Two fractions are equivalent (have equal value) if their cross-products are equal. (Sec. 307)

To raise a fraction to higher terms, multiply both numerator and denominator by the same number (greater than one). (Sec. 308)

To reduce a fraction to lower terms, divide its numerator and denominator by a number (greater than one) which is a divisor of both. (Sec. 309)

To reduce a fraction to lowest terms: (Sec. 311)

1. Divide both numerator and denominator by all factors common to both.
2. Use Euclid's Algorithm if you think it may be reduced further.

A *prime number* is a natural number greater than 1 that can only be divided by itself and 1. A prime number has no factors other than itself and 1. (Sec. 314)

A *composite number* is a natural number that can be divided by some number other than itself and 1. A composite number has factors other than itself and 1. (Sec. 314)

Commutative and Associative Properties (Sec. 323)

Addition and multiplication of fractions and mixed numbers are both commutative and associative.

Subtraction and division of fractions and mixed numbers are *not* commutative or associative.

Order of operations: (Sec. 324)

1. If there are any parentheses in the expression, that part of the expression within a pair of parentheses is evaluated first, then the entire expression.
2. Any evaluation always proceeds in three steps:
   First: Powers and roots are done in any order.
   Second: Multiplication and division are done in order from left to right.
   Third: Addition and subtraction are done in order from left to right.

All numbers that can be represented by points on the number line are called *real numbers*. Whole numbers, fractions, and mixed numbers are real numbers. (Sec. 326)

REVIEW EXERCISES 327, SET I. In Exercises 1 and 2, change the improper fractions to mixed numbers.

1. (a) $\frac{5}{2}$ (b) $\frac{11}{8}$ (c) $\frac{18}{13}$ (d) $\frac{26}{16}$ (e) $\frac{63}{32}$

2. (a) $\frac{7}{3}$ (b) $\frac{21}{19}$ (c) $\frac{47}{25}$ (d) $\frac{78}{33}$ (e) $\frac{124}{57}$

In Exercises 3 and 4, change the mixed numbers to improper fractions.

3. (a) $2\frac{3}{5}$ (b) $3\frac{7}{8}$ (c) $9\frac{2}{11}$ (d) $13\frac{5}{16}$ (e) $23\frac{27}{31}$

4. (a) $1\frac{3}{4}$ (b) $2\frac{7}{9}$ (c) $8\frac{5}{13}$ (d) $16\frac{7}{18}$ (e) $27\frac{13}{35}$

In Exercises 5-12, perform the indicated operations.

5. (a) $2 + 4\frac{2}{5}$ (b) $2\frac{2}{3} + 1\frac{3}{5}$ (c) $4\frac{5}{16} + \frac{5}{8}$ (d) $153\frac{2}{5} + 135\frac{3}{4}$

6. (a) $5 + 2\frac{1}{3}$ (b) $3\frac{1}{4} + 4\frac{1}{2}$ (c) $4\frac{2}{3} + 3\frac{1}{2}$ (d) $113\frac{7}{8} + 98\frac{5}{6}$

7. (a) $5\frac{4}{5} - 3\frac{7}{10}$ (b) $5\frac{1}{3} - 2\frac{3}{4}$ (c) $16 - 5\frac{7}{8}$ (d) $351\frac{2}{3} - 272\frac{7}{9}$

8. (a) $4\frac{5}{6} - 2\frac{2}{3}$ (b) $3\frac{2}{7} - 1\frac{9}{14}$ (c) $25 - 12\frac{5}{16}$ (d) $286\frac{5}{8} - 196\frac{3}{4}$

9. (a) $4\frac{1}{5} \cdot 2\frac{3}{7}$ (b) $3\frac{2}{3} \cdot 6\frac{3}{5}$ (c) $\frac{3}{5} \times 105$ (d) $\frac{12}{13} \times 8\frac{1}{3}$

10. (a) $2\frac{1}{3} \times 4\frac{1}{2}$ (b) $2\frac{4}{5} \cdot 2\frac{6}{7}$ (c) $\frac{5}{8} \times 56$ (d) $4\frac{12}{13} \times \frac{13}{16}$

11. (a) $2\frac{2}{5} \div 1\frac{1}{15}$ (b) $1\frac{1}{6} \div 4\frac{2}{3}$ (c) $16 \div \frac{8}{13}$ (d) $1\frac{9}{16} \div 10$

12. (a) $2\frac{1}{4} \div 3\frac{3}{8}$ (b) $1\frac{1}{12} \div 8\frac{2}{3}$ (c) $15 \div 1\frac{1}{8}$ (d) $3\frac{3}{5} \div 9$

In Exercises 13 and 14, reduce to lowest terms.

13. (a) $\frac{42}{54}$ (b) $\frac{45}{105}$ (c) $\frac{28}{57}$ (d) $\frac{207}{253}$

14. (a) $\frac{50}{60}$ (b) $\frac{60}{84}$ (c) $\frac{225}{360}$ (d) $\frac{465}{403}$

In Exercises 15 and 16, state whether the given fractions are equivalent. If the fractions are not equivalent, state which one is larger.

15. (a) $\frac{10}{35}, \frac{6}{21}$ (b) $\frac{12}{15}, \frac{44}{55}$ (c) $\frac{14}{30}, \frac{18}{45}$ (d) $\frac{12}{21}, \frac{20}{35}$

16. (a) $\frac{14}{21}, \frac{10}{15}$ (b) $\frac{27}{18}, \frac{18}{14}$ (c) $\frac{28}{35}, \frac{36}{45}$ (d) $\frac{39}{91}, \frac{6}{14}$

In Exercises 17 and 18, arrange the given fractions in order of size—the largest first.

17. $\frac{4}{9}, \frac{5}{12}, \frac{7}{18}$ 18. $\frac{2}{3}, \frac{7}{8}, \frac{5}{6}$

When doing Exercises 19 and 20, refer to Section 315 if necessary.

19. (a)   (b)   20. (a)   (b)

$\dfrac{7}{45} + \dfrac{25}{36}$ $\dfrac{11}{35} - \dfrac{7}{30}$ $\dfrac{5}{12} + \dfrac{13}{28}$ $\dfrac{23}{30} - \dfrac{7}{18}$

In Exercises 21 and 22, simplify the complex fractions.

21. (a) $\dfrac{\frac{5}{8}}{\frac{5}{6}}$ (b) $\dfrac{\frac{12}{}}{\frac{6}{11}}$ (c) $\dfrac{\frac{3}{4}}{6}$ (d) $\dfrac{\frac{2}{3} + \frac{1}{2}}{\frac{5}{6}}$

22. (a) $\dfrac{\frac{3}{5}}{\frac{6}{5}}$ (b) $\dfrac{\frac{15}{}}{\frac{5}{3}}$ (c) $\dfrac{\frac{5}{8}}{10}$ (d) $\dfrac{\frac{3}{8} + \frac{5}{4}}{4\frac{1}{3}}$

In working Exercises 23 and 24, be sure to perform the operations in the correct order.

23. (a) $\left(3\frac{1}{2} - 2\frac{3}{4}\right) \div 6 \cdot \dfrac{4}{7}$ (b) $\left(\dfrac{3}{4}\right)^2 + 2\frac{2}{3} \cdot \dfrac{3}{16}$

24. (a) $\dfrac{7}{3} \div \dfrac{14}{9} - 2 \cdot \dfrac{5}{12}$ (b) $\left(\dfrac{2}{3}\right)^2 + \dfrac{5}{9} + 2\frac{2}{5} \div \dfrac{24}{25}$

25. If you sleep 8 hours out of every 24, what fraction of the time do you sleep?

26. A family used $10\frac{1}{2}$ square yards of carpet for the living room, and $8\frac{3}{4}$ square yards for the hall and bedroom combined.
    (a) What was the total amount of carpet used?
    (b) If carpet cost $8 per square yard, how much did it cost them to carpet the hall and bedroom?

27. In driving across the country, a man drove $4\frac{1}{2}$ hours on Monday, $12\frac{3}{4}$ hours on Tuesday, $8\frac{1}{3}$ hours on Wednesday, and $15\frac{1}{6}$ hours on Thursday. What was his total driving time for the trip?

28. A bookshelf is 42 inches long. How many books that are $\frac{3}{4}$ inch thick can be stood on that shelf?

29. It costs a family $$4\frac{3}{4}$ for each round trip to their vacation cabin. If they make 24 trips in a year, what is the transportation cost?

30. A board that is 73 inches long is cut into three pieces of equal length. If $\frac{1}{8}$ inch is wasted each time the board is sawed, how long is each of the three finished pieces?

31. A man has a 16-foot board he is going to use for shelves. If the shelves are to be $3\frac{1}{2}$ feet long:

(a) How many shelves can be cut from the board? ($\frac{1}{8}$ inch is wasted each time the board is sawed.)

(b) What length will be left from the original board after he cuts as many shelves from it as he can?

32. <u>Brain Teaser</u>. An ancient bedouin willed one-half of his camels to his oldest son, one-third to his second son, and one-ninth to his youngest son. When the father died, he had 17 camels. The sons could not see how to divide this herd according to their inheritance without killing some of the camels. They called on an uncle to solve this problem for them. The uncle added one of his own camels to the herd, making it 18 camels in all, then gave each son his share of the 18 camels.

oldest son: $\frac{1}{2}$ of 18 = 9 camels

second son: $\frac{1}{3}$ of 18 = 6 camels

youngest son: $\frac{1}{9}$ of 18 = <u>2 camels</u>

Total = 17 camels

How was it possible for each son to receive more than his share of the 17 camels, and at the same time have the uncle get his camel back?

REVIEW EXERCISES 327, SET II

1. Change the improper fractions to mixed numbers.

(a) $\frac{5}{3}$ (b) $\frac{9}{2}$ (c) $\frac{25}{12}$ (d) $\frac{23}{19}$ (e) $\frac{125}{41}$

2. Change the mixed numbers to improper fractions.

(a) $3\frac{2}{5}$ (b) $4\frac{5}{6}$ (c) $5\frac{2}{3}$ (d) $8\frac{3}{5}$ (e) $10\frac{5}{8}$

In Exercises 3-6, perform the indicated operations.

3. (a) $4 + 2\frac{3}{5}$ (b) $2\frac{5}{8} + 5\frac{3}{4}$ (c) $112\frac{5}{6} + 218\frac{2}{3}$ (d) $115\frac{1}{2} + 94\frac{7}{12}$

4. (a) $6\frac{5}{8} - 2\frac{1}{4}$ (b) $18 - 5\frac{3}{4}$ (c) $18\frac{1}{6} - 11\frac{1}{3}$ (d) $208\frac{3}{5} - 140\frac{7}{10}$

5. (a) $5\frac{1}{4} \cdot 2\frac{2}{7}$ (b) $1\frac{3}{5} \times 2\frac{3}{16}$ (c) $\frac{3}{8} \cdot 248$ (d) $2\frac{4}{7} \times 7\frac{7}{12}$

6. (a) $3\frac{5}{6} \div 7\frac{2}{3}$ (b) $18 \div \frac{9}{16}$ (c) $1\frac{5}{9} \div \frac{7}{27}$ (d) $9\frac{5}{8} \div 11$

7. Reduce to lowest terms.

   (a) $\dfrac{48}{54}$ (b) $\dfrac{57}{93}$ (c) $\dfrac{115}{120}$ (d) $\dfrac{258}{301}$

8. State whether the given fractions are equivalent.

   (a) $\dfrac{9}{11}, \dfrac{11}{13}$ (b) $\dfrac{13}{15}, \dfrac{39}{45}$ (c) $\dfrac{68}{76}, \dfrac{17}{19}$ (d) $\dfrac{47}{61}, \dfrac{142}{183}$

9. When doing these exercises, refer to Section 315 if necessary.

   (a) $\dfrac{5}{18} + \dfrac{7}{30}$ (b) $\dfrac{7}{21} - \dfrac{5}{28}$

10. Simplify the complex fractions.

    (a) $\dfrac{\frac{3}{5}}{\frac{9}{10}}$ (b) $\dfrac{\frac{14}{7}}{\frac{7}{8}}$ (c) $\dfrac{\frac{4}{5}}{12}$ (d) $\dfrac{\frac{6}{7} - \frac{3}{4}}{\frac{3}{14}}$

11. In doing these exercises, be sure to perform the operations in the correct order.

    (a) $\left(2\dfrac{1}{3} - 1\dfrac{1}{2}\right) \div \dfrac{5}{18} \cdot \dfrac{5}{24}$ (b) $\left(1\dfrac{2}{3}\right)^2 - \dfrac{1}{3} + 3\dfrac{1}{3} \div 3$

12. A family used $29\dfrac{1}{2}$ square yards of carpet for their living room, $12\dfrac{1}{3}$ square yards for a bedroom, and $15\dfrac{1}{4}$ square yards for the hall.
    (a) What was the total amount of carpet used?
    (b) If the carpet cost them $12 a square yard laid, how much did it cost them for the complete job?

13. A bookshelf is 36 inches long. How many books that are $1\dfrac{1}{2}$ inches thick can be stood on that shelf?

14. A board $96\dfrac{1}{2}$ inches long is cut into three pieces of equal length. If $\dfrac{1}{4}$ inch is wasted each time the board is sawed, how long is each of the three finished pieces?

Chapter 3: Diagnostic Test

Name_____

_____

The purpose of this test is to see how well you know fractions. We recommend that you work this diagnostic test *before* your instructor tests you on this chapter. Allow yourself about an hour to do this test.

Complete solutions for all the problems on this test, together with section references, are given in the Answer Section. You should study the sections referred to for the problems you do incorrectly.

---

1. Change the improper fractions to mixed numbers.

   (a) $\dfrac{7}{4}$ (b) $\dfrac{63}{29}$

   (1a)_____

   (1b)_____

---

2. Change the mixed numbers to improper fractions.

   (a) $2\dfrac{3}{4}$ (b) $15\dfrac{7}{12}$

   (2a)_____

   (2b)_____

---

3. Determine which of the following pairs of fractions are equivalent (have equal values). If the fractions are not equivalent, state which one is larger.

   (a) $\dfrac{7}{9}$, $\dfrac{21}{27}$ (b) $\dfrac{5}{9}$, $\dfrac{41}{73}$

   (3a)_____

   (3b)_____

---

4. State whether each of the following numbers is prime or composite.

(a) 31   (b) 51   (c) 71

(4a) _____

(4b) _____

(4c) _____

5. Perform the indicated operations.

(a) $2 + 3\frac{1}{5}$ (b) $\begin{array}{r} 4\frac{2}{3} \\ + \ 3\frac{1}{2} \\ \hline \end{array}$

(5a) _____

(5b) _____

(c) $\frac{5}{6} \times \frac{3}{20}$ (d) $\frac{2}{3} \times 111$

(5c) _____

(5d) _____

(e) $\frac{3}{8} \div \frac{9}{16}$ (f) $4\frac{1}{5} \times 2\frac{1}{7}$

(5e) _____

(5f) _____

(g) $2\frac{2}{9} \div 3\frac{1}{3}$ (h) $\dfrac{\frac{5}{8}}{\frac{5}{6}}$

(5g) _____

(5h) _____

(i) $\dfrac{\dfrac{3}{8} + \dfrac{5}{8}}{4\dfrac{1}{3}}$

(5i) _____

6. Reduce to lowest terms.

   (a) $\dfrac{30}{42}$ (b) $\dfrac{180}{540}$

(6a) _____

(6b) _____

7. Use Euclid's Algorithm as an aid to reduce this fraction to lowest terms.

   $\dfrac{217}{248}$

(7) _____

8. Add: $27\dfrac{2}{9}$

   $18\dfrac{1}{2}$

   $4\dfrac{3}{4}$

   $10\dfrac{7}{8}$

(8) _____

9. Subtract: $124\dfrac{2}{3}$

   $- 17\dfrac{4}{5}$

(9) _____

10. Perform the operations in the correct order.

(a) $6 + 2 \times \dfrac{3}{4}$

(10a) _____

(b) $4 \cdot \left(\dfrac{3}{4}\right)^2 - \dfrac{1}{2}$

(10b) _____

(c) $\dfrac{2}{3} \div \dfrac{4}{3} \cdot \dfrac{2}{5}$

(10c) _____

11. When a $1\dfrac{1}{8}$-pound steak was trimmed of fat, it weighed $\dfrac{3}{4}$ pound. How much of the steak was fat?

(11) _____

12. How many tablets, each containing 3 milligrams of medicine, must be used to make up a $7\dfrac{1}{2}$-milligram dosage?

(12) _____

13. How many square yards of carpet will be needed to carpet a rectangular room that is $7\dfrac{1}{2}$ yards long and $4\dfrac{2}{3}$ yards wide?

(13) _____

FOUR
Decimal Fractions

Basic Concepts

DECIMAL FRACTIONS. A *decimal fraction* is a fraction whose denominator is a power of 10.

Example 1

(a) $\dfrac{3}{10^1} = \dfrac{3}{10}$ is read "three tenths."

(b) $\dfrac{3}{10^2} = \dfrac{3}{100}$ is read "three hundredths."

(c) $\dfrac{25}{10^3} = \dfrac{25}{1,000}$ is read "twenty-five thousandths."

(d) $\dfrac{5}{10^0} = \dfrac{5}{1} = 5$

(e) $\dfrac{251}{10^0} = \dfrac{251}{1} = 1$ (f) $\dfrac{0}{10^0} = \dfrac{0}{1} = 0$

In Examples (d), (e), and (f) we show that all whole numbers can be written as decimal fractions. However, not all decimal fractions are whole numbers [see Examples (a), (b), and (c)].

A decimal fraction can be written in two ways:

(1) *Fraction form:* $\dfrac{3}{10}$, $\dfrac{17}{1,000}$, $\dfrac{347}{100}$, $\dfrac{2,359}{10}$, etc.

(2) *Decimal form:* 0.3, 0.017, 3.47, 235.9, etc.

This new symbol "." is called a *decimal point*.

Example 2

(a) $\dfrac{4}{10}$ = 0.4   is read "four tenths."

(b) $\dfrac{5}{100}$ = 0.05   is read "five hundredths."

(c) $\dfrac{6}{1,000}$ = 0.006   is read "six thousandths."

Hereafter, decimal fractions will be called *decimals*. In most cases, we will use the term *fraction* for fractions having denominators other than powers of 10.

Reading and Writing Decimal Numbers

In Figure 402, we show place values of decimals. Note that the decimal point is written between the units place and the tenths place.

Place	Power	Value
millions	10^6	= 1,000,000
hundred-thousands	10^5	= 100,000
ten-thousands	10^4	= 10,000
thousands	10^3	= 1,000
hundreds	10^2	= 100
tens	10^1	= 10
ones	10^0	= 1
decimal point →		
tenths	0.1	= 1/10
hundredths	0.01	= 1/100
thousandths	0.001	= 1/1,000
ten-thousandths	0.0001	= 1/10,000
hundred-thousandths	0.00001	= 1/100,000
millionths	0.000001	= 1/1,000,000

Place-Values of Decimals

Figure 402

In examining the numbers of Figure 402, we see that the value of each place is one-tenth the value of the first place to its left.

To Read a Decimal

1. Read the number to the left of the decimal point as you read a whole number.
2. Say "and" for the decimal point.
3. Read the number to the right of the decimal point as a whole number, then say the name of the place occupied by the right-hand digit of the number.

READING AND WRITING DECIMALS LESS THAN ONE

Example 1

(a) 0.5   is read "five tenths."
    ↑————tenths place

(b) 0.06   is read "six hundredths."
    ↑————hundredths place

(c) 0.007   is read "seven thousandths."
    ↑————thousandths place

(d) 0.50   is read "fifty hundredths."
    ↑————hundredths place

(e) 0.500   is read "five hundred thousandths."
   ↑————thousandths place

(f) 0.567   is read "five hundred sixty-seven thousandths."
   ↑————thousandths place

(g) 0.62,354 is read "sixty-two thousand, three hundred
    ↑ ↑   fifty-four hundred-thousandths."
         └—hundred-thousandths place
     └————————Commas to the right of the decimal point are
                  placed in the proper positions to help read
                  the number to the right of the decimal point
                  as though it were a whole number.

In writing decimals less than 1, we usually write a 0 to the
left of the decimal point to call attention to the decimal
point so that it is not overlooked. For example, seventy-five
hundredths is written 0.75. However, both 0.75 and .75 are
correct ways of writing seventy-five hundredths.

READING AND WRITING DECIMALS GREATER THAN ONE

Example 2

   (a) 6.27   is read "six and twenty-seven hundredths."

   (b) 175.006   is read "one hundred seventy-five and
                   six thousandths."

   (c) 8.00004   is read "eight and four hundred-thousandths."
      ↑————hundred-thousandths place

   (d) 8.400   is read "eight and four hundred thousandths."

   (e) 107,060.756 is read "one hundred seven thousand, sixty,
                     and seven hundred fifty-six thousandths."

   (f) 6,000.5,437 is read "six thousand, and five thousand,
            ↑ four hundred thirty-seven ten-thousandths."
          └—ten-thousandths place

ANOTHER WAY OF READING AND WRITING DECIMALS GREATER THAN ONE

> To Read a Decimal
>
> 1. Ignore the decimal point and read the
>   number as a whole number.
> 2. Say the name of the place occupied by the
>   right digit of the given decimal.

Example 3. Read 3.4
              ↑————tenths place

   (1) Ignoring the decimal point and reading the number as
       a whole number, we say, "thirty-four."

   (2) The right-hand digit, 4, is in the tenths place;
       therefore, we say, "thirty-four tenths." Written $\frac{34}{10}$.

Example 4. Read 12.8

  ↑——— tenths place

Say "one hundred twenty-eight tenths."

$$\text{Written } \frac{128}{10} .$$

Example 5. Read 312.42

  ↑—— hundredths place

Say "thirty-one thousand, two hundred forty-two hundredths."

$$\text{Written } \frac{31,242}{100} .$$

AN EASY WAY TO REMEMBER THE NAME OF A DECIMAL PLACE. To find
the name of a decimal place, write a 0 for each decimal place
in the number, then precede them by a 1.

Example 6. Find the name of the place marked with the X.

(a) 0.0X
   |↑——— hundredths
   100   place

(b) 0.X
   ↑——— tenths
   10   place

(c) 0.00X
   ||↑——— thousandths
   1,000   place

(d) 0.000 00X
   ||| ||↑——— millionths
   1,000,000   place

EXERCISES 402, SET I. In Exercises 1-10, write the numbers in
decimal notation.

1. Six and four tenths
2. Seven and three thousandths
3. Twelve and two hundredths
4. Seven and twenty-one hundredths
5. Thirty-five hundredths
6. One hundred twenty-two and six tenths
7. Five thousand, eighty-six and seven hundredths
8. One hundred thirty-five thousandths
9. Seven hundred thousand, fifty-two and nine ten-thousandths
10. Eight million, forty thousand, five, and two thousand,
    seven hundred forty-six ten-thousandths

In Exercises 11-20, write the numbers in words.

11. 0.8   12. 0.95   13. 4.375   14. 20.6

15. 15.65   16. 137.95   17. 1,115   18. 6.0045

19. 5.3756   20. 47,028.05361

EXERCISES 402, SET II. In Exercises 1-5, write the numbers in
decimal notation.

1. Seventy-two hundredths
2. Fifteen and six hundredths

3. Four and seven thousandths
4. Nine thousand, four hundred seventy, and eight hundredths
5. Six million, nine thousand, eighty, and five thousand, fourteen ten-thousandths

In Exercises 6-10, write the numbers in words.

6. 0.47   7. 3.625   8. 65.39   9. 12.0028

10. 39,042.0807

ANOTHER METHOD OF READING DECIMALS. The method of reading decimals we have just used is a formal one which we use in this book, unless otherwise indicated. In common usage, we read decimals in the following way:

Example 7

    (a) 0.72   is read "zero point seven two."

    (b) 12.573   is read "one two point five seven three."

    (c) 6874.1935 is read "six eight seven four point one nine three five."

We start at the left of the number, read each digit in the order it comes, saying "point" when we come to the decimal point.

THE DECIMAL POINT IN WHOLE NUMBERS. Usually the decimal point is not written in whole numbers. If it is written, it is placed to the right of the units digit:

Without the decimal point	*With the decimal point*
4	4.
135	135.
1,000	1,000.

If a decimal point is not written in a number, it is understood that it follows the last digit of the number. Any whole number can be considered a decimal.

Adding Decimals

> To Add Decimals
>
> 1. Write the numbers under one another with their decimal points in the same vertical line.
> 2. Add the numbers like whole numbers.
> 3. Place the decimal point in your answer (sum) in the same vertical line as the other decimal points.

Writing the numbers clearly and keeping the columns straight helps to reduce the number of addition errors.

Example 1. Add 75.4 + 186 + 0.056 + 1.207 + 2,350.

Solution:

$$
\begin{array}{r}
\text{Thousands}\quad\text{Hundreds}\quad\text{Tens}\quad\text{Units}\quad\text{Decimal point}\quad\text{Tenths}\quad\text{Hundredths}\quad\text{Thousandths}\\[4pt]
7\;5\;.\;4\\
1\;8\;6\;.\\
.\;0\;5\;6\\
1\;.\;2\;0\;7\\
2,\;3\;5\;0\;.\\
\hline
2,\;6\;1\;2\;.\;6\;6\;3
\end{array}
$$

You may find it easier to keep the columns straight when zeros are written in the open spaces, as shown below.

$$
\begin{array}{r}
7\;5\;.\;4\;0\;0\\
1\;8\;6\;.\;0\;0\;0\\
0\;0\;0\;.\;0\;5\;6\\
0\;0\;1\;.\;2\;0\;7\\
2,\;3\;5\;0\;.\;0\;0\;0\\
\hline
2,\;6\;1\;2\;.\;6\;6\;3
\end{array}
$$

The method for adding decimals shown in Example 1 works because:

Only like things can be added directly.

Example 2. Add: 1 apple + 2 pears. These cannot be added because they are *not* like things.

Example 3. Add: 3 apples + 5 apples = 8 apples. These can be added because they *are* like things.

Example 4.
$$
\begin{array}{r}
12.8\\
3.24
\end{array}
$$
←——— 8 tenths + 4 hundredths These cannot be added directly.

Notice the decimal points are not in a vertical line.

Example 5.
$$
\begin{array}{r}
\overset{1}{1}2.8\\
3.24\\
\hline
0
\end{array}
$$
←——— 8 tenths + 2 tenths = 10 tenths These can be added directly.

This is why we place the decimal points in the same vertical line.

EXERCISES 403, SET I. Find the indicated sums.

1. 6.5 + 0.66 + 80.75 + 287 + 0.078

2. 100 + 20 + 7 + 0.6 + 0.09 + 0.008

3. $0.35 + $24.79 + $127.50 + $18.84 + $96

4. $0.85 + $286.83 + $7.89 + $46 + $19.95

5. 75.5 + 3.45 + 180 + 0.0056

6. 185 + 35.06 + 0.186 + 0.0007

7. 987.46 + 35.778 + 1,750.46 + 706.188 + 7,556.189

8. 75,000 + 398.46 + 79.06 + 5.0789 + 186,300 + 35.45

9. Mrs. Ramirez spent the following for food: Monday $5.33, Tuesday $7.47, Wednesday $3.89, Thursday $6.28, Friday $4.96, Saturday $11.24. What was the week's food bill?

10. Frank checked his gasoline credit slips after making a short trip and found that he had used the following amounts of gasoline: 11.2 gallons, 10.8 gallons, 14.1 gallons, 6.7 gallons, 9.4 gallons. How many gallons did he use for the trip?

11. Find the sum of the following numbers: three thousand, fifty, and thirty-seven hundredths; five and two hundred-thousandths; seventy and one hundred fifty ten-thousandths.

12. Find the perimeter of the figure shown.

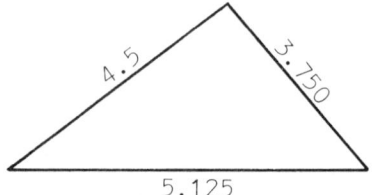

EXERCISES 403, SET II. Find the indicated sums.

1. 9.4 + 0.47 + 60.35 + 326 + 0.082

2. $0.43 + $375.92 + $8.96 + $68 + $17.84

3. 63.8 + 0.0087 + 9.46 + 120

4. 754.83 + 49.665 + 1,068.7 + 90.466 + 250.38

5. Find the sum of the following numbers: six thousand, thirty, and ninety-five hundredths; eight and four hundred-thousandths; ninety and five hundred ten ten-thousandths.

6. Mrs. Leoni spent the following for food: Monday $6.84, Tuesday $9.53, Wednesday $4.78, Thursday $8.09, Friday $5.64, Saturday $13.17. What was the week's food bill?

Subtracting Decimals

```
To Subtract Decimals

1. Write the number being subtracted (subtrahend)
   under the number it is being subtracted from (minu-
   end) with their decimal points in the same vertical
   line.
2. Subtract the numbers as whole numbers.
3. Place the decimal point in your answer (difference)
   in the same vertical line as the other decimal
   points.
```

Example 1. Subtract 6.07 from 75.14.

Solution:

$$7^15 \,.\, 1^14$$
$$6\,.\,07$$
$$\underline{11}$$
$$69\,.\,07$$

Example 2. Subtract 121.6 from 304.178.

Solution:

$$3^10\,4\,.^11\,7\,8$$
$$1\,2\,1\,.\,6$$
$$\underline{11}$$
$$1\,8\,2\,.\,5\,7\,8$$

Example 3. Subtract 1.654 from 38.6.

Solution:

$$3\,8\,.^16^10^10 \longleftarrow \text{Add zeros to the right of 6 in}$$
$$1\,.\,6\,5\,4 \qquad \text{the tenths and hundredths places.}$$
$$\underline{111}$$
$$3\,6\,.\,9\,4\,6$$

The method for subtracting decimals shown in Examples 1, 2, and 3 works·because:

Only like things can be subtracted directly.

Example 4. Subtract: 2 pears - 1 apple. These cannot be sub-tracted directly because they are *not* like things.

Example 5. Subtract: 5 apples - 2 apples = 3 apples. These can be subtracted directly because they *are* like things.

Example 6. Subtract: 12.8 - 3.24

Solution:

$$\overset{1}{12.8}\boxed{0}$$
$$\underline{-\ 3.2\boxed{4}} \longleftarrow 10 \text{ hundredths - 4 hundredths}$$
$$\overset{1}{6} \qquad = 6 \text{ hundredths}$$

EXERCISES 404, SET I. Find the indicated differences.

1. 7.85 - 3.44   2. 84.07 - 0.66   3. 208.5 - 7.16

4. 715.75 - 28.19   5. 300 - 0.145   6. 7,000 - 3.68

7. 81,284.56 - 2,784.8   8. 2,000,046 - 30,015.8

9. 5.785 - 0.9665   10. 6.005 - 0.8476

11. Mrs. Geller's bank statement showed a balance of $254.39 at the beginning of the month. During the month she made the following deposits: $183.50, $233.75, and $78.86. During the month she wrote the following checks: $27.15, $86.94, $123.47, $167.66, $122.20, and $38.67. Find her balance at the end of the month.

12. Mr. Baca's bank statement showed a balance of $175.57 at the beginning of the month. During the month he made the following deposits: $539.18, $177.57, and $358.58. During the month he wrote the following checks: $83.39, $12.66, $236.16, $15.68, $12.59, $33.68, $150.21, $37.73, $47.26, $41.29, $82.04, $93.05, and $15.74. Find his bank balance at the end of the month.

EXERCISES 404, SET II. Find the indicated differences.

1. 67.09 - 0.93   2. 817.64 - 92.57   3. 500 - 0.426

4. 800,073 - 60,905.2   5. 14.874 - 0.6559

6. Mr. Naslund's bank statement showed a balance of $164.38 at the beginning of the month. During the month he made the following deposits: $429.17, $126.53, and $295.46. During the month he wrote the following checks: $37.94, $86.26, $204.35, $17.49, $21.78, $55.16, $182.54, $41.83, $29.12, $7.03, $95.41, and $10.99. Find his bank balance at the end of the month.

The Number of Decimal Places in a Number

The number of decimal places in a number is the number of digits written to the right of the decimal point. A whole number has no decimal places.

Example 1

(a) 0.25   has two decimal places.

(b) 0.354   has three decimal places.

(c) 14.5   has one decimal place.

(d) 0.5000 has four decimal places.

(e) 167.   has no decimal places.

EXERCISES 405, SET I. How many decimal places are there in each of the following numbers?

1. 7.5   2. 18.04   3. 3.750   4. 256   5. 0.014

6. 1.1237   7. 0.444   8. 0.007   9. .007   10. 100.0006

EXERCISES 405, SET II. How many decimal places are there in each of the following numbers?

1. 6.125   2. 35.4   3. 8.09   4. 0.0256   5. 871

Multiplication of Decimals

To Multiply Decimals

1. Multiply the numbers as whole numbers.
2. Add the number of decimal places in the two numbers being multiplied.
3. Place the decimal point in your answer (product) so that your answer has as many decimal places as the sum found in (2).

Example 1. Multiply: 0.035 × 0.25

Solution:

$$
\begin{array}{rl}
0.25 & (\text{2 decimal places}) \\
0.035 & (\underline{\text{3}} \text{ decimal places}) \\
\hline
125 & 5 \\
75 & \\
\hline
0.00875 & (\text{5 decimal places})
\end{array}
$$

Example 2. Multiply: 0.276 × 358.4

Solution:

$$
\begin{array}{rl}
3\,58.4 & (\text{1 decimal place}) \\
0\,.276 & (\underline{\text{3}} \text{ decimal places}) \\
\hline
2\,1504 & 4 \\
25\,088 & \\
71\,68 & \\
\hline
98.9184 & (\text{4 decimal places})
\end{array}
$$

We use the following examples to show why the number of decimal places in the product is equal to the sum of the decimal places of the numbers being multiplied.

Example 3. Multiply: 0.2 × 0.003

Solution: $0.2 \times 0.003 = \dfrac{2}{10} \times \dfrac{3}{1{,}000} = \dfrac{6}{10{,}000} = 0.0006$

$$
\underbrace{0.2}_{\substack{1 \\ \text{decimal} \\ \text{place}}} \; + \; \underbrace{0.003}_{\substack{3 \\ \text{decimal} \\ \text{places}}} \qquad = \underbrace{0.0006}_{\substack{4 \\ \text{decimal} \\ \text{places}}}
$$

Example 4. Multiply: 3.24×12.8.

Solution: $3.\underset{\underset{\text{decimal places}}{\underset{2}{\downarrow}}}{24} \times 12.\underset{\underset{\text{decimal place}}{\underset{1}{\downarrow}}}{8} = \frac{324}{100} \times \frac{128}{10} = \frac{41,472}{1,000} = 41.\underset{\underset{\text{decimal places}}{\underset{3}{\downarrow}}}{772}$

$+$ $=$

EXERCISES 406, SET I. Since changing the order of the numbers being multiplied does not change the product, it is usually easier to use the number with fewer nonzero digits as the multiplier. We show this by working the first problem two different ways.

1. 4.6×3.749

First solution:

```
    3.7 4 9
        4.6
    2 2 4 9 4
  1 4 9 9 6
  1 7.2 4 5 4
```

Second solution:
```
          4.6
        3.7 4 9
        4 1 4
        1 8 4
        3 2 2
      1 3 8
      1 7.2 4 5 4
```

2. 2.96×3.8

3. 1.54×27.9

4. $37.8 \times 1,056$

5. 8.412×0.25

6. 1.35×95.67

7. 0.0056×386.45

8. 800.6×0.096

9. 0.0128×3.2

10. 0.1086×3.5

11. 8.7×5.607

12. 7.805×0.86

13. 0.002568×0.85

14. $7.86 \times 18,000$

Solution: We handle final zeros the same way we did in Section 116.

```
    7.8 6|
      1 8|0 0 0
    6 2 8 8|
    7 8 6  |
  1 4 1,4 8|0.0 0
```

15. $2.56 \times 93,000$

16. $230,000 \times 0.075$

17. Ann makes monthly payments of $17.63 for a stereo set. If she takes 18 months to pay for it, what is the total cost?

18. Irv makes $2.68 an hour.
    (a) What is his salary for a regular 40-hour week?
    (b) Irv receives time-and-a-half for each hour over 40 that he works in 1 week. How much does he make for working a 50-hour week?

In order to do Exercises 19 and 20, you need to know that

$$Pressure = \text{force on a unit of area}$$

$$Total~force = \text{pressure} \times \text{surface area}$$

Example 5. If the air pressure at sea level is 14.7 pounds per square inch, there is a force of 14.7 pounds on each square inch of surface area at sea level. Therefore, the total force on a surface area of 10 square inches is 147 pounds.

$$\text{Total force} = 14.7 \times 10 = 147 \text{ pounds}$$

19. Air pressure on each square inch of surface at sea level is 14.7 pounds. Find the total force of the air on a man's body at sea level. Assume the surface area of his body is 2,150 square inches.
20. In sea water the water pressure increases 0.444 pounds per square inch for each foot of depth. Remember, air pressure at sea level is 14.7 pounds per square inch.
   (a) Find the total pressure per square inch at a depth of 37 feet.

$$\text{Total pressure} = \text{air pressure} + \text{water pressure}.$$

   (b) Find the total force on a man's body in sea water at a depth of 37 feet. Assume the surface area of his body is 2,150 square inches.
   (c) Find the total force on a man's body (2,150 square inches) in sea water at a depth of 120 feet.
   (d) Find the force on an area of 1 square foot (144 square inches) at the greatest known depth (36,198 feet) of the Pacific Ocean.

EXERCISES 406, SET II. In Exercises 1-8, find the products.

1. 8.4 × 5.267
2. 46.9 × 1,058
3. 1.85 × 74.66
4. 700.8 × 0.069
5. 0.2049 × 7.5
6. 0.05287 × 0.63
7. 9.73 × 16,000
8. 360,000 × 0.045

9. Olga makes monthly payments of $37.45 for a quadriphonic hi-fi system. If she takes 24 months to pay for it, what is the total cost?
10. Rick makes $8.17 an hour.
   (a) What is his salary for a regular 40-hour week?
   (b) Rick gets time-and-a-half for each hour over 40 that he works in one week. How much would he make for working a 46-hour week?

Rounding off Decimals

In Section 122 we explained rounding off whole numbers and why it is done. The same need for rounding off decimals exists. For example, since the smallest coin we have is the 1¢ piece, any calculations done with numbers representing money are usually rounded off to the nearest cent. When figuring federal income taxes, you are permitted to round off your calculations to the nearest dollar to make it easier to figure and pay your taxes. Some measuring instruments are accurate to thousandths of an inch or tenths of a foot, etc. For this reason, we usually round off measurements to the accuracy of the instrument used.

IDENTIFYING THE PLACE WE ARE ROUNDING OFF TO

Example 1. When rounding off 75.17 to one decimal place (tenths), the place we are rounding off to is shown shaded.

tenths place

75.17

Example 2. When rounding off 2.4756 to two decimal places (hundredths), the place we are rounding off to is shown shaded.

hundredths place

2.4756

To Round Off a Decimal

CASE I. When the first digit to be dropped is less than 5.

unchanged

3.149
3.1

Therefore, 3.149 $\doteq$ 3.1 rounded off to tenths.

CASE II. When the first digit to be dropped is greater than 5.

increased by 1

56.427
56.43

Therefore, 56.427 $\doteq$ 56.43 rounded off to hundredths.

CASE III. When the first digit to be dropped is 5, and there are some nonzero digits to the right of 5.

increased by 1

0.126502
0.127

Therefore, 0.126502 $\doteq$ 0.127 rounded off to thousandths.

(continued on next page)

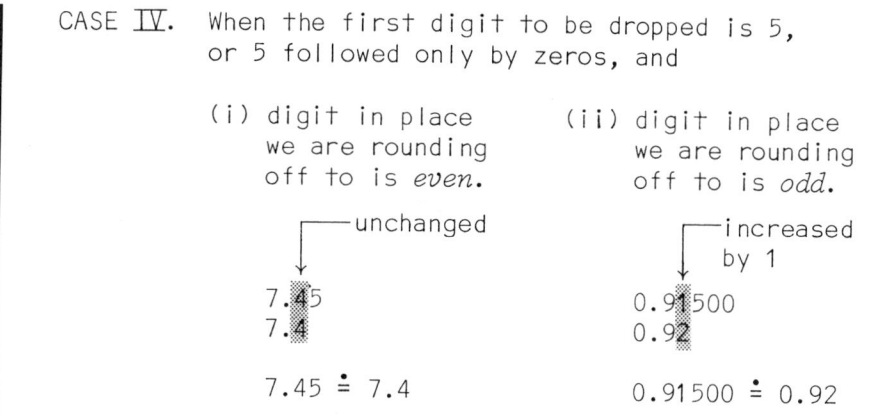

CASE IV. When the first digit to be dropped is 5, or 5 followed only by zeros, and

(i) digit in place we are rounding off to is *even*.

(ii) digit in place we are rounding off to is *odd*.

unchanged

7.45
7.4

7.45 ≐ 7.4

increased by 1

0.91500
0.92

0.91500 ≐ 0.92

Example 3. Examples of Case I.

(a) Round off 75.23 to one decimal place (tenths).

tenths

75.2③ ←—Since the first digit to be dropped is less
75.2   than 5, the part retained is unchanged.

Therefore, 75.23 ≐ 75.2, rounded off to one decimal place.

(b) Round off 186.4 to units.

units

186.④ ←—Since the first digit to be dropped is less
186   than 5, the part retained is unchanged.

Therefore, 186.4 ≐ 186, rounded off to units.

(c) 7.692 ≐ 7.69, rounded off to two decimal places (hundredths).

(d) 5,740.5 ≐ 5,700, rounded off to hundreds.

Example 4. Examples of Case II.

(a) Round off 75.17 to one decimal place (tenths).

tenths

75.1⑦ ←—Since the first digit to be dropped is
75.2   greater than 5, increase the preceding
             digit by one.

Therefore, 75.17 ≐ 75.2, rounded off to one decimal place.

(b) Round off 186.8 to units.

$$186.\overset{\text{units}}{\underset{\downarrow}{8}} \enspace \circled{8} \longleftarrow \text{Since the first digit to be dropped is}$$
187

186. ⑧ ←—Since the first digit to be dropped is
187   greater than 5, increase the preceding
          digit by one.

Therefore, 186.8 ≐ 187, rounded off to units.

(c) 7.698 ≐ 7.70, rounded off to two decimal places (hundredths).

(d) 5,180.5 ≐ 5,200, rounded off to hundreds.

Example 5. Examples of Case IV.

(a) Round off 23.45 to one decimal place (tenths).

23.**4** ⑤ ←—Since the first digit to be dropped is 5
23.**4**   and the digit in the place we are rounding
               off to is even, it remains unchanged.

Therefore, 23.45 ≐ 23.4, rounded off to one decimal place.

(b) Round off 23.35 to one decimal place (tenths).

23.**3** ⑤ ←—Since the first digit to be dropped is 5
23.**4**   and the digit in the place we are rounding
               off to is odd, increase it by one.

Therefore, 23.35 ≐ 23.4, rounded off to one decimal place.

(c) Round off 72.500 to units.

72. ⑤ 00 ←—Since the first digit to be dropped is 5
72.   followed only by zeros and the digit in
             the place we are rounding off to is even,
             it remains unchanged.

Therefore, 72.500 ≐ 72, rounded off to units.

Example 6. Round off 72.5001 to units. When the first digit to be dropped is 5 and there are nonzero digits to the right of the 5, increase the preceding digit by one. Therefore, 72.5001 ≐ 73, rounded off to units. (Case III)

A word of caution: Do not accumulate rounding offs. For example, to round off 1.7149 to two decimal places, do not round off 1.7149 to 1.715, then round off 1.715 to 1.72. Actually, 1.7149 ≐ 1.71, rounded off to two decimal places.

EXERCISES 407, SET I. In Exercises 1-10, round off each number to one decimal place.

1. 7.16   2. 3.24   3. 6.250   4. 3.150

5. 0.064   6. 0.051   7. 13.055   8. 5.049

9. 3.149   10. 18.009

In Exercises 11-20, round off each number to the nearest unit.

11. 7.5   12. 8.4   13. 9.5   14. 10.5

15. 10.51   16. 3.499   17. 0.67   18. 0.49

19. 5.09   20. 140.5

In Exercises 21-30, round off each number to the indicated place.

21. 1.236   2 decimal places

22. 0.045   2 decimal places

23. 0.035   2 decimal places

24. 1.37564   3 decimal places

25. 5.00716   thousandths

26. 0.05678   4 decimal places

27. 88.85   tens

28. 8.85   tens

29. 6.7445   thousandths

30. 0.5005   thousandths

EXERCISES 407, SET II. In Exercises 1-7, round off each number to one decimal place.

1. 8.27   2. 9.54   3. 7.650   4. 5.947

5. 14.007   6. 22.35   7. 6.2501

In Exercises 8-14, round off each number to the nearest unit.

8. 30.5   9. 41.38   10. 18.49   11. 16.501

12. 27.5   13. 39.7   14. 198.5

In Exercises 15-21, round off each number to the indicated place.

15. 5.728   2 decimal places

16. 10.047   1 decimal place

17. 0.32549   3 decimal places

18. 86.14   tenths

19. 86.14   tens

20. 7.106   hundredths

21. 2,560.0902   thousands

408 Division of Decimals

To Divide a Decimal by a Whole Number

1. Place a decimal point above the quotient line directly above the decimal point in the dividend.
2. Divide the numbers as whole numbers.

Example 1. Divide: 150.4 ÷ 47

Solution:

```
        3.2
  47 ) 150.4
       141
        9 4
        9 4
```

Example 2. Divide: 48.4 ÷ 83. Round off your answer to two decimal places.

Solution:

```
         .583 ≐ 0.58
  83 ) 48.400
       41 5
        6 90
        6 64
          260
          249
           11
```

EXERCISES 408A, SET I. In Exercises 1-10, the quotients are exact. Do not round off the quotients.

1. 86.96 ÷ 8   2. 249.2 ÷ 7   3. 93.6 ÷ 6

4. 673.2 ÷ 9   5. 52.8 ÷ 32   6. 51.66 ÷ 21

7. 6.825 ÷ 39   8. 28.13 ÷ 58   9. 472.5 ÷ 63

10. 311.1 ÷ 85

In Exercises 11-20, divide and round off the quotient to the indicated place.

11. 8.56 ÷ 7   2 decimal places

12. 456.7 ÷ 9   1 decimal place

13. 376.3 ÷ 8   3 decimal places

14. 514.7 ÷ 6   3 decimal places

15. 58.6 ÷ 42   2 decimal places

16. 75.4 ÷ 51   2 decimal places

17. 3.86 ÷ 76   3 decimal places

18. 5.77 ÷ 84   4 decimal places
19. 76.5 ÷ 208   2 decimal places
20. 90.6 ÷ 555   3 decimal places

EXERCISES 408A, SET II. In Exercises 1-5, the quotients are
exact. Do not round off the quotients.

1. 85.04 ÷ 8   2. 94.2 ÷ 6   3. 57.35 ÷ 31
4. 9.506 ÷ 49   5. 452.6 ÷ 73

In Exercises 6-10, divide and round off the quotient to the
indicated place.

6. 346.9 ÷ 9   2 decimal places
7. 91.4375 ÷ 7   3 decimal places
8. 2840.5 ÷ 46   1 decimal place
9. 7.6104 ÷ 84   3 decimal places
10. 178.12 ÷ 305   2 decimal places

To Divide a Decimal by a Decimal

1. Place a caret (∧) to the right of the last
   nonzero digit of the divisor.
2. Count the number of places between the decimal
   point and the caret in the divisor.
3. Place a caret in the dividend the same number
   of places to the right (or left) of its decimal point as in (2). Add zeros to the dividend
   when needed.
4. Place a decimal point in the quotient directly
   above the caret in the dividend.
5. Divide the numbers as whole numbers.

Example 3. Divide: 2.368 ÷ 0.32

Solution:

$$
\begin{array}{r}
7.4 \\
0.3\,2_\wedge \overline{)2.3\,6_\wedge 8} \\
2\,2\,4 \\
\hline
1\,2\,8 \\
1\,2\,8 \\
\end{array}
$$

Example 4. Divide: 166.4 ÷ 400. (A whole number can be considered a decimal).

Solution:

$$
\begin{array}{r}
.4\,1\,6 \\
4_\wedge 0\,0.\overline{)1_\wedge 6\,6.4}
\end{array}
$$

The divisor has become 4 instead of 400.

Example 5. Divide: 0.144 ÷ 1.20

Solution:

$$1.2_\wedge 0 \overline{\smash{\big)}\ 0.1_\wedge 4\ 4} \qquad .1\ 2$$

The divisor has become 12.

Examples 6 and 7 show that *we place the caret so that we make the divisor the smallest possible whole number.*

Example 6. Divide: 3.51 ÷ 0.065

Solution:

```
                    5 4.
0.0 6 5∧ ) 3.5 1 0     ←——This zero was added so that
           3 2 5          three places come between
           2 6 0          caret and decimal point
           2 6 0
```

Example 7. Divide: 0.2394 ÷ 5.7

Solution:

```
                        ——This zero was added to
              ↓           hold the place above the 3.
          .0 4 2
5.7∧ ) 0.2∧3 9 4
       2 2 8
       1 1 4
       1 1 4
```

Example 8. Divide: 197.2 ÷ 0.29

Solution:

```
                      ——This zero was added to
            ↓           hold the place above the 0
        6 8 0.          of the dividend.
0.2 9∧ ) 1 9 7.2 0∧
        1 7 4      ←——This zero was added so that
        2 3 2          two places come between
        2 3 2          caret and decimal point.
```

Example 9. Divide: 0.09267 ÷ 2.4. Round off your answer to three decimal places.

Solution:

```
        .0 3 8 6 ≐ 0.039
2.4∧ ) 0.0∧9 2 6 7
       7 2
       2 0 6
       1 9 2
         1 4 7
         1 4 4
             3
```

Explanation: In our study of fractions, we learned that the value of a fraction is not changed when the numerator and denominator are multiplied by the same number.

Example 10 $\qquad \dfrac{13}{8} = \dfrac{13 \times 10}{8 \times 10} = \dfrac{130}{80}$

Example 11. Since a fraction is equivalent to a division:

$$3.7\overline{)12.5} = 12.5 \div 3.7 = \frac{12.5 \times 10}{3.7 \times 10} = \frac{125}{37} = 37.\overline{)125.}$$

This shows that $3.7\overline{)12.5} = 37.\overline{)125.}$ Therefore, the quotient is unchanged when the decimal point in the divisor and dividend are both moved the same number of places to the right. Using the caret, we would write the last example:

$$3.7\overline{)12.5} = 3.7_\wedge\overline{)12.5_\wedge}$$

EXERCISES 408B, SET I. In Exercises 1-14, the quotients are exact. Do not round off the quotients.

1. 9.612 ÷ 2.7   2. 17.898 ÷ 3.8

3. 478.24 ÷ 6.1   4. 12.42 ÷ 9.2

5. 78.3 ÷ 300   6. 40.5 ÷ 500

7. 0.196 ÷ 1.40   8. 0.242 ÷ 1.10

9. 129.105 ÷ 0.45   10. 3.1304 ÷ 0.65

11. 0.020292 ÷ 0.0057   12. 0.0260484 ÷ 5.88

13. 1.04595 ÷ 0.367   14. 2.89896 ÷ 0.771

In Exercises 15-20, divide and round off the quotient to the indicated place.

15. 38.6 ÷ 170   3 decimal places

16. 7.66 ÷ 2.90   2 decimal places

17. 800 ÷ 35.7   1 decimal place

18. 756 ÷ 86.3   2 decimal places

19. 0.961 ÷ 7.55   3 decimal places

20. 0.0498 ÷ 4.95   5 decimal places

21. The balance owed on a car amounted to $2,467.14. What would the monthly payments be in order to pay it off in 36 months? Round off the payment to the nearest cent (two decimal places).

22. At the beginning of a trip, Raul's odometer reading was 65,479 miles. At the end of the trip the reading was 67,784 miles. He used 147 gallons of gasoline. How many miles did he get to the gallon of gasoline? Round off the answer to one decimal place.

23. If Rosie paid $16.27 a month on her department store charge account, how long would it take her to pay off a balance of $390.48?

24. If Al paid $13.92 a month on his finance company loan, how long would it take him to pay off a balance of $250.56?

EXERCISES 408B, SET II. In Exercises 1-7, the quotients are exact. Do not round off the quotients.

1. 15.036 ÷ 2.8   2. 785.33 ÷ 9.1

3. 224.8 ÷ 400   4. 0.435 ÷ 1.50

5. 245.1 ÷ 0.75   6. 0.041976 ÷ 0.0088

7. 6.18035 ÷ 0.661

In Exercises 8-10, divide and round off the quotient to the indicated place.

8. 100.656 ÷ 180   3 decimal places

9. 1600 ÷ 49.6   1 decimal place

10. 1.1023 ÷ 6.35   2 decimal places

11. The cost of a room-addition to Mr. Ota's home came to $10,573.86 after the interest had been added. What would the monthly payment be in order to pay it off in 5 years? Round off the payment to the nearest cent (2 decimal places).

Checking Operations with Decimals by Casting out Nines

All operations with decimals are carried out as we did with whole numbers—the only difference being how we handled the decimal point. Since all operations with whole numbers can be checked by casting out nines, we can use the same method for checking operations with decimals.

Example 1. Add:

$$
\begin{array}{rcl}
2.73 &\longrightarrow& 3 \\
15.2 &\longrightarrow& 8 \\
9.64 &\longrightarrow& 1 \\
\hline
27.57 && 12 \to \circled{3}
\end{array}
$$

Add 9-rems giving 12.

Check

9-rem for answer to problem

Note: This method checks the accuracy of the digits (see Section 107). It does not check the position of the decimal point.

Example 2. Multiply:

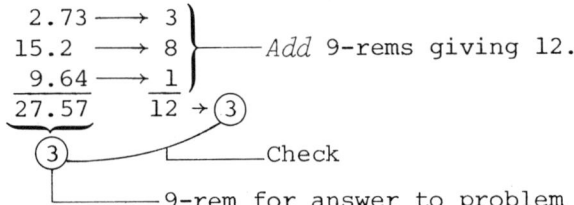

Multiply 9-rems giving 14.

Check

Division of decimals can only be checked if the quotient is not rounded off (see Example 3).

Example 3. Divide: 0.7960 ÷ 2.85. Carry your answer out to four decimal places. Do *not* round off.

Solution:

```
                          . 2  7  9  2
       2.8 5ʌ| 0.7  9ʌ6  0  0  0
                5  7  0
                2  2  6  0
                1  9  9  5
                   2  6  5  0
                   2  5  6  5
                         8  5  0
                         5  7  0
                         2  8  0
```

Check: Dividend = divisor × quotient + remainder

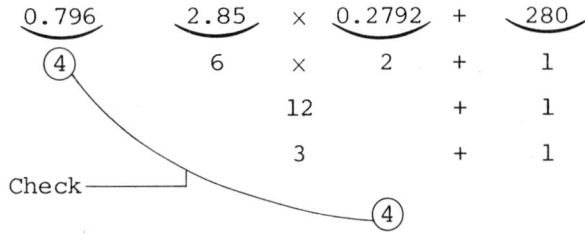

In order to have this check work, do not round off your quotient.

EXERCISES 409, SET I. Perform the indicated operations and check by casting out nines.

1. Add: 26.4 + 0.072 + 138. + 5.6

2. Add: 39. + 0.43 + 6.8 + 2.57

3. Multiply: 3.746 × 0.52

4. Multiply: 7.6 × 0.5463

5. Divide: 4.729 ÷ 5.8. Carry your answer out to three decimal places.

6. Divide: 3.051 ÷ 0.82. Carry your answer out to two decimal places.

EXERCISES 409, SET II. Perform the indicated operations and check by casting out nines.

1. Add: 58.3 + 0.069 + 147. + 9.2

2. Multiply: 4.857 × 0.62

3. Divide: 75.112 ÷ 9.2. Carry your answer out to two decimal places.

Multiplying and Dividing Decimals by Powers of Ten

Here are some powers of 10:

$$10^1 = 10, \quad 10^2 = 100, \quad 10^3 = 1,000$$

Refer to Section 120 for a more complete explanation of powers of 10.

To Multiply a Decimal by a Power of Ten

Move the decimal point to the *right* as many places as the number of zeros in the power of 10.

Example 1

(a) $24.7 \times 10 = 24 {_\wedge} 7. = 247$

1 zero → 1 place

(b) $24.7 \times 100 = 2,4 {_\wedge} 70. = 2,470$

2 zeros → 2 places

(c) $1.0567 \times 1,000 = 1 {_\wedge} 056.7 = 1,056.7$

3 zeros → 3 places

(d) $0.0973 \times 10^2 = 0 {_\wedge} 09.3 = 9.73$

exponent 2 → 2 places

(e) $34 \times 10^4 = 34 {_\wedge} 0000. = 340,000$

exponent 4 → 4 places

To Divide a Decimal by a Power of Ten

Move the decimal point to the *left* as many places as the number of zeros in the power of 10.

Example 2

(a) $395 \div 100 = 3.95 {_\wedge} = 3.95$

2 zeros ← 2 places

(b)
$$75.6 \div 10 = 7.5{\scriptstyle\wedge}6 = 7.56$$

1 zero ← 1 place

(c)
$$\frac{0.315}{100} = 0.00{\scriptstyle\wedge}315 = 0.00315$$

2 zeros ← 2 places

(d)
$$\frac{4,165.2}{10^3} = 4.165{\scriptstyle\wedge}2 = 4.1652$$

exponent 3 ← 3 places

(e)
$$75.6 \div 10^4 = 0.0075{\scriptstyle\wedge}6 = 0.00756$$

exponent 4 ← 4 places

EXERCISES 410, SET I. Perform the indicated operations.

1. $\dfrac{95.6}{10}$ 2. $7.98 \div 10$ 3. 27.8×100

4. 8.95×10 5. $\dfrac{750.2}{10^2}$ 6. $0.0614 \div 10$

7. 9.846×10^2 8. $0.0837 \times 1,000$ 9. $\dfrac{100}{10^3}$

10. $200 \cdot 10^3$ 11. $10^4 \times 27.4$ 12. $10,000 \times 0.457$

13. The $146.35 cost of a party was shared by 10 people. How much did each person have to pay? (Be sure to round your answer off to the nearest cent.)
14. A club charged $1.50 for each lottery ticket. If 10,000 tickets were sold, how much money did the club raise from this lottery?
15. 537 people attended a $100-a-plate fund-raising dinner given for a political candidate. How much campaign money did this dinner raise?
16. The $23,758 cost of putting in electric service along a rural road was shared equally by 100 home owners. How much did each owner have to pay?

EXERCISES 410, SET II. Perform the indicated operations.

1. $6.34 \div 10$ 2. 28.7×10 3. $\dfrac{905.3}{10^2}$

4. 0.0268×10^3 5. $300 \div 1,000$ 6. $10^4 \times 85.61$

7. The $167.49 cost of a week-end boat rental was shared by 10 people. How much did each person have to pay? (Be sure to round your answer off to the nearest cent.)

Some Shortcuts in Division

Example 1. $24\overline{)36}$

Solution:

$$24\overline{)36} = \frac{36}{24} = \frac{3}{2} = 2\overline{)3.0}^{1.5}$$

└─ At this point, reduce the fraction to lowest terms.

Example 2. $25\overline{)10.5}$

Solution:

$$25\overline{)10.5} = \frac{\overset{2.1}{\cancel{10.5}}}{\underset{5}{\cancel{25}}} = 5\overline{)2.10}^{.42}$$

─ Since canceling involves division, this decimal point must be directly over the other.

Now that you have seen that both numerator and denominator can be divided by the same number, we can arrange the work as follows:

Example 3.　$1.6\overline{)4.84}$

$= 16.\overline{)48.4}$ 　Both dividend and divisor were multiplied by 10.

$= 4\overline{)12.1}$ 　Both dividend and divisor were divided by 4. Then divide in the usual way.

Example 4. $1,300\overline{)564.3}$

$= 13\overline{)5.643}$ 　Both dividend and divisor were divided by 100.

Example 5.　$350\overline{)26.75}$

$= 35\overline{)2.675}$ 　Both divided by 10.

$= 7\overline{)0.535}$ 　Both divided by 5.

EXERCISES 411, SET I. Use shortcuts in doing the following divisions.

1. $18\overline{)27}$ 　　　　2. $33\overline{)714}$ 　　　　3. $25\overline{)2.45}$

4. $24\overline{)14.4}$ 　　　5. $1.2\overline{)4.26}$ 　　　6. $3.5\overline{)1.75}$

7. $30\overline{)71.1}$ 　　　8. $200\overline{)785.6}$ 　　9. $3,000\overline{)729.3}$

10. $500\overline{)8,500}$ 　11. $200\overline{)0.0248}$ 　12. $130\overline{)0.52}$

13. $4.44\overline{)88.80}$

14. $10.5\overline{)3.90}$ 　(Round off answer to two decimal places.)

15. $4.80\overline{)270.0}$ 　(Round off answer to nearest tenth.)

EXERCISES 411, SET II. Use shortcuts in doing the following divisions.

1. 15‾402

2. 32‾24.4

3. 4.5‾3.60

4. 20‾825.4

5. 300‾60,600

6. 120‾0.72

7. 3.60‾480.0 (Round off answer to nearest tenth.)

8. 10.5‾39.0 (Round off answer to two decimal places.)

Changing a Fraction to a Decimal

Since

$$\frac{a}{b} = a \div b$$

therefore,

> To Change a Fraction to a Decimal
>
> Divide the numerator by the denominator.

Sometimes a fraction can be changed to an exact decimal (Example 1).

Example 1. Change $\frac{3}{16}$ to a decimal.

Solution: $\frac{3}{16}$ = 3 ÷ 16 = 16‾3.0000 → .1875

```
     .1875
16 | 3.0000
     1 6
     1 40
     1 28
       120
       112
        80
        80
```

Therefore, $\frac{3}{16}$ = 0.1875

Not all fractions can be changed to exact decimals (Example 2). In this case, we must know the number of decimal places desired in the decimal representation of the fraction. The division of numerator by denominator must be carried out to at least one more decimal place than that required for the decimal representation of the fraction. We then round off the quotient to the required decimal place. (See Section 407 for rounding off.)

Example 2. Change $\frac{3}{7}$ to a decimal fraction. Round off to four decimal places.

Solution:
$$
\begin{array}{r}
.42857 \\
7\overline{)3.00000} \\
\underline{2\ 8} \\
20 \\
\underline{14} \\
60 \\
\underline{56} \\
40 \\
\underline{35} \\
50 \\
\underline{49} \\
1
\end{array}
$$

Therefore, $\frac{3}{7} \doteq 0.4286$

Example 3. Change $3\frac{5}{6}$ to a decimal. Round off to two decimal places.

First solution: Change $\frac{5}{6}$ to a decimal, then add its value to 3.

$$
\begin{array}{r}
.833 \\
6\overline{)5.000} \\
\underline{4\ 8} \\
20 \\
\underline{18} \\
20 \\
\underline{18} \\
2
\end{array}
$$

Therefore, $\frac{5}{6} \doteq 0.83$

Then $3\frac{5}{6} \doteq 3 + 0.83 = 3.83$

Second solution: Change $3\frac{5}{6}$ to the improper fraction $\frac{23}{6}$, then change $\frac{23}{6}$ to a decimal.

$$
3\frac{5}{6} = \frac{23}{6} =
\begin{array}{r}
3.833 \\
6\overline{)23.000} \\
\underline{18} \\
5\ 0 \\
\underline{4\ 8} \\
20 \\
\underline{18} \\
20 \\
\underline{18} \\
2
\end{array}
$$

Therefore, $3\frac{5}{6} \doteq 3.83$

EXERCISES 412, SET I. In Exercises 1-10, change the given fractions and mixed numbers to exact decimals.

1. $\frac{3}{4}$ 2. $\frac{5}{8}$ 3. $2\frac{1}{2}$ 4. $5\frac{3}{4}$ 5. $\frac{3}{8}$

6. $4\frac{3}{5}$ 7. $\frac{5}{16}$ 8. $\frac{5}{4}$ 9. $\frac{9}{32}$ 10. $4\frac{1}{40}$

11. Change $\frac{7}{11}$ to a decimal rounded off to two decimal places.

12. Change $4\frac{5}{7}$ to a decimal rounded off to three decimal places.

13. Change $\frac{2}{9}$ to a decimal rounded off to two decimal places.

14. Change $2\frac{7}{12}$ to a decimal rounded off to three decimal places.

15. Change $7\frac{7}{8}$ to a decimal rounded off to two decimal places.

16. Change $5\frac{5}{32}$ to a decimal rounded off to three decimal places.

17. A drill press operator uses a $\frac{5}{16}$-inch drill to put a hole through a steel bracket. Would a 0.315-inch diameter pin fit in the hole he drilled?

18. What is $\frac{3}{7}$ of a dollar, rounded off to the nearest cent?

EXERCISES 412, SET II. In Exercises 1-5, change the given fractions and mixed numbers to exact decimals.

1. $\frac{7}{8}$ 2. $6\frac{3}{5}$ 3. $\frac{9}{16}$ 4. $9\frac{3}{4}$ 5. $3\frac{5}{32}$

6. Change $\frac{5}{9}$ to a decimal rounded off to two decimal places.

7. Change $4\frac{5}{12}$ to a decimal rounded off to three decimal places.

8. Change $1\frac{3}{7}$ to a decimal rounded off to three decimal places.

Changing a Decimal to a Fraction

> To Change a Decimal to a Fraction
>
> 1. *Read* the decimal, then *write* it as shown in Section 402.
> 2. Reduce the fraction to lowest terms.

Example 1. Change 0.4 to a fraction in lowest terms.
         ↑————————tenths place

Read the decimal "four tenths."

Write $\frac{4}{10} = \frac{2}{5}$ reduced to lowest terms

Example 2. Change 0.25 to a fraction in lowest terms.
         ↑————————hundredths place

Read the decimal "twenty-five hundredths."

Write $\dfrac{25}{100} = \dfrac{1}{4}$ reduced to lowest terms

Example 3. Change 17.4 to a mixed number in lowest terms.

$\uparrow$
tenths place

Read the decimal "one hundred seventy-four tenths."

Write $\dfrac{174}{10} = \dfrac{87}{5} = 17\dfrac{2}{5}$

Example 4. Change 5.241 to a mixed number. We show *an easy way to write a decimal as a fraction:*

Remove decimal point.

$\downarrow$

5.241
$|\,|\,|\longrightarrow \dfrac{5241}{1,000} = 5\dfrac{241}{1,000}$

Write 1 $\longrightarrow$ 1,000
to the left
of the zeros.

Under each digit to the right of the decimal point, write a zero.

Example 5. Change 0.00785 to a fraction.

0.0 0 7 8 5
$|\,|\,|\,|\,|\longrightarrow \dfrac{785}{100,000} = \dfrac{157}{20,000}$
1 0 0,0 0 0

EXERCISES 413A, SET I. In Exercises 1-18, change the decimals to fractions or mixed numbers reduced to lowest terms.

1. 0.6	2. 0.8	3. 0.05	4. 0.65
5. 0.075	6. 0.750	7. 0.875	8. 1.8
9. 2.5	10. 3.7	11. 5.9	12. 4.3
13. 0.0625	14. 2.125	15. 37.5	16. 2.1875
17. 65.625	18. 0.000875		

19. A machinist must drill a 0.875-inch hole through a metal brace. What fractional size drill should he use?
20. A carpenter must drill a 0.625-inch hole through a rafter. What fractional size drill should he use?

EXERCISES 413A, SET II. Change the decimals to fractions or mixed numbers reduced to lowest terms.

1. 0.7	2. 0.55	3. 0.025	4. 2.4
5. 4.6	6. 2.625	7. 3.4375	8. 0.00625

COMPLEX DECIMALS. Decimals such as $0.33\dfrac{1}{3}$ and $0.67\dfrac{1}{2}$ are called *complex decimals.**

———————

*These are complex decimal *fractions*, but we continue to shorten the name "decimal fraction" to "decimal." (See Section 401.)

Example 6. Change $0.12\frac{1}{2}$ to a fraction in lowest terms.

$$0.1\underbrace{\left(2\frac{1}{2}\right)}_{1\ 0\ 0} \xrightarrow{\text{2nd decimal place}} \frac{12\frac{1}{2}}{100} = \frac{\frac{25}{2}}{100} = \frac{25}{2} \div \frac{100}{1} = \frac{25}{2} \cdot \frac{1}{\underset{4}{100}} = \frac{1}{8}$$

Example 7. Change $0.66\frac{2}{3}$ to a fraction in lowest terms.

$$0.6\underbrace{\left(6\frac{2}{3}\right)}_{1\ 0\ 0} \xrightarrow{\text{2nd decimal place}} \frac{66\frac{2}{3}}{100} = \frac{\frac{200}{3}}{100} = \frac{200}{3} \div \frac{100}{1} = \frac{200}{3} \cdot \frac{1}{\underset{1}{100}} = \frac{2}{3}$$

Example 8. Change $2.16\frac{2}{3}$ to a fraction in lowest terms.

$$2.16\frac{2}{3} = 2 + .16\frac{2}{3}$$

$$0.1\ 6\frac{2}{3} \xrightarrow{} \frac{16\frac{2}{3}}{100} = \frac{\frac{50}{3}}{100} = \frac{50}{3} \div \frac{100}{1} = \frac{50}{3} \cdot \frac{1}{\underset{2}{100}} = \frac{1}{6}$$

Therefore, $2.16\frac{2}{3} = 2 + \frac{1}{6} = 2\frac{1}{6}$.

EXERCISES 413B, SET I. Change the complex decimals to fractions or mixed numbers reduced to lowest terms.

1. $0.37\frac{1}{2}$ 2. $0.62\frac{1}{2}$ 3. $0.33\frac{1}{3}$ 4. $0.1\frac{1}{4}$

5. $0.5\frac{3}{4}$ 6. $2.062\frac{1}{2}$ 7. $1.0\frac{1}{5}$ 8. $0.0\frac{2}{3}$

9. $0.00\frac{5}{12}$ 10. $2.00\frac{1}{4}$ 11. $0.001\frac{1}{6}$ 12. $1.05\frac{3}{8}$

EXERCISES 413B, SET II. Change the complex decimals to fractions or mixed numbers reduced to lowest terms.

1. $0.87\frac{1}{2}$ 2. $0.6\frac{1}{4}$ 3. $3.03\frac{3}{4}$ 4. $0.0\frac{1}{3}$

5. $2.00\frac{1}{2}$ 6. $1.09\frac{5}{8}$

 **Multiplying Decimals by Fractions and Mixed Numbers**

Example 1. Multiply: $\frac{2}{3} \times 3.84$

$$\underset{1}{\frac{2}{\cancel{3}}} \times \frac{\overset{1.28}{\cancel{3.84}}}{1} = 2 \times 1.28 = 2.56$$

$$\begin{array}{r} 1.28 \\ \times\ \ 2 \\ \hline 2.56 \end{array}$$

Example 2. Multiply: $\frac{3}{14} \times 6.52$

$$\underset{7}{\frac{3}{\cancel{14}}} \times \frac{\overset{3.26}{\cancel{6.52}}}{1} = \frac{3 \times 3.26}{7} = \frac{9.78}{7} = 1.39\frac{5}{7} \doteq 1.40$$

$$\begin{array}{r} 1.39\frac{5}{7} \\ 7\overline{)9.78} \end{array}$$

Example 3. Multiply: $2\frac{3}{5} \times 2.55$

First solution: Change $2\frac{3}{5}$ to $\frac{13}{5}$, then multiply.

$$\underset{1}{\frac{13}{\cancel{5}}} \times \frac{\overset{0.51}{\cancel{2.55}}}{1} = 13 \times 0.51 = 6.63$$

$$\begin{array}{r} .51 \\ \times\ 13 \\ \hline 153 \\ 51\ \ \\ \hline 6.63 \end{array}$$

Second solution: Change $2\frac{3}{5}$ to a decimal, then multiply.

$$5\overline{)3.0}^{\ .6}$$

$$2\frac{3}{5} = 2 + \frac{3}{5} = 2 + .6 = 2.6$$

$$\begin{array}{r} 2.55 \\ \times\ 2.6 \\ \hline 15\ 30 \\ 51\ 0\ \ \\ \hline 6.630 \end{array}$$

Therefore, $2\frac{3}{5} \times 2.55 = 2.6 \times 2.55 = 6.63$

Example 4. Find the cost of $2\frac{3}{4}$ yards of silk at $3.59 a yard.

Solution: First change $2\frac{3}{4}$ to a decimal, then multiply it by $3.59.

$$2\frac{3}{4} = \frac{11}{4} = 2.75$$

$$\begin{array}{r} 3.59 \\ \times\ 2.75 \\ \hline 17\ 95 \\ 251\ 3\ \ \\ 718\ \ \ \ \\ \hline 9.8725 \end{array} \doteq \$9.87,\ \text{rounded off to the nearest cent.}$$

EXERCISES 414, SET I. In Exercises 1-8, perform the indicated operations, then express your answer as a decimal.

1. $\frac{3}{4} \times 5.24$ 2. $\frac{5}{6} \times 23.4$ 3. $2\frac{1}{2} \times 7.4$ 4. $4\frac{1}{3} \times 1.83$

5. $2\frac{3}{8} \times 13.6$ 6. $1\frac{7}{8} \times 2.84$ 7. $\frac{7}{10} \times 56.5$ 8. $\frac{3}{10} \times 62.4$

In Exercises 9-12, perform the indicated operations, then express your answer as a decimal rounded off to the indicated place.

9. $\frac{7}{12} \times 56.4$ 2 decimal places

10. $\frac{5}{6} \times 83.8$ 3 decimal places

11. $1\frac{9}{14} \times 7.72$ 3 decimal places

12. $2\frac{5}{22} \times 6.54$ 3 decimal places

13. Carmen makes $2.45 an hour. How much does she earn in a $7\frac{1}{2}$ - hour day?

14. Debbie needs $2\frac{1}{3}$ yards of material to make a dress. If she buys a fabric costing $3.95 a yard, what is the total cost of the material?

EXERCISES 414, SET II. In Exercises 1-4, perform the indicated operations, then express your answer as a decimal.

1. $\frac{5}{6} \times 28.8$ 2. $4\frac{2}{3} \times 1.71$ 3. $1\frac{5}{8} \times 4.28$ 4. $\frac{9}{10} \times 78.3$

In Exercises 5 and 6, perform the indicated operations, then express your answer as a decimal rounded off to the indicated place.

5. $\frac{5}{12} \times 105.2$ 2 decimal places

6. $1\frac{7}{15} \times 2.61$ 3 decimal places

7. Jaime makes $7.80 an hour when he works on Saturday. How much does he earn for $5\frac{1}{2}$ hours of work at that rate?

Comparing the Size of Decimals

Sometimes we need to recognize which of a group of decimals is largest. *We can compare the sizes of decimals by writing them all with the same number of decimal places.*

Example 1. Arrange the following decimals in order of size—the largest first: .27, .205, .2, .250.

Solution: Since the largest number of decimal places in any of the given decimals is three, we add final zeros wherever necessary to change all the given decimals to three decimal places.

.270, .205, .200, .250

Now arrange the 3-decimal place numbers in order of size—the largest first.

.270, .250, .205, .200

Therefore, the *original* decimals arranged according to size are:

0.27, 0.250, 0.205, 0.2

Example 2. Arrange the following decimals in order of size—the largest first: 0.06, 2.1, 1.20, 2.

Solution: 0.06, 2.10, 1.20, 2.00 { all numbers converted to 2-decimal places.

2.10, 2.00, 1.20, 0.06 { 2-decimal-place numbers arranged according to size.

2.1, 2, 1.20, 0.06 { original decimals arranged according to size.

EXERCISES 415, SET I. In Exercises 1-8, arrange the decimal numbers in order of size—the largest first.

1. 0.409	0.49	0.41		
2. 0.35	0.3	0.305		
3. 3.075	3.1	3.05	3.009	
4. 7.0	7.1	7.08	7.099	
5. 0.075	0.07501	0.0749	0.7	
6. 0.06	0.1998	0.6	0.059	
7. 5.05	5.5	5.0501	5	5.0496
8. 3.0505	3.051	3.0695	3.199	3

EXERCISES 415, SET II. In Exercises 1-4, arrange the decimals in order of size—the largest first.

1. 0.735	0.7	0.74	
2. 5.2	5.09	5.199	
3. 0.08	0.8	0.0796	0.095
4. 6.0505	6	6.055	6.1009

Chapter Summary

A _decimal_ is a fraction whose denominator is a power of 10. (Sec. 401)

To read a decimal, ignore the decimal point and read the number as a whole number. Then say the name of the place occupied by the right digit of the given decimal. (Sec. 402)

The number of decimal places in a number is the number of digits written to the right of the decimal point. (Sec. 405)

To add decimals, write the numbers under one another with their decimal points in the same vertical line. Then add the numbers like whole numbers. Place the decimal point in your answer in the same vertical line as the other decimal points. (Sec. 403)

To subtract decimals, write the number being subtracted under the number it is being subtracted from, with their decimal points in the same vertical line. Subtract the numbers as whole numbers. Place the decimal point in the answer in the same vertical line as the other decimal points. (Sec. 404)

To multiply decimals, multiply the numbers as whole numbers. Your answer has as many decimal places as the sum of the decimal places in the numbers being multiplied. (Sec. 406)

To divide decimals:

1. Place a caret (∧) in the divisor to make it the smallest possible whole number.
2. Place a caret in the dividend the same number of places to the right (or left) of its decimal point as was done in the divisor.
3. Place the decimal point in the quotient directly above the caret in the dividend.
4. Divide the numbers as whole numbers.

To multiply a decimal by a power of 10, move the decimal point to the _right_ as many places as the number of zeros in the power of 10. (Sec. 410)

To divide a decimal by a power of 10, move the decimal point to the _left_ as many places as the number of zeros in the power of 10. (Sec. 410)

To change a fraction to a decimal, divide the numerator by the denominator. (Sec. 412)

REVIEW EXERCISES 416, SET I.

1. Add: 75.23 + 186.56 + 7,896,448 + 8.007 + 386.759 + 0.0058

2. Add: 247.3 + 0.0856 + 4,582 + 10.907 + 6.84

3. Subtract: 509.6 - 81.34   4. Subtract: 47.93 - 28.5

In Exercises 5-26, perform the indicated operations.

5. 100×7.78 6. 10×6.5

7. $789 \div 100$ 8. $4.7 \div 10$

9. $1,000 \times 75$ 10. $10^2 \times 4.25$

11. $0.064 \div (10)^2$ 12. $4.25 \div 100$

13. $10^4 \times 0.0056$ 14. $10^3 \times 0.0075$

15. $786,000 \div 10,000$ 16. $85,600 \div 1,000$

17. $2 \times 5 \times 78$ 18. $8 \times 125 \times 45$

19. $5 \times 85.4 \times 2$ 20. $5 \times 1.67 \times 2$

21. $4 \times 9.6 \times 25$ 22. $2 \times 5 \times 10 \times 3.5$

23. $600\overline{)865.2}$ 24. $400\overline{)81.36}$

25. $260\overline{)67.6}$ 26. $340\overline{)173.4}$

In Exercises 27-36, change the given fractions and mixed numbers to exact decimals.

27. $\frac{7}{8}$ 28. $\frac{5}{16}$ 29. $2\frac{3}{4}$ 30. $5\frac{3}{8}$ 31. $\frac{17}{5}$ 32. $\frac{31}{20}$

33. Change $\frac{5}{6}$ to a decimal rounded off to two decimal places.

34. Change $\frac{7}{12}$ to a decimal rounded off to two decimal places.

35. Change $3\frac{2}{3}$ to a decimal rounded off to three decimal places.

36. Change $4\frac{9}{16}$ to a decimal rounded off to three decimal places.

In Exercises 37-42, change the decimals to fractions or mixed numbers reduced to lowest terms.

37. 0.68 38. 0.54 39. 3.85

40. 5.26 41. $8.7\frac{1}{2}$ 42. $0.06\frac{1}{4}$

In Exercises 43 and 44 arrange the decimals in order of size— the largest first.

43. 0.1075 0.09 0.11 0.2

44. 3 3.05 2.99 3.1 3.0501

In Exercises 45-52, perform the indicated operations, then round off your answer to the indicated place.

45. 70.9×94.78 1 decimal place

46. 40.8×68.59 1 decimal place

47. $6.007 \div 7.25$ 2 decimal places

48. $8.009 \div 4.67$ 2 decimal places

49. Add: 56.75 + 186.3 + 8.388   2 decimal places

50. Add: 4.864 + 795.3 + 25.81   2 decimal places

51. Subtract: 387.61 - 246.593   2 decimal places

52. Subtract: 94.073 - 76.48   2 decimal places

In Exercises 53-56, find the value of each expression, being careful to use the correct order of operations.

53. $100 \cdot 5 \div 4 + 7.25$ 54. $100 \div 5 \cdot 5 + 7.25$

55. $3(2^3) + 2.3 \times 5.7 - 7.55$ 56. $5 \cdot \sqrt{16} + 10 \times 5.6 - 3.75$

In Exercises 57-60, perform the indicated operations, then express your answer as a decimal.

57. $\frac{2}{3} \times 23.1$ 58. $\frac{3}{5} \times 2.95$ 59. $2\frac{1}{4} \times 3.48$ 60. $1\frac{3}{8} \times 21.6$

In Exercises 61 and 62, perform the indicated operations, then express your answer as a decimal rounded off to the indicated place.

61. $\frac{11}{12} \times 8.74$ 1 decimal place

62. $3\frac{1}{6} \times 7.41$ 1 decimal place

63. Carlos bought one pair of shoes for $19.95, two neckties for $3.95 each, three pairs of socks for $1.25 a pair, and one suit for $89.95. What was his total bill?

64. At the beginning of the month, Jim's bank balance was $275.38. During the month he wrote the following checks: $15.98, $46.75, $87.45, $135.46, $21.98, $174.89, $68, and $57.76. He made deposits of $250 and $350. Find his bank balance at the end of the month.

65. Find the cost of $4\frac{2}{3}$ yards of material at $3.49 a yard. Round off your answer to the nearest cent.

66. Nora made 18 equal monthly payments on her new stereo set. If the total cost of the set was $355, what was her monthly payment? (Round off to the nearest cent.)

67. The total pressure on 1 square inch at a depth of 50 feet in the sea is 36.9 pounds. Find the total force on a man's body at that depth if his body has an area of 2,345 square inches.

68. Rudy drove his car 9,600 miles last year. His total car expenses were $625 for the year. Find the average cost per mile. Round off your answer to the nearest tenth of a cent.

69. A car travels 196 miles. If it gets 14 miles per gallon of gas, how many gallons did it use for the trip? If gas is 42.9¢ per gallon, how much was spent for gas?

70. <u>Brain Teaser</u>. A tie and a pin cost $1.10. The tie costs $1 more than the pin. What is the cost of each?

71. <u>Brain Teaser</u>. Would you give a $10 reward for 1924 pennies?

72. <u>Brain Teaser</u>. If eggs cost 26¢ per dozen, how many can be

purchased for a cent and a quarter?

<u>REVIEW EXERCISES 416, SET II</u>. In Exercises 1-12, perform the
indicated operations.

1. Add: 59.43 + 278.46 + 9,725 + 6.008 + 176.953 + 0.0084

2. Subtract: 6005.9 - 70.83

3. 100×8.54

4. $96.5 \div 10$

5. 32.7×10^2

6. $5.62 \div 100$

7. $10^3 \times 0.0029$

8. $\dfrac{824,000}{10^4}$

9. $2 \times 6.39 \times 5$

10. $25 \times 39.2 \times 4$

11. $700\overline{)1496.6}$

12. $520\overline{)202.8}$

In Exercises 13-15, change the given fractions and mixed numbers to exact decimals.

13. $\dfrac{11}{16}$

14. $3\dfrac{5}{8}$

15. $\dfrac{27}{20}$

16. Change $\dfrac{5}{12}$ to a decimal rounded off to 2 decimal places.

17. Change $2\dfrac{15}{16}$ to a decimal rounded off to 3 decimal places.

In Exercises 18-20, change the decimals to fractions or mixed
numbers reduced to lowest terms.

18. 0.86

19. 4.65

20. $0.08\dfrac{3}{4}$

In Exercises 21-24, perform the indicated operations, then
round off your answer to the indicated place.

21. 60.9×87.46 1 decimal place

22. 7.008×5.79 2 decimal places

23. $85.63 + 294.5 + 7.689$ 2 decimal places

24. $708.91 - 563.692$ 2 decimal places

In Exercises 25 and 26, find the value of each expression,
being careful to use the correct order of operations.

25. $1000 \div 20 \cdot 5 + 9.74$ 26. $(10)^2 \cdot \sqrt{36} + 100 \times 4.3 - 973.8$

In Exercises 27 and 28, perform the indicated operations; then
express your answer as a decimal.

27. $\dfrac{4}{5} \times 6.85$

28. $2\dfrac{5}{8} \times 37.6$

29. Perform the indicated operations, then express your answer
as a decimal rounded off to two decimal places: $\dfrac{7}{12} \times 5.36$

30. Marion bought 2 pairs of shoes at $22.99 a pair, 5 pairs of hose at $2.98 a pair, and a pantsuit for $74.95. What was her total bill?

31. At the beginning of the month, Morey's bank balance was $174.29. During the month he made deposits of $323.74 and $186.55. During the month he wrote checks for $59.83, $112.09, $23.75, $84.37, $25, $4.99, $98.75, and $46.52. Find his bank balance at the end of the month.

32. Marilyn drove her car 21,000 miles last year. Her total car expenses for the year were $1,640. Find the average cost per mile of driving her car during the year. Round off your answer to the nearest tenth of a cent.

33. Find the cost of $24\frac{2}{3}$ feet of copper pipe selling for 98 cents a foot. Round off your answer to the nearest cent.

Chapter 4: Diagnostic Test

Name_____

_____

The purpose of this test is to see how well you know decimals. We recommend that you work this diagnostic test *before* your instructor tests you on this chapter. Allow yourself about an hour to do this test.

Complete solutions for all the problems on this test, together with section references, are given in the Answer Section. You should study the sections referred to for the problems you do incorrectly.

1. Write the following numbers in decimal notation:

   (a) One hundred twenty-five thousandths.

   (1a)_____

   (b) Eight thousand, fifty, and seven hundredths.

   (1b)_____

   (c) Sixteen and nine hundred seventy-three ten-thousandths.

   (1c)_____

2. Write the following numbers in words:

   (a) 0.67   (2a)_____

   _____

   (b) 81.012   (2b)_____

   _____

3. Add: 7.8 + 56 + 0.017 + 500.94

   (3)_____

4. Subtract: 40.6 - 8.54

   (4)_____

5. Subtract: 9.073 - 0.87

   (5)_____

6. Multiply: 3.75×0.058

(6) _____

7. Divide: $27.51 \div 3.5$

(7) _____

8. Round off each of the following numbers to the indicated place:

   (a) 2.817   (2 decimal places)   (8a) _____

   (b) 54.749 (tenths)   (8b) _____

   (c) 0.0465 (3 decimal places)   (8c) _____

9. Divide, and round off to one decimal place:

   $4.38 \div 0.56$

(9) _____

10. Change $\dfrac{5}{6}$ to a decimal rounded off to two decimal places.

(10) _____

11. Change $5\dfrac{3}{16}$ to an exact decimal.

(11) _____

12. Change 0.78 to a fraction reduced to lowest terms.

(12) _____

13. Perform the indicated operations:

   (a) 100×5.816 (13a) _____

   (b) $764.1 \div 10$ (13b) _____

   (c) 3.9×10^3 (13c) _____

   (d) $\dfrac{41.8}{10^2}$ (13d) _____

14. Perform the indicated operations, then round off your answer to the indicated place.

(a) $\frac{2}{3} \times 67.2$ (This gives an exact decimal.)

(14a)_____

(b) $2\frac{3}{5} \times 8.41$ (1 decimal place)

(14b)_____

(c) $5(2^3) - 2.4 \div 0.8$ (units)

(14c)_____

15. The total pressure on 1 square inch at a depth of 35 feet in the sea is 30.2 pounds. Find the total force on a man's body at that depth if his body has an area of 2,325 square inches.

(15)_____

16. The balance owed on a car amounts to $2,870, including interest. What would the monthly payment be in order to pay it off in 36 months? Round off the payment to the nearest cent (two decimal places).

(16)_____

17. At the beginning of the month, Jeff's bank balance was $346.52. During the month he wrote the following checks: $17.75, $64.57, $91.35, $135.46, and $186.40. He made a deposit of $325. Find his bank balance at the end of the month.

(17)_____

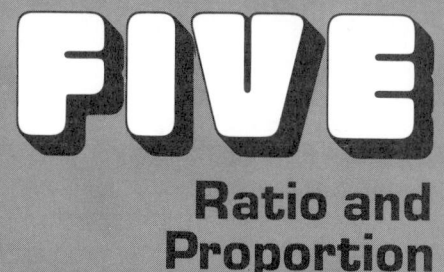

FIVE

Ratio and
Proportion

Meaning of Ratio

A *ratio* is another name for fraction. In fact, fractions in algebra are called *ratio*nal numbers.

$$\frac{a}{b}$$ is called "the ratio of a to b."

a and b are called the *terms* of the ratio. In older books, the ratio a to b is sometimes written $a:b$, but this hides the fact that a ratio is a fraction. We will emphasize the fraction meaning of ratio.

The Three Meanings of $\frac{a}{b}$. We have now given three different meanings to $\frac{3}{4}$:

1. 3 of the 4 equal parts a unit has been divided into. (Fraction meaning)
2. $3 \div 4$. (Division meaning)
3. The ratio of 3 to 4. (Ratio meaning)

The meaning chosen depends upon how it is used. The ratio meaning is used when *comparing* two numbers.

Example 1. In an arithmetic class there are 13 men and 17 women.

  (a) The ratio of men to women is $\frac{13}{17}$.

  (b) The ratio of women to men is $\frac{17}{13}$.

Example 2. A ball player makes 11 hits for every 30 times at bat.

The ratio of hits to times at bat is $\frac{11}{30}$.

Example 3. A man earns $35 for every 8 hours he works.

The ratio of dollars to hours is $\frac{35}{8}$.

EXERCISES 501, SET I. Express the following as ratios.

1. 40 Frenchmen to 27 Englishmen
2. 117 radios to 38 families
3. 53 students to 307 nonstudents
4. 3 wins to 11 losses
5. 240 miles to 13 gallons
6. A college football team won 7 out of 9 games played. There were no tie games.
   (a) What is the ratio of wins to games played?
   (b) What is the ratio of wins to losses?
   (c) What is the ratio of losses to wins?

EXERCISES 501, SET II. Express the following as ratios.

1. 5 wins to 7 losses
2. 127 sheep to 211 cattle
3. 43 women to 47 men
4. $1,049 to $283
5. A college basketball team won 23 out of 30 games played. There were no tie games.
   (a) What is the ratio of wins to games played?
   (b) What is the ratio of wins to losses?
   (c) What is the ratio of losses to wins?

Reducing a Ratio to Lowest Terms

Since a ratio is a fraction, it can be reduced to lowest terms by either of the two methods used with fractions.

First Method. *Repeated canceling*. Reduce the ratio of 48 to 30 to lowest terms.

$$\frac{48}{30} = \frac{\overset{8}{\cancel{\overset{24}{\cancel{48}}}}}{\underset{5}{\cancel{\underset{15}{\cancel{30}}}}} = \frac{8}{5}$$

Second Method. Divide the numerator and denominator by the largest number that divides both of them (found by *Euclid's Algorithm*). Reduce the ratio of 273 to 351 to lowest terms.

$$273\overline{\smash)351}$$
$$\underline{273} \quad 3$$
$$78\overline{\smash)273}$$
$$\underline{234} \quad 2$$
$$39\overline{\smash)78}$$

largest number that
divides 273 and 351 ──────→ $\underline{78}$
$$0$$

$$39\overline{\smash)273} \qquad 39\overline{\smash)351}$$
$$\underline{273} \qquad\quad \underline{351}$$

Therefore, $\dfrac{273}{351} = \dfrac{273 \div 39}{351 \div 39} = \dfrac{7}{9}$

Example 1. Reduce the ratio of 18 to 12 to lowest terms.

$$\frac{\overset{3}{\cancel{18}}}{\underset{2}{\cancel{12}}} = \frac{3}{2}$$

Example 2. Reduce the ratio of 155 to 217 to lowest terms.

(continued on next page)

$$155\overline{)217}$$ ...
$$\begin{array}{r} 1 \\ 155\overline{)217} \\ 155 \\ \hline 62\overline{)155} \quad 2 \\ 124 \quad 2 \\ \hline 31\overline{)62} \\ 62 \\ \hline 0 \end{array}$$

$$31\overline{)155} \quad \begin{array}{r} 5 \\ 155 \\ \hline \end{array} \qquad 31\overline{)217} \quad \begin{array}{r} 7 \\ 217 \\ \hline \end{array}$$

largest number that
divides 155 and 217 ——→

Therefore, $\dfrac{155}{217} = \dfrac{155 \div 31}{217 \div 31} = \dfrac{5}{7}$

A ratio can sometimes be reduced by expressing both numerator and denominator in terms of a common unit. In many cases this reduced form of the ratio is more meaningful.

Example 3. Reduce the ratio of 6 inches to 1 foot to lowest terms.

$$\frac{6 \text{ inches}}{1 \text{ foot}} = \frac{6 \text{ inches}}{12 \text{ inches}} = \frac{\overset{1}{\cancel{6}}}{\underset{2}{\cancel{12}}} = \frac{1}{2}$$

————— because 1 foot = 12 inches

Example 4. Reduce the ratio of 5 cents tax per dollar to lowest terms.

$$\frac{5 \text{ cents}}{1 \text{ dollar}} = \frac{5 \text{ cents}}{100 \text{ cents}} = \frac{\overset{1}{\cancel{5}}}{\underset{20}{\cancel{100}}} = \frac{1}{20}$$

————— because 1 dollar = 100 cents

Example 5. Al works 16 hours every 3 days. Reduce this ratio.

$$\frac{16 \text{ hours}}{3 \text{ days}} = \frac{16 \text{ hours}}{3 \cdot 24 \text{ hours}} = \frac{\overset{2}{\cancel{16}}}{\underset{9}{\cancel{72}}} = \frac{2}{9}$$

————— because 1 day = 24 hours

EXERCISES 502, SET I. Reduce the following ratios to lowest terms.

1. 28 to 14	2. 116 to 48	3. 52 to 39
4. 135 to 60	5. 119 to 153	6. 69 to 92

7. 85 cents to 15 cans
8. 20 quarts of ice cream to 44 children
9. 42 yards to 12 dresses
10. 15 radios to 3 students
11. 6 hours to 40 minutes
12. 40 inches to 3 feet
13. 88 cents to 2 dollars
14. 45 cents to 3 dollars
15. Write the ratio that shows the comparison of 1 hour to 10 minutes.
16. Write the ratio that shows the comparison of 1 week to 1 day.
17. Write the ratio that shows the comparison of a 24-hour day to 2 hours.

18. Write the ratio that shows the comparison of 5 minutes to 1 hour.
19. Compare 3 inches to 1 foot by a ratio.
20. Compare 9 inches to 1 foot by a ratio.
21. Find the ratio that shows the comparison of 50¢ to 5¢.
22. One school bus carries 52 passengers while a smaller bus carries 32 passengers. Write a ratio that compares the number of passengers in the larger bus to the number of passengers in the smaller bus.
23. Mr. Lee weighs 175 pounds, and his daughter weighs 25 pounds. Compare Mr. Lee's weight to his daughter's weight by a ratio.
24. Car A (an economy car) averages 25 miles to the gallon of gasoline. A heavier car, B, averages 15 miles to the gallon of gasoline. Compare the mileage of Car A to Car B by a ratio.
25. Olympic sprinters run about 22 miles per hour. The African cheetah has been clocked up to 70 miles per hour. Use a ratio to compare the speed of an Olympic sprinter to that of a cheetah.

EXERCISES 502, SET II. Reduce the following ratios to lowest terms.

1. 36 to 12
2. 48 to 32
3. 90 cents to 20 cans
4. 150 dollars to 25 people
5. 5 hours to 40 minutes
6. 60 inches to 2 feet
7. 70 cents to 2 dollars
8. Write the ratio that shows the comparison of 1 hour to 12 minutes.
9. Write the ratio that shows the comparison of 1 week to 7 hours.
10. Compare 8 inches to 1 foot by a ratio.
11. Mr. Wong's age is 35 years, and his daughter is 5 years old. Compare Mr. Wong's age to his daughter's age by a ratio.
12. Compare 2 yards to 9 inches by a ratio.

Ratios Whose Terms Are Not Whole Numbers

In all the ratios studied so far, the terms have been whole numbers. This is not always the case. The terms of a ratio can be any kind of number, the only restriction being that the denominator cannot be 0.

Example 1. Find the ratio of $1\frac{3}{8}$ to 11.

$$\frac{1\frac{3}{8}}{11} = 1\frac{3}{8} \div 11 = \frac{11}{8} \div \frac{11}{1} = \frac{\cancel{11}}{8} \cdot \frac{1}{\cancel{11}} = \frac{1}{8}$$

Example 2. Find the ratio of $1\frac{3}{4}$ to $3\frac{1}{2}$.

$$\frac{1\frac{3}{4}}{3\frac{1}{2}} = 1\frac{3}{4} \div 3\frac{1}{2} = \frac{7}{4} \div \frac{7}{2} = \frac{\cancel{7}}{\cancel{4}_2} \cdot \frac{\cancel{2}^1}{\cancel{7}} = \frac{1}{2}$$

Example 3. Find the ratio of $\frac{3}{5}$ to $1\frac{4}{5}$.

$$\frac{\frac{3}{5}}{1\frac{4}{5}} = \frac{3}{5} \div 1\frac{4}{5} = \frac{3}{5} \div \frac{9}{5} = \frac{\cancel{3}^1}{\cancel{5}} \cdot \frac{\cancel{5}}{\cancel{9}_3} = \frac{1}{3}$$

Example 4. Find the ratio of the volume of a 15-cubic-foot refrigerator to one having a volume of 3.25 cubic feet.

$$\frac{15}{3.25} = \frac{1500}{325} = \frac{\overset{60}{\overset{\cancel{300}}{\cancel{1500}}}}{\underset{13}{\underset{\cancel{65}}{\cancel{325}}}} = \frac{60}{13}$$

This ratio has more meaning if we write it with a denominator of 1.

$$\frac{60}{13} = 60 \div 13 \stackrel{\bullet}{=} 4.6 = \frac{4.6}{1}$$

This means that the large refrigerator holds approximately 4.6 times as much as the small one.

Example 5. A 100-foot flag pole casts a 40-foot shadow. Use a ratio to compare the height of the pole to the length of its shadow.

$$\frac{\text{height of pole}}{\text{length of shadow}} = \frac{100 \text{ feet}}{40 \text{ feet}} = \frac{\cancel{100}^5}{\cancel{40}_2} = \frac{5}{2}$$

Because $\frac{5}{2} = 2\frac{1}{2} = \frac{2\frac{1}{2}}{1}$, the pole is $2\frac{1}{2}$ times as long as its shadow.

EXERCISES 503, SET I. In Exercises 1-12, find the ratio of the first number to the second number.

1. $\frac{1}{2}$ to 2

2. $\frac{1}{4}$ to 8

3. 4 to $\frac{3}{8}$

4. 3 to $\frac{7}{9}$

5. $\frac{3}{4}$ to $\frac{5}{8}$

6. $\frac{2}{3}$ to $\frac{4}{15}$

7. $2\frac{1}{2}$ to 5

8. $3\frac{1}{3}$ to 10

9. 6 to $1\frac{3}{5}$

10. 7 to $2\frac{4}{5}$ 11. $2\frac{2}{3}$ to $1\frac{1}{15}$ 12. $2\frac{1}{2}$ to $3\frac{1}{3}$

13. A 6-foot man casts a $4\frac{1}{2}$-foot shadow.

   (a) Compare his height to the length of his shadow.
   (b) The man's height is how many times the length of his shadow?

14. A building that is 26 feet high casts a $6\frac{1}{2}$-foot shadow. Find the ratio of its height to its shadow.

15. The gasoline tank on Mel's sport car holds $9\frac{3}{4}$ gallons. Tom's compact car has a tank holding $16\frac{1}{2}$ gallons. Using a ratio, compare the size of the sport car's tank to that of the compact.

16. The small refrigerator in the Novak family's camper has a volume of $4\frac{1}{2}$ cubic feet. Their kitchen refrigerator has a volume of 14.4 cubic feet.

   (a) Compare the size of the two refrigerators by a ratio.
   (b) The volume of the larger refrigerator is how many times that of the smaller one?

EXERCISES 503, SET II. In Exercises 1-6, find the ratio of the first number to the second number.

1. $\frac{1}{3}$ to 2   2. $\frac{1}{5}$ to 4   3. $\frac{3}{4}$ to $\frac{9}{8}$

4. $2\frac{1}{5}$ to 22   5. 6 to $2\frac{2}{5}$ 6. $3\frac{1}{4}$ to $2\frac{7}{16}$

7. A $5\frac{1}{2}$-foot woman casts a $2\frac{1}{2}$-foot shadow.

   (a) Compare her height to the length of her shadow.
   (b) The woman's height is how many times the length of her shadow?

8. The area of North America is approximately 9,400,000 square miles and the area of South America is approximately 6,800,000 square miles.

   (a) Find the ratio of the area of North America to the area of South America.
   (b) The area of North America is how many times the area of South America?

Meaning of a Proportion

A *proportion* is a statement that two ratios are equal.

Common notation for a proportion:

$$\frac{a}{b} = \frac{c}{d}$$ Read: "a is to b as c is to d."
           or: "a over b equals c over d."

Another notation for a proportion:

$$a : b :: c : d \quad \text{Read: "}a \text{ is to } b \text{ as } c \text{ is to } d.\text{"}$$

The *terms* of a proportion:

1st term   3rd term

$$\frac{a}{b} = \frac{c}{d}$$

2nd term   4th term

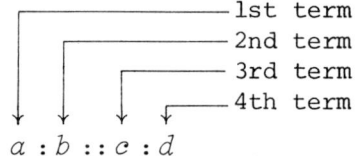

The *means and extremes* of a proportion:

means

$$\frac{a}{b} = \frac{c}{d}$$

extremes

extremes

$$a : b :: c : d$$

means

The means of this proportion are "b" and "c."

The extremes of this proportion are "a" and "d."

Example 1. In the proportion $\dfrac{2}{7} = \dfrac{6}{21}$,

    read: "2 is to 7 as 6 is to 21,"

    the first term is 2,   the second term is 7,
    the third term is 6,   the fourth term is 21,
    the means are 7 and 6, the extremes are 2 and 21.

Example 2. In the proportion $\dfrac{x}{3} = \dfrac{5}{15}$,

    read: "x is to 3 as 5 is to 15,"

    the first term is x,   the second term is 3,
    the third term is 5,   the fourth term is 15,
    the means are 3 and 5, the extremes are x and 15.

EXERCISES 504, SET I

1. In the proportion $\dfrac{12}{5} = \dfrac{24}{10}$, find the following:
   (a) the first term   (b) the second term
   (c) the third term   (d) the fourth term
   (e) the means   (f) the extremes

2. In the proportion $\dfrac{8}{14} = \dfrac{16}{x}$, find the following:
   (a) the first term   (b) the second term
   (c) the third term   (d) the fourth term
   (e) the means   (f) the extremes

1. In the proportion $\frac{2}{3} = \frac{4}{6}$, find the following:

   (a) the first term   (b) the second term
   (c) the third term   (d) the fourth term
   (e) the means   (f) the extremes

Product of Means Equals Product of Extremes

In any proportion, *the product of the means equals the product of the extremes.*

In the proportion $\frac{2}{3} = \frac{4}{6}$,

$$3 \times 4 \qquad = \qquad 2 \times 6$$

product of means = product of extremes

Again, in the proportion $\frac{3}{5} = \frac{6}{10}$,

$$5 \times 6 \qquad = \qquad 3 \times 10$$

product of means = product of extremes

In the proportion

$$\frac{a}{b} = \frac{c}{d}$$

$$b \cdot c \qquad = \qquad a \cdot d$$

product of means = product of extremes

This is sometimes called the *cross-multiplication rule.* This was the method shown in Section 307 for checking to see if two fractions are equivalent. In this section, cross-multiplication makes it possible to see if two ratios are equal.

If the product of the means equals the product of the extremes, then the ratios form a proportion. If the product of the means does not equal the product of the extremes, then the ratios do not form a proportion.

<u>Example 1.</u> Do the given ratios form a proportion?

(a) $\frac{5}{12}$, $\frac{3}{7}$. No, because $5 \cdot 7 \neq 12 \cdot 3$.   This is not a
$35 \neq 36$ proportion.

(b) $\frac{16}{6}$, $\frac{8}{3}$. Yes, because $16 \cdot 3 = 6 \cdot 8$.   This is a
$48 = 48$ proportion.

(c) $\frac{15}{6}$, $\frac{45}{18}$. Yes, because $15 \cdot 18 = 6 \cdot 45$. This is a
$270 = 270$ proportion.

(d) $\frac{6}{31}$, $\frac{5}{26}$. No, because $6 \cdot 26 \neq 31 \cdot 5$.   This is not a
$156 \neq 155$ proportion.

<u>EXERCISES 505, SET I</u>. Do the given ratios form a proportion?

1. $\frac{3}{5}$, $\frac{6}{10}$

2. $\frac{6}{4}$, $\frac{3}{2}$

3. $\frac{2}{3}$, $\frac{5}{7}$

4. $\frac{3}{7}$, $\frac{4}{9}$

5. $\frac{6}{9}$, $\frac{4}{6}$

6. $\frac{12}{9}$, $\frac{4}{3}$

7. $\frac{21}{17}$, $\frac{19}{15}$

8. $\frac{16}{35}$, $\frac{17}{37}$

9. $\frac{36}{39}$, $\frac{24}{26}$

10. $\frac{25}{15}$, $\frac{32}{19}$

11. $\frac{12}{18}$, $\frac{28}{42}$

12. $\frac{40}{100}$, $\frac{22}{55}$

13. $\frac{114}{162}$, $\frac{95}{130}$

14. $\frac{253}{891}$, $\frac{161}{567}$

<u>EXERCISES 505, SET II</u>. Do the given ratios form a proportion?

1. $\frac{3}{4}$, $\frac{9}{12}$

2. $\frac{5}{7}$, $\frac{15}{20}$

3. $\frac{6}{8}$, $\frac{9}{12}$

4. $\frac{10}{15}$, $\frac{12}{18}$

5. $\frac{220}{330}$, $\frac{2}{3}$

6. $\frac{117}{121}$, $\frac{128}{133}$

7. $\frac{207}{311}$, $\frac{142}{213}$

Solving a Proportion for an Unknown Term

When three of the four terms of a proportion are known, it is always possible to find the value of the unknown term.

<u>Example 1</u>. In the proportion $\frac{2}{3} = \frac{x}{51}$, the third term is unknown and is represented by the letter x. Since the product of the means equals the product of the extremes,

$$\frac{2}{3} = \frac{x}{51}$$

$$3 \cdot x = 2 \cdot 51$$

$$3 \cdot x = 102$$

The "=" sign tells us that $3 \cdot x$ and 102 are the same number. Therefore, if $3 \cdot x$ and 102 are both divided by 3, the resulting numbers will be equal. That is,

$$\frac{\overset{1}{\cancel{3}} \cdot x}{\underset{1}{\cancel{3}}} = \frac{\overset{34}{\cancel{102}}}{\underset{1}{\cancel{3}}}$$

$$x = 34$$

Check: In the original proportion, $\frac{2}{3} = \frac{x}{51}$, replace x by 34. Then,

$$\frac{2}{3} = \frac{34}{51}$$

$$3 \cdot 34 = 2 \cdot 51$$

$$102 = 102$$

Method of Solving a Proportion for an Unknown Letter

1. Set the product of the means equal to the product of the extremes.
2. Divide each product by the number that the unknown letter is multiplied by.

To Check Your Solution

1. Replace the unknown letter in the proportion by the value you obtained for it.
2. Cross-multiply, and verify that the cross-products are equal.

Example 2. Solve $\frac{x}{25} = \frac{18}{15}$ for x.

Solution:

$$\frac{x}{25} = \frac{18}{15}$$

$$15 \cdot x = 18 \cdot 25 \qquad \text{(product of extremes =}$$
$$15 \cdot x = 450 \qquad\qquad \text{product of means)}$$

$$\frac{\overset{1}{\cancel{15}} \cdot x}{\underset{1}{\cancel{15}}} = \frac{\overset{30}{\cancel{450}}}{\underset{1}{\cancel{15}}} \qquad \text{(dividing both products by the number that } x \text{ is multiplied by)}$$

$$x = 30$$

Check:

$$\frac{30}{25} = \frac{18}{15} \qquad (x \text{ was replaced by 30})$$

$$30 \cdot 15 = 25 \cdot 18$$

$$450 = 450$$

Often the work in solving a proportion can be simplified by reducing a ratio to its lowest terms. We use this procedure to simplify the solution of Example 2.

$$\frac{x}{25} = \frac{18}{15}$$

$$\frac{x}{25} = \frac{6}{5} \qquad \text{(since } \frac{18}{15} = \frac{6}{5}, \text{ when reduced to}$$
$$\text{lowest terms)}$$

$$5 \cdot x = 25 \cdot 6$$

$$\frac{\cancel{5} \cdot x}{\cancel{5}} = \frac{\overset{5}{\cancel{25}} \cdot 6}{\cancel{5}}$$

$$x = 30$$

Example 3. Solve $\frac{8}{14} = \frac{16}{x}$ for x.

$$\left(\begin{array}{c} \text{reducing} \\ \frac{8}{14} \text{ to } \frac{4}{7} \end{array} \right) \longrightarrow \quad \frac{4}{7} = \frac{16}{x}$$

$$4 \cdot x = 7 \cdot 16$$

$$\frac{\cancel{4} \cdot x}{\cancel{4}} = \frac{7 \cdot \overset{4}{\cancel{16}}}{\cancel{4}}$$

$$x = 28$$

Check: $\frac{8}{14} = \frac{16}{28}$ ←in place of x

$$\frac{4}{7} = \frac{4}{7}$$

obtained by reducing both ratios to lowest terms

Example 4. Solve $\frac{20}{30} = \frac{x}{150}$ for x.

$$\left(\begin{array}{c} \text{reducing} \\ \frac{20}{30} \text{ to } \frac{2}{3} \end{array} \right) \longrightarrow \quad \frac{2}{3} = \frac{x}{150}$$

$$2 \cdot 150 = 3 \cdot x$$

$$\frac{2 \cdot \overset{50}{\cancel{150}}}{\cancel{3}} = \frac{\cancel{3} \cdot x}{\cancel{3}}$$

$$100 = x$$

Check: $\frac{20}{30} = \frac{100}{150}$ ← in place of x

$$\frac{2}{3} = \frac{2}{3}$$

by reducing both ratios to lowest terms

Example 5. Solve $\frac{49}{x} = \frac{42}{18}$ for x.

$$\left(\begin{array}{c} \text{reducing} \\ \frac{42}{18} \text{ to } \frac{7}{3} \end{array} \right) \longrightarrow \quad \frac{49}{x} = \frac{7}{3}$$

$$49 \cdot 3 = 7 \cdot x$$

$$\frac{\overset{7}{\cancel{49}} \cdot 3}{\cancel{7}} = \frac{\cancel{7} \cdot x}{\cancel{7}}$$

$$21 = x$$

Check: $\frac{49}{21} = \frac{42}{18}$

in place of x

$$\frac{7}{3} = \frac{7}{3}$$

by reducing both ratios to lowest terms

1. $\frac{x}{4} = \frac{2}{3}$ 2. $\frac{x}{5} = \frac{6}{4}$ 3. $\frac{8}{x} = \frac{4}{5}$

4. $\frac{10}{x} = \frac{15}{4}$ 5. $\frac{4}{7} = \frac{x}{21}$ 6. $\frac{15}{12} = \frac{x}{9}$

7. $\frac{4}{13} = \frac{16}{x}$ 8. $\frac{28}{18} = \frac{14}{x}$ 9. $\frac{x}{18} = \frac{24}{30}$

10. $\frac{26}{x} = \frac{39}{14}$ 11. $\frac{15}{22} = \frac{x}{33}$ 12. $\frac{x}{21} = \frac{80}{42}$

13. $\frac{55}{x} = \frac{35}{28}$ 14. $\frac{44}{77} = \frac{x}{14}$ 15. $\frac{100}{x} = \frac{40}{30}$

16. $\frac{144}{36} = \frac{96}{x}$ 17. $\frac{x}{100} = \frac{75}{125}$ 18. $\frac{24}{98} = \frac{x}{147}$

19. $\frac{39}{x} = \frac{104}{48}$ 20. $\frac{60}{84} = \frac{45}{x}$

EXERCISES 506, SET II. Solve for x in each of the following
proportions.

1. $\frac{x}{3} = \frac{6}{9}$ 2. $\frac{5}{x} = \frac{7}{3}$ 3. $\frac{5}{4} = \frac{x}{6}$

4. $\frac{5}{3} = \frac{2}{x}$ 5. $\frac{4}{x} = \frac{7}{11}$ 6. $\frac{x}{6} = \frac{8}{15}$

7. $\frac{35}{25} = \frac{x}{8}$ 8. $\frac{18}{x} = \frac{12}{9}$ 9. $\frac{x}{27} = \frac{54}{36}$

10. $\frac{111}{x} = \frac{69}{46}$

In solving these proportions for the unknown letter, you have
been using algebra. We will learn more about algebra in Chapter 11.

Proportions Whose Terms Are Not Whole Numbers

In all the ratios studied so far, the terms have been whole
numbers. This is not always the case. The terms of a proportion can be any kind of number, the only restriction being that
the denominator in either ratio cannot be 0.

Letters other than x are often used to represent the unknown
term.

Example 1. Solve for P.

$$\frac{P}{3} = \frac{\frac{5}{6}}{5}$$

$$5 \cdot P = \frac{\cancel{3}^{1}}{1} \cdot \frac{5}{\cancel{6}_{2}}$$

$$5 \cdot P = \frac{5}{2}$$

$$\frac{\cancel{5} \cdot P}{\cancel{5}} = \frac{\frac{5}{2}}{5}$$

$$P = \frac{5}{2} \div 5 = \frac{\cancel{5}^{1}}{2} \cdot \frac{1}{\cancel{5}_{1}} = \frac{1}{2}$$

Example 2. Solve for x.   Example 3. Solve for B.

$$\frac{3\frac{1}{2}}{5\frac{1}{4}} = \frac{x}{4} \qquad\qquad \frac{0.24}{2.7} = \frac{4}{B}$$

$$5\frac{1}{4} \cdot x = 3\frac{1}{2} \cdot 4 \qquad\qquad \frac{\overset{4}{\cancel{24}}}{\underset{45}{\cancel{270}}} = \frac{4}{B}$$

$$\frac{21}{4} \cdot x = \frac{7}{\cancel{2}_{1}} \cdot \frac{\cancel{4}^{2}}{1} = 14 \qquad\qquad \frac{4}{45} = \frac{4}{B}$$

$$\frac{\frac{\cancel{21}}{4} \cdot x}{\frac{\cancel{21}}{4}} = \frac{14}{\frac{21}{4}} \qquad\qquad 45 \cdot 4 = 4 \cdot B$$

$$\qquad\qquad\qquad\qquad \frac{45 \cdot \cancel{4}}{\cancel{4}} = \frac{\cancel{4} \cdot B}{\cancel{4}}$$

$$x = 14 \div \frac{21}{4} \qquad\qquad\qquad 45 = B$$

$$x = \frac{\cancel{14}^{2}}{1} \cdot \frac{4}{\cancel{21}_{3}} = \frac{8}{3} = 2\frac{2}{3}$$

EXERCISES 507, SET I. Solve for the letter in each of the following proportions.

1. $\dfrac{\frac{3}{4}}{6} = \dfrac{P}{16}$ 2. $\dfrac{\frac{2}{5}}{4} = \dfrac{P}{25}$ 3. $\dfrac{A}{9} = \dfrac{3\frac{1}{3}}{5}$

4. $\dfrac{A}{8} = \dfrac{2\frac{1}{4}}{18}$ 5. $\dfrac{7.7}{B} = \dfrac{3.5}{5}$ 6. $\dfrac{6.8}{B} = \dfrac{17}{57.4}$

7. $\dfrac{P}{100} = \dfrac{\frac{3}{2}}{15}$ 8. $\dfrac{P}{100} = \dfrac{\frac{7}{5}}{35}$ 9. $\dfrac{12\frac{1}{2}}{100} = \dfrac{A}{48}$

10. $\dfrac{16\frac{2}{3}}{100} = \dfrac{9}{B}$

11. $\dfrac{2.54}{1} = \dfrac{X}{7.5}$

12. $\dfrac{12.5}{W} = \dfrac{7.8}{16}$

(Round off to two decimal places)

EXERCISES 507, SET II. Solve for the letter in each of the following proportions.

1. $\dfrac{\frac{3}{5}}{6} = \dfrac{P}{25}$

2. $\dfrac{A}{12} = \dfrac{2\frac{1}{2}}{5}$

3. $\dfrac{16.2}{B} = \dfrac{3}{5.5}$

4. $\dfrac{P}{100} = \dfrac{4.8}{1.5}$

5. $\dfrac{12\frac{1}{2}}{100} = \dfrac{A}{96}$

6. $\dfrac{57.4}{39.6} = \dfrac{7.4}{B}$

(Round off to one decimal place)

Solving Word Problems Using Proportions

> **Method for Solving Word Problems Using Proportions**
>
> 1. Represent the unknown quantity by x.
> 2. Be sure to put the units next to the numbers when writing the proportion.
> 3. Be sure the same units occupy corresponding positions in the two ratios of the proportion.
>
> | _Correct Arrangements_ | _Incorrect Arrangements_ |
> |---|---|
> | $\dfrac{\text{miles}}{\text{hours}} = \dfrac{\text{miles}}{\text{hours}}$ | $\dfrac{\text{dollars}}{\text{weeks}} = \dfrac{\text{weeks}}{\text{dollars}}$ |
> | $\dfrac{\text{hours}}{\text{miles}} = \dfrac{\text{hours}}{\text{miles}}$ | $\dfrac{\text{dollars}}{\text{weeks}} = \dfrac{\text{dollars}}{\text{days}}$ |
> | $\dfrac{\text{miles}}{\text{miles}} = \dfrac{\text{hours}}{\text{hours}}$ | |
>
> 4. Once the numbers have been correctly entered in the proportion by using the units as a guide, drop the units when cross-multiplying to solve for the unknown.

Example 1. A man used 10 gallons of gas on a 180-mile trip. How many gallons of gas can he expect to use on a 300-mile trip?

Solution: Think: "gallons is to miles" as "gallons is to miles."

$$\frac{x \text{ gallons}}{300 \text{ miles}} = \frac{10 \text{ gallons}}{180 \text{ miles}}$$

Note: the ratios used on each side have gallons in the numerator and miles in the denominator.

Therefore,

$$180x = 300 \cdot 10$$

$$x = \frac{\overset{50}{\cancel{300}} \cdot \cancel{10}}{\underset{3}{\cancel{180}}} = \frac{50}{3} = 16\frac{2}{3} \text{ gallons}$$

Example 2. A market is selling four cans of beets for 53 cents. How much will 12 cans cost at the same rate?

Solution:

$$\frac{x \text{ cents}}{12 \text{ cans}} = \frac{53 \text{ cents}}{4 \text{ cans}}$$

$$4 \cdot x = 12(53)$$

$$x = \frac{\overset{3}{\cancel{12}}(53)}{\cancel{4}} = 159 \text{ cents} = \$1.59$$

Example 3. A baseball team wins seven of its first 12 games. How many would you expect it to win out of its first 36 games if the team continues to play with the same degree of success?

Solution:

$$\frac{7 \text{ wins}}{12 \text{ games}} = \frac{x \text{ wins}}{36 \text{ games}}$$

$$7 \cdot 36 = 12 \cdot x$$

$$\frac{7 \cdot \overset{3}{\cancel{36}}}{\cancel{12}} = \frac{\cancel{12} \cdot x}{\cancel{12}}$$

$$21 = x$$

Therefore, the team can expect to win 21 out of its first 36 games.

Example 4. There are 25 men in a college class containing 38 students. Assuming this is typical of all classes, how many of the college's 7,500 students would be men?

Solution:

$$\frac{25 \text{ men}}{38 \text{ students}} = \frac{x \text{ men}}{7,500 \text{ students}}$$

$$38 \cdot x = 25(7,500)$$

$$\frac{\cancel{38} \cdot x}{\cancel{38}} = \frac{25(\overset{3,750}{\cancel{7,500}})}{\underset{19}{\cancel{38}}}$$

$$x = \frac{25(3,750)}{19} = \frac{93,750}{19} \doteq 4,934 \text{ men}$$

Example 5. At a soda fountain 8 quarts of ice cream were used to make 100 milk shakes. How many quarts are needed to make 550 milk shakes?

Solution:

$$\frac{8 \text{ quarts}}{100 \text{ milk shakes}} = \frac{x \text{ quarts}}{550 \text{ milk shakes}}$$

$$100 \cdot x = 8(550)$$

$$\frac{\cancel{100} \cdot x}{\cancel{100}} = \frac{\overset{4}{\cancel{8}}(\overset{11}{\cancel{550}})}{\underset{2}{\cancel{100}}}$$

$$x = 4(11) = 44 \text{ quarts}$$

Example 6. If a 6-foot man casts a $4\frac{1}{2}$-foot shadow, how tall is a tree that casts a 30-foot shadow?

Solution:

$$\frac{6\text{-foot tall}}{4\frac{1}{2}\text{-foot shadow}} = \frac{x\text{-foot tall}}{30\text{-foot shadow}}$$

$$\frac{6}{4\frac{1}{2}} = \frac{x}{30}$$

$$4\frac{1}{2} \cdot x = 6 \cdot 30$$

$$\frac{9}{2} \cdot x = 180$$

$$\frac{\cancel{\frac{9}{2}} \cdot x}{\cancel{\frac{9}{2}}} = \frac{180}{\frac{9}{2}}$$

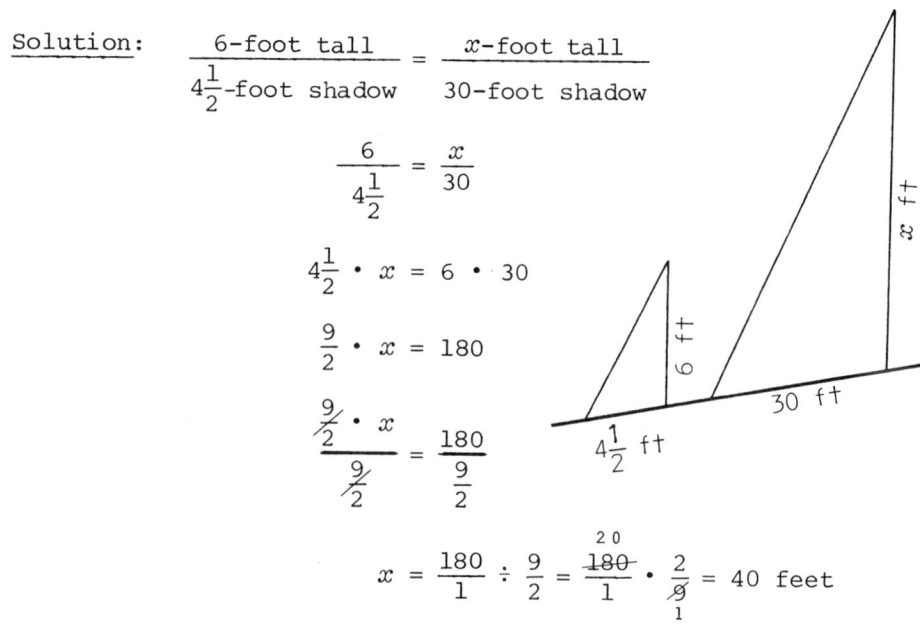

$$x = \frac{180}{1} \div \frac{9}{2} = \frac{\overset{20}{\cancel{180}}}{1} \cdot \frac{2}{\underset{1}{\cancel{9}}} = 40 \text{ feet}$$

EXERCISES 508, SET I

1. A painter uses about 3 gallons of paint in doing 2 rooms. How many gallons would he need to paint 20 rooms?

2. A man drives 600 miles in $1\frac{1}{2}$ days. How long would it take him to drive 3,000 miles?

3. A 6-foot man has a 4-foot shadow when a tree casts a 20-foot shadow. How tall is the tree?

4. Seven men finish 10 houses in a month. How many houses could 35 men finish in the same time?

5. The scale in an architectural drawing is 1 inch equals 8 feet. What are the dimensions of a room that measures $2\frac{1}{2}$ by 3 inches on the drawing?

6. The property tax on a $15,000 home is $450. What would be the tax on a $25,000 home?

7. An investment of $3,000 earned $180 for a year. How much would have to be invested to earn $540 in the same time?

8. A store has a bargain price of 85¢ for three jars of grape jelly. How many jars could someone buy for $5.95?

9. A man's car burns $2\frac{1}{2}$ quarts of oil on a 1,800-mile trip.

How many quarts of oil can he expect to use on a 12,000-mile trip?

10. Fifteen defective axles were found in 100,000 cars of a particular model. How many defective axles would you expect to find in the 2 million cars made of that same model?

EXERCISES 508, SET II

1. A 6-foot man has a $3\frac{1}{2}$-foot shadow when a tree casts a $9\frac{1}{3}$-foot shadow. How tall is the tree?

2. The scale in an architectural drawing is 1 inch equals 8 feet. What are the dimensions of a room that measures $2\frac{1}{2}$ by $3\frac{1}{4}$ inches on the drawing?

3. A man's car averages $1\frac{1}{4}$ quarts of oil for each 1,000 miles driven. How many quarts of oil can he expect to use on an 8,000-mile trip?

4. The property tax on a $25,000 home is $450. What would be the tax on a $35,000 home?

5. A painter uses 3 gallons of paint to paint 2 rooms. How many gallons would he need to paint 7 rooms?

Chapter Summary

A _ratio_ is a fraction. It is a means of comparing two numbers. (Sec. 501)

Always reduce any ratio to its lowest terms to simplify your work. (Sec. 502)

A _proportion_ is a statement that two ratios are equal. (Sec. 504)

In a proportion, _the product of the means is equal to the product of the extremes_. (Sec. 505)

To Solve a Proportion for an Unknown Letter (Sec. 506)

1. Set the product of the means equal to the product of the extremes.
2. Divide each product by the number that the unknown letter is multiplied by.

To Check Your Solution of a Proportion (Sec. 506)

1. Replace the unknown letter in the original proportion by the value you obtained for it.
2. Cross-multiply, and verify that the cross-products are equal.

To Solve a Word Problem By Using a Proportion (Sec. 508)

1. Read the word problem completely, then use x to represent the unknown quantity.

2. Be sure that the same units occupy corresponding positions in both ratios of your proportion.
3. Solve your proportion for the unknown x.

REVIEW EXERCISES 509, SET I. In Exercises 1-4, reduce to lowest terms.

1. 27 to 12

2. 63 to 49

3. 2 yards to 9 feet

4. 35 quarters to 42 nickels

In Exercises 5-8, do the given ratios form a proportion?

5. $\dfrac{50}{120}$, $\dfrac{125}{300}$

6. $\dfrac{46}{32}$, $\dfrac{18}{12}$

7. $\dfrac{63}{81}$, $\dfrac{28}{36}$

8. $\dfrac{117}{97}$, $\dfrac{82}{68}$

In Exercises 9-14, solve for the unknown letter.

9. $\dfrac{45}{70} = \dfrac{x}{56}$

10. $\dfrac{39}{x} = \dfrac{130}{210}$

11. $\dfrac{24}{15} = \dfrac{18}{x}$

12. $\dfrac{x}{\frac{2}{3}} = \dfrac{\frac{9}{16}}{\frac{5}{6}}$

13. $\dfrac{2.8}{5.2} = \dfrac{A}{6.5}$

14. $\dfrac{2\frac{1}{2}}{B} = \dfrac{9}{2\frac{7}{10}}$

15. If 5 pounds of apples cost 62¢, how much does $2\frac{1}{2}$ pounds cost?

16. The Gibson family uses 2 pounds of margarine per week. How much will the family use in 4 days?

17. Fred knows his car needs a quart of oil every 1,500 miles. How many quarts will he need for a 12,000-mile trip?

18. Four students finish a class for every five students who begin. For 252 students to finish, how many must have begun the class?

19. If a 6-foot man casts an 8-foot shadow, find the height of a tree that casts a 96-foot shadow.

20. A speed of 60 miles per hour is equal to 88 feet per second. If sound travels at a speed of 1,100 feet per second, what is the speed of sound in miles per hour?

21. Kenneth rode his bicycle 2.8 miles in 15 minutes ($\frac{1}{4}$ hour). What was his speed in miles per hour? (To determine the speed in miles per hour, find the miles traveled in 60 minutes.)

22. A doctor's prescription calls for $\frac{1}{8}$ ounce of a particular ingredient for every 35 pounds of body weight.
    (a) How many ounces of this ingredient should a 105-pound woman take?
    (b) How many ounces of this ingredient should a 200-pound man take?

23. <u>Brain Teaser</u>. In July 1972 the race horse Quack beat the record in the $1\frac{1}{4}$ mile race. The old record was set in 1956 by the famous horse Swaps. The new record time set by

Quack for the $1\frac{1}{4}$ mile distance was 1 minute and 58.1 seconds. What was Quack's speed for this distance in miles per hour? (Hints: 1 minute 58.1 seconds is equal to $\left[1 + \frac{58.1}{60}\right]$ minutes. To determine the speed in miles per hour, find the miles traveled in 60 minutes.)

REVIEW EXERCISES 509, SET II. In Exercises 1 and 2, reduce to lowest terms.

1. 35 to 45

2. 6 yards to 2 feet

In Exercises 3 and 4, do the given ratios form a proportion?

3. $\frac{113}{339}$, $\frac{27}{81}$

4. $\frac{214}{785}$, $\frac{89}{325}$

In Exercises 5-7, solve for the unknown letter.

5. $\frac{4}{21} = \frac{x}{1\frac{3}{4}}$

6. $\frac{4.8}{x} = \frac{16}{6.5}$

7. $\frac{\frac{1}{5}}{8} = \frac{2\frac{3}{4}}{x}$

8. If 5 pounds of oranges cost 85¢, how much should 3 pounds cost?

9. Four students finish a class for every six students who begin. For 48 students to finish, how many must have begun the class?

10. If a $5\frac{1}{2}$-foot woman casts an $8\frac{1}{4}$-foot shadow, find the height of a tree that casts a 30-foot shadow.

11. Mike rode his bicycle 4 miles in 15 minutes. What was his speed in miles per hour?

Chapter 5: Diagnostic Test

Name _____

_____

The purpose of this test is to see how well you understand ratio and proportion. We recommend that you work this diagnostic test *before* your instructor tests you on this chapter. Allow yourself about 40 minutes to do this test.

Complete solutions for all the problems on this test, together with section references, are given in the Answer Section. You should study the sections referred to for the problems you do incorrectly.

---

1. A college football team won 7 out of 10 games played. There were no tie games.
   (a) What is the ratio of wins to games played?   (1a) _____

   (b) What is the ratio of wins to losses?   (1b) _____

---

2. Reduce the following ratios to lowest terms.

   (a) 15 to 18   (2a) _____

   (b) 12 cars to 8 families   (2b) _____

   (c) 9 inches to 2 feet   (2c) _____

---

3. Do the given ratios form a proportion?

   (a) $\dfrac{21}{37}$, $\dfrac{13}{23}$ (3a) _____

   (b) $\dfrac{24}{54}$, $\dfrac{36}{81}$ (3b) _____

---

4. Solve for the unknown letter.

   (a) $\dfrac{x}{12} = \dfrac{40}{60}$ (b) $\dfrac{24}{75} = \dfrac{P}{100}$

   (c) $\dfrac{3\frac{1}{2}}{B} = \dfrac{21}{40}$

   (4a) _____

   (4b) _____

   (4c) _____

5. If 5 pounds of oranges cost 84¢, what will $7\frac{1}{2}$ pounds cost?

(5) _____

6. If a 6-foot man casts a 4-foot shadow, find the height of a flagpole that casts an 18-foot shadow.

(6) _____

7. Henry's car used 2 quarts of oil on a 1,500-mile trip. How many quarts can he expect to use on a 6,000-mile trip?

(7) _____

8. A doctor's prescription calls for $\frac{1}{4}$ ounce of a particular ingredient for every 25 pounds of body weight. How many ounces of this ingredient would be needed by someone weighing 175 pounds?

(8) _____

SIX
Percent

The Meaning of Percent

Consider a 5-gallon paint can containing 2 gallons of paint. We can say, "The can is $\frac{2}{5}$ full." But you have learned that $\frac{2}{5} = \frac{2 \cdot 20}{5 \cdot 20} = \frac{40}{100}$, so we can also say, "The can is $\frac{40}{100}$ full."

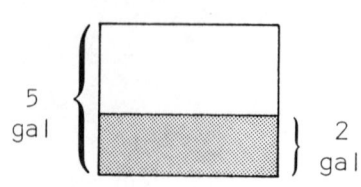

5 gal

2 gal

The decimal representation for $\frac{40}{100}$ is 0.40. Since $\frac{40}{100}$ is 40 per hundred, we can say, "40 *per cent*." (*Cent*um means 100 in Latin.) The can is 40 percent full. 40 percent can also be written 40%.

% is the symbol for percent.

This discussion leads us to the following definition of percent.

Percent is the number of parts per hundred.

Today the meaning of percent is much broader than the original concept of "by the hundred." 5% still means 5 parts out of 100, but we can extend the use of percent to show, for example, that 500% means five times the original number, or that 5,000% means fifty times the original number, or that 5% means 1/20th of the original number. The following clipping from the *Los Angeles Times* is an example of this broader use of percent. In this clipping, 400% is used to show that by 1990 the increase in the amount of electricity used will be four times as much as we are using now.

400% RISE IN NATION'S ELECTRIC DEMANDS SEEN
Power Commission Projection for 1990 Says Consumer Cuts May Be Needed

Figure 601

Example 1. In a class of 20 students, 3 get A. We can describe this by saying:

(a) "3 out of 20 get A." This can be written:

"$\frac{3}{20}$ of the class get A." (*fraction form*)

(b) Since $\frac{3}{20} = \frac{3 \cdot 5}{20 \cdot 5} = \frac{15}{100}$, we can say "15 per 100 get A," written "15 percent get A." Also written

"15% get A." (*percent form*)

——— Note that the percent symbol (%) contains two zeros, one for each zero in 100.

(c) $\frac{15}{100} \doteq 0.15$ (*decimal form*) Therefore, 15% = 0.15.

$\frac{3}{20}$, 15%, 0.15, are three different ways of saying the same thing. We need to know how to change from one form to another.

Changing a Fraction to a Decimal

> To Change a Fraction to a Decimal
>
> Divide the numerator by the denominator (see Section 412).

Example 1.

(a) Change $\frac{3}{5}$ to a decimal. $\frac{3}{5} = 3 \div 5 = 0.6$

(b) Change $\frac{7}{8}$ to a decimal. $\frac{7}{8} = 7 \div 8 = 0.875$

(c) Change $1\frac{2}{7}$ to a decimal. (Round off to three decimal places.) $1\frac{2}{7} = \frac{9}{7} = 9 \div 7 \doteq 1.286$

EXERCISES 602, SET I. Change the following fractions or mixed numbers to decimals. If it is not an exact decimal, round off to two decimal places.

1. $\frac{1}{4}$ 2. $\frac{1}{2}$ 3. $\frac{3}{8}$ 4. $\frac{1}{5}$ 5. $\frac{5}{6}$

6. $\frac{4}{5}$ 7. $\frac{5}{16}$ 8. $\frac{3}{16}$ 9. $\frac{5}{7}$ 10. $\frac{5}{3}$

11. $\frac{4}{25}$ 12. $\frac{7}{50}$ 13. $1\frac{2}{5}$ 14. $2\frac{5}{8}$ 15. 2

EXERCISES 602, SET II. Change the following fractions, or mixed numbers, to decimals. If it is not an exact decimal, round off to two decimal places.

1. $\dfrac{3}{4}$ 2. $\dfrac{5}{8}$ 3. $\dfrac{7}{12}$ 4. $\dfrac{9}{16}$

5. $\dfrac{3}{7}$ 6. $\dfrac{9}{40}$ 7. $1\dfrac{8}{25}$ 8. $3\dfrac{11}{20}$

Changing a Decimal to a Percent

Percent means the number of hundredths. Since $0.25 = \dfrac{25}{100} =$ 25 hundredths = 25%, changing a decimal to a percent simply means reading how many hundredths there are in the decimal. Therefore, _to change a decimal to a percent_, determine the number of hundredths contained in the decimal, and then write the percent (%) symbol.

Example 1

(a) $0.2 = 0.20 = \dfrac{20}{100} = 20$ hundredths = 20%

(b) $0.03 = \dfrac{3}{100} = 3$ hundredths = 3%

(c) $1.5 = 1.50 = \dfrac{150}{100} = 150$ hundredths = 150%

(d) $0.756 = \dfrac{75.6}{100} = 75.6$ hundredths = 75.6%

Notice in Example 1 that:

> To Change a Decimal to a Percent
>
> Move the decimal point two places to the right and write the percent symbol (%).

Example 2

(a) $0.136 = 13.6\%$ (b) $0.035 = 3.5\%$ (c) $2.30 = 230\%$

EXERCISES 603, SET I. Change each decimal to a percent.

1. 0.27   2. 0.35   3. 0.06   4. 0.125   5. 1.4

6. 2.05   7. 0.186   8. 0.015   9. 0.075   10. 0.175

11. 2.9   12. 3.8   13. 2.005   14. 3.015   15. 1.36

16. 2.11   17. 4   18. 3   19. 5.74   20. 7.15

EXERCISES 603, SET II. Change each decimal to a percent.

1. 0.45   2. 0.08   3. 0.375   4. 1.06   5. 0.065

6. 4.6   7. 2.009   8. 5   9. 7.473   10. 13.8

 **Changing a Percent to a Decimal**

Percent means the number of hundredths.

Since 27% = 27 hundredths, then $27\% = \dfrac{27}{100} = 0.27$.

Example 1

    (a) $54\% = 54$ hundredths $= \dfrac{54}{100} = 0.54$

    (b) $7\% = \dfrac{7}{100} = 0.07$

    (c) $105\% = \dfrac{105}{100} = 1.05$

Notice in Example 1 that:

To Change a Percent to a Decimal

Move the decimal point two places to the left
and remove the percent symbol (%).

Example 2

    (a) $38.5\% = 0.385$ (b) $2.8\% = 0.028$ (c) $130\% = 1.30$

    (d) $400\% = 4$ (e) $5\frac{1}{2}\% = 5.5\% = 0.055$

EXERCISES 604, SET I. Change the following percents to decimals.
If the decimal is not exact, round off to four decimal places.

  1. 45% 2. 78% 3. 125% 4. 150% 5. 6.5%

  6. 8.6% 7. 2.35% 8. 3.85% 9. $2\frac{1}{2}\%$ 10. $4\frac{3}{4}\%$

11. $3\frac{1}{4}\%$ 12. $5\frac{2}{5}\%$ 13. 10.05% 14. 2.08% 15. $\frac{3}{4}\%$

16. $\frac{1}{2}\%$ 17. $66\frac{2}{3}\%$ 18. $33\frac{1}{3}\%$ 19. $12\frac{1}{2}\%$ 20. $37\frac{1}{2}\%$

EXERCISES 604, SET II. Change the following percents to decimals. If the decimal is not exact, round off to four decimal
places.

  1. 58% 2. 175% 3. 4.8% 4. 5.37% 5. $2\frac{1}{4}\%$

  6. $4\frac{3}{5}\%$ 7. 6.09% 8. $62\frac{1}{2}\%$ 9. $\frac{5}{4}\%$ 10. $83\frac{1}{3}\%$

Changing a Decimal to a Fraction

We showed how to change a decimal to a fraction in Section 413. We review the method here.

To Change a Decimal to a Fraction (see Section 413)

1. Read the decimal, then write it in fraction form.
2. Reduce the fraction to lowest terms.

Example 1

(a) Change 0.5 to a fraction in lowest terms.

$$0.5 = \frac{5}{10} = \frac{1}{2}$$

(b) Change 0.38 to a fraction in lowest terms.

$$0.38 = \frac{38}{100} = \frac{19}{50}$$

(c) Change 1.75 to a mixed number in lowest terms.

$$1.75 = \frac{175}{100} = \frac{7}{4} = 1\frac{3}{4}$$

We have now shown that *when we take a fractional part of something, we can express it in three ways: as a fraction, as a decimal, or as a percent.*

Example 2. In a class of 20 students, 3 get A. We can describe this by saying:

(a) $\frac{3}{20}$ of the class get A. (*fraction*)

(b) 0.15 of the class get A. (*decimal*)

(c) 15 percent of the class get A. (*percent*)

We have also shown how each of these forms can be changed to either of the other forms.

EXERCISES 605, SET I. In Exercises 1-12, change each decimal to a proper fraction or mixed number in lowest terms.

1. 0.7	2. 0.8	3. 0.64	4. 0.88
5. 0.125	6. 0.375	7. 3.5	8. 2.8
9. 1.75	10. 1.25	11. 0.1875	12. 0.5625

In Exercises 13-38, change the given form into the two missing forms.

	Fraction	Decimal	Percent			Fraction	Decimal	Percent
13.	$\frac{1}{2}$			14.	$\frac{1}{4}$		
15.		0.6		16.		0.4	
17.			10%	18.			20%
19.	$\frac{3}{4}$			20.		0.36	
21.			6%	22.	$\frac{4}{5}$		
23.		0.48		24.			8%
25.	$1\frac{1}{8}$			26.		0.075	
27.			44%	28.	$2\frac{3}{8}$		
29.		0.025		30.			28%
31.			350%	32.			275%
33.		6.25		34.			$4\frac{1}{2}\%$
35.			$5\frac{1}{4}\%$	36.		8.75	
37.			$\frac{3}{4}\%$	38.			$\frac{6}{12}\%$

In Exercises 39-44, change the given form into the two missing forms. Round off decimals to three decimal places. Round off percents to one decimal place.

	Fraction	Decimal	Percent
39.	$\frac{2}{3}$		
41.		$0.5\frac{3}{8}$	
43.			$24\frac{1}{4}\%$

	Fraction	Decimal	Percent
40.	$\frac{5}{6}$		
42.		$0.45\frac{1}{4}$	
44.			$6\frac{5}{8}\%$

45. Due to illness a student was absent from school 10 days out of 40 days.
(a) What fraction of the time was he absent?
(b) What decimal part of the time was he absent?
(c) What percent of the time was he absent?

46. If $\frac{1}{3}$ of a man's salary is deducted from his paycheck, what percent of his check does he get?

47. The price of a car is discounted 20%. What fraction of the original price must the buyer pay?

48. A student takes an examination in mathematics and solves 9 problems correctly out of 12.
(a) What fraction of the problems did he solve correctly?
(b) What decimal part of the problems did he solve correctly?
(c) What was his percent grade?

49. A student spends $20\frac{5}{6}\%$ of his time working in a market. What fraction of his time is spent working at the market?

EXERCISES 605, SET II. In Exercises 1-6, change each decimal to a proper fraction or mixed number in lowest terms.

1. 0.6   2. 0.36   3. 0.625   4. 2.4   5. 3.75   6. 0.6875

In Exercises 7-22, change the given form to the two missing forms. Round off decimals to three decimal places. Round off percents to one decimal place.

	Fraction	Decimal	Percent
7.	$\frac{3}{4}$		
9.			40%
11.		0.54	
13.	$1\frac{5}{8}$		

	Fraction	Decimal	Percent
8.		0.8	
10.	$\frac{3}{5}$		
12.			12%
14.		0.05	

	Fraction	Decimal	Percent
15.			68%
17.		6.75	
19.			$\frac{1}{2}$%
21.		$0.2\frac{5}{8}$	

	Fraction	Decimal	Percent
16.			325%
18.			$13\frac{1}{4}$%
20.	$\frac{5}{3}$		
22.			$4\frac{1}{4}$%

23. A student gets 48 answers correct out of 60 on a multiple choice exam. What percent of the questions did he do incorrectly?
24. If the price on a piece of furniture is reduced 45%, what fraction of the original price must the buyer pay?
25. Ted had deductions totaling $140 taken from his $350 paycheck. What percent of his check is take-home pay?

Finding a Fractional Part of a Number

In the last section we showed that a fractional part of a number can be expressed as a percent, a decimal, or as a fraction. For example:

$$25\% = 0.25 = \frac{1}{4}$$

$$50\% = 0.50 = \frac{1}{2}$$

$$10\% = 0.10 = \frac{1}{10}$$

$$200\% = 2.00 = \frac{2}{1}$$

To Find a Fractional Part of a Number

1. If the fractional part is expressed *as a fraction*, multiply the fraction times the number.
2. If the fractional part is expressed *as a decimal*, multiply the decimal times the number.
3. If the fractional part is expressed *as a percent*, change the percent to a decimal or fraction, then multiply.

Example 1. In a class of 30 students, $\frac{3}{5}$ of them are women. How many are women?

Solution: $\frac{3}{5}$ of $30 = \frac{3}{5} \times \frac{30}{1} = \frac{3}{\cancel{5}} \times \frac{\cancel{30}^{6}}{1} = 18$ women

"of" means "to multiply" in problems of this type.

Example 2. Find 0.13 of $75.

Solution: 0.13 of $75 = 0.13 \times 75. = \$9.75$

$$\begin{array}{r} 75 \\ .13 \\ \hline 225 \\ 75 \\ \hline 975 = 9.75 \end{array}$$

Example 3. A man's monthly salary is $725. His total deductions amount to 23% of his monthly check.
   (a) How much is deducted from his check each month?
   (b) What is his net take-home pay?

Solution: 23% = 0.23. Therefore, 23% of $725

$$= 0.23 \times \$725 = \$166.75$$

$$\begin{array}{r} \$725 \\ .23 \\ \hline 2175 \\ 1450 \\ \hline \$166.75 \end{array} = \text{amount deducted from his paycheck}$$

$$\begin{array}{r} \$725.00 \\ \$166.75 \\ \hline \$558.25 \end{array} = \text{his take-home pay}$$

Example 4. Find 25% of 12.

Solution: $25\% = \frac{25}{100} = \frac{1}{4}$. Therefore, 25% of $12 = \frac{1}{4} \times \frac{12}{1} = 3$

Example 5. Find $3\frac{1}{3}\%$ of 240.

Solution: $3\frac{1}{3}\% = 3\frac{1}{3}$ hundredths $= \dfrac{3\frac{1}{3}}{100} = 3\frac{1}{3} \div 100$

$$= \frac{\cancel{10}^{1}}{3} \times \frac{1}{\cancel{100}_{10}} = \frac{1}{30}$$

Therefore, $3\frac{1}{3}\%$ of $240 = \frac{1}{\cancel{30}_{1}} \times \frac{\cancel{240}^{8}}{1} = 8$

EXERCISES 606, SET I.

1. Find $\frac{1}{6}$ of 48.

2. Find $\frac{1}{2}$ of 38.

3. Find 0.25 of 36.

4. Find 0.20 of 15.

5. Find 15% of 32.

6. Find 75% of 12.

7. Find $\frac{3}{4}$ of 52.

8. Find $\frac{5}{8}$ of 64.

9. Find 0.225 of 140.

10. Find 0.375 of 150.

11. Find 200% of 56.

12. Find 300% of 72.

13. Find $\frac{5}{6}$ of 27.

14. Find $\frac{7}{12}$ of 32.

15. Find 0.03125 of 960.

16. Find 0.0625 of 480.

17. Find $13\frac{1}{3}$% of 702.

18. Find $8\frac{2}{3}$% of 504.

19. Find $\frac{1}{2}$% of 300.

20. Find $\frac{1}{4}$% of 200.

21. Find 0.35% of 550.

22. Find 0.55% of 750.

23. A man's two-week salary is $375. His total deductions amount to 27% of his check. How much is deducted from his check?

24. On a mathematics examination of 20 problems, John solved 85% of the problems correctly. How many problems did he have correct?

25. During an 88-day spring semester, a student was absent $\frac{1}{11}$ of the time. How many days was he absent?

26. A man's weekly salary check is $165. His deductions amount to $\frac{1}{5}$ of his check. Find his "take-home" pay.

27. In purchasing a car, a 15% down payment is required. Find the down payment on a car that sells for $2,150.

28. The present value of a house is 150% of its 1965 cost. If the house sold for $15,500 in 1965, what is its present value?

29. If a student sleeps $\frac{1}{3}$ of each 24-hour day, how many hours does he sleep in a week?

30. If $\frac{3}{4}$ of the total weight of a steer will produce usable products, how many pounds of usable products can be taken from a 1,250-pound steer?

31. An automobile listed at $2,490 was sold at a 15% discount. What was the selling price?

32. Mr. Miller, with a salary of $12,500 a year, was given a 4.5% raise. What was his salary after the raise?

33. A dealer had a sale and marked several items down 22%. Calculate the sale price of each of the items which had been selling for $237.50, $94.50, $248.50, and $89.50.

34. The total surface area of the earth is 196,940,000 square miles, of which 29% is land and 71% water. Find the land area and the water area of the earth.

EXERCISES 606, SET II.

1. Find $\frac{1}{3}$ of 54.

2. Find 0.40 of 35.

3. Find 35% of 46.

4. Find $\frac{3}{8}$ of 72.

5. Find 0.325 of 160.

6. Find 400% of 83.

7. Find $\frac{5}{12}$ of 42.

8. Find 0.09375 of 64.

9. Find $9\frac{2}{3}$% of 405.

10. Find $\frac{1}{5}$% of 800.

11. Find 0.65% of 450.

12. On a statistics examination of 50 questions, Penny answered 64% of the questions correctly. How many questions did she have correct?

13. During an 84-day fall semester, Jeff was absent $\frac{2}{7}$ of the time. How many days was he absent?

14. The present value of some municipal bonds is 130% of their cost in 1970. If Mrs. Kaplan paid $12,500 for them in 1970, what is their present value?

15. If $\frac{7}{8}$ of the total weight of a steer will produce usable products, how many pounds of usable products can be obtained from a 1,340-pound steer?

16. A catamaran listed at $5,650 was sold at a 12% discount. What was the actual sales price?

17. The surface of the moon is about 7.43% of the surface of the earth. If the total surface area of the earth is 196,940,000 square miles, what is the surface area of the moon?

The Percent Proportion

Take a second look at the example of the paint can discussed in Section 601. We described that situation in three different ways.

1. The can is $\frac{2}{5}$ full.
2. The can is 0.40 (40 hundredths) full.
3. The can is 40 percent full.

We have talked about three quantities:

1. 40 *percent* (*P*)
2. 5, which represents the *whole thing*, and is called the *base* (*B*)
3. 2, which is called the *amount* (*A*)

These three numbers are related by the proportion:

$$\frac{2}{5} = \frac{40}{100}$$

This proportion can be written:

$$\frac{A}{B} = \frac{P}{100}$$

and is called "*the percent proportion*."

In Section 506 you learned how to solve a proportion for an unknown term.

<u>Example</u>. Suppose we want to find P in the above example, where $A = 2$ and $B = 5$.

<u>Solution</u>: $\dfrac{A}{B} = \dfrac{P}{100}$

$\dfrac{2}{5} = \dfrac{P}{100}$ Here we replaced A with 2 and B with 5.

Then, $5 \cdot P = 2 \cdot 100$

$\dfrac{\cancel{5}^{1} \cdot P}{\cancel{5}_{1}} = \dfrac{\cancel{200}^{40}}{\cancel{5}_{1}}$ ⎫
⎬ —See Section 506.
$P = 40$ ⎭

<u>EXERCISES 607, SET I</u>. Solve $\dfrac{A}{B} = \dfrac{P}{100}$:

1. for P when $A = 15$ and $B = 75$.

2. for P when $A = 12$ and $B = 25$.

3. for A when $B = 20$ and $P = 60$.

4. for A when $B = 15$ and $P = 40$.

5. for B when $P = 75$ and $A = 48$.

6. for B when $A = 9$ and $P = 45$.

7. for P when $A = 2.84$ and $B = 40$.

8. for P when $A = 1.05$ and $B = 35$.

9. for A when $P = 125$ and $B = 78.6$.

10. for A when $B = 78.6$ and $P = 3.5$. Round off the answer to one decimal place.

11. for B when $A = 7.4$ and $P = 37.5$. Express the answer accurately to tenths.

12. for B when $A = 37.4$ and $P = 12.5$.

13. for P when $A = 14.7$ and $B = 37\frac{1}{2}$.

14. for P when $A = 3.94$ and $B = 58.6$. Round off the answer

to two decimal places.

15. for A when $B = 58.6$ and $P = 7\frac{1}{2}$. Round off the answer to two decimal places.

16. for A when $B = 95.7$ and $P = 16\frac{2}{3}$. Round off the answer to one decimal place.

EXERCISES 607, SET II. Solve $\frac{A}{B} = \frac{P}{100}$:

1. for P when $A = 14$ and $B = 35$.

2. for A when $B = 45$ and $P = 80$.

3. for B when $A = 42$ and $P = 24$.

4. for P when $A = 6.3$ and $B = 42$.

5. for A when $B = 59.4$ and $P = 6.5$. Round off the answer to one decimal place.

6. for B when $A = 4.8$ and $P = 73.5$. Express the answer accurately to tenths.

7. for P when $A = 2.68$ and $B = 47.2$. Round off the answer to two decimal places.

8. for A when $B = 62.4$ and $P = 83\frac{1}{3}$.

Identifying the Numbers in a Percent Problem

There are three unknown terms in the percent proportion. They are:

1. the amount A
2. the base B
3. the percent P

In the last section, we learned how to solve the percent proportion for one of the three terms when the other two terms are known. In every percent problem, two numbers must be given. We must be able to identify which of the three terms the given numbers represent. The easiest term to identify is the percent (P). "P" is the number written with the word "percent," or the symbol $(\%)$.

Example 1

(a) What number is $\underset{P}{\boxed{8 \text{ percent}}}$ of 40?

(b) 16 is $\underset{P}{\boxed{40\%}}$ of what number?

(c) 15 is $\underset{P}{\boxed{\text{what percent}}}$ of 60?

We next identify the base (B). "B" is the number that represents 100 percent, and usually follows the words "percent of."

Example 2

    (a) What number is 8 *percent of* 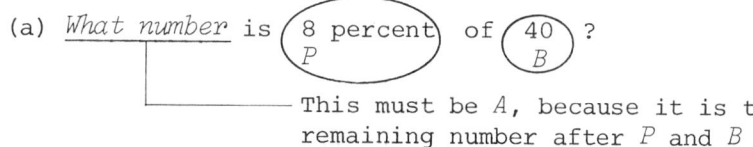 ?

    (b) 16 is 40% *of* what number ?

    (c) 15 is what *percent of* 60 ?

"A" is the remaining number after P and B have been identified. "A" is the *amount*.

Example 3

    (a) *What number* is 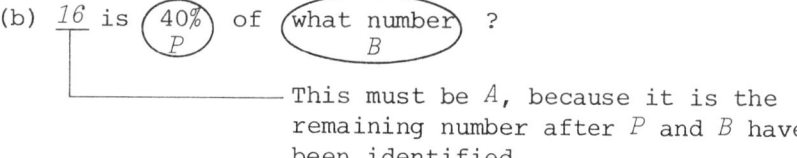 8 percent of 40 ?

               This must be A, because it is the remaining number after P and B have been identified.

    (b) *16* is 40% of what number ?

               This must be A, because it is the remaining number after P and B have been identified.

    (c) *15* is 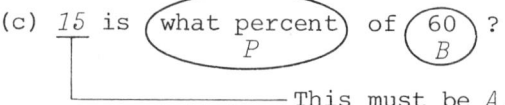 what percent of 60 ?

               This must be A.

We identify all three terms in the following percent problems.

Example 4

    (a) What number is 175% of 80 ?

    (b) 35 is what percent of 105 ?

    (c) 400 is 200% of what number ?

    (d) A team wins 75% of its games. If it wins 30 games, how many games has it played?

Solution: The problem can be restated:

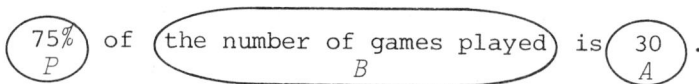

    75% of the number of games played is 30 .

```
┌─────────────────────────────────────────────────────────┐
│                                                         │
│   To Identify the Numbers in a Percent Problem          │
│                                                         │
│   1. P is the number written with the word "percent" (%).│
│   2. B is the number that follows the words "percent of."│
│   3. A is the remaining number after P and B have been  │
│      identified.                                        │
│                                                         │
└─────────────────────────────────────────────────────────┘
```

EXERCISES 608, SET I. In the following exercises, identify the percent P, base B, and amount A. One of these will be unknown. Tell which is the unknown. Do not solve the exercises at this time.

1. 15 is 30% of what number?
2. 16 is 20% of what number?
3. 115 is what percent of 250?
4. 330 is what percent of 225?
5. What is 25% of 40?
6. What is 45% of 65?
7. 15% of what number is 127.5?
8. 32% of what number is 256?
9. What percent of 8 is 17?
10. What percent of 6 is 12?
11. 63% of 48 is what number?
12. 87% of 49 is what number?
13. 750 is 125% of what number?
14. 325 is 130% of what number?
15. 23 is what percent of 16?
16. 57 is what percent of 23?
17. What is 200% of 12?
18. What is 300% of 9?
19. 500 grams of a solution contain 27 grams of a drug. Find the percent of drug strength.
20. 15% of a number is 37.5. What is the number?
21. A team wins 80% of its games. If it wins 68 games, how many games has it played?
22. What is 27% of $135?
23. In a class of 42 students, 7 students received a grade of B. What percent of the class received a grade of B?
24. John's weekly gross pay is $110 but 23% of his check is withheld. How much is withheld?

EXERCISES 608, SET II. In the following exercises, identify the percent P, base B, and amount A. One of these will be unknown. Tell which is the unknown. Do not solve the exercises at this time.

1. 18 is 40% of what number?
2. 950 is what percent of 325?
3. What is 55% of 82?
4. 23% of what number is 347?
5. What percent of 12 is 18?
6. 91% of 64 is what number?
7. 558 is 134% of what number?
8. 39 is what percent of 27?

9. What is 400% of 19?
10. There are 43 grams of sulfuric acid in 500 grams of solution. Find the percent of acid in the solution.
11. A team wins 105 games. This is 70% of the games played. How many games were played?
12. Rosie's weekly salary is $225. If her deductions amount to $68, what percent of her salary is deducted?

The Three Types of Percent Problems

Since there are only three letters in the percent proportion,

$$\frac{A}{B} = \frac{P}{100}$$

There can be only three kinds of percent problems.

The Three Types of Percent Problems

Type I: Solving for A
in $\frac{A}{B} = \frac{P}{100}$ when B and P are known.

Type II: Solving for P
in $\frac{A}{B} = \frac{P}{100}$ when A and B are known.

Type III: Solving for B
in $\frac{A}{B} = \frac{P}{100}$ when A and P are known.

TYPE I PERCENT PROBLEM SOLUTION

Example 1. What is 12% of 85?

Solution: (What) is (12%) of (85)?
      A P B

$$\frac{A}{B} = \frac{P}{100}$$

$$\frac{A}{85} = \frac{12}{100}$$

$$\frac{A}{85} = \frac{\overset{3}{\cancel{12}}}{\underset{25}{\cancel{100}}}$$

$$25 \cdot A = 85 \cdot 3$$

$$\frac{\cancel{25} \cdot A}{\cancel{25}} = \frac{255}{25}$$

$$A = 10.2$$

Therefore, 10.2 is 12% of 85.

We have already worked Type I percent problems in Section 606 using a different method. We suggest that you rework some of the exercises of Section 606 using the percent proportion. We leave it up to you to decide which method you prefer to use in solving Type I problems. Our main concern is that you use the percent proportion in solving Type II and Type III problems.

TYPE II PERCENT PROBLEM SOLUTION

Example 2. 5 is what percent of 20?

Solution:

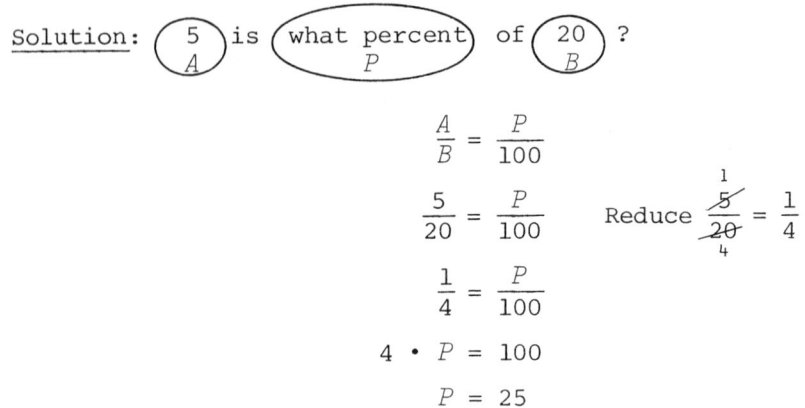

$$\frac{A}{B} = \frac{P}{100}$$

$$\frac{5}{20} = \frac{P}{100} \qquad \text{Reduce } \frac{\cancel{5}^{1}}{\cancel{20}_{4}} = \frac{1}{4}$$

$$\frac{1}{4} = \frac{P}{100}$$

$$4 \cdot P = 100$$

$$P = 25$$

Therefore, 5 is 25% of 20.

TYPE III PERCENT PROBLEM SOLUTION

Example 3. In an examination a student worked 15 problems correctly. This was 75% of the problems. Find the total number of problems on the examination.

Solution: In this problem we are saying:

75% of some number is 15.
P   B   A

$$\frac{A}{B} = \frac{P}{100}$$

$$\frac{15}{B} = \frac{75}{100} \qquad \text{Reduce } \frac{\cancel{75}^{3}}{\cancel{100}_{4}} = \frac{3}{4}.$$

$$\frac{15}{B} = \frac{3}{4}$$

$$3 \cdot B = 4 \cdot 15$$

$$B = \frac{60}{3} = 20$$

Therefore, 20 problems were on the examination.

Example 4. Mr. Delgado, a salesman, makes a 6% commission on all items he sells. One week he made $390. What were his gross sales for the week?

<u>Solution</u>. In this problem we are saying:

$$\left(\begin{array}{c}6\%\\P\end{array}\right) \text{of} \left(\begin{array}{c}\text{some number}\\B\end{array}\right) \text{is} \left(\begin{array}{c}\$390\\A\end{array}\right)$$

$$\frac{A}{B} = \frac{P}{100}$$

$$\frac{390}{B} = \frac{6}{100} \qquad \text{Reduce } \frac{\overset{3}{\cancel{6}}}{\underset{50}{\cancel{100}}} = \frac{3}{50}.$$

$$\frac{390}{B} = \frac{3}{50}$$

$$3 \cdot B = 50 \cdot 390$$

$$\frac{\cancel{3} \cdot B}{\cancel{3}} = \frac{\overset{6,500}{\cancel{19,500}}}{\cancel{3}}$$

$$B = \$6,500$$

Therefore, his gross sales for the week were $6,500.

<u>EXERCISES 609, SET I</u>. Solve the following percent problems.

1. 15 is 30% of what number?
2. 16 is 20% of what number?
3. 115 is what percent of 250?
4. 330 is what percent of 225?
5. What is 25% of 40?
6. What is 45% of 65?
7. 15% of what number is 127.5?
8. 32% of what number is 256?
9. What percent of 8 is 17?
10. What percent of 6 is 12?
11. 63% of 48 is what number?
12. 87% of 49 is what number?
13. 750 is 125% of what number?
14. 325 is 130% of what number?
15. 23 is what percent of 16?
16. 57 is what percent of 23?
17. What is 200% of 12?
18. What is 300% of 9?
19. 42 is $66\frac{2}{3}\%$ of what number?
20. 36 is $16\frac{2}{3}\%$ of what number?
21. 500 grams of a solution contain 27 grams of a drug. Find the percent of drug strength.
22. 15% of a number is 37.5. What is the number?
23. A team wins 80% of its games. If it wins 68 games, how many games has it played?
24. What is 27% of $135?
25. In a class of 42 students, 7 students received a grade of B. What percent of the class received a grade of B?
26. John's weekly gross pay is $110, but 23% of his check is withheld. How much is withheld?
27. 54 out of 210 civil service applicants pass their exams. What percent of the applicants passes?

28. A 4,200-pound automobile contains 462 pounds of rubber. What percent of the car's total weight is rubber?
29. 5.85% of a person's salary is withheld for social security. How much is deducted from Leon's $460 salary for social security?
30. Every quarter an employer must send 11.7% of the total amount earned by his employees to the collector of internal revenue. If the employees' earnings are $178,475 for a quarter, how much must he send to the collector of internal revenue? (Round off your answer to the nearest cent.)

EXERCISES 609, SET II. Solve the following percent problems.

1. 24 is 40% of what number?
2. 650 is what percent of 325?
3. What is 55% of 82?
4. 23% of what number is 13.11?
5. What percent of 12 is 18?
6. 91% of 64 is what number?
7. 335 is 134% of what number?
8. 39 is what percent of 27? (Round off your answer to one decimal place.)
9. What is 400% of 19?
10. There are 43 grams of sulfuric acid in 500 grams of solution. Find the percent of acid in the solution.
11. A team wins 105 games. This is 70% of the games played. How many games were played?
12. Rosie's weekly salary is $225. If her deductions amount to $63, what percent of her salary is take-home pay?

Interest

A common use of percent is in working interest problems. When we borrow money, we pay *interest* for the use of the money. Naturally, the more money we borrow the more interest we pay. The amount of money we borrow is called the *principal*. The interest we pay is a *fractional part* of the amount we borrow. That is, the interest is a percent of the principal. The percent is called the *rate* of interest.

Example 1. Find the *yearly* interest on $1,000 when the yearly interest rate is 9%.

Solution: Find 9% of $1,000.   (We are finding a *fractional part* of $1,000. See Section 606.)

$$9\% = 0.09$$
$$0.09 \times \$1,000 = \$90.00$$

Therefore, the yearly interest is $90.00. To find the interest for 3 years, multiply the yearly interest by 3.

$$\text{Interest for 3 years} = 3 \times \$90 = \$270$$

Example 2. On his bank credit card account, Mr. Garcia had a balance due of $250. The *monthly* rate charged by the bank was 1.5%. Find the interest charged for the month.

Solution: Find 1.5% of $250.

$$1.5\% = 0.015$$

$$0.015 \times \$250 = \$3.75$$

Therefore, the monthly interest charged by the bank was $3.75. To find the interest for 4 months, multiply the monthly interest by 4.

$$\text{Interest for 4 months} = 4 \times \$3.75 = \$15.00$$

Example 3. The United California Bank charges its credit card customers a *daily* rate of 0.04932%. (This was the rate in 1973.) Find the finance charge on $350 for a period of 25 days.

Solution: 0.04932% = 0.0004932.

The interest for 1 day is 0.0004932 × $350 = $0.17262. To find the interest for 25 days, we multiply the daily interest ($0.17262) by 25.

$$25 \times \$0.17262 = \$4.3155 \doteq \$4.32$$

A basic formula used to work problems like the above examples is:

Interest = principal × rate × time

$$I = Prt$$

r and *t* must be expressed in the *same* time units.

In the interest formula:

1. With a *yearly rate*, the time must be expressed in *years*.
2. With a *monthly rate*, the time must be expressed in *months*.
3. With a *daily rate*, the time must be expressed in *days*.

Example 4. A student borrows $1,800 from the government at $3\frac{1}{2}\%$ yearly interest to pay his school expenses. If he agrees to pay it back at the end of 5 years, what will the interest be?

Solution: $3\frac{1}{2}\% = 0.035$.

$$\text{Interest} = \text{principal} \times \text{rate} \times \text{time}$$
$$\text{Interest} = \quad 1{,}800 \quad \times\ 0.035 \times \quad 5$$
$$\text{Interest} = \quad \$315$$

EXERCISES 610, SET I. Round off answers to the nearest cent.

	Principal	Rate	Find interest for: (a)	(b)
1.	$1,400	9% per yr.	1 year	3 years
2.	$2,500	8% per yr.	1 year	2 years
3.	$ 350	1.5% per mo.	1 month	2 months
4.	$ 175	1.5% per mo.	1 month	3 months
5.	$2,000	0.04932% per day	1 day	30 days
6.	$2,150	0.04932% per day	1 day	20 days

In Exercises 7-12, express answers to the nearest cent.

7. Find the interest on $750 for $2\frac{1}{2}$ years when the yearly interest rate is 7%.

8. Find the interest on $1,250 for $1\frac{1}{2}$ years when the yearly interest rate is 8%.

9. Mr. Medrano's credit card account has a balance due of $175. The monthly rate charged by the bank is 1.5%. Find the interest charged for 2 months.

10. Mr. Chan's credit card account had a balance due of $129. The monthly rate charged by the bank is $1\frac{1}{2}$%. Find the interest charged for 3 months.

11. Miss Jung's credit card account has a balance due of $38.37. The daily rate of interest is 0.04932%. Find the finance charge for 23 days.

12. Mrs. Fry's credit card account has a balance due of $56.92. The daily rate of interest is 0.04932%. Find the finance charge for 19 days.

EXERCISES 610, SET II. Round off answers to the nearest cent.

	Principal	Rate	Find interest for: (a)	(b)
1.	$1,800	7% per yr.	1 year	3 years
2.	$ 224	1.5% per mo.	1 month	4 months
3.	$ 680	0.04932% per day	1 day	20 days

4. Find the interest on $925 for $2\frac{1}{2}$ years when the yearly interest rate is 9%.

5. Mr. Wong's credit card account had a balance due of $237. The monthly rate charged by the bank is $1\frac{1}{2}$%. Find the interest charged for 2 months.

6. Mrs. Volkmann's credit card account has a balance due of $91.24. The daily rate of interest is 0.04932%. Find the finance charge for 22 days.

Add-on-Interest and Monthly Payments

With *add-on-interest*, the interest as found in the last section is added to the principal to get the *total amount owed*.

$$\text{Total amount owed} = \text{principal} + \text{interest}$$

To find the amount of each *monthly payment*, divide the total amount owed by the number of months over which the payments are made.

$$\text{Monthly payment} = \frac{\text{total amount owed}}{\text{number of months}}$$
$$= \frac{\text{principal} + \text{interest}}{\text{number of months}}$$

Example 1. After trading his car in on a new car, Mr. Jones had a balance due of $2,654. He was told that he could pay the balance in equal monthly payments over a period of 36 months. The yearly interest rate was 6.5%. To find the monthly payment, proceed as follows:

Step 1. Find the interest.

Rate = 6.5% = 0.065 (Yearly rate)
Time = 36 months = 3 years
Interest = principal × rate × time
        = $2,654 × 0.065 × 3 = $517.53

Step 2. Find the total amount owed.

Total amount owed = principal + interest
                  = $2,654 + $517.53
                  = $3,171.53

Step 3. Find the monthly payment.

$$\text{Monthly payment} = \frac{\text{total amount owed}}{\text{number of months}}$$

$$= \frac{\$3,171.53}{36} \doteq \$88.10$$

Example 2. JoAnne makes an $86 down payment on a new $860 spinet piano. The yearly interest rate is 7.5%. Find her monthly payment if she takes 36 months to pay for it.

Step 1. Find the principal.

$$\text{Principal} = \text{cost} - \text{down payment}$$
$$= \$860 - \$86 = \$774$$

Step 2. Find the interest.

Rate = 7.5% = 0.075 (Yearly rate)
Time = 36 months = 3 years
Interest = principal × rate × time
$\qquad = \$774 \times 0.075 \times 3 = \$174.15$

Step 3. Find the total amount owed.

Total amount owed = principal + interest
$\qquad = \$774 + \$174.15$
$\qquad = \$948.15$

Step 4. Find the monthly payment.

$$\text{Monthly payment} = \frac{\text{total amount owed}}{\text{number of months}}$$

$$= \frac{\$948.15}{36} \doteq \$26.34$$

The method of add-on-interest is one method used in calculating monthly payments that has been, and is still being, used in installment accounts in many places. There are other methods for calculating monthly payments that will not be discussed in this book.

EXERCISES 611, SET I. In the following exercises, round off the monthly payment to the nearest cent.

1. Miss Duran made a $40 down payment on a $375 stereo set. The yearly interest rate was 7%. Find her monthly payment if she takes 24 months to pay.
2. Mr. Gee bought a console color TV set for $450. He made a $45 down payment. The payment contract calls for 24 monthly payments. The yearly interest rate is 8%. Find the monthly payment.
3. After trading in his car on a new car, Mr. Muller had a balance due of $2,300. His payment contract calls for 36 monthly payments. The yearly interest rate is $6\frac{1}{2}\%$. Find his monthly payment.
4. After trading in his car on a new car, Mr. Matsuda had a balance due of $4,250. His payment contract calls for 36

monthly payments. The yearly interest rate is $6\frac{3}{4}\%$. Find his monthly payment.

5. Mrs. Avila made a $25 down payment on a $250 refrigerator. The yearly interest rate on the balance due was $7\frac{3}{4}\%$. Find her monthly payment if she pays it off in 12 months.

6. Mrs. Myles made a $35 down payment on a washer-dryer costing $325. The yearly interest rate was 7.25%. Find her monthly payment if she pays it off in 24 months.

7. Mr. Norton made a $350 down payment on a camper costing $3,500. The yearly interest rate was $6\frac{1}{4}\%$. Find his monthly payment if he pays for it in 36 months.

8. Mr. Siebert made a $250 down payment on a boat costing $2,500. The yearly interest rate was 8%. Find his monthly payment if he pays for it in 36 months.

EXERCISES 611, SET II. In the following exercises, round off the monthly payment to the nearest cent.

1. Mr. and Mrs. Wells made an $800 down payment on a mini-motorhome costing $7,950. Find their monthly payment if it is to be paid off in five years. The yearly rate of interest is $7\frac{1}{2}\%$.

2. Greg made a $200 down payment on a $2,150 motorcycle. Find his monthly payment if he pays it off in two years. The yearly rate of interest is 8%.

3. Sarah bought a catamaran for $3,500. She made a down payment of $350. The yearly interest rate is 7%. If she agrees to pay it off in three years, what will her monthly payment be?

4. After trading in his car on a new van, David had a balance due of $4,650. His payment contract calls for 36 monthly payments. The yearly interest rate is $6\frac{1}{2}\%$. Find his monthly payment.

 # Chapter Summary

Percent is the number of hundredths in a number. (Sec. 603)
Percent is the number of parts per hundred. (Sec. 601)

We can express a fractional part of something in three ways: as a fraction, as a decimal, or as a percent. (Sec. 605)

To change a fraction to a decimal, divide the numerator by the denominator. (Sec. 602)

To change a decimal to a percent, move the decimal point two places to the right, then write the percent symbol (%). (Sec. 603)

To change a percent to a decimal, move the decimal point two places to the left, and remove the percent symbol (%). (Sec. 604)

To change a decimal to a fraction: (Sec. 605)

1. Read the decimal, then write it in fraction form.
2. Reduce the fraction to lowest terms.

To find a fractional part of a number: (Sec. 606)

1. If the fractional part is expressed _as a fraction_, multiply the fraction times the number.
2. If the fractional part is expressed _as a decimal_, multiply the decimal times the number.
3. If the fractional part is expressed _as a percent_, change the percent to a decimal or fraction, then multiply.

The percent proportion is $\dfrac{A}{B} = \dfrac{P}{100}$. (Sec. 607)

To identify the numbers in a percent problem: (Sec. 608)

1. P is the number written with the word "percent" (%).
2. B is the number that follows the words "percent of."
3. A is the remaining number after P and B have been identified.

The three types of percent problems: (Sec. 609)

Type I: Solving for A in $\dfrac{A}{B} = \dfrac{P}{100}$ when B and P are known.

Type II: Solving for P in $\dfrac{A}{B} = \dfrac{P}{100}$ when A and B are known.

Type III: Solving for B in $\dfrac{A}{B} = \dfrac{P}{100}$ when A and P are known.

Interest = principal × rate × time. Rate and time must be expressed in the same time units. (Sec. 610)

To find the monthly payment when the _add-on-interest_ method is used: (Sec. 611)

$$\underline{Monthly~payment} = \frac{\text{total amount owed}}{\text{number of months}} = \frac{\text{principal + interest}}{\text{number of months}}$$

REVIEW EXERCISES 612, SET I. In Exercises 1-6, change the given form into the two remaining forms.

	Fraction	Decimal	Percent
1.	$\dfrac{2}{5}$		
3.			30%
5.		1.75	

	Fraction	Decimal	Percent
2.		0.45	
4.	$2\dfrac{1}{2}$		
6.			$12\dfrac{1}{2}\%$

7. What is 35% of $275?
8. 45 is what percent of 300?
9. 77.5 is 31% of what number?
10. What is 245% of $450?
11. 7 is what percent of 8?
12. 366 is 150% of what number?

13. Find $2\frac{1}{2}$% of $400.

14. Find $4\frac{3}{4}$% of $200.

15. At a certain college the fall enrollment was 5% more than the spring enrollment. The spring enrollment was 3,560. Find the fall enrollment.
16. If you work nine problems correctly on an examination of 12 problems, what is your percent grade?
17. A $95 suit of clothes is marked down 20%. Find the selling price of the suit.
18. The rent on a $90-a-month apartment is raised $2\frac{1}{2}$%. Find the new rent.
19. 500 grams of a solution contain 75 grams of a drug. Find the percent of drug.
20. Find the monthly interest on $280 when the monthly interest rate is 1.5%.
21. Find the yearly interest on $1,150 when the yearly interest rate is 9.6%.
22. One week a salesman working on a 15% commission made $255. Find his total sales for the week.
23. Mr. Garcia, with a salary of $9,500, was given a 5.6% raise. What was his salary after the raise?
24. Use the method of add-on-interest to find the monthly payment when the balance due is $1,500. The yearly rate of interest is 7%, and payments are to run for 24 months.

REVIEW EXERCISES 612, SET II. In Exercises 1-3, change the given form into the two remaining forms.

	Fraction	Decimal	Percent
1.	$\frac{3}{8}$		
2.		1.25	
3.			$16\frac{2}{3}$%

4. What is 175% of $350?
5. 93 is what percent of 124?
6. 62.4 is 26% of what number?

7. Find $7\frac{3}{4}$% of $500.

8. The evening enrollment at a certain college is 6% less than the day enrollment. If the day enrollment is 2,450, find the evening enrollment.

9. The rent on a $180-a-month apartment was raised $4\frac{1}{2}\%$. Find the new rent.

10. Find the yearly interest on $940 when the yearly interest rate is 8.4%.

11. Mr. Edmonson's new contract calls for a 12% raise in salary. His present salary is $11,250. What will his new salary be?

12. Use the method of add-on-interest to find the monthly payment, if the balance due is $4,500. The yearly rate of interest is 8%, and payments are to run for 36 months.

Chapter 6: Diagnostic Test

Name_____

_____

The purpose of this test is to see how well you understand percent. We recommend that you work this diagnostic test *before* your instructor tests you on this chapter. Allow yourself about an hour to do this test.

Complete solutions for all the problems on this test, together with section references, are given in the Answer Section. You should study the sections referred to for the problems you do incorrectly.

---

1. Change $\frac{4}{5}$ to a decimal.

(1)_____

2. Change $4\frac{2}{3}$ to a decimal. (Round off to two decimal places.)

(2)_____

3. Change 0.74 to percent.

(3)_____

4. Change 2.05 to percent.

(4)_____

5. Change 48% to a decimal.

(5)_____

6. Change 565% to a decimal.

(6)_____

7. Change 0.58 to a fraction reduced to lowest terms.

(7)_____

8. Change 2.75 to a mixed number in lowest terms.

(8) _____

9. Find $\frac{3}{8}$ of 56.

(9) _____

10. Find 0.32 of 480.

(10) _____

11. Find 78% of 350.

(11) _____

12. 12 is what percent of 30?

_____ (12) _____

13. 28 is 42% of what number? (Round off your answer to the nearest unit.)

(13) _____

_____

14. Bill worked 17 problems correctly on a test having 20 problems. Find his percent score.

(14) _____

_____

15. After trading his car in on a new car, Tony had a balance due of $2,000. He agreed to pay the balance in 36 equal monthly payments. The yearly interest rate was 7%.

(a) Find the interest for 36 months (3 years).

(15a) _____

(b) Find the total amount to be paid.

(15b)_____

(c) Find the monthly payment. (Round off to the nearest cent.)

(15c)_____

SEVEN

The English System of Measurement

The English System of Measurement

From the earliest times, different systems have been used for measuring lengths, weights, time, volumes, etc. The two major systems in use today are the *English system* and the *metric system*. In this section, we discuss the English system. The metric system is discussed in Chapter 8.

The English system that we use every day includes such units as inches, feet, yards, miles, pounds, gallons, etc. The relationships between some of these commonly used English system units are given in the following table.

Some Common Equivalent English Units of Measure		
Length Units:	1 quart (qt)	= 2 pints (pt)
	1 gallon (gal)	= 4 quarts (qt)
	1 gallon (gal)	= 231 cubic inches (cu in)
Time Units:	1 minute (min)	= 60 seconds (sec)
	1 hour (hr)	= 60 minutes (min)
	1 day (da)	= 24 hours (hr)
	1 week (wk)	= 7 days (da)
	1 year (yr)	= 52 weeks (wk)
	1 year (yr)	= 365 days (da)
Volume Units:	1 foot (ft)	= 12 inches (in)
	1 yard (yd)	= 3 feet (ft)
	1 mile (mi)	= 5,280 feet (ft)
Weight Units:	1 pound (lb)	= 16 ounces (oz)
	1 ton	= 2,000 pounds (lb)

Figures 701A and 701B show how two English units were derived.

Figure 701A.—King Henry I decreed: "A yard should be the distance from the tip of my nose to the end of my thumb."

Figure 701B.—King Edward I decreed: "One inch should equal three barley corns laid end to end."

 **Changing Units of Measure by Means of Unit Fractions**

It is often necessary to change from one unit of measure to another. This can be done by means of ratios called *unit fractions*. We now show the unit fraction method most commonly used in science. We think you will find it helpful.

You *can* change from one unit of measure to another *without* using unit fractions, but students are sometimes confused as to whether they should divide or multiply in making the conversion. The method of unit fractions minimizes this confusion.

<u>Example 1</u>. Change 27 yards to feet.

<u>Solution</u>: 1 yd = 3 ft

If we form the ratio $\boxed{\dfrac{3\ \text{ft}}{1\ \text{yd}}}$ ———— unit fraction this ratio must equal 1 because

the numerator and denominator are equal.

So that $27 \text{ yd} = 27 \text{ yd} \times (1) = \dfrac{27\ \cancel{\text{yd}}}{1}\left(\dfrac{3\ \text{ft}}{1\ \cancel{\text{yd}}}\right) = (27 \times 3) \text{ ft}$

$= 81 \text{ ft}$

Notice that units can be canceled as well as numbers.

↑——— This is a unit fraction because 1 yd = 3 ft.

Therefore, 27 yd = 81 ft

You know that the value of the unit fraction must be 1. We now explain how to select the correct unit fraction.

To Select the Correct Unit Fraction

1. The unit fraction must have two units:
   (a) the unit you want in your answer
   (b) the unit you want to get rid of
2. The unit fraction must be written so that the unit of measure you want to get rid of can be canceled.
   (a) *If the unit you want to get rid of is in the numerator* of the given expression, that same unit must be in the denominator of the unit fraction chosen. (See Example 2.)
   (b) *If the unit you want to get rid of is in the denominator* of the given expression, that same unit must be in the numerator of the unit fraction chosen. (See Example 10.)

Example 2. Change 7 inches to feet.

Solution:

Feet placed in the numerator so we get feet in the answer.

$$\frac{7 \text{ in}}{1} \left(\frac{1 \text{ ft}}{12 \text{ in}} \right)$$

Inches placed in the denominator to cancel with inches in 7 inches.

$$\frac{7 \text{ in}}{1} \left(\frac{1 \text{ ft}}{12 \text{ in}} \right) = \frac{7}{12} \text{ ft}$$

Therefore,
$$7 \text{ in} = \frac{7}{12} \text{ ft}$$

If you try to use the ratio the other way, it won't work.

$$\frac{7 \text{ in}}{1} \left(\frac{12 \text{ in}}{1 \text{ ft}} \right)$$

Here you see that the inches will not cancel, so you will not be left with just feet in the answer.

Example 3. 5 lb = _?_ oz

Solution:

Because 1 lb = 16 oz

$$\frac{5 \text{ lb}}{1} \left(\frac{16 \text{ oz}}{1 \text{ lb}} \right) = (5 \times 16) \text{ oz} = 80 \text{ oz}$$

Therefore,
$$5 \text{ lb} = 80 \text{ oz}$$

Example 4. 7,920 ft = _?_ mi

Solution:

Because 1 mi = 5,280 ft

$$\frac{7,920 \text{ ft}}{1} \left(\frac{1 \text{ mi}}{5,280 \text{ ft}} \right) = \frac{7,920}{5,280} \text{ mi} = \frac{3}{2} \text{ mi} = 1\frac{1}{2} \text{ mi}$$

Therefore,
$$7,920 \text{ ft} = 1\frac{1}{2} \text{ mi}$$

Example 5. 84 hr = _?_ da

Solution:

Because 1 da = 24 hr

$$\frac{84 \text{ hr}}{1} \left(\frac{1 \text{ da}}{24 \text{ hr}} \right) = \frac{84}{24} \text{ da} = \frac{7}{2} \text{ da} = 3\frac{1}{2} \text{ da}$$

Therefore,
$$84 \text{ hr} = 3\frac{1}{2} \text{ da}$$

Example 6. $7\frac{1}{2}$ gal = _?_ qt

Solution:

Because 4 qt = 1 gal

$$\frac{7\frac{1}{2} \text{ gal}}{1} \left(\frac{4 \text{ qt}}{1 \text{ gal}} \right) = (7\frac{1}{2} \times 4) \text{ qt} = 30 \text{ qt}$$

Therefore,
$$7\frac{1}{2} \text{ gal} = 30 \text{ qt}$$

Note: Sometimes it is necessary to use two or more unit fractions to change to the required units. (See Example 7.)

Example 7. 2 hr = __?__ sec

Solution: Step 1. Change hours to minutes.

$$\frac{2 \cancel{hr}}{1} \left(\frac{60 \text{ min}}{1 \cancel{hr}} \right) = (2 \times 60) \text{ min} = 120 \text{ min}$$

Step 2. Change minutes to seconds.

$$\frac{120 \cancel{min}}{1} \left(\frac{60 \text{ sec}}{1 \cancel{min}} \right) = (120 \times 60) \text{ sec} = 7{,}200 \text{ sec}$$

Therefore, 2 hr = 7,200 sec

This problem could be worked in one step as follows:

$$\frac{2 \cancel{hr}}{1} \left(\frac{60 \cancel{min}}{1 \cancel{hr}} \right) \left(\frac{60 \text{ sec}}{1 \cancel{min}} \right) = (2 \times 60 \times 60) \text{ sec} = 7{,}200 \text{ sec}$$

A word of caution: The unit fraction must be written so that the unit of measure you want to get rid of can be canceled.

Sometimes the unit you want to get rid of is in the numerator (Examples 8, 9).

Example 8. 26 mi = __?__ ft

To get rid of miles and be left with feet:

Correct choice	*Incorrect choice*
because miles cancel	because miles do *not* cancel

$$\frac{26 \cancel{mi}}{1} \left(\frac{5280 \text{ ft}}{1 \cancel{mi}} \right) \qquad\qquad \frac{26 \text{ mi}}{1} \left(\frac{1 \text{ mi}}{5280 \text{ ft}} \right)$$

Therefore, 26 mi = (26 × 5280) ft = 137,280 ft

Example 9. 13.6 hr = __?__ min

To get rid of hours and be left with minutes:

Correct choice	*Incorrect choice*
because hours cancel	because hours do *not* cancel

$$\frac{13.6 \cancel{hr}}{1} \left(\frac{60 \text{ min}}{1 \cancel{hr}} \right) \qquad\qquad \frac{13.6 \text{ hr}}{1} \left(\frac{1 \text{ hr}}{60 \text{ min}} \right)$$

Therefore, 13.6 hr = (13.6 × 60) min = 816 min

Sometimes the unit you want to get rid of is in the denominator (Example 10).

Example 10. 2000 mi per hr = 2000 $\dfrac{\text{mi}}{\text{hr}}$ = $\dfrac{?}{\text{min}}$ $\dfrac{\text{mi}}{}$ = __?__ mi per min

To get rid of hours and be left with minutes:

Correct choice	*Incorrect choice*
because hours cancel	because hours do *not* cancel

$$\frac{2000 \text{ mi}}{\cancel{\text{hr}}} \left(\frac{1 \ \cancel{\text{hr}}}{60 \text{ min}} \right) \qquad\qquad \frac{2000 \text{ mi}}{\text{hr}} \left(\frac{60 \text{ min}}{1 \text{ hr}} \right)$$

Therefore, 2000 mi per hr = $\left(\dfrac{2000}{60} \right) \dfrac{\text{mi}}{\text{min}} \doteq 33.33$ mi per min

Example 11. Fred drives 10 miles in 20 minutes. What is his average speed in miles per hour?

Solution: $\qquad \dfrac{10 \text{ mi}}{\cancel{20 \text{ min}}} \left(\dfrac{\overset{3}{\cancel{60 \text{ min}}}}{1 \text{ hr}} \right) = 30 \ \dfrac{\text{mi}}{\text{hr}} = 30 \text{ mph}$

EXERCISES 702, SET I. In the following exercises, find the missing numbers.

1. 5 yd = __?__ ft

2. 7 yd = __?__ ft

3. 3 ft = __?__ in

4. 5 ft = __?__ in

5. $3\frac{1}{3}$ yd = __?__ ft

6. $4\frac{1}{3}$ yd = __?__ ft

7. $5\frac{2}{3}$ ft = __?__ in

8. $3\frac{1}{6}$ ft = __?__ in

9. 8 gal = __?__ qt

10. 7 gal = __?__ qt

11. $2\frac{3}{4}$ gal = __?__ qt

12. $5\frac{1}{4}$ gal = __?__ qt

13. 7 qt = __?__ pt

14. 11 qt = __?__ pt

15. 2.5 qt = __?__ pt

16. 8.5 qt = __?__ pt

17. $1\frac{1}{2}$ hr = __?__ min

18. $3\frac{1}{3}$ hr = __?__ min

19. 7.75 min = __?__ sec

20. 3.25 min = __?__ sec

21. $1\frac{3}{4}$ mi = __?__ ft

22. $2\frac{1}{4}$ mi = __?__ ft

23. $3\frac{1}{4}$ da = __?__ hr

24. $5\frac{1}{6}$ da = __?__ hr

25. 10 yd = __?__ ft = __?__ in

26. 7 yd = __?__ ft = __?__ in

27. 8 gal = __?__ qt = __?__ pt

28. $2\frac{1}{2}$ gal = __?__ qt = __?__ pt

29. 24 in = __?__ ft

30. 36 in = __?__ ft

31. 84 in = __?__ ft

32. 96 in = __?__ ft

33. 90 sec = __?__ min

34. 75 sec = __?__ min

35. 150 min = __?__ hr

36. 105 min = __?__ hr

37. 5,280 ft = __?__ mi

38. 10,560 ft = __?__ mi

39. 730 da = __?__ yr

40. 511 da = __?__ yr

41. 104 wk = __?__ yr

42. 65 wk = __?__ yr

43. 2 gal = __?__ cu in

44. $2\frac{1}{3}$ gal = __?__ cu in

45. $2\frac{1}{2}$ lb = __?__ oz

46. $3\frac{1}{4}$ lb = __?__ oz

47. 48 oz = __?__ lb

48. 36 oz = __?__ lb

49. $2\frac{1}{2}$ tons = __?__ lb

50. $3\frac{1}{4}$ tons = __?__ lb

51. 2,200 lb = __?__ tons

52. 3,600 lb = __?__ tons

53. Change 1.25 miles to feet.
54. Change 8,820 feet to miles.
55. Change 2 weeks to hours.
56. Change 5 miles to yards.
57. Change 2,160 minutes to days.
58. Change 60 ounces to pounds.
59. Louise drove 8 miles in 12 minutes. What was her average speed in miles per hour?
60. Abe walks 2 miles in 50 minutes. What is his average walking speed in miles per hour?
61. Sound travels about 1,100 feet per second. What is the speed of sound in miles per hour?
62. A certain glacier moves 100 yards in a year. What is its speed in feet per month?

EXERCISES 702, SET II. In the following exercises, find the missing numbers.

1. 5 yd = __?__ in

2. $3\frac{1}{4}$ ft = __?__ in

3. 5 gal = __?__ qt

4. 9 pt = __?__ gal

5. 7 qt = __?__ pt

6. $1\frac{3}{4}$ hr = __?__ min

7. 320 sec = __?__ min

8. $2\frac{1}{2}$ da = __?__ hr

9. $2\frac{1}{2}$ mi = __?__ ft

10. 60 hr = __?__ da

11. 13,200 ft = __?__ mi

12. $1\frac{3}{4}$ lb = __?__ oz

13. $1\frac{1}{4}$ tons = __?__ lb

14. 1,600 lb = __?__ tons

15. 3 wk = __?__ hr

16. 1,728 min = __?__ da

17. 15 miles per hour is how many feet per minute?
18. 88 feet per second is how many miles per hour?

Common Household Measures

Here is a table of common household measures:

Common Household Measures
1 tablespoon (tbsp) = 3 teaspoons (tsp)
1 cup = 16 tablespoons (tbsp)
1 cup = 8 ounces (oz)
1 pint (pt) = 2 cups
1 pint (pt) = 16 ounces (oz)

In Figure 703 we show a comparison of cups, pints, and ounces.

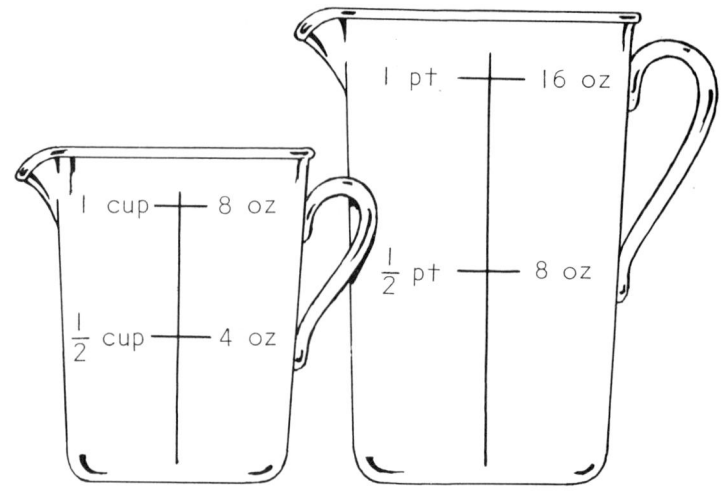

Figure 703

Unit fractions can be used to change from one unit of measure to another.

Example 1. 12 tablespoons = __?__ cups

Solution: $\dfrac{12 \text{ tbsp}}{1} \left(\dfrac{1 \text{ cup}}{16 \text{ tbsp}} \right) = \dfrac{12}{16} \text{ cup} = \dfrac{3}{4} \text{ cup}$

Therefore, $\qquad\qquad$ 12 tbsp = $\dfrac{3}{4}$ cup

EXERCISES 703, SET I. In Exercises 1-10, change to the units indicated.

1. $\dfrac{1}{4}$ cup = __?__ oz

2. $\dfrac{1}{4}$ cup = __?__ tbsp

3. $\dfrac{3}{8}$ cup = __?__ tbsp

4. $\dfrac{5}{8}$ cup = __?__ tsp

5. 1 gal = __?__ oz

6. 4 cups = __?__ pt

7. $1\dfrac{1}{3}$ tbsp = __?__ tsp

8. 3 pt = __?__ cups

9. 5 pt = __?__ cups   10. $2\frac{2}{3}$ tbsp = __?__ tsp

11. The ingredients for one lemon pie filling are as follows:

$1\frac{1}{4}$ cups sugar $\frac{1}{4}$ cup butter or margarine, melted

2 tablespoons flour   3 eggs

$\frac{1}{8}$ teaspoon salt   1 teaspoon grated lemon peel

$\frac{1}{2}$ cup water   1 lemon, peeled and sliced thin

Mr. Gomez plans to bake three of these pies. How much of each ingredient will he need?

12. The recipe for a Brownie Dessert Royal is as follows:

$\frac{3}{4}$ cup butter or margarine   3 eggs

3 oz unsweetened chocolate   1 cup all-purpose flour

1 cup sugar $\frac{1}{2}$ teaspoon baking powder

$1\frac{1}{2}$ teaspoons vanilla

In preparing dessert for a church social, Mrs. Smith needs to make five times the amount of the recipe. How much of each ingredient will she need?

EXERCISES 703, SET II. Change to the units indicated.

1. $\frac{3}{4}$ cup = __?__ oz   2. 18 tbsp = __?__ cup

3. $1\frac{1}{2}$ gal = __?__ oz   4. 9 tsp = __?__ tbsp

5. 7 pt = __?__ cups   6. 12 oz = __?__ cups

7. 9 cups = __?__ pt   8. $2\frac{1}{2}$ cups = __?__ oz

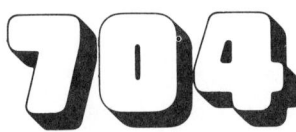

Denominate Numbers

If you wish to describe how tall you are, how much you weigh, how old you are, or how much money you have in your pocket, you use certain *standard units of measure*. For example, you may be 69 *inches* tall, weigh 160 *pounds*, be 19 *years* old, and have 15 *dollars* in your pocket. Here the standard units of measure are inches, pounds, years, and dollars.

Numbers expressed in standard units of measure, such as inches, pounds, years, dollars, and so on, are called *denominate numbers*. A denominate number is a number with a name (*nomen* means name in Latin). When we say 3 feet, the "3 feet" is a denominate number. When we say 5 hours, the "5 hours" is a denominate number. Numbers such as 5, 25, $\frac{3}{4}$, etc., which are not given with units, are called *abstract numbers*. When we say 35 + 10 = 45, we are using abstract numbers, but when we say 35 feet + 10 feet = 45 feet, we are using denominate numbers.

When we say *like numbers*, we mean denominate numbers expressed in the same units.

Example 1. Examples of like numbers.

    (a) 7 feet and 20 feet

    (b) 15 pounds and 10 pounds

Example 2. Examples of unlike numbers.

    (a) 5 feet and 10 inches

    (b) 2 hours and 15 minutes

EXERCISES 704, SET I

1. Arrange the following numbers into sets so that all like numbers are in the same set: 7 inches; 5 pounds; $1; 4 gallons; 6 inches; 3 gallons; $10; 2 inches.
2. Arrange the following numbers into two sets so that all the abstract numbers are in one set and all the denominate numbers are in the other set: 8 miles; 15; 7 gallons; $\frac{5}{8}$; 4 pounds; 6.
3. Make a list of the different units of measure used in the following sentences: John drove 4 miles to the store, where he bought 3 pounds of hamburger, 2 dozen eggs, 13 ounces of cookie mix, 283 grams of slivered almonds, 2 quarts of milk, and 1 pint of mayonnaise. His bill came to exactly $7.
4. Arrange the following into two sets of numbers so that all abstract numbers are in one set and all denominate numbers are in the other set: $5; 7 inches; 21; $\frac{3}{4}$; 8 miles; 5 pounds; 6; 3 dozen.
5. Arrange the following into sets so that all like numbers are in the same set: 2 pints; 5 gallons; 3 quarts; $5; 75¢; 3 pints; 2 gallons; $4.

EXERCISES 704, SET II

1. Arrange the following numbers into sets so that all like numbers are in the same set: 3 feet; 2 teaspoons; 5 cups; 4 feet; $8; 2 cups; 4 yards; $60; 1 yard.
2. Make a list of the different units of measure used in the following sentences: Janet drove 11 miles to the village to shop. She purchased 5 pounds of meat, 2 dozen eggs, 10 pounds of oranges, 1 dozen apples, and 1 gallon of milk.
3. Arrange the following into two sets of numbers so that all abstract numbers are in one set and all denominate numbers are in the other set: 2 pints; $3; 35; $2\frac{1}{2}$ feet; 8 inches; 7; 2 gallons.

Simplifying Denominate Numbers

To simplify a denominate number having two or more different units, change a quantity of smaller units to larger units whenever possible.

Example 1. Simplify: 2 ft 18 in

Solution:
$$2 \text{ ft} + 18 \text{ in}$$
$$= 2 \text{ ft} + 1 \text{ ft} + 6 \text{ in}$$
$$= 3 \text{ ft} + 6 \text{ in}$$
$$= 3 \text{ ft } 6 \text{ in}$$

Another way of writing the solution:

2 ft   18 in

1 ft 6 in

1 ft
3 ft   6 in

Therefore, 2 ft 18 in = 3 ft 6 in.

Example 2. Simplify: 5 hr 76 min

Solution:
$$5 \text{ hr} + 76 \text{ min}$$
$$= 5 \text{ hr} + 1 \text{ hr} + 16 \text{ min}$$
$$= 6 \text{ hr} + 16 \text{ min}$$
$$= 6 \text{ hr } 16 \text{ min}$$

Another way of writing the solution:

5 hr   76 min

1 hr 16 min

1 hr
6 hr   16 min

A further simplification:

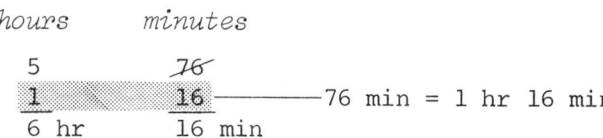

hours	minutes
5	~~76~~
1	16
6 hr	16 min

76 min = 1 hr 16 min

Example 3. Simplify: 14 hr 126 min

Solution:

hours	minutes	
14	~~126~~	
2	6	——126 min = 2 hr 6 min
16 hr	6 min	

Therefore, 14 hr 126 min = 16 hr 6 min.

Example 4. Simplify: 2 da 23 hr 150 min

Solution:

days	hours	minutes	
2	23	~~150~~	
	2	30	——150 min = 2 hr 30 min
	~~25~~		
1	1		——25 hr = 1 da 1 hr
3 da	1 hr	30 min	

Therefore, 2 da 23 hr 150 min = 3 da 1 hr 30 min.

Example 5. Simplify: 4 gal 11 qt 5 pt

Solution:

gallons	quarts	pints	
4	11	~~5~~	
	2	1	—— 5 pt = 2 qt 1 pt
	~~13~~		
3	1		—— 13 qt = 3 gal 1 qt
7 gal	1 qt	1 pt	

Example 6. Simplify: 5 yd 7 ft 15 in

Solution:

yards	feet	inches	
5	7	~~15~~	
	1	3	—— 15 in = 1 ft 3 in
	~~8~~		
2	2		—— 8 ft = 2 yd 2 ft
7 yd	2 ft	3 in	

EXERCISES 705, SET I. Simplify the following denominate numbers.

1. 4 ft 15 in   2. 7 yd 5 ft

3. 2 wk 9 da   4. 2 da 36 hr

5. 3 gal 15 qt   6. 7 qt 11 pt

7. 5 yd 4 ft 27 in   8. 10 yd 2 ft 34 in

9. 2 hr 73 min 110 sec   10. 3 hr 82 min 125 sec

11. 3 gal 7 qt 5 pt   12. 4 gal 5 qt 6 pt

13. 2 mi 6,000 ft   14. 3 mi 6,400 ft

15. 2 yr 48 wk 75 da   16. 3 yr 51 wk 55 da

17. 2 tons 3,500 lb   18. 3 tons 2,250 lb

19. 4 lb 20 oz   20. 5 lb 56 oz

21. 2 da 23 hr 75 min   22. 3 yd 5 ft 15 in

EXERCISES 705, SET II. Simplify the following denominate numbers.

1. 3 ft 19 in   2. 3 wk 11 da

3. 4 gal 6 qt   4. 3 yd 5 ft 18 in

5. 4 hr 70 min 84 sec   6. 2 gal 5 qt 3 pt

7. 2 mi 6,380 ft   8. 3 yr 51 wk 75 da

9. 6 lb 54 oz   10. 3 da 22 hr 250 min

Adding Denominate Numbers

Only like numbers can be added.

$$5 \text{ dollars} + 3 \text{ dollars} = 8 \text{ dollars}$$
$$2 \text{ feet} + 4 \text{ feet} = 6 \text{ feet}$$
$$6 \text{ days} + 11 \text{ days} = 17 \text{ days}$$

This shows that only addition of like numbers makes sense.

```
To Add Denominate Numbers

1. Write the denominate numbers under one another
   with like units in the same vertical line.
2. Add the numbers in each vertical line.
3. Simplify the denominate number found in (2).
```

Example 1. Add: 7 ft 3 in, 3 ft 5 in, and 2 ft 7 in.

Solution:
```
          7 ft   3 in
          3      5
          2      7
          1
          ─      ──
                 15

        13 ft   3 in
```

Example 2. Add: 2 gal 1 qt 1 pt, 3 gal 2 qt 1 pt, and 5 gal 3 qt 1 pt.

Solution:
```
                2 gal 1 qt 1 pt
                3     2    1
                5     3    1
                1     1
                ─    ─̶7̶̶   ─̶3̶̶
```

11 gal 3 qt 1 pt

Example 3. Add: 3 da 18 hr 42 min, 5 da 9 hr 29 min, and 1 da 14 hr 51 min.

Solution:
```
                3 da 18 hr  42 min
                5     9     29
                1    14     51
                1     2
               ─̶1̶0̶̶  ─̶4̶3̶̶   ─̶1̶2̶2̶̶
```

1 wk 3 da 19 hr 2 min

EXERCISES 706, SET I. Find the following sums, then simplify.

1. 4 ft 5 in
   3 ft 6 in
   5 ft 2 in

2. 2 yd 2 ft
   3 yd 2 ft
   4 yd 1 ft

3. 3 hr 15 min
   2 hr 50 min
   7 hr 24 min

4. 35 min 54 sec
   48 min 27 sec

5. 3 gal 2 qt 1 pt
   5 gal 3 qt 1 pt
   8 gal 1 qt 1 pt

6. 3 gal 2 qt 1 pt
   5 gal 3 qt 1 pt
   4 gal 1 qt 1 pt

7. 3 yd 2 ft 10 in
   1 yd 1 ft 9 in
   8 yd 2 ft 7 in

8. 1 mi 4,000 ft
   2 mi 3,800 ft

9. 1 da 12 hr 15 min
   5 da 23 hr 54 min
   2 da 18 hr 47 min

10. 2 yr 41 wk 5 da
    1 yr 18 wk 4 da
    3 yr 27 wk 3 da

11. 3 tons 1,500 lb
    5 tons   450 lb
    7 tons 1,850 lb

12. 3 lb 5 oz
     8 lb 15 oz
    13 lb 9 oz

13. 7 lb 3 oz
    15 lb 9 oz
    22 lb 17 oz

14. 5 hr 17 min 35 sec
    3 hr 44 min 47 sec
    2 hr 53 min 24 sec

EXERCISES 706, SET II. Find the following sums, then simplify.

1. 5 ft 7 in
   2 ft 5 in
   3 ft 8 in

2. 2 hr 50 min
   4 hr 15 min
   5 hr 20 min

```
3.  4 gal 2 qt 1 pt              4.  2 yd 2 ft  7 in
    3 gal 1 qt 2 pt                  4 yd 3 ft  5 in
    1 gal 3 qt 3 pt                  3 yd 1 ft 10 in

5.  6 lb 8 oz                    6.  2 wk 13 da 18 hr 50 min
    7 lb 6 oz                        1 wk  5 da 15 hr 40 min
    3 lb 9 oz                              6 da  3 hr 20 min

7.  3 tons 1,400 lb
    2 tons   600 lb
    5 tons   800 lb
```

Subtracting Denominate Numbers

Only like numbers can be subtracted.

> To Subtract Denominate Numbers
>
> 1. Write the number being subtracted under the number it is being subtracted from, writing like units in the same vertical line.
> 2. Subtract the numbers in each vertical line, borrowing when necessary from the first nonzero number to the left.
> 3. Simplify your answer.

Example 1. Subtract 2 ft 4 in from 10 ft 6 in.

Solution:

$$
\begin{array}{rr}
10 \text{ ft} & 6 \text{ in} \\
- 2 & 4 \\
\hline
8 \text{ ft} & 2 \text{ in}
\end{array}
$$

Example 2. Subtract 4 gal 3 qt from 7 gal 1 qt.

$$
\begin{array}{rr}
{}^{6}\!\!\not{7} \text{ gal} & {}^{5}\!\!\not{1} \text{ qt} \\
- 4 & 3 \\
\hline
2 \text{ gal} & 2 \text{ qt}
\end{array}
$$

← The 1 gallon "borrowed" makes 4 quarts, which, when added to the 1 quart already there, gives 5 quarts.

Example 3. Subtract 1 hr 50 min 40 sec from 3 hr 21 min.

Solution:

$$
\begin{array}{rrr}
 & {}^{80}\!\!\!\not{20} & \\
{}^{2}\!\!\not{3} \text{ hr} & \not{21} \text{ min} & {}^{60}\!\!\not{0} \text{ sec} \\
- 1 & 50 & 40 \\
\hline
1 \text{ hr} & 30 \text{ min} & 20 \text{ sec}
\end{array}
$$

Example 4. Subtract 1 yd 2 ft 10 in from 3 yd 4 in.

Solution:

$$\begin{array}{c c c}
\overset{2}{\cancel{3}} \text{ yd} & \overset{\overset{2}{\cancel{2}}}{\cancel{0}} \text{ ft} & \overset{16}{\cancel{4}} \text{ in} \\
- \ 1 & 2 & 10 \\
\hline
1 \text{ yd} & 0 \text{ ft} & 6 \text{ in} = 1 \text{ yd } 6 \text{ in}
\end{array}$$

EXERCISES 707, SET I. Subtract and simplify.

1. 8 ft 10 in
    − 3 ft 4 in

2. 9 ft 6 in
    − 7 ft 2 in

3. 13 yd 2 ft
    − 7 yd 1 ft

4. 7 yd 2 ft
    − 3 yd 1 ft

5. 8 gal 2 qt
    − 3 gal 3 qt 1 pt

6. 8 hr 15 min
    − 3 hr 50 min

7. 5 lb 3 oz
    − 2 lb 8 oz

8. 3 lb 7 oz
    − 1 lb 10 oz

9. 3 tons   700 lb
    − 1 ton 1,200 lb

10. 4 tons   500 lb
     − 2 tons 1,600 lb

11. 3 da 5 hr
     − 1 da 15 hr

12. 5 gal   1 pt
     − 3 gal 3 qt

13. 5 yd 2 ft 6 in
     − 2 yd 2 ft 9 in

14. 3 yd 1 ft 9 in
     − 1 yd 2 ft 10 in

15. 4 mi 3,000 ft
     − 1 mi 4,700 ft

16. 3 mi 2,000 ft
     − 2 mi 2,350 ft

17. 35 min 40 sec
     − 20 min 55 sec

18. 27 min 20 sec
     − 15 min 32 sec

19. 5 da 13 hr 22 min
     − 2 da 18 hr 45 min

20. 5 yr 43 wk 3 da
     − 2 yr 50 wk 5 da

21. Brain Teaser. If a puppy in a box weighs 2 pounds 10 ounces, and the box alone weighs 10 ounces, what does the puppy weigh?

EXERCISES 707, SET II. Subtract and simplify.

1. 6 ft 8 in
    − 4 ft 5 in

2. 8 yd 2 ft
    − 2 yd 1 ft

3. 10 gal 2 qt
    − 4 gal 3 qt 1 pt

4. 5 lb 9 oz
    − 2 lb 10 oz

5. 3 tons 400 lb
    − 1 ton   900 lb

6. 7 gal   1 pt
    − 3 gal 3 qt

7. 8 yd 2 ft 4 in
    − 2 yd 3 ft 6 in

8. 8 mi 1,500 ft
    − 7 mi 1,800 ft

9. 7 yr 25 wk 18 da
    − 5 yr 30 wk 20 da

10. 5 wk 3 da 10 hr
     − 1 wk 8 da 12 hr

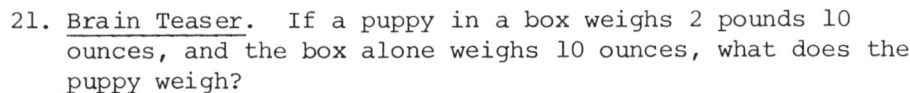

Multiplying Denominate Numbers

To Multiply a Denominate Number by an Abstract Number

1. Multiply each part of the denominate number by the abstract number.
2. Simplify the product.

Example 1. Multiply 4 × (2 ft 8 in).

Solution:

```
        2 ft   8 in
             ×  4
        8 ft  32 in  ←
        2      8         ⎤ Simplifying the product
       10 ft   8 in  ←
```

Example 2. Multiply 5 × (4 gal 3 qt 1 pt).

Solution:

```
        4 gal   3 qt 1 pt
                    ×  5
       20 gal  15 qt  5 pt  ←
                2      1        ⎤ Simplifying
               17               ⎥ the product
        4       1
       24 gal   1 qt 1 pt  ←
```

EXERCISES 708A, SET I. Multiply and simplify.

1. 4 × (3 wk 5 da)   2. 3 × (2 yr 225 da)

3. 6 × (5 mi 2,850 ft)   4. 5 × (3 mi 1,550 ft)

5. 5 × (2 yd 1 ft 3 in)   6. 3 × (5 gal 2 qt 1 pt)

7. 4 × (1 hr 25 min 11 sec)   8. 7 × (2 yd 2 ft 6 in)

9. 8 × (3 gal 3 qt 1 pt)   10. 6 × (2 hr 15 min 21 sec)

EXERCISES 708A, SET II. Multiply and simplify.

1. 3 × (6 wk 4 da)   2. 4 × (5 mi 750 ft)

3. 5 × (4 gal 3 qt 1 pt)   4. 6 × (7 yd 2 ft 9 in)

5. 8 × (3 hr 20 min 25 sec)

MULTIPLYING A DENOMINATE NUMBER BY A DENOMINATE NUMBER. Sometimes it is possible to multiply a denominate number by another denominate number.

Example 3. Find the area of a rectangle that is 3 yards long and 2 yards wide.

Solution:

$$\begin{aligned}\text{Area} &= \text{length} \times \text{width}\\ &= 3 \text{ yards} \times 2 \text{ yards}\\ &= 6 \text{ square yards (sq yd)}\end{aligned}$$

The reason we can multiply these two denominate numbers is that the following has meaning:

$$1 \text{ yd} \times 1 \text{ yd} = 1 \text{ sq yd}$$

Example 4. A crew of 10 men works for 8 hours.

Solution: 10 men × 8 hours = 80 man-hours

The reason we can multiply these two denominate numbers is that the following has meaning:

$$1 \text{ man} \times 1 \text{ hour} = 1 \text{ man-hour}$$

Example 5. 5 apples × 3 chairs.

Solution: 5 apples × 3 chairs = 15 apple-chairs

Since apple-chair has no meaning that we can think of, multiplying these denominate numbers does not make sense.

Multiplying Denominate Numbers

1. A denominate number can always be multiplied by an abstract number. (See Examples 1 and 2.)
2. A denominate number can be multiplied by another denominate number only when the product of their units has meaning.

EXERCISES 708B, SET I

1. Find the area of a rectangle that is 176 feet long and 48 feet wide.
2. Find the area of a rectangle that is 28 inches by 17 inches.
3. A baseball diamond is a square 90 feet on each side. Find the area enclosed in a baseball diamond.
4. A football field measures 100 yards between goal lines and measures 55 yards between side lines. Find the area enclosed by these lines.
5. A construction crew of 17 men worked 12 eight-hour days to construct a building. Find the total man-hours used to construct the building. Find the cost to construct the building if the cost per man-hour is $6.50.
6. In designing a new vacuum cleaner, 956.4 man-hours were used. Find the cost of designing this machine if the average cost of a man-hour was $14.23.

1. Find the area of a rectangle that is 204 feet long and 75 feet wide.
2. Find the area of a playing field that is 90 yards long and 52 feet 6 inches wide.
3. In making a wing assembly on an airplane 875.5 man-hours were used. Find the cost of assembling this wing if the average cost of each man-hour was $12.50.

Dividing Denominate Numbers

To divide a denominate number expressed in unlike units by an abstract number, we divide the largest unit first, then the next largest unit, and so on until the division is completed. Our first example is simple and has no remainder.

Example 1. Divide (10 yd 2 ft 8 in) by 2.

Solution:

$$\begin{array}{r} 5 \text{ yd } 1 \text{ ft } 4 \text{ in} \\ 2\overline{\smash{\big)}10 \text{ yd } 2 \text{ ft } 8 \text{ in}} \end{array}$$

To Divide a Denominate Number by an Abstract Number

1. Divide the largest unit by the abstract number.
2. If a remainder is left from (1), change it to the next smaller unit and add it to those units already there.
3. Divide the sum found in (2) by the abstract number.
4. Repeat steps (2) and (3) until the division is complete.

Example 2. Divide (11 yd 1 ft 8 in) by 3.

Solution:

$$\begin{array}{r} \underline{3 \text{ yd}} \quad 2 \text{ ft} \quad 6\frac{2}{3} \text{ in} \\ 3\overline{\smash{\big)}11 \text{ yd} \quad 1 \text{ ft} \quad 8 \text{ in}} \\ \underline{9} \\ 2 \text{ yd} = 6 \text{ ft} \\ \overline{7 \text{ ft}} \end{array}$$

$$\begin{array}{r} \underline{2 \text{ ft}} \\ 3\overline{\smash{\big)}7} \\ \underline{6} \\ 1 \text{ ft} = 12 \text{ in} \\ \overline{20 \text{ in}} \end{array}$$

This 2 inches has not been divided by 3. When it is divided by 3, we get $\frac{2}{3}$ inch. ⟶

$$\begin{array}{r} 6\frac{2}{3} \text{ in} \\ 3\overline{\smash{\big)}20 \text{ in}} \\ \underline{18} \\ 2 \text{ in} \end{array}$$

The writing of this solution can be shortened as follows:

$$
\begin{array}{r}
3 \text{ yd} \quad\quad 2 \text{ ft} \quad\quad 6\frac{2}{3} \text{ in} \\
\hline
3\,\overline{)\,11 \text{ yd} \quad\quad 1 \text{ ft} \quad\quad 8 \text{ in}} \\
\underline{9} \\
2 \text{ yd} = \underline{6 \text{ ft}} \\
7 \text{ ft} \\
\underline{6} \\
1 \text{ ft} = \underline{12 \text{ in}} \\
20 \text{ in} \\
\underline{18} \\
2 \text{ in} \longrightarrow \frac{2 \text{ in}}{3} = \frac{2}{3} \text{ in}
\end{array}
$$

__Example 3.__ Divide (3 gal 2 qt 1 pt) by 4.

__Solution:__

$$
\begin{array}{r}
0 \text{ gal} \quad\quad 3 \text{ qt} \quad\quad 1\frac{1}{4} \text{ pt} \\
\hline
4\,\overline{)\,3 \text{ gal} \quad\quad 2 \text{ qt} \quad\quad 1 \text{ pt}} \\
\underline{0} \\
3 \text{ gal} = \underline{12 \text{ qt}} \\
14 \text{ qt}
\end{array}
$$

$$
\begin{array}{r}
3 \text{ qt} \\
\hline
4\,\overline{)\,14 \text{ qt}} \\
\underline{12} \\
2 \text{ qt} = \underline{4 \text{ pt}} \\
5 \text{ pt}
\end{array}
$$

This 1 pint has not been divided by 4. When it is divided by 4, we get $\frac{1}{4}$ pint. $\longrightarrow$

$$
\begin{array}{r}
1\frac{1}{4} \text{ pt} \\
\hline
4\,\overline{)\,5 \text{ pt}} \\
\underline{4} \\
1 \text{ pt}
\end{array}
$$

The writing of this solution can be shortened as follows:

$$
\begin{array}{r}
0 \text{ gal} \quad\quad 3 \text{ qt} \quad\quad 1\frac{1}{4} \text{ pt} \\
\hline
4\,\overline{)\,3 \text{ gal} \quad\quad 2 \text{ qt} \quad\quad 1 \text{ pt}} \\
\underline{0} \\
3 \text{ gal} = \underline{12 \text{ qt}} \\
14 \text{ qt} \\
\underline{12} \\
2 \text{ qt} = \underline{4 \text{ pt}} \\
5 \text{ pt} \\
\underline{4} \\
1 \text{ pt} \longrightarrow \frac{1 \text{ pt}}{4} = \frac{1}{4} \text{ pt}
\end{array}
$$

EXERCISES 709A, SET I. Divide and simplify.

1. (2 ft 6 in) ÷ 2   2. (3 qt 1 pt) ÷ 3

3. (2 qt 1 pt) ÷ 4   4. (6 ft 8 in) ÷ 5

5. (5 hr 30 min) ÷ 3   6. (6 hr 40 min) ÷ 4

7. (4 gal 3 qt 1 pt) ÷ 3   8. (5 yd 2 ft 6 in) ÷ 3

9. (8 yd 2 ft 10 in) ÷ 5   10. (5 gal 3 qt 1 pt) ÷ 6

11. (5 lb 8 oz) ÷ 7   12. (4 lb 10 oz) ÷ 8

13. (13 wk 5 da 15 hr) ÷ 3 14. (12 wk 4 da 10 hr) ÷ 5

15. (8 mi 4,500 ft) ÷ 6   16. (4 mi 4,000 ft) ÷ 12

EXERCISES 709A, SET II. Divide and simplify.

1. (5 ft 8 in) ÷ 2   2. (7 hr 40 min) ÷ 4

3. (8 gal 2 qt 1 pt) ÷ 3   4. (10 lb 8 oz) ÷ 5

5. (8 wk 15 da 10 hr) ÷ 3   6. (8 yd 2 ft 10 in) ÷ 4

7. (20 wk 25 da 15 hr) ÷ 6 8. (10 mi 2,752 ft) ÷ 8

DIVIDING A DENOMINATE NUMBER BY A DENOMINATE NUMBER. Sometimes it is possible to divide a denominate number by another denominate number.

Example 1. How many 3-foot shelves can be cut from a 12-foot board?

Solution:
$$\frac{12 \text{ ft}}{3 \text{ ft}} = \frac{\overset{4}{\cancel{12} \cancel{\text{ft}}}}{\underset{1}{\cancel{3} \cancel{\text{ft}}}} = 4$$
Notice that the answer is an abstract number.

Example 2. How many $2 ties can be bought with $18?

Solution:
$$\frac{18 \text{ dollars}}{2 \text{ dollars}} = \frac{\overset{9}{\cancel{18 \text{ dollars}}}}{\underset{1}{\cancel{2 \text{ dollars}}}} = 9$$

Example 3. If a sponsor buys enough TV time to permit a total of one hour for commercials, how many 3-minute commercials can he put on?

Solution:
$$\frac{1 \text{ hour}}{3 \text{ min}} = \frac{\overset{20}{\cancel{60 \text{ min}}}}{\underset{1}{\cancel{3 \text{ min}}}} = 20$$

1. How many 2-foot fence posts can be cut from a 16-foot board?

2. How many $1\frac{3}{4}$-foot stakes can be cut from a 14-foot board?

3. How many 15¢ postcards can $1.75 buy?

4. How many special-addressed envelopes costing 13 cents each can be bought for $17.50?

5. How many 2-minute radio commercials can be fitted into 1 hour and 30 minutes available just for commercials?

6. It takes 32 minutes to machine a special fitting. How many of these fittings can be made in an 8-hour work day?

EXERCISES 709B, SET II

1. How many $1\frac{1}{2}$-foot stakes can be cut from a 12-foot board? Disregard the waste in the cutting.

2. How many 35¢ cards can be bought for $3.85? Disregard the sales tax.

3. How many 14¢ stamped envelopes can be bought for $3.50?

Chapter Summary

The *English system* of measurement is the one in common use in this country today. (Sec. 701)

Changing Units of Measure by Means of Unit Fractions. (Sec. 702) The unit fraction must be written so that the unit of measure you want to get rid of can be canceled.

1. *If the unit you want to get rid of is in the numerator* of the given expression, that same unit must be in the denominator of the unit fraction chosen.

2. *If the unit you want to get rid of is in the denominator* of the given expression, that same unit must be in the numerator of the unit fraction chosen.

Denominate numbers have specific names such as 7 feet, 5 gallons, etc., while numbers such as 3, 4, 5, are *abstract numbers*. (Sec. 704)

To add denominate numbers, write like units under one another. Then add each set of like units and simplify. (Sec. 706)

To subtract denominate numbers, write like units under one another. Then subtract each set of units, borrowing when necessary, and simplify. (Sec. 707)

To multiply a denominate number by an abstract number, multiply each part of the denominate number by the abstract number and simplify. A denominate number can always be multiplied by an abstract number. A denominate number can be multiplied by another denominate number only when the product of their units has meaning. (Sec. 708)

For division of denominate numbers, see Section 709.

REVIEW EXERCISES 710, SET I. In Exercises 1-12, find the missing numbers.

1. 1.5 mi = ? ft   2. 18,480 ft = ? mi

3. 3.75 yd = ? in   4. 216 in = ? yd

5. 11 gal = ? pt   6. 44 pt = ? gal

7. $2\frac{3}{4}$ da = ? hr   8. 280 min = ? hr

9. 240 oz = ? lb   10. $3\frac{1}{2}$ tons = ? lb

11. $1\frac{1}{2}$ cups = ? oz   12. $1\frac{3}{4}$ qt = ? cups

13. Add: 3 hr 25 min 45 sec   14. Add: 3 tons 1,500 lb
          4 hr 50 min 30 sec   2 tons 1,200 lb

15. Subtract: 7 gal 1 qt 1 pt   16. Subtract: 5 lb 9 oz
              - 3 gal 3 qt 1 pt   - 2 lb 13 oz

17. Multiply: 18. Multiply:

    7 × (3 yd 2 ft 4 in)   3 × (5 hr 22 min 45 sec)

19. Divide: 20. Divide:

    (7 hr 20 min 48 sec) ÷ 4   (5 yd 2 ft 9 in) ÷ 3

21. The greatest known oceanic depth is 36,198 feet. What is this depth in miles? Express your answer to the nearest tenth of a mile.
22. The height of Mt. Everest is 29,028 feet. What is this height in miles? Express your answer to the nearest tenth of a mile.
23. At a certain place on the Colorado River the speed of the current is $4\frac{1}{2}$ miles per hour. What is this speed in feet per minute?
24. Find the number of feet per minute you are traveling when you are driving 45 miles per hour.
25. A *light-year* is the *distance* traveled by light in one year. If light travels approximately 186,000 miles per second, express 4.3 light-years in millions of miles. This is the distance to the nearest star, Alpha Centauri.

26. Brain Teaser. When does 960 equal 1,000?

REVIEW EXERCISES 710, SET II. In Exercises 1-6, find the missing numbers.

1. 2.5 mi = ? ft   2. 2.75 yd = ? in

3. 64 pt = ? gal   4. 112 oz = ? lb

5. $2\frac{1}{2}$ tons = ? lb   6. 33 pt = ? qt

7. Add: 5 hr 42 min 17 sec
         3 hr 35 min 52 sec

8. Subtract: 9 gal 1 qt 1 pt
             - 6 gal 2 qt 2 pt

9. Multiply:

   6 × (3 yd 2 ft 8 in)

10. Divide:

    (5 wk 4 da 10 hr) ÷ 6

11. The height of Mt. McKinley is 20,320 feet. What is this height in miles? Express your answer to the nearest tenth of a mile.

12. Find the number of feet per minute you are traveling when you are driving 70 miles per hour.

Chapter 7: Diagnostic Test

Name_____

_____

The purpose of this test is to see how well you know English weights and measures. We recommend that you work this diagnostic test *before* your instructor tests you on this chapter. Allow yourself about 50 minutes to do this test.

Complete solutions for all the problems on this test, together with section references, are given in the Answer Section. You should study the sections referred to for the problems you do incorrectly.

---

1. Fill in the missing numbers.

   (a) 4 yd = _____ ft = _____ in

   (b) 2 gal = _____ qt = _____ pt

   (c) 3 hr = _____ min = _____ sec

   (d) 5 wk = _____ da = _____ hr

   (e) 10 lb = _____ oz

---

2. Simplify the following denominate numbers.

   (a) 2 hr 85 min 97 sec

   (2a)_____ hr _____ min _____ sec

   (b) 1 yr 49 wk 67 da

   (2b)_____ yr _____ wk _____ da

   (c) 7 mi 6,080 ft

   (2c)_____ mi _____ ft

---

3. Add and simplify:

   5 yd 2 ft 8 in
   3 yd   10 in
   6 yd 1 ft 9 in
   _____

   (3)_____ yd _____ ft _____ in

---

4. Subtract and simplify:

```
  7 gal     1 pt
- 3 gal 2 qt
```

(4) _____ gal _____ qt _____ pt

5. Multiply and simplify:

6 × (2 lb 7 oz)

(5) _____ lb _____ oz

6. Find the area of a rectangle that is 55 feet by 24 feet.

(6) _____ sq ft

7. Divide and simplify:

(7 yd 2 ft 9 in) ÷ 5

(7) _____ yd _____ ft _____ in

8. If beans cost 21¢ a can, how many cans can be bought for $3.15?

(8) _____

9. Fill in the missing numbers.

   (a) 3 cups = _____ pt

   (b) 6 tsp = _____ tbsp

   (c) 8 tbsp = _____ cup

   (d) 2 pt = _____ oz

_____

10. Change 60 miles per hour to feet per second.

                                    (10)_____

EIGHT

The Metric System of Measurement

The metric system is a decimal system of measurement that originated in France during the French Revolution. Today, most of the world uses the metric system or will use it in the near future. The metric system is already used in our country in medicine, nursing, the pharmaceutical industry, the National Aeronautics and Space Administration, photography, the military, and sports. For example, you are probably familiar with the 100-meter race, 35-millimeter film and slides, the liter bottle, the 105-millimeter gun, and so on. See Figure 800.

Figure 800
Metric Uses in the United States

Astronauts on the moon use meters to describe their positions. In many countries—Mexico, for example—metric units are used for speeds and distances on road signs, weight and volume, clothing sizes, etc. The United States is committed to changing to the metric system in the near future. So, the metric system is a necessary part of your education.

Basic Metric Units of Measure

There are four commonly used basic units in the metric system that we will consider.

1. The *meter* (m). Used to measure *length*.
2. The *liter* (ℓ). Used to measure *volume*.
3. The *gram* (g). Used to measure *weight*.
4. The *centigrade degree* (°C). Used to measure *temperature*. (Also called Celsius degree.)

1. THE METER (m). The basic unit for measuring length in the metric system is the meter (m). It corresponds roughly to the yard in the English system of measurement. (See Figure 801A.)

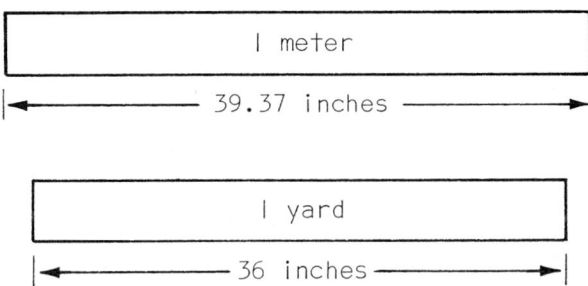

| 1 meter

|←——— 39.37 inches ———→|

| 1 yard

|←——— 36 inches ———→|

Figure 801A

2. THE LITER (ℓ). The basic unit for measuring volume (capacity) in the metric system is the liter (ℓ). It corresponds roughly to the quart in the English system of measurement. We buy gasoline by the gallon in this country. In metric countries gasoline is sold by the liter. (See Figure 801B.)

1 liter
= 1.06 qt

Figure 801B

3. THE GRAM (g). The basic unit for measuring weight in the metric system is the gram (g). One gram is a small quantity of weight. It takes approximately 454 grams to equal one pound. (See Figure 801C.)

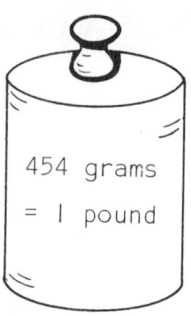

Figure 801C

4. THE CENTIGRADE DEGREE (°C). The basic unit for measuring temperature in the metric system is the centigrade degree (often called Celsius degree).

Some Equivalent Centigrade-Fahrenheit Temperatures		
Boiling point of water	100°C	= 212°F
Normal body temperature	37°C	= 98.6°F
Freezing point of water	0°C	= 32°F

Changing Units in the Metric System

PREFIXES. Basic units can be changed to larger or smaller units by means of *prefixes*. We will consider the three most commonly used prefixes.

1. Kilo means 1,000.

   Therefore, 1 *kilo*meter (km) = 1,000 meters (m)
   1 *kilo*liter (kl) = 1,000 liters (ℓ)
   1 *kilo*gram (kg) = 1,000 grams (g)

2. Centi means $\frac{1}{100}$. (Remember, 1 cent = $\frac{1}{100}$ dollar.)

   Therefore, 1 *centi*meter (cm) = $\frac{1}{100}$ meter *or* 100 cm = 1 m

   1 *centi*liter (cl) = $\frac{1}{100}$ liter *or* 100 cl = 1 ℓ

   1 *centi*gram (cg) = $\frac{1}{100}$ gram *or* 100 cg = 1 g

3. Milli means $\frac{1}{1,000}$.

   Therefore, 1 *milli*meter (mm) = $\frac{1}{1,000}$ meter

   or   1,000 mm = 1 m

$$1 \text{ } \textit{milli}\text{liter (ml)} = \frac{1}{1,000} \text{ liter}$$

or $\qquad$ $1,000 \text{ ml} = 1 \text{ } \ell$

$$1 \text{ } \textit{milli}\text{gram (mg)} = \frac{1}{1,000} \text{ gram}$$

or $\qquad$ $1,000 \text{ mg} = 1 \text{ g}$

CHANGING UNITS. All the prefixes involve either a multiplication or a division by a power of ten. Multiplying or dividing a number by a power of ten can be carried out just by moving the decimal point. Therefore, *to change to larger or smaller units in the metric system, it is only necessary to move the decimal point.* This is one of the main advantages to using the metric system.

Example 1 $\qquad\qquad$ 2.4 m = _?_ cm

$\qquad\qquad\qquad\qquad$ 1 m = 100 cm

$$2.4 \text{ m} \left(\frac{100 \text{ cm}}{1 \text{ m}} \right) = 2.4 \times 100 \text{ cm} = 2\wedge40. \text{ cm}$$

unit fraction _____↑ $\qquad\qquad\qquad$ └── 2 places to right

Therefore, $\qquad\qquad$ 2.4 m = 240 cm

For more help in using *unit fractions* to change units, see Section 702.

Example 2 $\qquad\qquad$ 5.67 kg = _?_ g

$\qquad\qquad\qquad\qquad$ 1 kg = 1,000 g

$$5.67 \text{ kg} \left(\frac{1,000 \text{ g}}{1 \text{ kg}} \right) = 5.67 \times 1,000 \text{ g} = 5\wedge670. \text{ g}$$

$\qquad\qquad\qquad\qquad\qquad\qquad\qquad\qquad$ └─3 places to right

Therefore, $\qquad\qquad$ 5.67 kg = 5,670 g

Example 3 $\qquad\qquad$ 352 ml = _?_ ℓ

$\qquad\qquad\qquad\qquad$ 1 ℓ = 1,000 ml

$$352 \text{ ml} \left(\frac{1 \text{ } \ell}{1,000 \text{ ml}} \right) = \frac{352}{1,000} \text{ } \ell = .352\wedge \text{ } \ell$$

$\qquad\qquad\qquad\qquad\qquad\qquad\qquad\qquad$ └── 3 places to left

Therefore, $\qquad\qquad$ 352 ml = 0.352 ℓ

Example 4 $\qquad\qquad$ 150 mg = _?_ g

$\qquad\qquad\qquad\qquad$ 1 g = 1,000 mg

$$150 \text{ mg} \left(\frac{1 \text{ g}}{1,000 \text{ mg}} \right) = \frac{150}{1,000} \text{ g} = .150\wedge \text{ g}$$

$\qquad\qquad\qquad\qquad\qquad\qquad\qquad\qquad$ └── 3 places to left

Therefore, $\qquad\qquad$ 150 mg = 0.150 g

We now study each prefix in more detail.

KILO (means 1,000)

```
┌──────────────────────────────────────────────────────────────┐
│                                                                │
│   Changing Units Involving the Prefix Kilo                     │
│                                                                │
│           Large units to small units                           │
│                                                                │
│              ⎧ kilometers to meters ⎫ move the decimal point   │
│   To change ⎨ kiloliters to liters  ⎬ 3 places to the right.   │
│              ⎩ kilograms to grams    ⎭ (See Example 5.)        │
│   ─ ─ ─ ─ ─ ─ ─ ─ ─ ─ ─ ─ ─ ─ ─ ─ ─ ─ ─ ─ ─ ─ ─ ─ ─ ─ ─ ─    │
│           Small units to large units                           │
│                                                                │
│              ⎧ meters to kilometers ⎫ move the decimal point   │
│   To change ⎨ liters to kiloliters  ⎬ 3 places to the left.    │
│              ⎩ grams to kilograms    ⎭ (See Example 6.)        │
│                                                                │
└──────────────────────────────────────────────────────────────┘
```

Example 5. Changing from large units to small units involving kilo.

(a) 0.42 km = 0ʌ420. m = 420 m
                  ⟶
                  3

(b) 6.039 kl = 6ʌ039. ℓ = 6,039 ℓ   move decimal
                  ⟶   three places
                  3   to the right

(c) 8.7 kg = 8ʌ700. g = 8,700 g
                ⟶
                3

Example 6. Changing from small units to large units involving kilo.

(a) 9,025 m = 9.025ʌ km = 9.025 km   move decimal
                  ⟵   three places
                  -3   to the left

(b) 640 ℓ = .640ʌ kl = 0.640 kl   (The -3 means a
                ⟵   movement of 3
                -3   places to the
                                             left.)
(c) 62,300 g = 62.300ʌ kg = 62.3 kg
                     ⟵
                     -3

EXERCISES 802A, SET I

1. 1.8 km = _?_ m   2. 34 kl = _?_ ℓ

3. 0.249 kg = _?_ g   4. 5.71 km = _?_ m

5. 60.5 ℓ = _?_ kl   6. 322 g = _?_ kg

7. 275 g = _?_ kg   8. 56.4 ℓ = _?_ kl

9. 0.78 km = _?_ m   10. 9.3 kg = _?_ g

11. 72,350 g = _?_ kg   12. 2,365 m = _?_ km

13. A pharmaceutical house's orders for hydrogen peroxide
    average 125 liters per month. How many kiloliters is this

per year?

14. A rectangular alfalfa field measures 0.90 km by 0.20 km. Find the area of this field in square meters (m^2).

EXERCISES 802A, SET II

1. 0.532 kg = ? g
2. 305 ℓ = ? kl
3. 0.68 km = ? m
4. 2.4 kl = ? ℓ
5. 10.9 g = ? kg
6. 4,375 m = ? km
7. If 235 grams of a particular chemical are used every hour, how many kilograms of this chemical will be needed per week?

CENTI (means $\frac{1}{100}$)

Changing Units Involving the Prefix *Centi*

Small units to large units

To change { centimeters to meters / centiliters to liters / centigrams to grams } move the decimal point 2 places to the *left*. (See Example 7.)

Large units to small units

To change { meters to centimeters / liters to centiliters / grams to centigrams } move the decimal point 2 places to the *right*. (See Example 8.)

Example 7. Changing from small units to large units involving centi.

(a) 155 cm = 1.55∧ m = 1.55 m (← −2)

(b) 76 cl = .76∧ ℓ = 0.76 ℓ (← −2)

(c) 4.9 cg = .04∧9 g = 0.049 g (← −2)

move decimal two places to the *left*

Example 8. Changing from large units to small units involving centi.

(a) 6.8 m = 6∧80. cm = 680 cm (→ 2)

(b) 0.47 ℓ = 0∧47. cl = 47 cl (→ 2)

(c) 5.873 g = 5∧87.3 cg = 587.3 cg (→ 2)

move decimal two places to the *right*

1. 279 cm = __?__ m   2. 54 cl = __?__ ℓ

3. 8.3 cg = __?__ g   4. 4,090 cm = __?__ m

5. 2.5 m = __?__ cm   6. 0.72 ℓ = __?__ cl

7. 3.906 g = __?__ cg   8. 0.842 m = __?__ cm

9. 632 cl = __?__ ℓ   10. 58.1 cg = __?__ g

11. 0.0263 g = __?__ cg   12. 0.092 ℓ = __?__ cl

13. Bob's height is 1.82 meters. Express his height in centimeters.

14. Hilda's gift weighed 1,430 grams. Express the weight of the gift in centigrams.

EXERCISES 802B, SET II

1. 26 cg = __?__ g   2. 0.506 ℓ = __?__ cl

3. 85.3 cm = __?__ m   4. 2,500 cl = __?__ ℓ

5. 0.38 g = __?__ cg   6. 0.046 m = __?__ cm

7. At 18 months, Susie was 71 centimeters tall. Express her height in meters.

MILLI (means $\frac{1}{1,000}$)

Changing Units Involving the Prefix *Milli*

Small units to large units

To change { millimeters to meters / milliliters to liters / milligrams to grams } move the decimal point 3 places to the *left*. (See Example 9.)

- -

Large units to small units

To change { meters to millimeters / liters to milliliters / grams to milligrams } move the decimal point 3 places to the *right*. (See Example 10.)

Example 9. Changing from small units to large units involving milli.

(a) 56 mm = .056∧ m = 0.056 m ⎤
         ←   ⎥
        −3   ⎥
                                 ⎥ move decimal
(b) 4,800 ml = 4.800∧ ℓ = 4.8 ℓ ⎥ three places
            ←   ⎥ to the *left*
           −3   ⎥
                                 ⎥
(c) 250 mg = .250∧ g = 0.25 g   ⎦
          ←
         −3

Example 10. Changing from large units to small units involving milli.

(a) 1.4 m = 1ᴧ400. mm = 1,400 mm
       ⎯⎯→
        3

(b) 0.68 ℓ = 0ᴧ680. ml = 680 ml
          ⎯⎯→
           3

(c) 0.2050 g = 0ᴧ205.0 mg = 205.0 mg
            ⎯⎯→
             3

move decimal
three places
to the *right*

Another commonly used metric unit is the *cubic centimeter*.

1 cubic centimeter (cc) = 1 milliliter (ml)

In this section we have only considered the metric units that are most commonly used. There are many more units and prefixes used in the metric system. We include a more complete table of metric units and prefixes on page 301 for your reference.

EXERCISES 802C, SET I

1. 91 mm = __?__ m

2. 5,600 ml = __?__ ℓ

3. 470 mg = __?__ g

4. 4,300 mm = __?__ m

5. 2.6 m = __?__ mm

6. 0.39 ℓ = __?__ ml

7. 0.1080 g = __?__ mg

8. 0.0827 m = __?__ mm

9. 230 ml = __?__ ℓ

10. 9,160 mg = __?__ g

11. 7.04 g = __?__ mg

12. 21.6 ℓ = __?__ ml

13. 2 ℓ = __?__ cc

14. 3.55 ℓ = __?__ cc

15. 175 cc = __?__ ℓ

16. 2,500 cc = __?__ ℓ

17. A doctor recommends that Jackie take 1.5 g of vitamin C a day. How many milligrams would she take in a week?
18. Water evaporates from a swimming pool at the rate of 52 ml per minute. How many liters of water would have to be added each day to make up for the water lost in evaporation? (Round off to the nearest liter.)

EXERCISES 802C, SET II

1. 0.18 m = __?__ mm

2. 4,200 cc = __?__ ℓ

3. 0.0236 g = __?__ mg

4. 3,500 ml = __?__ ℓ

5. 860 mm = __?__ m

6. 57.2 ℓ = __?__ ml

7. 90.1 mg = __?__ g

8. 0.6 ℓ = __?__ cc

9. It was recommended that Helen take 5 grams of vitamin C per week. If she takes three 250-mg capsules each day, is she taking more or less than the recommended dosage? By how much?

Changing English System Units to Metric Units (and Vice Versa)

Have you ever had the experience of trying to use an American-made wrench on a bolt on a foreign-made car? The American-made wrench does not fit the metric-made bolt. Since over 90 percent of the people in the world today use the metric system, we must learn how to change the English system units that we use into metric units, and vice versa.

Following is a short table listing commonly used conversions between the metric and the English systems of measurement.

COMMON ENGLISH—METRIC CONVERSIONS

```
        1 inch (in) = 2.54 centimeters (cm)
  39.4 inches (in) = 1 meter (m)
   0.621 miles (mi) = 1 kilometer (km)
        1 mile (mi) = 1.61 kilometers (km)
       1 pound (lb) = 454 grams (g)
    2.20 pounds (lb) = 1 kilogram (kg)
    1.06 quarts (qt) = 1 liter (ℓ)
      2.47 acres = 1 hectare (ha)
```

It is possible to get slightly different answers if different conversion factors are used.

Example 1. Mrs. Peralta weighs 62 kg. What is her weight in pounds?

Solution: 1 kg = 2.20 lb

$$62 \text{ kg} \left(\frac{2.20 \text{ lb}}{1 \text{ kg}} \right) = 62 \times 2.20 \text{ lb} \doteq 136 \text{ lb}$$

Therefore, 62 kg $\doteq$ 136 lb

Example 2. A road sign in Mexico reads "85 kilometers to Ensenada." How far is this in miles?

Solution: 1 km = 0.621 mi

$$85 \text{ km} \left(\frac{0.621 \text{ mi}}{1 \text{ km}} \right) = 85 \times 0.621 \text{ mi} \doteq 52.8 \text{ mi}$$

Therefore, 85 km $\doteq$ 52.8 mi

Example 3. 125 ha = __?__ acres

Solution: $125 \text{ ha} \left(\frac{2.47 \text{ acres}}{1 \text{ ha}} \right) = 125 \times 2.47 \text{ acres} \doteq 309 \text{ acres}$

Therefore, 125 ha $\doteq$ 309 acres

Example 4. 22 gal = __?__ ℓ

Solution:

$$22 \; \cancel{gal} \left(\frac{4 \; \cancel{qt}}{1 \; \cancel{gal}} \right) \left(\frac{1 \; \ell}{1.06 \; \cancel{qt}} \right) = \frac{\cdot 22 \times 4}{1.06} \; \ell \doteq 83.0 \; \ell$$

Therefore, 22 gal $\doteq$ 83.0 ℓ

Example 5. Change 104 grams to ounces.

Solution: 104 g = $\dfrac{104 \; \cancel{g}}{1} \left(\dfrac{1 \; \cancel{lb}}{454 \; \cancel{g}} \right) \left(\dfrac{16 \; oz}{1 \; \cancel{lb}} \right)$

This is 1 because ⟶   ⟵ This is 1 because
1 lb = 454 g   1 lb = 16 oz

Therefore, 104 g = $\dfrac{104 \times 16}{454}$ oz $\doteq$ 3.7 ft

Example 6. Change 228 centimeters to feet.

Solution: 228 cm = $\dfrac{228 \; \cancel{cm}}{1} \left(\dfrac{1 \; \cancel{in}}{2.54 \; \cancel{cm}} \right) \left(\dfrac{1 \; ft}{12 \; \cancel{in}} \right)$

This is 1 because ⟶   ⟵ This is 1 because
1 in = 2.54 cm   1 ft = 12 in

Therefore, 228 cm = $\dfrac{228}{2.54 \times 12}$ ft $\doteq$ 7.5 ft

Example 7. Change 750 feet per second to kilometers per minute.

Solution: $\dfrac{750 \; \cancel{ft}}{\cancel{sec}} \left(\dfrac{1 \; \cancel{mi}}{5,280 \; \cancel{ft}} \right) \left(\dfrac{1.61 \; km}{1 \; \cancel{mi}} \right) \left(\dfrac{60 \; \cancel{sec}}{1 \; min} \right) \doteq 13.7 \; km \; per \; min$

This is 1 because ⟶
1 mi = 5,280 ft   ⟵ This is 1 because
                               1 min = 60 sec
                               This is 1 because
                               1 mi = 1.61 km

Therefore, 750 ft per sec $\doteq$ 13.7 km per min

EXERCISES 803, SET I. In the following exercises, if quotients are not exact, round off to two decimal places.

1. 10 in = __?__ cm   2. 18 in = __?__ cm

3. 2.12 qt = __?__ ℓ 4. 3.18 qt = __?__ ℓ

5. 0.55 kg = __?__ lb   6. 4.7 kg = __?__ lb

7. 82 km = __?__ mi   8. 140 km = __?__ mi

9. 10 ℓ = __?__ qt   10. 7 ℓ = __?__ qt

11. 31 mi = __?__ km   12. 155 mi = __?__ km

13. 17.6 lb = __?__ kg   14. 26.4 lb = __?__ kg

15. 150 ha = __?__ acres   16. 65 acres = __?__ ha

17. 66 in = __?__ m   18. 74 in = __?__ m

19. 2 m = __?__ in
20. 3 m = __?__ in
21. 908 g = __?__ lb
22. 1,362 g = __?__ lb
23. 0.75 lb = __?__ g
24. 1.25 lb = __?__ g
25. 1.5 yd = __?__ cm
26. 100 cm = __?__ yd
27. 227 g = __?__ oz
28. 12 oz = __?__ g

29. A road sign in France reads "120 kilometers to Paris." Find this distance in miles.
30. A speed control sign reads 30 kilometers per hour. What is this in miles per hour?
31. Mr. Dubois steps on a scale in Orly Airport. The scale reads 85.7 kilograms. What is his weight in pounds?
32. A crate of transistor radios arrives from Japan marked "67.5 kilograms net weight." Find this weight in pounds.
33. In the Olympics, there is a 1,500-meter race which is about the same distance as our 1-mile race. Which race is longer and by how much? (Express the difference in feet.)
34. In the Olympics, there is a 400-meter race and a quarter-mile race. Which race is longer and by how much? (Express the difference in feet.)
35. A Howitzer muzzle measures 175 millimeters. What is this measurement in inches?
36. What is the width in inches of a 35-millimeter roll of film?
37. How many gallons does a 20-liter container hold?
38. A moon lander is decending at the rate of 1,000 miles per hour. Express this rate of descent in meters per second.

EXERCISES 803, SET II. In the following exercises, if the quotients are not exact, round off to two decimal places.

1. 1.56 ℓ = __?__ qt
2. 50 cm = __?__ in
3. 20 kg = __?__ lb
4. 160 km = __?__ mi
5. 15 qt = __?__ ℓ
6. 80 mi = __?__ km
7. 50 lb = __?__ kg
8. 200 ha = __?__ acres
9. 50 in = __?__ m
10. 3 m = __?__ in
11. 850 g = __?__ lb
12. 1.25 lb = __?__ g
13. 0.5 yd = __?__ cm
14. 300 g = __?__ oz

15. On the Autobahn in Germany, the off-ramp speed signs read (100). This speed is understood to be in kilometers per hour. To what speed in miles per hour must the driver slow when leaving the Autobahn?
16. A backpacker bought a tent made in Switzerland. Its weight was listed as 2.3 kilograms. Find its weight in pounds.
17. Find the number of centimeters in 1 foot.
18. 100 meters equals how many feet and inches?
19. The distance from the north pole to the equator along the surface of the earth has been taken as 10 million meters. Find this distance in miles.
20. Water is flowing through a pipe at the rate of 500 gallons per minute. How many liters per second is this?

Changing Fahrenheit to Centigrade Temperature (and Vice Versa)

When planning what clothing to wear, or what activity to engage in, we usually check the temperature first. When traveling in metric countries this can be confusing unless you can change centigrade to Fahrenheit. For example, if you hear that the temperature will be 30°C tomorrow, do you plan on going to the beach or going ice skating?

CHANGING CENTIGRADE TO FAHRENHEIT. This can be done by using the following formula.

To Change Centigrade to Fahrenheit

$$F = \frac{9}{5}C + 32$$

where F = number of °F

and C = number of °C

Example 1. 30°C = __?__ °F

$$F = \frac{9}{\cancel{5}}(\cancel{30}^{6}) + 32 = 54 + 32 = 86$$

Therefore, 30°C = 86°F

Example 2. 7°C = __?__ °F

$$F = \frac{9}{5}(7) + 32 = \frac{63}{5} + 32 = 12.6 + 32 = 44.6 \doteq 45$$

Therefore, 7°C $\doteq$ 45°F

CHANGING FAHRENHEIT TO CENTIGRADE. This can be done by using the following formula.

To Change Fahrenheit to Centigrade

$$C = \frac{5}{9}(F - 32)$$

where C = number of °C

and F = number of °F

<u>Example 3.</u> $68°F = \underline{\ ?\ }\ °C$

$$C = \frac{5}{9}(68 - 32) = \frac{5}{\cancel{9}}(\cancel{36}^{4}) = 20$$

Therefore, $68°F = 20°C$

<u>Example 4.</u> $97°F = \underline{\ ?\ }\ °C$

$$C = \frac{5}{9}(97 - 32) = \frac{5}{9}(65) \doteq 36$$

Therefore, $97°F \doteq 36°C$

<u>EXERCISES 804, SET I.</u> Round off answers to the nearest degree.

1. $20°C = \underline{\ ?\ }\ °F$ 2. $15°C = \underline{\ ?\ }\ °F$

3. $50°F = \underline{\ ?\ }\ °C$ 4. $59°F = \underline{\ ?\ }\ °C$

5. $8°C = \underline{\ ?\ }\ °F$ 6. $17°C = \underline{\ ?\ }\ °F$

7. $72°F = \underline{\ ?\ }\ °C$ 8. $85°F = \underline{\ ?\ }\ °C$

9. A Frenchman traveling in the U.S. notes that a Fahrenheit thermometer reads $41°$. What is the equivalent centigrade reading?

10. Is $10°C$ warmer or colder than $48°F$? By how much?

<u>EXERCISES 804, SET II.</u> Round off answers to the nearest degree.

1. $40°C = \underline{\ ?\ }\ °F$ 2. $59°F = \underline{\ ?\ }\ °C$

3. $24°C = \underline{\ ?\ }\ °F$ 4. $56°F = \underline{\ ?\ }\ °C$

5. Is $98°F$ warmer or colder than $35°C$? By how much?

Chapter Summary

The <u>*English system*</u> of measurement is the one in common use in this country today. (Sec. 701)

The <u>*metric system*</u> is a decimal system used in almost every country in the world.

<u>Changing units</u> in the English system usually involves long calculations. Changing units in the metric system only involves moving the decimal point. (Sec. 802)

To remember the most commonly used units in the metric system, you need to know only four basic units and three prefixes. (Secs. 801, 802, 804)

BASIC METRIC UNITS AND PREFIXES

```
           Basic Units              Prefixes

   Meter measures length.    Kilo means 1,000.
                                            1
   Liter measures volume.    Milli means -------.
                                          1,000
                                            1
   Gram measures weight.     Centi means -----.
                                           100
   Centigrade degree measures temperature.
```

USEFUL ENGLISH—METRIC CONVERSIONS (Sec. 803)

```
        1 inch (in) = 2.54 centimeters (cm)
    39.4 inches (in) = 1 meter (m)
   0.621 miles (mi) = 1 kilometer (km)
        1 mile (mi) = 1.61 kilometers (km)
        1 pound (lb) = 454 grams (g)
    2.20 pounds (lb) = 1 kilogram (kg)
    1.06 quarts (qt) = 1 liter (ℓ)
        2.47 acres = 1 hectare (ha)
```

REVIEW EXERCISES 805, SET I

1. 2 ℓ = ? ml   2. 240 ml = ? ℓ

3. 2,000 g = ? kg   4. 3.86 kg = ? g

5. 35 mm = ? in   6. 9 in = ? mm

7. 1.75 m = ? cm   8. 300 cm = ? m

9. 500 mg = ? g   10. 0.056 g = ? mg

11. 1,200 acres = ? ha   (Round off to nearest hectare.)

12. 640 ha = ? acres   (Round off to nearest acre.)

13. 27°C = ? °F   (Round off to nearest degree.)

14. 42°F = ? °C   (Round off to nearest degree.)

15. Our planet Earth measures 24,902 miles around at the
    equator. Express this distance to the nearest kilometer.
16. The distance from the north pole to the equator along the
    surface of the earth has been taken as 10 million meters.
    Find this distance in miles.
17. A speed control sign reads 80 kilometers per hour. What is
    this in miles per hour?

18. A speed control sign reads 70 mph. What is this in kilometers per hour?

19. A woman weighs 120 pounds on an American scale. What is her weight in kilograms? (Express answer to the nearest kilogram.)

20. A man steps on a metric scale and finds he weighs 79.6 kilograms. What is his weight in pounds? (Express answer to the nearest pound.)

21. When you travel by air the airlines allow you to take up to 20 kilograms of luggage. How much is this in pounds?

22. Temperatures in the 1,800-mile-thick mantle of the Earth are thought to be as high as 3,000°F. Express this temperature in centigrade degrees.

23. Find the volume in cubic centimeters of a container that holds 1.75 liters of water. (1 cubic centimeter [cc] = 1 ml)

24. One model of the Ford Pinto has a 2,000-cubic centimeter displacement engine. What is the equivalent cubic inch displacement? (1 cubic centimeter [cc] = 1 ml)

25. A light-year is the distance traveled by light in one year. If light travels approximately 186,000 miles per second, express 8.6 light-years in millions of kilometers. This is the distance to Sirius, the brightest star visible in the northern hemisphere.

26. To start the U.S. Bicentennial Celebration, July 4, 1976, the light from a star 200 light-years from Earth was used to trip a sensor that flipped a switch lighting the lantern in the Old North Church in Boston. That light had left the star when the U.S. was born in 1776. How many billion kilometers had that light traveled? (See Exercise 25.)

REVIEW EXERCISES 805, SET II

1. 800 ml = __?__ ℓ

2. 1.6 kg = __?__ g
   (Round off to one decimal place.)

3. 105 mm = __?__ in

4. 250 cm = __?__ m

5. 2.4 g = __?__ mg

6. 50 ha = __?__ acres   (Round off to nearest acre.)

7. 24°C = __?__ °F   (Round off to nearest degree.)

8. The Datsun 280Z has a 2,800-cubic centimeter engine displacement. What is the equivalent cubic inch displacement? (Round off to the nearest cubic inch.)

9. The luggage weight limit for overseas flights is 20 kilograms. Mrs. Helmes's luggage weighs 52 pounds. Is this over or under the limit? By how many pounds?

10. Change 66 feet per second to miles per hour. (Round off answer to nearest mile per hour.)

11. Ben can run 100 meters in 10 seconds. What is his average speed in miles per hour? (Round off answer to nearest mile per hour.)

12. Mauna Kea, the highest mountain in Hawaii, stands approximately 33,500 feet above the ocean floor. What is this height in meters? (Round off to nearest meter.)

Chapter 8: Diagnostic Test

Name_____

_____

The purpose of this test is to see how well you know metric weights and measures as well as converting metric to English and English to metric units. We recommend that you work this diagnostic test *before* your instructor tests you on this chapter. Allow yourself about fifty minutes to do this test.

Complete solutions for all the problems on this test, together with section references, are given in the Answer Section. You should study the sections referred to for the problems you do incorrectly.

_____

In Problems 1-17, fill in the missing numbers. Consider all given numbers to be exact.

1. 5.24 m = _____ cm

2. 25 kg = _____ lb

3. 500 cc = _____ ml = _____ ℓ

4. 100 km = _____ mi

5. 7.42 qt = _____ ℓ

6. 100 ha = _____ acres

7. 870 g = _____ kg

8. 35°C = _____ °F $\left(F = \dfrac{9}{5}C + 32\right)$

9. 1,260 cm = _____ m

10. 42.5 kl = _____ ℓ

11. 1,075 mg = _____ g

12. 1.65 ℓ = _____ ml = _____ cc

13. 1.3 m = _____ mm

14. 5 lb = _____ g

15. 5,094 m. = _____ km

16. 77°F = _____ °C $\left(C = \frac{5}{9}[F - 32] \right)$

17. 20 in = _____ cm

---

18. A Mexican speed sign reads (50) . This is understood to be in kilometers per hour. What is this speed limit in miles per hour? (Round off to nearest mile per hour.)

(18) _____

---

19. A prescription calls for taking 650 mg of aspirin per day. How many grams would this be per week?

(19) _____

---

20. Express the length of a 10-foot board in meters. Round off your answer to the nearest tenth of a meter.

(20) _____

NINE

Arithmetic in Geometry

In this chapter we consider some of the applications of arithmetic to the more common geometric figures. We live in *rectangular* rooms, we use *circular* plates, our roofs have *triangular* shapes, much of our food comes in *cylindrical* cans, we play with *spherical* basketballs, volleyballs, and so on.

We will discuss some of the properties of these figures in this chapter.

Rectangle

A *rectangle* is shown in Figure 901A. The *area* of the rectangle is the space inside the lines (see Section 118).

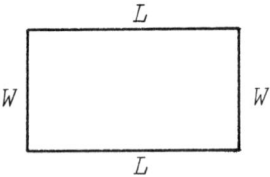

Figure 901A

Area of rectangle = length × width

$$A = L \times W$$

The *perimeter* of a rectangle is the sum of the lengths of all its sides. The word "perimeter" means "the measure around a figure."

Perimeter of rectangle = 2 × length + 2 × width

$$P = 2L + 2W$$

SQUARE. A *square* is shown in Figure 901B. A square is a rectangle in which the length and width are equal. In the same figure, we let *s* represent both length and width.

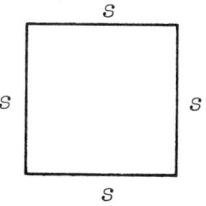

Figure 901B

$$\text{Area of square} = s \times s$$
$$A = s^2$$
$$\text{Perimeter of square} = s + s + s + s$$
$$P = 4s$$

Example 1. Find the area and perimeter of a rectangle that is 10 inches long and 6 inches wide. Give area in both square inches and square feet. Give perimeter in both inches and feet.

Solution:

$$\text{Area} = L \times W$$
$$= 10 \text{ in} \times 6 \text{ in} = 60 \text{ sq in}$$
$$= 60 \text{ sq in} \left(\frac{1 \text{ sq ft}}{144 \text{ sq in}} \right)$$
$$= \frac{5}{12} \text{ sq ft}$$

$$\text{Perimeter} = 2L + 2W$$
$$= 2 \times 10 \text{ in} + 2 \times 6 \text{ in}$$
$$= \quad 20 \text{ in} \quad + \quad 12 \text{ in} = 32 \text{ in}$$
$$= 32 \text{ in} \left(\frac{1 \text{ ft}}{12 \text{ in}} \right)$$
$$= \frac{8}{3} \text{ ft} = 2\frac{2}{3} \text{ ft}$$

Example 2. Find the area and perimeter of a square that measures $6\frac{1}{2}$ centimeters on a side.

First Solution:

	Area		Perimeter

$$A = s^2 = s \cdot s \qquad\qquad P = 4s$$

$$A = 6\frac{1}{2} \times 6\frac{1}{2} \qquad\qquad P = 4 \times 6\frac{1}{2}$$

$$A = \frac{13}{2} \times \frac{13}{2} = \frac{169}{4} = 42\frac{1}{4} \text{ sq cm} \qquad P = 4 \times \frac{13}{2}$$

$$P = \frac{4}{1} \times \frac{13}{2} = 26 \text{ cm}$$

Second Solution:

Area   Perimeter

$$A = s^2 = s \cdot s \qquad\qquad P = 4s$$

$$A = 6.5 \times 6.5 = 42.25 \text{ sq cm} \qquad P = 4 \times 6.5 = 26 \text{ cm}$$

Example 3. Find the cost to carpet a 9-foot × 12-foot room at $8.74 per square yard.

Solution: We first find the area in square yards, then multiply by the cost per square yard.

$$A = L \times W$$

$$A = 4 \text{ yd} \times 3 \text{ yd}$$

$$A = 12 \text{ sq yd}$$

$$\text{Cost} = 12 \times \$8.75 = \$105.00$$

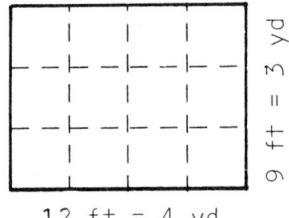

EXERCISES 901, SET I. In Exercises 1-10, find the area and perimeter of each rectangle.

	Length	Width	Express area in:	Express perimeter in:
1.	7 ft	5 ft	sq ft	ft
2.	13 ft	11 ft	sq ft	ft
3.	$4\frac{1}{2}$ ft	$2\frac{1}{2}$ ft	sq ft	ft
4.	25 cm	15 cm	sq cm	cm
5.	75 cm	35 cm	sq cm	m
6.	4.375 in	2.625 in	sq in (to nearest tenth)	in

	Length	Width	Express area in:	Express perimeter in:
7.	12 in	12 in	sq in & sq ft	ft
8.	3 ft	3 ft	sq ft & sq yd	ft
9.	3 yd	3 yd	sq yd & sq ft	yd
10.	$5\frac{1}{4}$ ft	$2\frac{3}{8}$ ft	sq ft	ft

11. The dimensions of a kitchen and family room are shown in Figure 901C. At 23¢ a square foot, find the cost to cover this floor with vinyl tile. By drawing lines, you can divide the figure into several rectangles. Then add the areas of the smaller rectangles to find the total area of the floor.

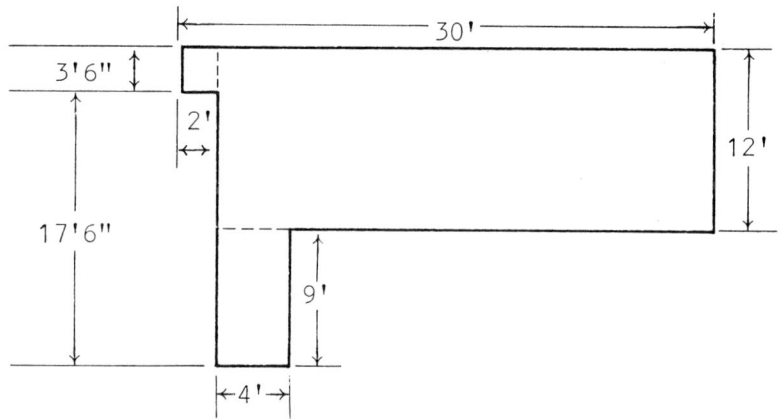

Figure 901C

12. A picture is 30 inches long and 20 inches wide. There is a 2-inch border around the picture. Find the number of square inches in this border.

EXERCISES 901, SET II. In Exercises 1-5, find the area and the perimeter of each rectangle.

	Length	Width	Express area in:	Express perimeter in:
1.	8 ft	6 ft	sq ft	ft
2.	$5\frac{1}{2}$ ft	$3\frac{1}{2}$ ft	sq ft	ft
3.	65 cm	45 cm	sq cm	m
4.	2 yd	4 ft	sq yd	yd
5.	$6\frac{3}{4}$ in	$5\frac{1}{3}$ in	sq in	in

6. Find the cost to cover this floor with a vinyl tile that costs 35¢ a square foot.

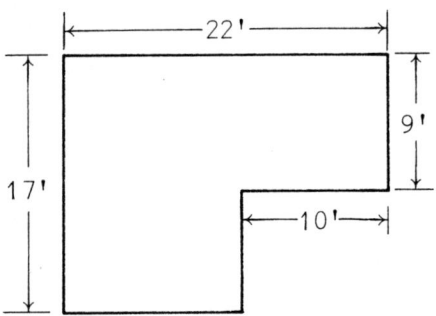

Triangle

A *triangle* is shown in Figure 902. The *base* (*b*) of the triangle is the side on which it appears to rest. The *height* (*h*) of the triangle is the distance from the base to the top of the triangle.

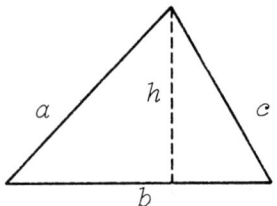

Figure 902

Area of triangle = $\frac{1}{2}$ base x height

$$A = \frac{1}{2} bh$$

Perimeter of triangle:

$$P = a + b + c$$

Example 1. Find the area and perimeter of the triangle shown at the right.

Solution: Area = $\frac{1}{2} bh$

$$= \frac{1}{2}(6)(3) = 9 \text{ sq cm}$$

Perimeter = $a + b + c$

$$= 5 + 6 + 3.6 = 14.6 \text{ cm}$$

Example 2. Find the area and perimeter of the triangle shown.

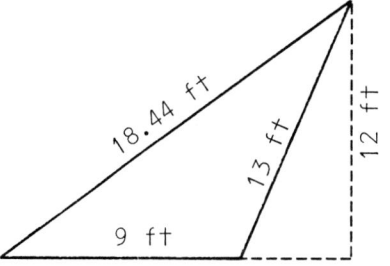

Solution: Notice that the height is 12 feet and not 13 feet, because the height is measured vertically from a level base.

$$\text{Area} = \frac{1}{2}\, bh$$

$$= \frac{1}{2}(9)(12) = 54 \text{ sq ft}$$

Perimeter $= a + b + c$

$$= 18.44 + 9 + 13 = 40.44 \text{ ft}$$

Example 3. Find the area and perimeter of the triangle shown.

Solution:

$$\text{Area} = \frac{1}{2}\, bh$$

$$= \frac{1}{2}(9 \text{ ft } 2 \text{ in})(5 \text{ ft})$$

$$= \frac{1}{2}(110 \text{ in})(60 \text{ in}) = 3,300 \text{ sq in}$$

$$= 3,300 \text{ sq in} \left(\frac{1 \text{ sq ft}}{144 \text{ sq in}}\right) = \frac{3,300}{144} \text{ sq ft}$$

$$= 22\frac{11}{12} \text{ sq ft}$$

Perimeter $= 8 \text{ ft } 4 \text{ in} + 5 \text{ ft } 7 \text{ in} + 9 \text{ ft } 2 \text{ in}$

```
      8 ft  4 in
      5     7
      9     2
   _____
      1    13 in
     23 ft  1 in
```

EXERCISES 902, SET I

1. Find the area and perimeter of the given figure.

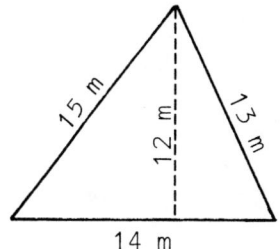

2. Find the area of the given triangle.
   (a) Express answer in square inches.
   (b) Express answer in square feet.

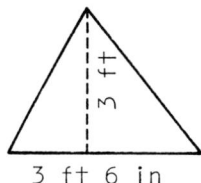

3. Find the area of the given triangle.
   (a) Express answer in square inches.
   (b) Express answer in square feet.

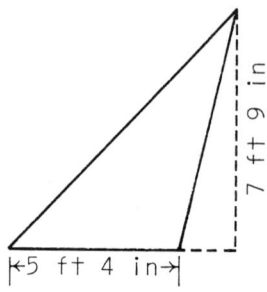

4. (a) Find the perimeter of the given figure in yards and feet.
   (b) Find the area of the figure in square feet.

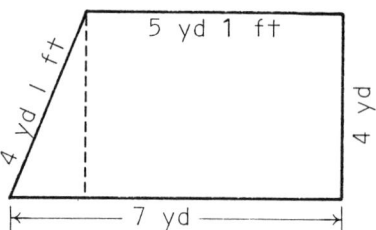

5. (a) Find the perimeter of this figure in feet and inches.
   (b) Find the total area of this figure in square feet.

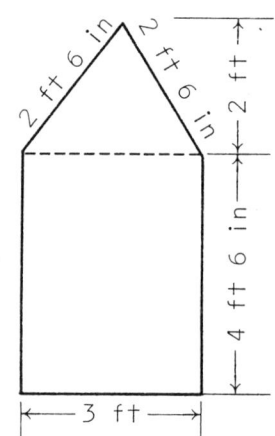

6. (a) Find the perimeter of this figure in feet and inches.
   (b) Find the total area of this figure in square feet.

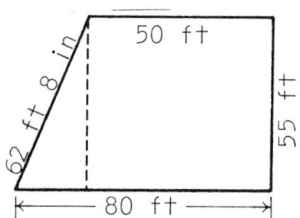

7. <u>Brain Teaser</u>. Make eight triangles with six matches.

EXERCISES 902, SET II

1. Find the area and perimeter of the given triangle.

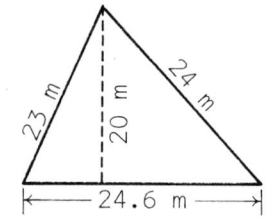

2. Find the area of the given
   triangle.
   (a) Express answer in square
       inches.
   (b) Express answer in square
       feet.

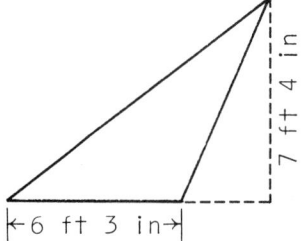

3. (a) Find the perimeter of
       this figure in feet
       and inches.
   (b) Find the total area of
       the figure in square
       feet.

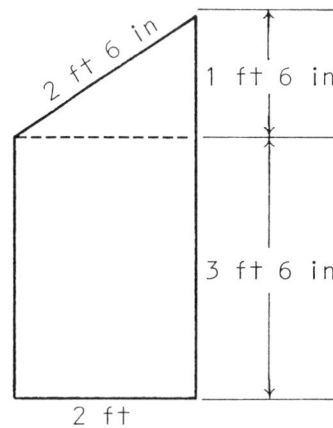

Circle

A *circle* is shown in Figure 903. All points
of a circle are the same distance from a point
within it called the *center* (0). The *radius*
(r) of the circle is the distance from the
center to any point on the circle. The
diameter (d) of the circle is the greatest
distance across the circle; it is twice the
radius. The *circumference* (C) of the circle
is its length; it is the perimeter of the
circle. If we divide the circumference of any
circle by its diameter, we always get the
same number. That number has been named *pi*
(a Greek letter) and is written π.

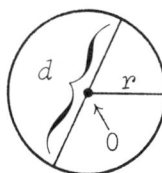

Figure 903

$$\pi = \frac{\text{circumference}}{\text{diameter}} = 3.141592653 \ldots$$

In using the number π, we round it off to whatever degree of
accuracy is called for in the problem.

Area of circle: $A = \pi r^2$

Circumference of circle: $C = \pi d = 2\pi r$

Example 1. Find the circumference and area of a circle having a radius of 6.25 inches. Use π = 3.14 and round off the answers to one decimal place.

Solution:

$$C = 2\pi r$$

$$= 2(3.14)(6.25) \doteq 39.2 \text{ in}$$

$$A = \pi r^2$$

$$= (3.14)(6.25)^2$$

$$= (3.14)(39.0625) \doteq 122.7 \text{ sq in}$$

EXERCISES 903A, SET I. Complete the following table using the given information. Use π = 3.14 and round off the answers to one decimal place.

	Radius	Diameter	Circumference	Area
1.	10 ft	___ ft	___ ft	___ sq ft
2.	___ in	8 in	___ in	___ sq in
3.	3 yd	___ yd	___ yd	___ sq yd
4.	2 ft 6 in	___ ft	___ ft	___ sq ft
5.	___ cm	20 cm	___ cm	___ sq cm
6.	___ ft	9 ft	___ ft	___ sq ft
7.	3.8 in	___ in	___ in	___ sq in
8.	___ in	7 in	___ in	___ sq in

EXERCISES 903A, SET II. Complete the following table using the given information. Use π = 3.14 and round off the answers to one decimal place.

	Radius	Diameter	Circumference	Area
1.	4 yd	___ yd	___ yd	___ sq yd
2.	___ cm	10 cm	___ cm	___ sq cm
3.	2.5 in	___ in	___ in	___ sq in
4.	___ ft	3 ft	___ ft	___ sq ft

A FACT ABOUT CIRCLES

Example 2. Compare the area of a 4-inch circle with the area of a 2-inch circle.

$d = 2$ in

$r = 1$ in

$A = \pi r^2 = \pi(1)^2 = \pi$

$d = 4$ in

$r = 2$ in

$A = \pi r^2$

$\quad = \pi(2)^2 = 4\pi$

You can see that: a diameter   2   times as large makes

the area   4   times as large

$2^2 = 4$

This means that a 4-inch pipe will carry four times as much water as a 2-inch pipe.

Example 3. Compare the area of a 6-inch circle with the area of a 2-inch circle.

$d = 2$ in

$r = 1$ in

$A = \pi r^2$

$\quad = \pi(1)^2 = \pi$

$d = 6$ in

$r = 3$ in

$A = \pi r^2$

$\quad = \pi(3)^2 = 9\pi$

You can see that: a diameter   3   times as large makes

the area   9   times as large.

$3^2 = 9$

This means that a 6-inch pipe will carry nine times as much water as a 2-inch pipe.

FLOW OF WATER IN A GARDEN HOSE. Did you know that a $\frac{3}{4}$-inch garden hose will carry $2\frac{1}{4}$ times as much water as a $\frac{1}{2}$-inch garden hose? Or that a $\frac{5}{8}$-inch garden hose will carry over $1\frac{1}{2}$ times as much water as a $\frac{1}{2}$-inch garden hose? Or that a 1-inch pipe will carry four times as much water as a $\frac{1}{2}$-inch pipe? The reason for this was shown in Examples 1 and 2.

The proportion that compares the flow of water in two different pipes is:

$$\frac{F}{f} = \left(\frac{D}{d}\right)^2 \quad \text{where} \begin{cases} F = \text{the flow in the large pipe} \\ f = \text{the flow in the small pipe} \\ D = \text{diameter of the large pipe} \\ d = \text{diameter of the small pipe} \end{cases}$$

Example 4. Compare the flow of water in a 3-inch pipe with that in a 1-inch pipe.

Solution:
$$\frac{F}{f} = \left(\frac{D}{d}\right)^2$$

$$\frac{F}{f} = \left(\frac{3}{1}\right)^2 = 9$$

This means the flow in the 3-inch pipe is nine times the flow in the 1-inch pipe.

Example 5. How many $\frac{3}{4}$ - inch pipe lines can be run from a $1\frac{1}{2}$ - inch line?

Solution:
$$\frac{F}{f} = \left(\frac{D}{d}\right)^2$$

$$\frac{F}{f} = \left(\frac{1\frac{1}{2}}{\frac{3}{4}}\right)^2 = \left(\frac{\frac{3}{2}}{\frac{3}{4}}\right)^2 = \left(\frac{\cancel{3}}{\cancel{2}} \cdot \frac{\overset{2}{\cancel{4}}}{\cancel{3}}\right)^2 = 2^2 = 4$$

This means you can run four $\frac{3}{4}$ - inch pipe lines from a single $1\frac{1}{2}$ - inch line.

EXERCISES 903B, SET I

1. Compare the flow of water in a $\frac{7}{8}$ - inch hose with that in a $\frac{1}{2}$ - inch hose.

2. Compare the flow of water in a $\frac{5}{8}$ - inch hose with that in a $\frac{1}{2}$ - inch hose.

3. Disregarding the friction in the lines, how many sprinklers with $\frac{1}{2}$ - inch pipes can be run from a 2-inch main?

4. It takes 16 hours to fill a pool with a $\frac{1}{2}$ - inch hose. How long will it take to fill the pool with a $\frac{7}{8}$ - inch hose?

5. Compare the flow of water in a $\frac{7}{8}$ - inch hose with that in a $\frac{1}{4}$ - inch hose.

6. Compare the flow of water in a 1-inch hose with that in a $\frac{3}{4}$ - inch hose.

EXERCISES 903B, SET II

1. Compare the flow of water in a $\frac{3}{4}$ - inch hose with that in a $\frac{1}{2}$ - inch hose.

2. Compare the flow of water in a 1-inch hose with that in a $\frac{5}{8}$ - inch hose.

3. Disregarding the friction in the line, how many sprinklers with $\frac{1}{2}$ - inch pipes can be run from a $1\frac{3}{4}$ - inch main?

Volume and Surface Area of a Rectangular Box

VOLUME. A square box having equal length, width, and height is called a *cube*. See Figure 904A. If the length, width, and height are each 1 foot, the cube is called a *cubic foot* (cu ft). If the length, width, and height are 1 inch, the cube is called a *cubic inch* (cu in). If the length, width, and height are each 1 yard, the cube is called a *cubic yard* (cu yd). The *volume* of any container is a measure of the space inside that container. Volume is often measured in cubic units: cubic feet, cubic inches, cubic yards, and so on.

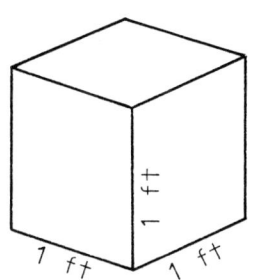

Figure 904A

We often need to know the volume of a *rectangular box*. See Figure 904B. The top, bottom, and all sides of a rectangular box are rectangles. Examples of rectangular boxes are most classrooms, most rooms in houses and apartments, shipping boxes and crates, Kleenex boxes, laundry soap boxes, and so on.

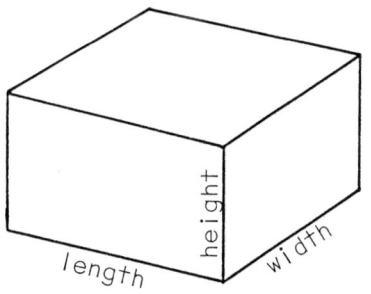

Figure 904B

A rectangular box is shown in Figure 904C. It has a length of 4 inches, a width of 3 inches, and a height of 1 inch.

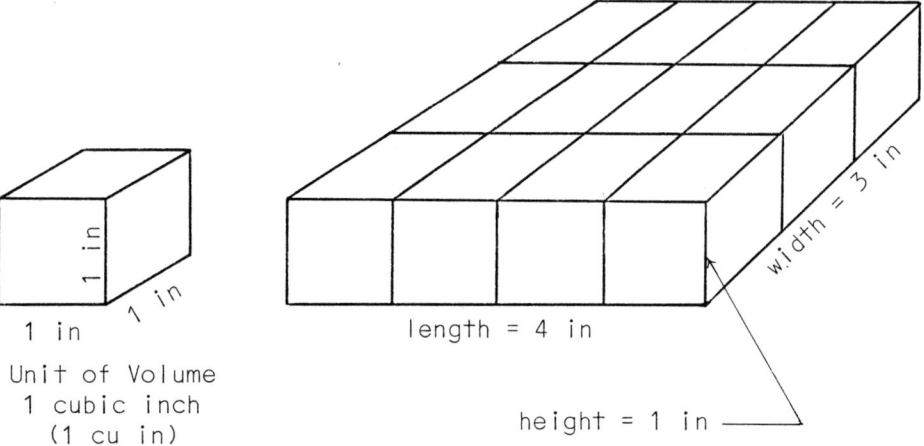

1 in
Unit of Volume
1 cubic inch
(1 cu in)

Figure 904C

Since the unit of volume (1 cu in) fits into the rectangular box 12 times, we say that the volume of the rectangular box is 12 cu in. If the length and width of the rectangular box remain the same and its height is increased to 2 inches, the volume is made up of 2 layers each containing 12 cu in. So that the volume of the box is 2 · 12 cu in = 24 cu in. If the height were 5 inches, there would be 5 layers each containing 12 cu in, so that the volume of the box would be 5 · 12 cu in = 60 cu in. This leads to the following formula for the volume of a rectangular box.

Volume of rectangular box = length × width × height

$$V = LWH$$

Example 1. Find the volume of a classroom having a length of 33 feet, a width of 30 feet, and a height of 12 feet.

Solution: Volume = length × width × height

= 33 ft × 30 ft × 12 ft

= 11,880 ft^3 (cu ft)

EXERCISES 904A, SET I

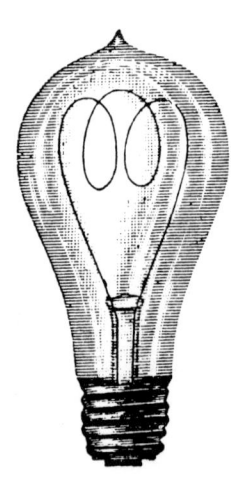

1. Find the volume of a living room that is 22 feet long, 15 feet wide, and 8 feet high.
2. A carton of light bulbs measures 15 inches by 11 inches by 13 inches. Find its volume.
3. Find the number of cubic feet in a cubic yard. (Hint: A cubic yard is a cube measuring 3 feet on each edge.)
4. Find the number of cubic inches in a cubic foot.
5. A classroom measures 10 yards long, 9 yards wide, and 4 yards high.
   (a) Find the volume of this classroom.
   (b) If air weighs about 2 pounds per cubic yard, find the weight of the air in this room.
6. An aquarium measures 24 inches long, 11 inches wide, and is filled to a depth of 13 inches.
   (a) Find the volume of the aquarium.
   (b) Find the weight of the water in the aquarium if 1 cubic inch of water weighs 0.0361 pounds. (Round off to the nearest pound.)
   (c) Find the number of gallons of water in this aquarium. (Hint: 1 gallon of water = 231 cubic inches.)
7. Find the number of cubic feet of dirt that can be removed from a rectangular hole that is 6 feet deep, $2\frac{1}{2}$ feet wide, and $6\frac{1}{3}$ feet long.

1. Find the volume of a room that is 24 feet long, 18 feet wide, and 8 feet high.
2. Find the number of cubic centimeters in a cubic meter. (Hint: A cubic meter is a cube measuring 100 centimeters on each edge.)
3. A box measures 1 meter long, 40 centimeters wide, and 30 centimeters high.
   (a) Find the volume of this box in cm^3.
   (b) Find the volume of this box in m^3.

SURFACE AREA. To find the *surface area* of a box, we must find the sum of the areas of the rectangles that form its top, bottom, and sides.

One of the rectangles that forms a *side* of a rectangular box is shown shaded in Figure 904D. To find its area, multiply its length times its width. The areas of the other 5 surface rectangles are found in the same way.

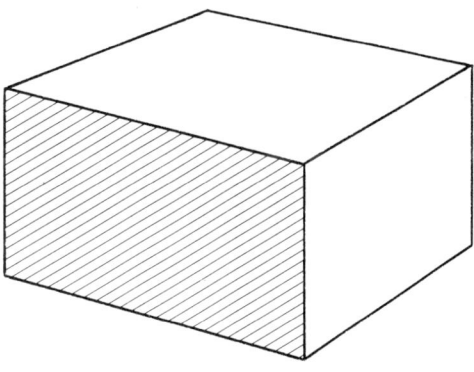

Figure 904D

Lateral area = sum of the rectangular
                         areas that form its sides

Total surface area = lateral area + top area
                                + bottom area

Example 2. Find the volume and total surface area of the rectangular box shown.

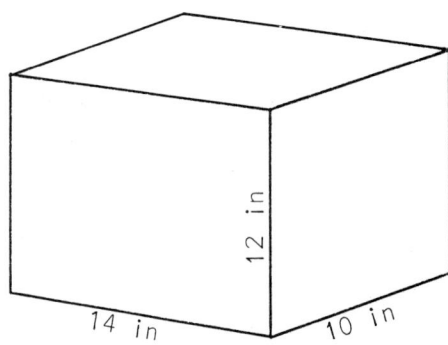

Solution:

$$\text{Volume} = LWH$$
$$= (14 \text{ in})(10 \text{ in})(12 \text{ in})$$
$$= 1{,}680 \text{ cu in}$$

$$\text{Lateral area} = 2(14 \text{ in} \times 12 \text{ in}) + 2(10 \text{ in} \times 12 \text{ in})$$
$$= 2(168 \text{ sq in}) + 2(120 \text{ sq in})$$
$$= 336 \text{ sq in} + 240 \text{ sq in}$$
$$= 576 \text{ sq in}$$

$$\text{Total surface area} = \text{lateral area} + \text{top area} + \text{bottom area}$$
$$= 576 \text{ sq in} + 2(14 \text{ in} \times 10 \text{ in})$$
$$= 576 \text{ sq in} + 280 \text{ sq in}$$
$$= 856 \text{ sq in}$$

CUBE. A cube is a rectangular box in which the length, width, and height are all equal. (See Figure 904E.) We represent the equal length, width, and height by the letter e.

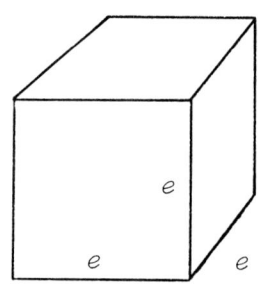

Figure 904E

$$\text{Volume of cube} = e \cdot e \cdot e$$
$$V = e^3$$
$$\text{Total surface area} = 6e^2$$

Example 3. Find the volume and total surface area of a cube that measures 10 inches on each edge.

Solution:

$$\text{Volume} = e^3$$
$$= 10^3 = 10 \cdot 10 \cdot 10$$
$$= 1{,}000 \text{ cu in}$$

$$\text{Total surface area} = 6e^2$$
$$= 6(10)^2 = 6 \cdot 10 \cdot 10$$
$$= 600 \text{ sq in}$$

1. Find the volume and total surface area of a rectangular box 10 inches long, 5 inches wide, and 3 inches high.
2. Find the volume and total surface area of a rectangular box 25 centimeters long, 17 centimeters wide, and 12 centimeters high.
3. Find the volume and total surface area of a cube that measures 8 centimeters on each edge.
4. Find the number of cubic feet in a cubic yard. (Hint: A cubic yard is a cube measuring 1 yard on each edge.)
5. (a) Find the number of cubic inches in a cubic foot. (Hint: A cubic foot is a cube measuring 12 inches on each edge.)
   (b) How many gallons will a cubic-foot container hold?
6. (a) Find the number of cubic centimeters in 1 cubic meter.
   (b) How many liters in 1 cubic meter?
7. Find the lateral area of a room 11 feet 6 inches wide, 16 feet long, and 8 feet high.
8. Find the total surface area of a box without a top cover. The box measures 22 inches long, 14 inches wide, and 12 inches high.
9. Find the volume and total surface area of a rectangular storage bin. The bin is 30 feet long, 10 feet 6 inches wide, and 11 feet 6 inches high. Express the volume in cubic feet and the area in square feet.
10. Find the volume and total surface area of a rectangular mixing tank in a chemical plant. The tank is 12 feet 6 inches long, 10 feet wide, and 8 feet 3 inches high. Express the volume in cubic feet and the area in square feet.

EXERCISES 904B, SET II

1. Find the volume and total surface area of a rectangular box 15 inches long, 9 inches wide, and 6 inches high.
2. Find the number of square inches of surface on a cubic foot.
3. Find the volume and total surface area of a rectangular storage bin. The bin is 25 feet long, 10 feet 6 inches wide, and 12 feet high. Express the volume in cubic feet and the area in square feet.
4. A rectangular tank has inside measurements of 6 feet by 4 feet by 5 feet. How many gallons of fluid will this tank hold? (A cubic foot equals 7.48 gallons.)
5. Find the lateral area of a box that is 18 inches long, 12 inches wide, and 14 inches high.

Cylinder

A right circular *cylinder* is shown in Figure 905. This cylinder is called "circular" because the top and bottom are circles. This cylinder is called "right" because the top and bottom form square corners with the sides.

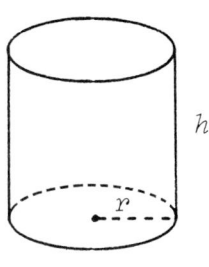

Figure 905

$$\boxed{\begin{array}{l} \text{Volume of cylinder} = \text{area of base} \times \text{height} \\ \qquad\qquad V = \pi r^2 h \\[4pt] \text{Lateral area} = 2\pi r h \\[4pt] \text{Total surface area} = \text{lateral area} + \text{top area} \\ \qquad\qquad\qquad\qquad\qquad\qquad + \text{bottom area} \\ \qquad\qquad\qquad = 2\pi r h + 2(\pi r^2) \end{array}}$$

<u>Example 1</u>. Find the volume and total surface area of the cylinder shown.

<u>Solution</u>:

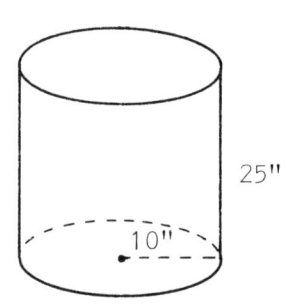

$$\begin{aligned} \text{Volume} &= \pi r^2 h \\ &= (3.14)(10)^2(25) \\ &= 7{,}850 \text{ cu in} \\[6pt] \text{Lateral area} &= 2\pi r h \\ &= 2(3.14)(10)(25) \\ &= 1{,}570 \text{ sq in} \end{aligned}$$

$$\begin{aligned} \text{Total surface area} &= 2\pi r h + 2(\pi r^2) = 1{,}570 + 2(3.14)(10)^2 \\ &= 1{,}570 + 628 = 2{,}198 \text{ sq in} \end{aligned}$$

<u>EXERCISES 905, SET I</u>. Use $\pi = 3.14$ and round off answers to one decimal place.

1. Find the volume and total surface area of a right circular cylinder 12 inches long and 10 inches in diameter.
2. Find the volume of a right circular cylinder that is 8 feet 6 inches long and 10 inches in diameter. Express the answer in cubic feet.
3. A cylindrical cistern is 16 feet deep and 12 feet in diameter. Calculate its volume in gallons. (1 cubic foot equals 7.48 gallons.)
4. A cylindrical water tank measures 20 inches in diameter and is 50 inches high. Find the volume of this tank in gallons. (1 gallon is equal to 231 cubic inches.)

<u>EXERCISES 905, SET II</u>. Use $\pi = 3.14$ and round off answers to one decimal place.

1. Find the volume of a right circular cylinder that is 4 feet 6 inches long and 10 inches in diameter. Express your answer in cubic feet.
2. A cylindrical water tank measures 24 inches in diameter and is 48 inches high. Find the volume of this tank in gallons. (1 gallon is equal to 231 cubic inches.)

Sphere

A *sphere* is shown in Figure 906. The *radius* (r) is the distance from the center to any point on the sphere.

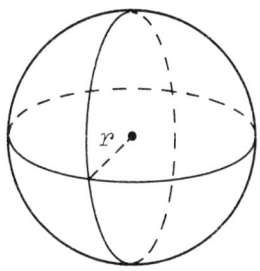

Figure 906

$$\text{Volume of sphere} = \frac{4}{3}\pi r^3$$

$$\text{Surface area of sphere} = 4\pi r^2$$

Example 1. Find the volume and surface area of a 20-inch sphere. (This means its diameter = 20 inches.)

Solution: Volume = $\frac{4}{3}\pi r^3$

$= \frac{4}{3}(3.14)(10^3) \doteq 4,186.7$ cu in

Surface = $4\pi r^2$

$= 4(3.14)(10^2) = 1,256$ sq in

EXERCISES 906, SET I. Use $\pi = 3.14$ and round off answers to one decimal place.

1. Find the surface area and volume of a 16-inch sphere.
2. Find the surface area and volume of an 8-inch sphere.
3. Find the ratio of the volumes found in Exercises 1 and 2. Can you discover what happens to the volume of a sphere when its diameter is doubled?
4. Using the results of Exercise 3, find the volume of a 24-inch sphere. (In Exercise 2 you found the volume of an 8-inch sphere. Use that answer in solving this exercise.)
5. Find the volume of the hemisphere shown below.

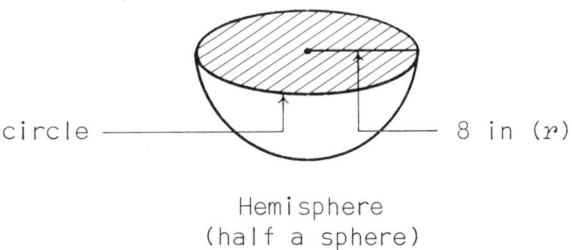

circle ———————— 8 in (r)

Hemisphere
(half a sphere)

6. A hemispherical water tank is 5 feet in diameter. Find its volume in gallons. (1 cubic foot = 7.48 gallons)

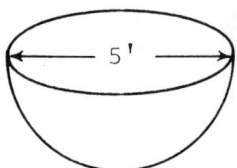

7. The figure shown to the right is a cylinder capped with a hemisphere.
   (a) Find the total volume of this figure.
   (b) Find the total surface area of this figure.

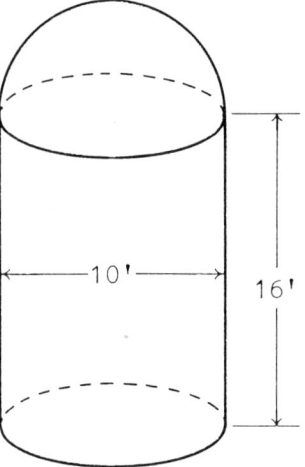

EXERCISES 906, SET II. Use π = 3.14 and round off answers to one decimal place.

1. Find the surface area and volume of a 24-inch sphere.
2. Find the surface area and volume of a 12-inch sphere.
3. Find the ratio of the volumes found in Exercises 1 and 2. Can you discover what happens to the volume of a sphere when its diameter is doubled?
4. Find the volume of a hemisphere having a diameter of 10 inches.
5. The figure shown to the right is a cylinder capped with a hemisphere.
   (a) Find the total volume of this figure.
   (b) Find the total surface area of this figure.

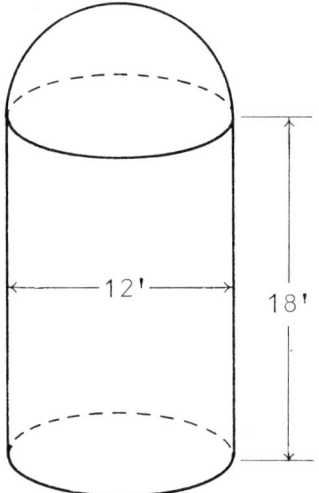

Similar Geometric Figures

Similar geometric figures are geometric figures that have exactly the *same shape* but usually different sizes.

Example 1

 (a) A basketball has the same shape as a baseball, but they certainly have different sizes. They are spheres with different radii.

 (b) A model ship has exactly the same shape as a real ship, but is much smaller in size.

 (c) The two triangles shown here have the same shape, but are different sizes.

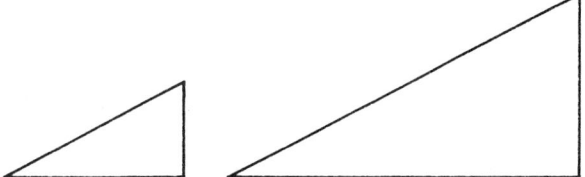

In the case of a model ship that is $\frac{1}{100}$ the size of the actual ship, any length on the model is $\frac{1}{100}$ the corresponding length on the actual ship. For example, if the actual ship is 100 feet long, then the length of the model would be $\frac{1}{100}$ of 100 feet, or 1 foot.

RATIO OF SIMILITUDE. The ratio of the corresponding lengths of similar geometric figures is called the *ratio of similitude*. In the example of the model boat just mentioned, the ratio of similitude is $\frac{1}{100}$.

Example 2. These two circles are similar geometric figures:
The ratio of similitude
of (a) to (b) = $\frac{2 \text{ in}}{1 \text{ in}} = \frac{2}{1}$.

The ratio of similitude
of (b) to (a) = $\frac{1}{2}$.

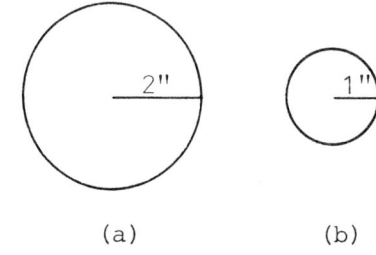

  (a)   (b)

Example 3. These two rectangles are similar geometric figures:

The ratio of similitude
of (a) to (b) = $\frac{6 \text{ m}}{2 \text{ m}} = \frac{3}{1}$.

The ratio of similitude
of (b) to (a) = $\frac{1}{3}$.

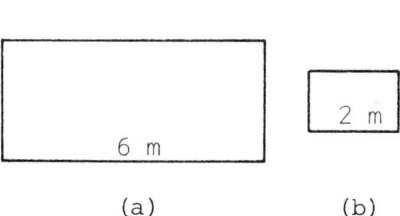

  (a)   (b)

Example 4. These two cubes are similar geometric figures:

The ratio of similitude
of (a) to (b) $= \dfrac{3 \, ft}{12 \, ft} = \dfrac{1}{4}$.

The ratio of similitude
of (b) to (a) $= \dfrac{4}{1}$.

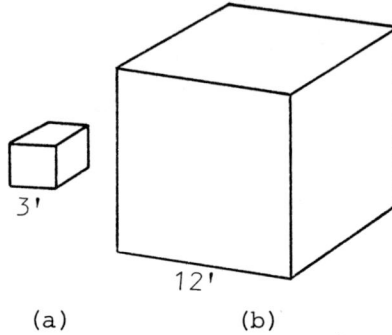

(a)   (b)

If we represent the ratio of similitude by the letter r:

1. The ratio of corresponding *lengths* = the ratio of similitude = r. (See Example 5.)
2. The ratio of corresponding *areas* = the *square* of the ratio of similitude = r^2. (See Example 6.)
3. The ratio of corresponding *volumes* = the *cube* of the ratio of similitude = r^3. (See Example 7.)

Example 5. Example of corresponding *lengths* in similar geometric figures.

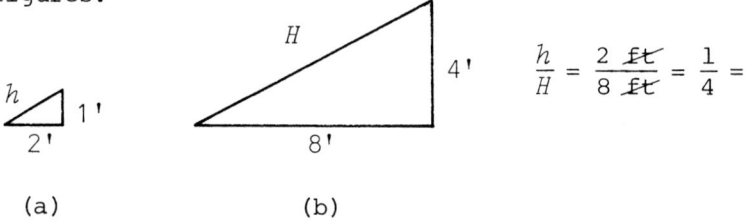

$\dfrac{h}{H} = \dfrac{2 \, ft}{8 \, ft} = \dfrac{1}{4} = r$

(a)   (b)

Example 6. Example of corresponding *areas* in similar geometric figures.

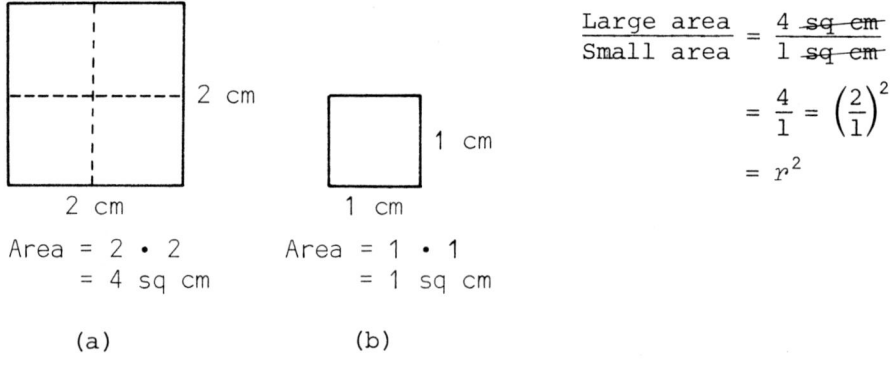

Area = 2 • 2
    = 4 sq cm

Area = 1 • 1
    = 1 sq cm

(a)   (b)

$\dfrac{\text{Large area}}{\text{Small area}} = \dfrac{4 \, sq \, cm}{1 \, sq \, cm}$

$= \dfrac{4}{1} = \left(\dfrac{2}{1}\right)^2$

$= r^2$

Example 7. Example of corresponding *volumes* of similar geometric figures.

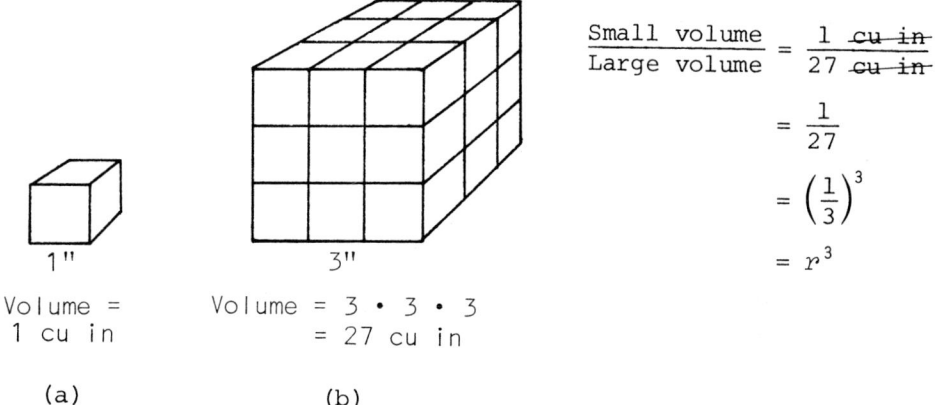

Volume =
1 cu in

(a)

Volume = 3 • 3 • 3
= 27 cu in

(b)

$$\frac{\text{Small volume}}{\text{Large volume}} = \frac{1 \text{ cu in}}{27 \text{ cu in}}$$

$$= \frac{1}{27}$$

$$= \left(\frac{1}{3}\right)^3$$

$$= r^3$$

Example 8. The use of the ratio of similitude is shown in Table 907.

Table 907

Ratio of similitude	Ratio of lengths	Ratio of areas	Ratio of volumes
2	2	4	8
3	3	9	27
4	4	16	64
5	5	25	125
⋮	⋮	⋮	⋮
10	10	100	1,000

From Table 907 we see that:

1. If a length in one figure is 2 times the corresponding length in a similar figure, the area is 4 times the area in the similar figure, and the volume is 8 times the volume of the similar figure.
2. If a length is 3 times as large, the area is 9 times as large, and the volume is 27 times as large.
3. If a length is 5 times as large, the area is 25 times as large, and the volume is 125 times as large.

Example 9

(a) The ratio of similitude
of cube (a) to cube (b),
$$r = \frac{12 \text{ in}}{3 \text{ in}} = \frac{4}{1}.$$

(b) The ratio of the shaded
areas of cube (a) to
cube (b) equals the *square*
of the ratio of similitude
$$r^2 = \frac{4}{1} \cdot \frac{4}{1} = \frac{16}{1}.$$
This means the large shaded
area is 16 times the small
shaded area.

(c) The ratio of the *volumes*
of cube (a) to cube (b)
equals the *cube* of the
ratio of similitude
$$r^3 = \frac{4}{1} \cdot \frac{4}{1} \cdot \frac{4}{1} = \frac{64}{1}.$$
This means that the volume
of cube (a) is 64 times the
volume of cube (b).

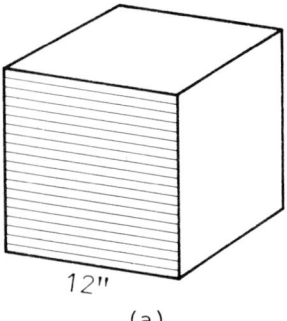

12"

(a)

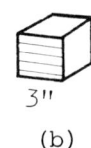

3"

(b)

Example 10. These are two similar cylinders. Ratio of similitude (large to small) $= \frac{10 \text{ in}}{2 \text{ in}} = \frac{5}{1}.$

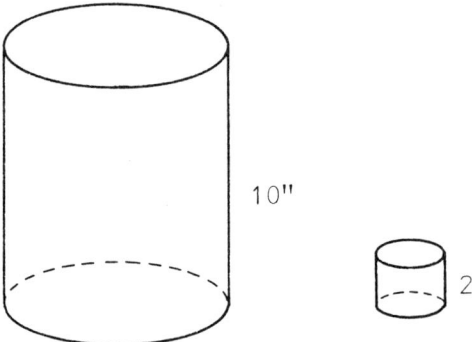

10"

2"

If the total surface area of the small cylinder $\doteq$ 20 sq in,
then the total surface area of the large cylinder $\doteq \left(\frac{5}{1}\right)^2 \cdot 20$
$= \frac{25}{1} \cdot 20 = 500$ sq in.

If the volume of the small cylinder $\doteq$ 6 cu in, then the volume
of the large cylinder $\doteq \left(\frac{5}{1}\right)^3 \cdot 6 = \frac{125}{1} \cdot 6 = 750$ cu in.

Example 11. These are two spheres. The ratio of similitude (large to small) = $\frac{15 \text{ in}}{5 \text{ in}} = \frac{3}{1}$.

Small sphere 5"

Surface area $\doteq 4(3.14)(5^2) = 314$ sq in

Volume $\doteq \frac{4}{3}(3.14)(5^3) = 524$ cu in

Large sphere 15"

Surface area = 9 · small sphere area
$\doteq 9 \cdot 314$
= 2,826 sq in

Volume = 27 · small sphere volume
$\doteq 27 \cdot 524$
= 14,148 cu in

EXERCISES 907, SET I

1. The diagonal of a 5-inch square is approximately 7.07 inches long. (A diagonal of a square is the line joining opposite corners.)
   (a) Find the length of the diagonal of a 10-inch square.
   (b) Find the length of the diagonal of a 15-inch square.
   (c) Find the length of the diagonal of a 20-inch square.
2. A triangle has a 10-inch base and a height of 4 inches.
   (a) Find the height of a similar triangle whose base is 5 inches.
   (b) Find the height of a similar triangle whose base is 15 inches.
3. A 6-foot man casts a 4-foot shadow. Find the height of a tree that casts a 32-foot shadow.
4. A 40-foot flagpole casts a 25-foot shadow. Find the height of a nearby building that casts a 20-foot shadow.
5. A cylindrical water tank is 4 feet high and holds 50 gallons. How many gallons will a similar tank hold that is 5 feet high?
6. A rectangular storage bin is 8 feet high and holds 12 tons of grain. How many tons will a similar bin hold that is 10 feet high?
7. It takes 16 gallons to paint a cylindrical oil storage tank that is 30 feet in diameter. How many gallons will it take to paint a similar tank that is 40 feet in diameter?
8. A 10-foot diameter bathysphere has a 1,000-pound force on it at a given depth in water. Find the force on an 8-foot diameter bathysphere at the same depth.

EXERCISES 907, SET II

1. These two rectangles are similar geometric figures.
   (a) Find the ratio of similitude of (A) to (B).
   (b) Find the ratio of the area of the large rectangle to that of the small rectangle.

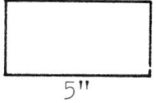

5"   2"

(A)   (B)

2. These are two similar triangles.
   (a) Find the ratio of
       similitude of (B) to (A).
   (b) Find the ratio of
       H to h.
   (c) Find the ratio of the
       area of the large
       triangle to that of
       the small triangle.

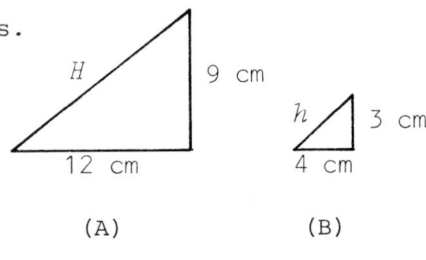

(A)   (B)

3. A 6-foot man casts a 2-foot shadow. Find the height of a
   tree that casts a 14-foot shadow.
4. A 12-foot diameter bathysphere has a 1,000-pound force on
   it at a given depth in water. Find the force on an 8-foot
   diameter bathysphere at the same depth.

Accuracy of Calculations

Measurements are obtained by reading instruments such as rulers,
steel tapes, micrometers, thermometers, scales, measuring cups,
clocks, etc. Measuring instruments all have limited *accuracy*.
For this reason, when using measurements in calculations, care
must be used in indicating the accuracy of answers.

SIGNIFICANT DIGITS

---

1. Every nonzero digit is significant.
2. Zeros *between* significant digits are significant.
3. *Final* zeros that follow the decimal point are
   significant.

Example 1

   (a) 265   3 significant digits

   (b) 7004   4 significant digits

   (c) 860   2 significant digits (See page 334,
                                              Example 6.)

   (d) 5.00   3 significant digits

   (e) 0.00043   2 significant digits

   (f) 9.005   4 significant digits

ACCURACY

> A number is considered accurate to its rightmost significant digit.

Example 2

   (a) 25.1   3-figure accuracy *(Accurate to nearest tenth)*

   (b) 9700   2-figure accuracy *(Accurate to nearest hundred)* (See page 334, Example 6.)

   (c) 20.76   4-figure accuracy *(Accurate to nearest hundredth)*

ROUNDING OFF. See Section 407 for the details of rounding off decimals.

Example 3

   (a) 876. ≐ 880   Rounded to tens

   (b) 44.62 ≐ 44.6   Rounded to 1 decimal place

   (c) 36.50 ≐ 36.   Rounded to units ⎱ When exactly in middle, round off

   (d) 35.50 ≐ 36.   Rounded to units ⎰ so last digit is *even*.

   (e) 0.16504 ≐ 0.17   Rounded to 2 decimal places

ACCURACY OF CALCULATED ANSWERS. We will use the following convention in expressing the accuracy of calculated answers, *although this accuracy is not always obtained.*

> When Calculating Using Numbers Obtained by Measurement
>
> 1. Your answer should be expressed to the same accuracy as the *least* accurate of the numbers you start with.
> 2. Only round off *after* you have obtained the result of *all* calculations.

Example 4

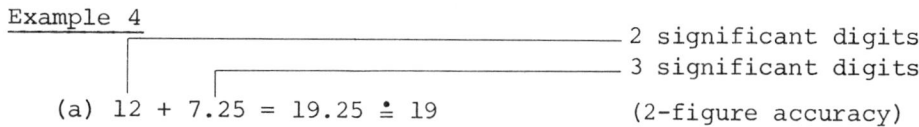

   (a) 12 + 7.25 = 19.25 ≐ 19   (2-figure accuracy)

```
                                              ┌────── 3 significant digits
                                              ├────── 4 significant digits
    (b)  64.7 × 129.3 = 8365.71 ≐ 8370        (3-figure accuracy)

                                              ┌────── 4 significant digits
         26.35
    (c)  ───────  ≐ 0.030746 ≐ 0.03075        (4-figure accuracy)
         857.00                     └─Since the answer is to be rounded
                                       off to 4 significant digits, the
                                       division is carried out to one more
                                       significant digit.
                                              └────── 5 significant digits
```

A word of caution: Zeros following a decimal point but preceding a nonzero digit are *not* significant. They are merely *placeholders*.

Example 5

    (a) 0.00089   2 significant digits. These three zeros are only placeholders.

    (b) 7.00089   6 significant digits. These three zeros *are* significant because they lie between significant digits 7 and 8.

FINAL ZEROS TO LEFT OF DECIMAL POINT. When we do not know the accuracy of a measurement, we do not consider final zeros preceding the decimal point as significant. (See Example 1[c].) In some cases, however, one or more final zeros may be significant, depending upon the accuracy of the measurement (Example 6).

Example 6. Suppose an actual measurement of 600 feet were accurate to the nearest 10 feet. How could we indicate this accuracy? Writing the number as 600 does *not* indicate the accuracy of 10 feet. The correct accuracy can be indicated by writing

       6.0×10^2 2 significant digits
       (600   1 significant digit)

EXACT NUMBERS. The discussion in this section has been concerned with the accuracy of numbers representing measurements, and calculations with such numbers. It is important to realize that some numbers are *exact*. A person has exactly $5, 2 dogs, 1 car, and so on. *Exact numbers can be considered to have an unlimited number of significant digits.*

EXERCISES 908, SET I. In Exercises 1-16, indicate the number of significant digits in each of the given numbers.

1. 47	2. 89	3. 3,506	4. 7,802
5. 750	6. 60	7. 85.0	8. 34.0
9. 0.006	10. 0.015	11. 0.0700	12. 0.0810
13. 2.03	14. 7.008	15. 50.40	16. 80.300

In Exercises 17-24, indicate the number of figures in the accuracy of each of the given numbers as well as the place the number is accurate to.

17. 42.3   18. 3.56   19. 840   20. 7,200

21. 6.00   22. 30.0   23. 50.24   24. 30.05

In Exercises 25-32, round off each number to the indicated place. (See Section 407.)

25. 534   tens   26. 386   tens

27. 56.97   1 decimal place   28. 24.84   1 decimal place

29. 27.50   units   30. 28.50   units

31. 0.028501   3 decimal places

32. 0.036499   3 decimal places

In Exercises 33-46, perform the indicated calculations. Express your answers to the accuracy of the least accurate of the numbers used in the calculation.

33. 23.75 + 66.25 - 3.60   34. 47.25 + 52.75 - 8.70

35. 6 × 7.8 × 35   36. 8 × 12.5 × 5.0

37. 1,400 ÷ 175   38. 1,350 ÷ 225

39. $\dfrac{0.0045 \times 8.05}{25}$ 40. $\dfrac{0.036 \times 6.39}{27}$

41. 127 × 0.43 × 85.6   42. 386 × 0.58 × 34.7

43. 3.14(12.5)8   44. 3.14(37.5)16

45. $3.14(9.8)^2$ 46. $3.14(7.9)^2$

EXERCISES 908, SET II. In Exercises 1-8, indicate the number of significant digits in each of the given numbers.

1. 24   2. 2,407   3. 70   4. 28.0

5. 0.028   6. 0.00230   7. 8.004   8. 90.200

In Exercises 9-12, indicate the number of figures in the accuracy of each of the given numbers as well as the place the number is accurate to.

9. 2.67   10. 2,300   11. 90.0   12. 20.06

In Exercises 13-16, round off each number to the indicated place. (See Section 407.)

13. 267   tens   14. 74.73   1 decimal place

15. 74.50   units   16. 0.028499   3 decimal places

In Exercises 17-23, perform the indicated calculations. Express your answers to the accuracy of the least accurate of the numbers used in the calculation.

17. 2.30 + 27.75 - 14.25

18. 7 × 15.6 ÷ 7.8

19. 2,520 ÷ 120

20. $\dfrac{0.027 \times 9.33}{36}$

21. 284 × 0.25 × 21.8

22. 3.14(24.5)(12)

23. 3.14(8.7)^2

Chapter Summary

Rectangle (Sec. 901)

Area = LW
Perimeter = $2L + 2W$

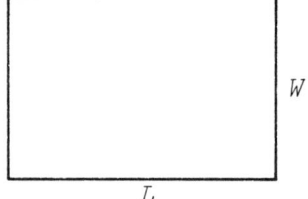

Square (Sec. 901)

Area = s^2
Perimeter = $4s$

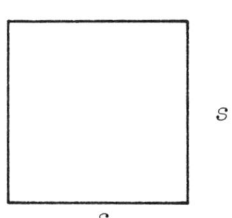

Triangle (Sec. 902)

Area = $\dfrac{1}{2}bh$
Perimeter = $a + b + c$

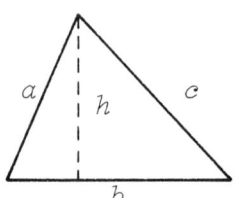

Circle (Sec. 903)

Area = πr^2
Circumference = $\pi d = 2\pi r$

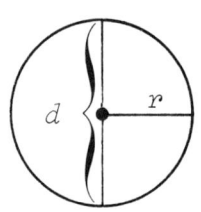

Rectangular Box (Sec. 904)

Volume = LWH
Lateral area = sum of the rectangular
                         areas that form its sides
Total surface area = lateral area
                              + top area
                              + bottom area

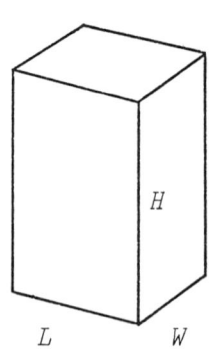

Cube (Sec. 904)

Volume = e^3
Total surface = $6e^2$

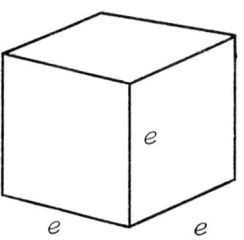

Cylinder (Sec. 905)

Volume = $\pi r^2 h$
Lateral surface = $2\pi rh$
Total surface = $2\pi rh + 2\pi r^2$

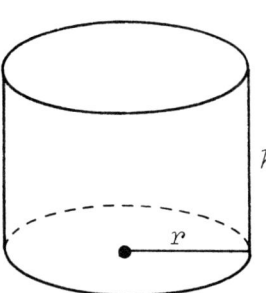

Sphere (Sec. 906)

Volume = $\dfrac{4}{3}\pi r^3$
Surface = $4\pi r^2$

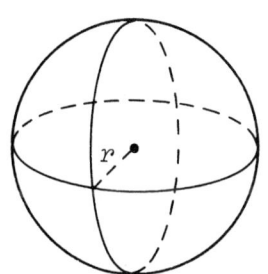

REVIEW EXERCISES 909, SET I. In Exercises 1-8, consider all given numbers to be exact numbers. Only round off answers when asked to do so in the problem.

1. Find the perimeter and area of this figure.

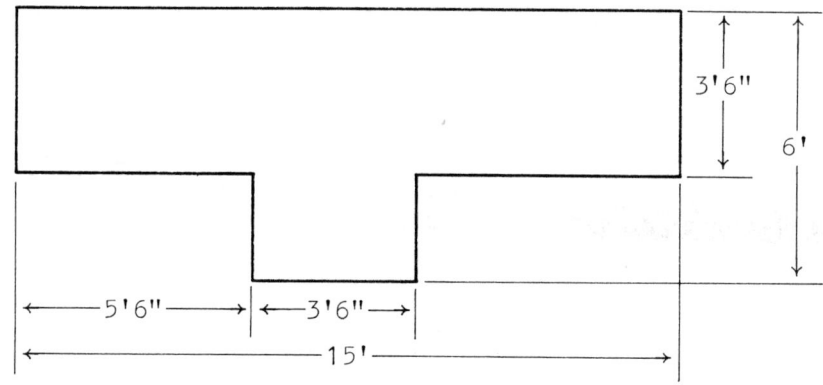

2. (a) Find the perimeter in yards and feet.
   (b) Find the area in square feet.

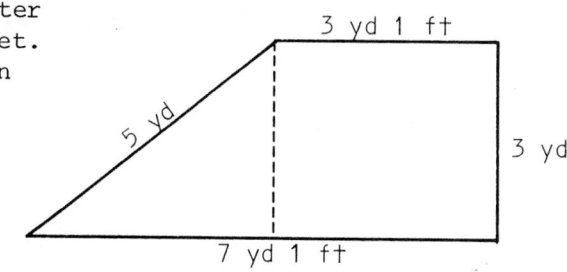

3. Find the circumference in feet and the area in square feet of a circle whose radius is 2 feet 6 inches. Use $\pi = 3.14$ and round off answers to one decimal place.

4. Find the volume of a cylinder that has a radius of 5 inches and a height of 15 inches. Use $\pi = 3.14$ and round off answer to the nearest unit.

5. Find the volume and total surface of a rectangular box with L = 3 feet 9 inches, W = 2 feet, and H = 4 feet 6 inches. Express the volume in cubic feet. Express the surface in square feet. Round off answers to one decimal place.

6. Find the volume of a cylindrical silo 6 yards in diameter and 14 yards high. Use $\pi = 3.14$ and round off to the nearest cubic yard.

7. Two circles have diameters of $\frac{1}{2}$ inch and $1\frac{1}{4}$ inches, respectively. The area of the larger circle is how many times greater than the area of the smaller circle?

8. A hemispherical water tank is 6 feet in diameter. Find its capacity in gallons. Use $\pi = 3.14$ and round off your answer to the nearest gallon. (1 cubic foot = 7.48 gallons)

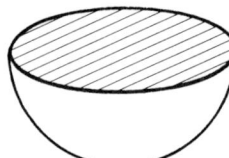

In Exercises 9-15, express your answers to the accuracy of the least accurate of the given numbers.

9. 28.96 + 5.01 - 4.70

10. 40.08 + 2.07 - 9.60

11. 2.06 × 35.0 × 0.08

12. 47.0 × 3.05 × 0.06

13. $\dfrac{4.80 \times 0.0564}{16}$

14. $\dfrac{5.6 \times 0.0048}{14}$

15. For the Bicentennial Celebration, July 4, 1976, Bob Older, a rancher, had a U.S. flag made that measured 67 feet by 102 feet. Find the area of this flag in square meters.

REVIEW EXERCISES 909, SET II. In Exercises 1-5, express your answers to the accuracy of the least accurate of the given numbers.

1. 85.14 + 5.07 - 2.80

2. 2.05 × 27.0 × 0.04

3. 28.0 × 2.06 × 0.07

4. $\dfrac{2.90 \times 0.0266}{28}$

5. $(3.14)(8.1)^2$

In Exercises 6-9, consider all given numbers to be exact numbers. Only round off answers when asked to do so in the problem.

6. Find the perimeter and area of this figure.

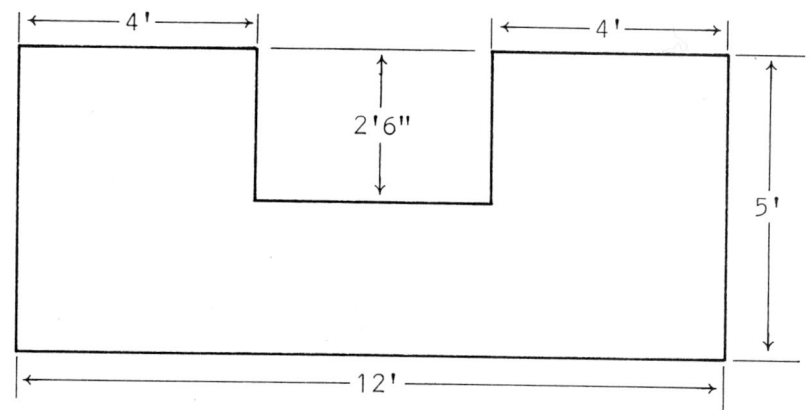

7. Find the volume of a cylinder that has a radius of 18 inches and a height of 3 feet. Use π = 3.14. Express your answer to the nearest tenth of a cubic foot.
8. Find the volume and surface of a sphere that has a diameter of 6 inches. Use dimensions in inches and express your answers to the nearest unit.
9. A hemispherical water tank is 8 feet in diameter. Find its capacity in gallons. Use π = 3.14 and round off your answer to the nearest gallon. (1 cubic foot = 7.48 gallons)

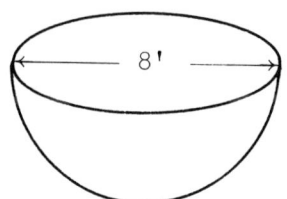

Chapter 9: Diagnostic Test

Name_____

_____

The purpose of this test is to see how well you can apply arithmetic to common geometric figures. We recommend that you work this diagnostic test *before* your instructor tests you on this chapter. Allow yourself about an hour to do this test.

Complete solutions for all the problems on this test, together with section references, are given in the Answer Section. You should study the sections referred to for the problems you do incorrectly.

In Problems 1-10, consider all given numbers to be exact numbers. Only round off answers when asked to do so in the problem.

1. (a) Find the area of the rectangle in square feet.

   (b) Find the perimeter in feet.

   17 ft 6 in

   12 ft

   (1a)_____

   (1b)_____

---

2. (a) Find the area of the square in square meters.

   (b) Find the perimeter in meters.

   1 m 27 cm

   (2a)_____

   (2b)_____

---

3. (a) Find the area of the
   triangle in square
   inches.

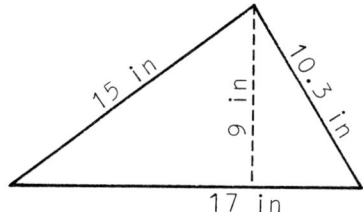

   (b) Find the perimeter
   in inches.

                                        (3a) _____

                                        (3b) _____

4. Find the circumference in feet and the area in square
   feet of a circle whose diameter is 5 feet. Use $\pi = 3.14$
   and round off answers to 3 significant digits.

                    (4) Circumference = _____ ft

                              Area = _____ sq ft

5. Compare the flow of water in a 1-inch hose with the flow
   of water in a $\frac{7}{8}$-inch hose.

                                        (5) _____

6. Find the volume and total surface area of a rectangular box 9 inches long, 5 inches wide, and 4 inches high.

(6) Volume = _____

Surface = _____

7. Find the volume and total surface area of a cube that measures 10 centimeters on each edge.

(7) Volume = _____

Surface = _____

8. Find the volume and the total surface area of a right circular cylinder 10 inches in diameter and 15 inches high. Use $\pi = 3.14$ and round off answers to the nearest unit.

(8) Volume = _____

Surface = _____

9. Find the surface area and volume of a 12-inch sphere. Use π = 3.14 and round off answers to three significant digits.

(9) Volume = _____

Surface = _____

10. A 5-foot-6-inch woman casts a 3-foot shadow. Find the height of a tree that casts an 18-foot shadow.

(10) _____

11. Round off your answer to the following product to the accuracy of the least accurate of the given numbers.

3.14 × 15.2 × 7.1

(11) _____

TEN
Systems of
Numeration

A *system of numeration* consists of (1) a set of symbols called digits and (2) a set of rules for combining the digits to make it possible to represent any number. In this chapter we will gain a clearer understanding of our own system of numeration, the decimal system, by studying several others as well.

The Decimal System (Base 10)

The set of digits = {0, 1, 2, 3, 4, 5, 6, 7, 8, 9}.

The base of the system = 10. (Note that this equals the number of digits.)

PLACE-VALUE. The decimal system is a *place-value system.* In Figure 1001A we show the place-values and their corresponding powers of 10. This figure is a modification of Figure 120, Chapter 1.

Written Number ⟶ 5 1 7, 0 2 6 .

Power of 10 ⟶ 10^5 10^4 10^3 10^2 10^1 10^0
(Base = 10)

Place-name ⟶

Place-value system:

| hundred-thousand | ten-thousand | one-thousand | hundred (unit) | ten (unit) | one (unit) |

Figure 1001A

EXPANDED FORM OF A NUMBER (Base 10). We use examples to show the meaning of the expanded form of a number in base 10.

Example 1. Write the expanded form (base 10) of 23.

$$23 = 20 \qquad + 3$$
$$= 2(10) \quad + 3(1)$$
$$\text{Expanded form} = 2 \cdot 10^1 + 3 \cdot 10^0$$

Remember $10^0 = 1$.

Example 2. Write the expanded form (base 10) of 504.

$$504 = 500 \qquad\qquad + \qquad\qquad 4$$
$$= 5(100) \quad + \quad 0(10) \quad + \quad 4(1)$$

Expanded form $= 5 \cdot 10^2 + \boxed{0 \cdot 10^1} + 4 \cdot 10^0$

└──────────────── Zero terms may

$$= 5 \cdot 10^2 + 4 \cdot 10^0 \qquad \text{be omitted}$$

when writing
expanded forms.

Example 3. Write the expanded form (base 10) of 8,724.

$$8,724 = 8,000 \quad + 700 \quad + 20 \quad + 4$$
$$= 8(1,000) + 7(100) \quad + 2(10) \quad + 4(1)$$

Expanded form $= 8 \cdot 10^3 \; + 7 \cdot 10^2 + 2 \cdot 10^1 + 4 \cdot 10^0$

Example 4. Write the expanded form (base 10) of 930,640.

$$930,640 = 9(100,000) + 3(10,000) + 0(1,000)$$
$$+ 6(100) + 4(10) + 0(1)$$

Expanded form $= 9 \cdot 10^5 + 3 \cdot 10^4 + 0 \cdot 10^3 + 6 \cdot 10^2$
$$+ 4 \cdot 10^1 + 0 \cdot 10^0$$

also $= 9 \cdot 10^5 + 3 \cdot 10^4 + 6 \cdot 10^2 + 4 \cdot 10^1$

ADDITIVE SYSTEM. Our decimal system is an *additive*, place-value system. The actual value of a number is found by *adding* the terms of the expanded form of the number.

Example 5. $27,036 = 2 \cdot 10^4 + 7 \cdot 10^3 + 0 \cdot 10^2 + 3 \cdot 10^1 + 6 \cdot 10^0$

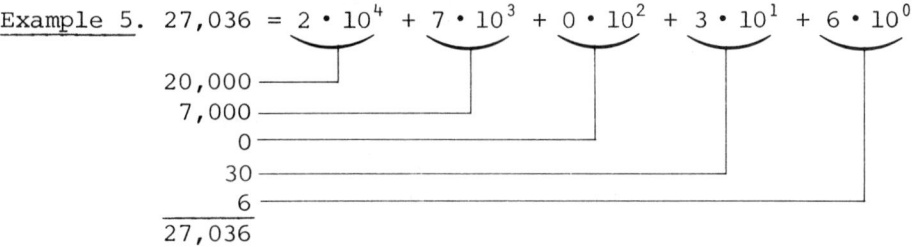

```
  20,000 ─────
   7,000 ─────
       0 ─────
      30 ─────
       6 ─────
  ──────
  27,036
```

In Figure 1001B, we show the relation between place-value and the expanded form of a number in an additive system of numeration (base 10).

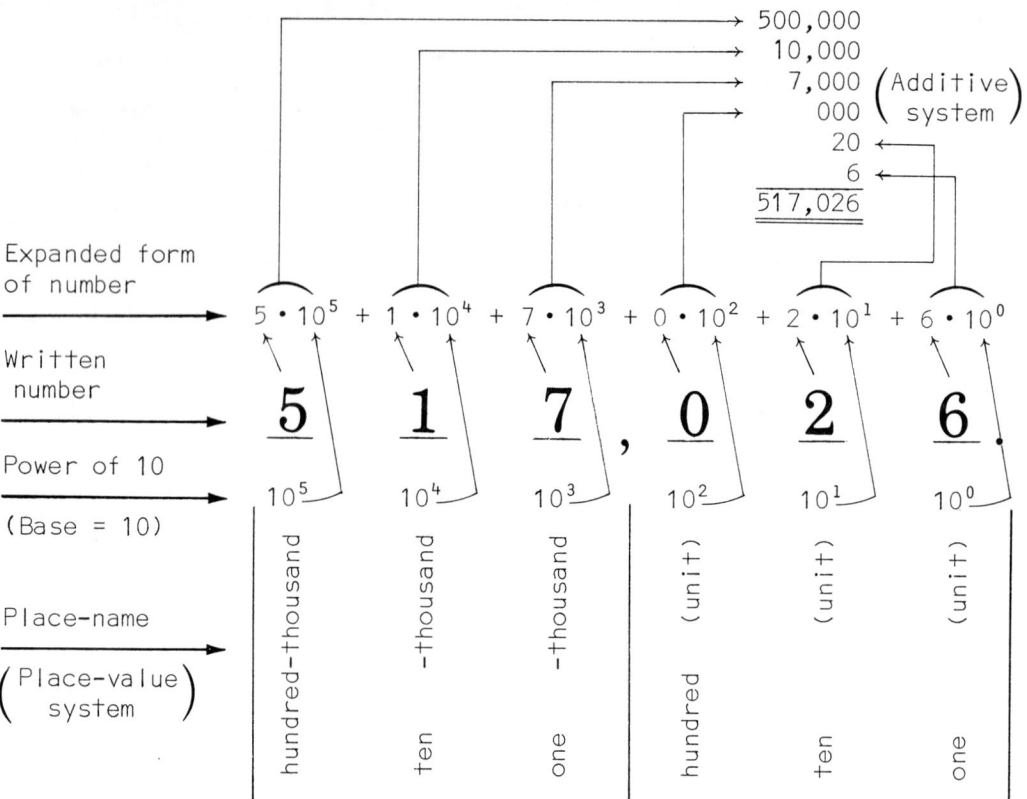

Figure 1001B

In Exercises 1-16, write the expanded form (base 10) for each of the given numbers.

1. 27	2. 98	3. 105	4. 503
5. 8,074	6. 7,105	7. 11,011	8. 10,101
9. 46,903	10. 65,024	11. 570,462	12. 805,902
13. 6,041,980		14. 4,703,068	
15. 10,609,048		16. 21,150,100	

In Exercises 17-22, write the number for each of the given expanded forms.

17. $3 \cdot 10^2 + 2 \cdot 10^1$

18. $5 \cdot 10^3 + 4 \cdot 10^1$

19. $6 \cdot 10^2 + 3 \cdot 10^0$

20. $7 \cdot 10^4 + 8 \cdot 10^2 + 2 \cdot 10^0$

21. $9 \cdot 10^4 + 3 \cdot 10^3 + 5 \cdot 10^2$

22. $3 \cdot 10^5 + 2 \cdot 10^2 + 4 \cdot 10^1$

EXERCISES 1001, SET II. In Exercises 1-8, write the expanded form (base 10) for each of the given numbers.

1. 34	2. 857	3. 5,864	4. 10,307

5. 45,013

6. 207,015

7. 3,708,072

8. 32,280,200

In Exercises 9-11, write the number for each of the expanded forms.

9. $4 \cdot 10^2 + 3 \cdot 10^1$

10. $7 \cdot 10^4 + 5 \cdot 10^2 + 3 \cdot 10^1 + 6 \cdot 10^0$

11. $2 \cdot 10^6 + 7 \cdot 10^4 + 5 \cdot 10^2 + 3 \cdot 10^1 + 0 \cdot 10^0$

The Binary System (Base 2)

To write any number in the base 10 (decimal) system, we need 10 digits. To write any number in the base 2 (*binary*) system, we need only two digits: 0 and 1. For this reason, base 2 arithmetic is used in many computers because "no current flow" can represent 0, while "current flow" can represent 1.

The set of digits in the binary system = { 0,1}.

The base of the system = 2. (Note that this equals the number of digits.)

<u>POWERS OF 2</u>. Before reading and writing binary numbers, we review the powers of 2.

$$2^0 = 1$$
$$2^1 = 2$$
$$2^2 = 2 \cdot 2 = 4$$
$$2^3 = 2 \cdot 2 \cdot 2 = 8$$
$$2^4 = 2 \cdot 2 \cdot 2 \cdot 2 = 16$$
$$2^5 = 2 \cdot 2 \cdot 2 \cdot 2 \cdot 2 = 32$$
$$2^6 = 64$$
$$2^7 = 128$$
$$2^8 = 256$$

To distinguish between a number written in base 10 and one written in base 2, we will write numbers in the following way:

$(53)_{10}$ to show the number is written 53 in base 10

$(110101)_2$ to show the number is written 110101 in base 2

This same method will be used when we learn other bases:

$(203)_5$ to show the number is written 203 in base 5

The base 5 (*quinary*) system will be discussed in Section 1003.

If no base is written, it will be understood to be base 10. In some cases we will write "base 10," so there can be no confusion.

PLACE-VALUE. The binary system is a *place-value* system. In Figure 1002A we show the place-values and their corresponding powers of 2.

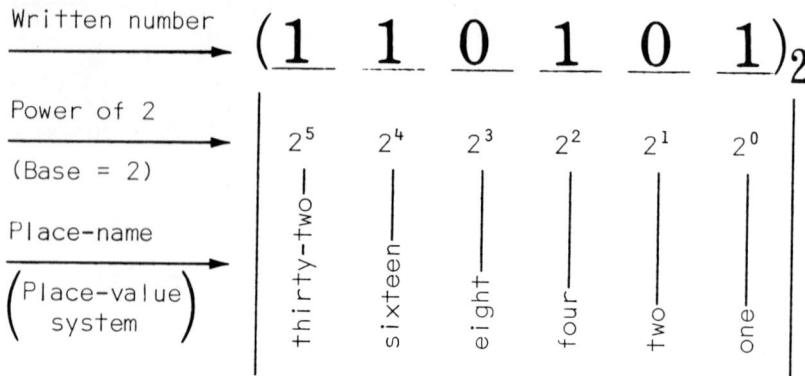

Figure 1002A

EXPANDED FORM OF A NUMBER (Base 2). The only difference in writing the expanded form in base 2 instead of base 10 is that we use powers of 2 instead of powers of 10. We use examples to show how to write the expanded form of a number in base 2.

Example 1. Write the expanded form of $(11)_2$, then convert it to base 10 form.

$$(11)_2 = 1 \cdot 2^1 + 1 \cdot 2^0 = 1(2) + 1(1) = 2 + 1 = (3)_{10}$$

Example 2. Write the expanded form of $(110)_2$, then convert it to base 10 form.

$$(110)_2 = 1 \cdot 2^2 + 1 \cdot 2^1 + 0 \cdot 2^0 = 1(4) + 1(2) + 0(1)$$
$$= 4 + 2 + 0 = (6)_{10}$$

In the next three examples, we compare base 10 and base 2 numbers by writing both in the expanded form.

Example 3

$$(11)_{10} = 1 \cdot 10^1 + 1 \cdot 10^0 = 10 + 1 = (11)_{10}$$

$$(11)_2 = 1 \cdot 2^1 + 1 \cdot 2^0 = 2 + 1 = (3)_{10}$$

Example 4

$$(111)_{10} = 1 \cdot 10^2 + 1 \cdot 10^1 + 1 \cdot 10^0 = 100 + 10 + 1 = (111)_{10}$$

$$(111)_2 = 1 \cdot 2^2 + 1 \cdot 2^1 + 1 \cdot 2^0 = 4 + 2 + 1 = (7)_{10}$$

Example 5

$$(1010)_{10} = 1 \cdot 10^3 + 0 \cdot 10^2 + 1 \cdot 10^1 + 0 \cdot 10^0$$
$$= 1000 \quad + \quad 0 \quad + \quad 10 \quad + \quad 0 \quad = (1010)_{10}$$

$$(1010)_2 = 1 \cdot 2^3 + 0 \cdot 2^2 + 1 \cdot 2^1 + 0 \cdot 2^0$$
$$= 8 \quad + \quad 0 \quad + \quad 2 \quad + \quad 0 \quad = (10)_{10}$$

We now show how to write some familiar base 10 numbers in the base 2 system.

Base 10	Base 2	Explanation
1	1	$(1)_2 = 1 \cdot 2^0 = 1(1) = (1)_{10}$
2	10	$(10)_2 = 1 \cdot 2^1 + 0 \cdot 2^0 = 1(2) + 0(1) = 2 + 0 = (2)_{10}$
3	11	$(11)_2 = 1 \cdot 2^1 + 1 \cdot 2^0 = 1(2) + 1(1) = 2 + 1 = (3)_{10}$
4	100	$(100)_2 = 1 \cdot 2^2 + 0 \cdot 2^1 + 0 \cdot 2^0 = 4 + 0 + 0 = (4)_{10}$
5	101	$(101)_2 = 1 \cdot 2^2 + 0 \cdot 2^1 + 1 \cdot 2^0 = 4 + 0 + 1 = (5)_{10}$
6	110	$(110)_2 = 1 \cdot 2^2 + 1 \cdot 2^1 + 0 \cdot 2^0 = 4 + 2 + 0 = (6)_{10}$
7	111	$(111)_2 = 1 \cdot 2^2 + 1 \cdot 2^1 + 1 \cdot 2^0 = 4 + 2 + 1 = (7)_{10}$
8	1000	$(1000)_2 = 1 \cdot 2^3 + 0 \cdot 2^2 + 0 \cdot 2^1 + 0 \cdot 2^0 = (8)_{10}$
9	1001	We leave the remaining explanations for the student.
10	1010	
16	10000	
17	10001	
20	10100	
32	100000	
60	111100	

ADDITIVE SYSTEM. The binary system is an *additive*, place-value system. The value of a number is found by *adding* the terms of the expanded form of the number as we have seen in the preceding examples. In Figure 1002B, we show the relation between place-value and the expanded form of a number in this additive system of numeration.

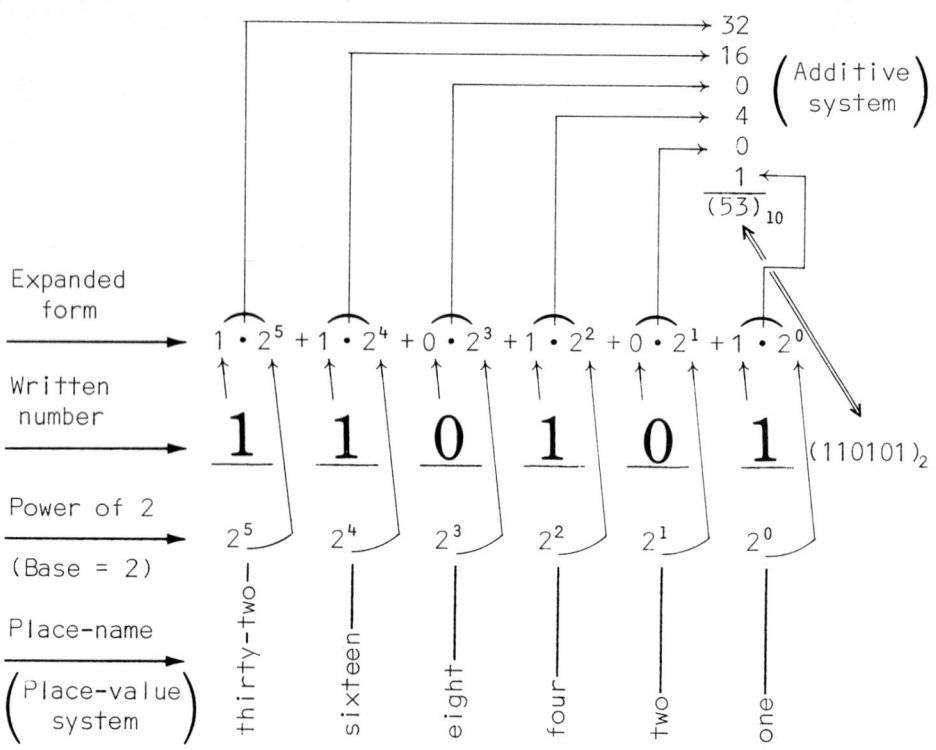

Figure 1002B

CHANGING A NUMBER FROM BASE 2 TO BASE 10

<u>Example 6</u>. Write $(1101)_2$ in base 10.

$$(1101)_2 = 1 \cdot 2^3 + 1 \cdot 2^2 + 0 \cdot 2^1 + 1 \cdot 2^0$$
$$= 1 \cdot 8 + 1 \cdot 4 + 0 \cdot 2 + 1 \cdot 1$$
$$= 8 + 4 + 0 + 1 = (13)_{10}$$

These small numbers indicate the power of 2 for each place.

<u>Example 7</u>. Write $(10110)_2$ in base 10.

$$(10110)_2 = 1 \cdot 2^4 + 0 \cdot 2^3 + 1 \cdot 2^2 + 1 \cdot 2^1 + 0 \cdot 2^0$$
$$= 16 + 0 + 4 + 2 + 0 = (22)_{10}$$

CHANGING A NUMBER FROM BASE 10 TO BASE 2

<u>Example 8</u>. Write $(19)_{10}$ in base 2.

<u>Method 1</u>. (i) Subtract the highest power of 2 possible.

$$19 - 16 = 3$$
$$2^4$$

(ii) Subtract the highest power of 2 from the remainder obtained in (i).

$$3 - \boxed{2} = 1$$
$$\uparrow \quad 2^1$$

(iii) Subtract the highest power of 2 possible from the remainder obtained in (ii).

$$1 - \boxed{1} = 0$$
$$\uparrow \quad 2^0$$

Therefore,

$$(19)_{10} = (1 \underline{\quad} 0 \underline{\quad} 0 \underline{\quad} 1 \underline{\quad} 1)_2$$

$\boxed{16}$	8	4	$\boxed{2}$	$\boxed{1}$
2^4	2^3	2^2	2^1	2^0

Notice that the ones are placed where that power of 2 was subtracted.

Method 2. An *easier way* to find the base 2 representation is by repeated division by 2.

```
2 | 19
2 |  9  R 1  ↑
2 |  4  R 1  |     (10011)₂
2 |  2  R 0  |
      1  R 0  ⌐
          READ
```

```
19 ÷ 2 = 9 R 1  ↑
 9 ÷ 2 = 4 R 1  |
 4 ÷ 2 = 2 R 0  |     (10011)₂
 2 ÷ 2 = 1 R 0  ⌐
          READ
```

Example 9. Write $(45)_{10}$ in base 2.

Method 1

```
   45
 - 32  = 2⁵
 ─────
   13
 -  8  = 2³
 ─────
    5
 -  4  = 2²
 ─────
    1
 -  1  = 2⁰
 ─────
    0
```

Therefore, $(45)_{10} =$

$$(1 \underline{\quad} 0 \underline{\quad} 1 \underline{\quad} 1 \underline{\quad} 0 \underline{\quad} 1)_2$$

32	16	8	4	2	1
$\boxed{2^5}$	2^4	$\boxed{2^3}$	$\boxed{2^2}$	2^1	$\boxed{2^0}$

Notice that the ones are put in each place where the power of 2 was subtracted.

Method 2

```
        2 | 45
        2 | 22  R 1
        2 | 11  R 0
        2 |  5  R 1        (101101)₂
        2 |  2  R 1
            1  R 0
              READ
```

EXERCISES 1002, SET I. Write the following base 2 numbers as base 10 numbers.

1. $(111)_2$

2. $(10101)_2$

3. $(11011)_2$

4. $(110110)_2$

5. $(1010101)_2$

6. $(11111111)_2$

Write the following base 10 numbers as base 2 numbers.

7. $(12)_{10}$

8. $(15)_{10}$

9. $(99)_{10}$

10. $(105)_{10}$

11. $(489)_{10}$

12. $(8625)_{10}$

In Exercises 13-16, write the expanded form (base 2) for each of the given numbers.

13. $(1101)_2$

14. $(1011)_2$

15. $(10101)_2$

16. $(11011)_2$

EXERCISES 1002, SET II. Write the following base 2 numbers as base 10 numbers.

1. $(101)_2$

2. $(1101)_2$

3. $(110111)_2$

Write the following base 10 numbers as base 2 numbers.

4. $(23)_{10}$

5. $(84)_{10}$

6. $(3852)_{10}$

In Exercises 7 and 8, write the expanded form (base 2) for each number.

7. $(11101)_2$

8. $(100111)_2$

The Quinary System (Base 5)

The set of digits in the quinary system = $\{0, 1, 2, 3, 4\}$.

The base of the system = 5. (Note that this equals the number of digits.)

$$5^0 = 1$$
$$5^1 = 5$$
$$5^2 = 5 \cdot 5 = 25$$
$$5^3 = 5 \cdot 5 \cdot 5 = 125$$
$$5^4 = 625$$
$$5^5 = 3125$$

The quinary (base 5) system is also an additive, place-value system. We can use the same methods with base 5 that we have learned for base 10 and base 2.

Example 1. Write $(342)_5$ in base 10.

$$(342)_5 = 3 \cdot 5^2 + 4 \cdot 5^1 + 2 \cdot 5^0 = 3 \cdot 25 + 4 \cdot 5 + 2 \cdot 1$$
$$= 75 + 20 + 2 = (97)_{10}$$

Example 2. Write $(2034)_5$ in base 10.

$$(2034)_5 = 2 \cdot 5^3 + 0 \cdot 5^2 + 3 \cdot 5^1 + 4 \cdot 5^0$$
$$= 2 \cdot 125 + 0 \cdot 25 + 3 \cdot 5 + 4 \cdot 1$$
$$= 250 + 0 + 15 + 4 = (269)_{10}$$

Example 3. Write $(952)_{10}$ in base 5.

```
5 | 952
5 | 190 R 2
5 |  38 R 0        (12302)₅
5 |   7 R 3
       1 R 2
        READ
```

$(12302)_5$

Notice we used only the simpler *Method 2* shown in Section 1002.

Example 4. Write $(2876)_{10}$ in base 5.

```
5 | 2876
5 |  575 R 1
5 |  115 R 0        (43001)₅
5 |   23 R 0
        4 R 3
         READ
```

$(43001)_5$

EXERCISES 1003, SET I. Write the following base 5 numbers as base 10 numbers.

1. $(104)_5$　　　　　2. $(234)_5$　　　　　3. $(1201)_5$

4. $(2110)_5$　　　　5. $(4231)_5$　　　　6. $(31402)_5$

Write the following base 10 numbers as base 5 numbers.

7. $(30)_{10}$　　　　　8. $(45)_{10}$　　　　　9. $(331)_{10}$

10. $(580)_{10}$　　　11. $(7906)_{10}$　　　12. $(78904)_{10}$

Complete the following table. (Hint: When necessary, convert a number to base 10 first, then change to the other bases from base 10.

	Base 2	Base 5	Base 10
13.	11011		
14.		134	
15.			46

EXERCISES 1003, SET II. Write the following base 5 numbers as base 10 numbers.

1. $(302)_5$　　　　　2. $(2104)_5$　　　　3. $(42030)_5$

Write the following base 10 numbers as base 5 numbers.

4. $(55)_{10}$　　　　　5. $(227)_{10}$　　　　6. $(45096)_{10}$

Complete the following table. (Hint: When necessary, convert a number to base 10 first, then change to the other bases from base 10.)

	Base 2	Base 5	Base 10
7.	10111		
8.		143	
9.			73

Roman Numerals

Roman numerals are still used. Uses include dates, first pages in books, some clock faces, names such as Elizabeth II and Louis XIV, and so on.

ROMAN NUMBER SYMBOLS

I	V	X	L	C	D	M
1	5	10	50	100	500	1,000

RULES FOR WRITING ROMAN NUMERALS. All numbers are written using the above symbols and four rules: (1) repetition, (2) addition, (3) subtraction, and (4) multiplication.

REPETITION RULE. A symbol can be repeated as many as three times. Occasionally a number such as 4 is written IIII, but in more recent usage, this is not common.

Example 1

   (a) I = 1; II = 2; III = 3

   (b) X = 10; XX= 20; XXX = 30

   (c) C = 100; CC = 200; CCC = 300

   (d) M = 1,000; MM = 2,000; MMM = 3,000

ADDITION RULE. The values of symbols are added when they are repeated, and when a lesser symbol appears to the right of a larger symbol. *Exception*: When the lesser symbol appears *between* two larger symbols.

Example 2

   (a) VI = 5 + 1 = 6

   (b) XVII = 10 + 5 + 1 + 1 = 17

   (c) LXV = 50 + 10 + 5 = 65

   (d) MDCCC = 1,000 + 500 + 100 + 100 + 100 = 1,800

SUBTRACTION RULE. When a lesser symbol appears to the left of a larger symbol, it is subtracted.

   (a) I can be subtracted only from V or X.
       [Example 3(a), 3(b)]

   (b) X can be subtracted only from L or C.
       [Example 3(c), 3(d)]

   (c) C can be subtracted only from D or M.
       [Example 3(e), 3(f)]

   (d) V, L, D are never subtracted.

   (e) When a lesser symbol appears between two larger symbols, it is subtracted from the symbol on its right instead of being added to the symbol on its left. [Example 3(g)]

Example 3

    (a) IV = 5 - 1 = 4

    (b) IX = 10 - 1 = 9

    (c) XL = 50 - 10 = 40

    (d) XC = 100 - 10 = 90

    (e) CD = 500 - 100 = 400

    (f) CM = 1,000 - 100 = 900

    (g) XIV = 10 + 4 = 14 (*Not* 11 + 5)

Notice that subtraction is used only to write fours and nines wherever they appear in the number.

MULTIPLICATION RULE. When a single bar appears over a symbol, its value is multiplied by 1,000. If a double bar appears over a symbol, its value is multiplied by 1 million (1,000 × 1,000).

Example 4

    (a) $\overline{V}$ = 5 · 1,000 = 5,000

    (b) $\overline{\overline{V}}$ = 5 · 1,000 · 1,000 = 5 · 1,000,000 = 5,000,000

    (c) $\overline{D}$ = 500 · 1,000 = 500,000

    (d) $\overline{\overline{D}}$ = 500 · 1,000,000 = 500,000,000

    (e) $\overline{IV}$ = 4 · 1,000 = 4,000

Example 5. Examples of Numbers Written in Roman Numerals. First we will show how all units, tens, and hundreds are written.

1	I	10	X	100	C
2	II	20	XX	200	CC
3	III	30	XXX	300	CCC
4	IV	40	XL	400	CD
5	V	50	L	500	D
6	VI	60	LX	600	DC
7	VII	70	LXX	700	DCC
8	VIII	80	LXXX	800	DCCC
9	IX	90	XC	900	CM
10	X	100	C	1,000	M

Now a variety of numbers will be written, using all the rules.

17	XVII	496	CDXCVI
19	XIX	943	CMXLIII
34	XXXIV	70,000	$\overline{LXX}$
52	LII	300,000	$\overline{CCC}$
95	XCV (not VC)	217,000	$\overline{CCXVII}$
99	XCIX (not IC)	1,325,729	$\overline{MCCCXXV}$DCCXXIX

EXERCISES 1004, SET I. In Exercises 1-26, write the given numbers in Roman numerals.

1. 17	2. 42	3. 19	4. 48
5. 64	6. 53	7. 59	8. 87
9. 145	10. 289	11. 414	12. 799
13. 1,973	14. 2,119	15. 4,994	16. 8,948
17. 86,045	18. 255,648	19. 157,095	20. 2,365,434

21. One hundred seventy-nine
22. Four hundred ninety-nine
23. Three thousand, nineteen
24. Nine thousand, eight hundred sixty-eight
25. Fifteen thousand, two hundred forty-nine
26. Six hundred twenty-three thousand, four hundred ninety-seven

In Exercises 27-44, write the given Roman numerals in our own system.

27. XIV	28. XVII	29. XXIII
30. XIX	31. XLV	32. XXII
33. XLVII	34. LXIX	35. CXXV
36. CDLXII	37. MMDCCXCII	38. LXXIX
39. CCXLV	40. CDLXXIV	41. MMMCDXCIII
42. $\overline{\text{VI}}$DCCLI	43. $\overline{\text{CLXXXVII}}$CCXV	44. $\overline{\text{CCLIV}}$CCCXIX

EXERCISES 1004, SET II. In Exercises 1-13, write the given numbers in Roman numerals.

1. 18	2. 14	3. 52	4. 93
5. 369	6. 599	7. 3,219	8. 7,942
9. 365,347		10. 3,439,276	

11. Two hundred twenty-nine
12. Eight thousand, six hundred nineteen
13. Five hundred forty thousand, two hundred ninety-nine

In Exercises 14-22, write the following Roman numerals in our own system.

14. XIX	15. XLV	16. XCIX
17. MCXIV	18. CMXXV	19. CDXCIV
20. MDCXXII	21. $\overline{\text{XXXVII}}$CCL	22. $\overline{\text{CCL}}$DXXX

 **Chapter Summary**

A *system of numeration* consists of:

1. A set of symbols called digits
2. A set of rules for combining the digits to make it possible to represent any number

In this chapter we studied *four different systems of numeration*:

1. The decimal system (base 10) (Sec. 1001)
2. The binary system (base 2) (Sec. 1002)
3. The quinary system (base 5) (Sec. 1003)
4. The Roman system (Sec. 1004)

REVIEW EXERCISES 1005, SET I. Complete the following table. (Hint: When necessary, convert a number to base 10 first, then change to the other systems.)

	Base 10	Base 2	Base 5	Roman
1.	17			
2.		1111		
3.			341	
4.				LXIX
5.	123			
6.		101101		
7.			1002	
8.				CXIV

9. Write 1974 using Roman numerals.

10. <u>Brain Teaser</u>. How can you take one from nine and get ten?

REVIEW EXERCISES 1005, SET II. Complete the following table.
(Hint: When necessary, convert a number to base 10 first, then
change to the other systems.)

	Base 10	Base 2	Base 5	Roman
1.	39			
2.		11001		
3.			423	
4.				CDXIV

5. Write 1978 using Roman numerals.

Chapter 10: Diagnostic Test

Name_____

_____

The purpose of this test is to see how well you understand other systems of numeration as well as our own base 10 system. We recommend that you work this diagnostic test *before* your instructor tests you on this chapter. Allow yourself about an hour to do this test.

Complete solutions for all the problems on this test, together with section references, are given in the Answer Section. You should study the sections referred to for the problems you do incorrectly.

_____

1. Write the expanded form (base 10) for each of the following numbers.

    (a) 475   (1a) _____

    (b) 20,306   (1b) _____

2. Write the number for each of the following expanded forms.

    (a) $5 \cdot 10^2 + 3 \cdot 10^1$ (2a)_____

    (b) $8 \cdot 10^4 + 9 \cdot 10^2 + 7 \cdot 10^0$ (2b)_____

3. Write $(1101)_2$ in the expanded form (base 2).

    (3) _____

4. Write $(10110)_2$ as a base 10 number.

    (4)_____

5. Write $(67)_{10}$ as a base 2 number.

    (5)_____

6. Write $(1204)_5$ in the expanded form (base 5).

(6) _____

7. Write $(342)_5$ as a base 10 number.

(7)_____

8. Write $(174)_{10}$ as a base 5 number.

(8)_____

9. Complete the following table.

	Base 2	Base 5	Base 10
(a)	1011		
(b)		143	
(c)			39

10. Write each of the following numbers in Roman numerals.

(a) 29   (10a)_____

(b) 144   (10b)_____

(c) 699   (10c)_____

(d) 1974   (10d)_____

11. Write each of the following Roman numerals in our base 10 system.

(a) XXVIII   (11a)_____

(b) CXLIX   (11b)_____

(c) MDCLXXV   (11c)_____

(d) $\overline{VI}$CCXXX   (11d)_____

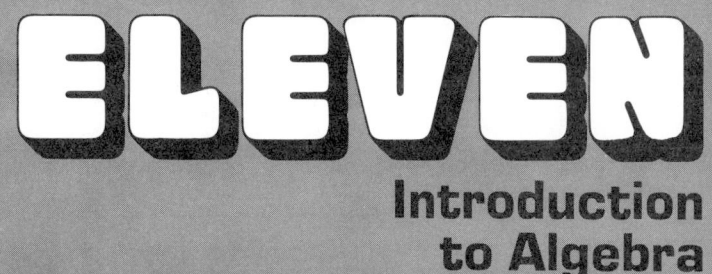

ELEVEN
Introduction to Algebra

In this chapter we give a brief introduction to beginning algebra. *Algebra* is a branch of mathematics dealing with relations and properties of numbers by means of symbols. We have already used symbols in several places in this book to show relations between numbers.

Example 1

    (a) Area of a rectangle = length × width

$$A = LW$$

    (b) Area of a triangle = $\frac{1}{2}$ base × height

$$A = \frac{1}{2}bh$$

    (c) Area of a circle = pi × (radius)^2

$$A = \pi r^2$$

Before going on with our introduction to algebra we must learn about negative numbers.

Negative Numbers

In Section 101 we showed how whole numbers could be represented by equally spaced points along the number line. We now extend the number line to the left and continue with the set of equally spaced points.

Numbers used to name the points to the left of 0 on the number line are called *negative numbers*. Numbers used to name the points to the right of 0 on the number line are called *positive numbers*. The positive and negative numbers are referred to as *signed numbers* (Figure 1101A).

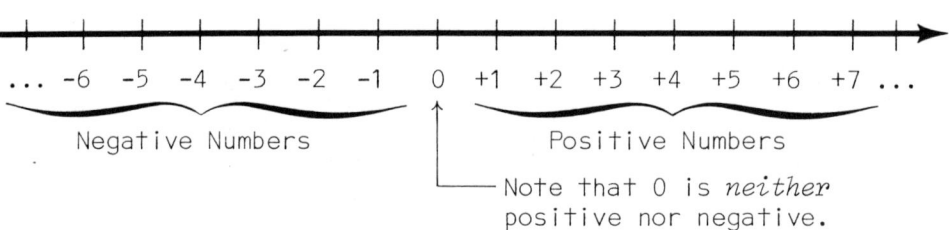

Figure 1101A

INTEGERS. The numbers used to name the points shown in Figure 1101A are called *integers*. The set of integers can be represented in the following way:

$$\{ \ \ldots \ , \ -3, \ -2, \ -1, \ 0, \ +1, \ +2, \ +3, \ \ldots \ \}$$

READING POSITIVE AND NEGATIVE INTEGERS

Example 1

    (a) -1   is read "negative one"

    (b) -575 is read "negative five hundred seventy-five"

    (c)   25 is read "twenty-five" or "positive twenty-five"

When reading or writing positive numbers, we usually omit the word "positive" and the + sign. Therefore, when there is no sign in front of a number, it is understood to be positive.

USING INEQUALITY SYMBOLS WITH INTEGERS. The arrowhead on the number line indicates the direction in which numbers get larger. Let X represent some number on the number line. Then numbers to the right of X on the number line are greater than X, written "$>X$." Numbers to the left of X on the number line are less than X, written "$<X$."

Example 2

    (a) $6 > 4$ (b) $0 > -1$ (c) $-2 > -5$

    (d) $-20 < -10$ (e) $-5 < 3$

In Section 101 we stated that a largest natural number could never be found because no matter how far we count there are always larger natural numbers. Similarly, no matter how far we count along the number line to the left of 0 we never reach a smallest negative number.

EXAMPLES SHOWING THE USE OF POSITIVE AND NEGATIVE INTEGERS

Example 3

On a cold day in Minnesota, the temperature could be -40°F. This, as you know, means that the temperature is 40°F below 0.

Example 4

The altitudes of some unusual places on Earth are as follows:

    (a) Mt. Everest . . . . 29,028 feet
        This means that the peak of Mt. Everest
        is 29,028 feet *above* sea level.

    (b) Mt. Whitney (California). . . . . 14,495 feet

    (c) Lowest point in Death Valley
        (California). . . . .   -282 feet
        This means that the lowest point
        in Death Valley is 282 feet *below*
        sea level.

    (d) Dead Sea (Jordan) . . . . -1,299 feet

    (e) World's deepest well (Oklahoma, 1974) . . -31,441 feet

    (f) Mariana Trench (Pacific Ocean). . . . . . -36,198 feet

Representing Some Negative Numbers Other Than Integers. Many
points exist on the number line to the left of 0 other than
those representing the negative integers. Some of these points
are shown in Figure 1101B.

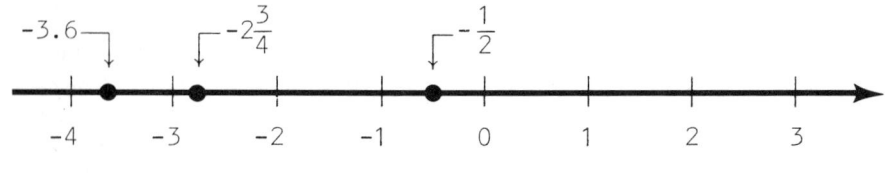

Figure 1101B

In Section 326 we explained that all the numbers that can be
represented by points on the number line are called real numbers. Since negative numbers can be represented by points on
the number line, they are a part of the real number system.

EXERCISES 1101, SET I

1. Write -75 in words.
2. Write -49 in words.
3. Use digits to write negative fifty-four.
4. Use digits to write negative one hundred nine.
5. Which is larger, -2 or -4?
6. Which is larger, 0 or -10?
7. A scuba diver descends to a depth of sixty-two feet.
   Represent this number by an integer.
8. The temperature at Fairbanks, Alaska, was forty-five degrees
   Fahrenheit below zero. Represent this number by an integer.

In Exercises 9-16, determine which of the two symbols < or >
should be used to make each statement true.

9. 8 _?_ 5   10. 7 _?_ 9   11. 0 _?_ -3

12. -10 _?_ 0   13. -17 _?_ -11   14. -10 _?_ -4

15. -5 _?_ -16   16. -3 _?_ -20

17. What is the largest negative integer?
18. What is the smallest negative integer?
19. What is the smallest two-digit integer?
20. What is the largest two-digit integer less than zero?

EXERCISES 1101, SET II

1. Write -57 in words.
2. Use digits to write negative one hundred twenty-eight.
3. Which is larger, -12 or -8?
4. The temperature in Butte, Montana, was eighteen degrees
   Fahrenheit below zero. Represent this temperature by an
   integer.

In Exercises 5-8, determine which of the two symbols < or >
should be used to make each statement true.

5. -15 _?_ 0   6. -7 _?_ -15

7. 12 _?_ 4   8. -24 _?_ -18

9. What is the largest negative integer?
10. What is the smallest one-digit integer?

Adding Signed Numbers

Now that we have introduced a new kind of number, we will do some of the operations with them that we learned for positive numbers. In this section we will learn how to add signed numbers. A signed number has two distinct parts, the number part* and the sign.

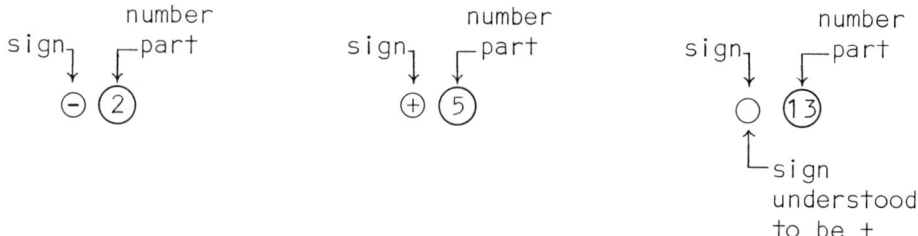

We now state without proof the rules for adding signed numbers.

```
To Add Signed Numbers

1. When the numbers      < Add the number parts.
   have the same sign      The sum has the same sign
                           as both numbers.

2. When the numbers      < Subtract the number parts.
   have different signs    The sum has the sign of
                           the larger number part.
```

We show how to add signed numbers by example.

Example 1. Add: 3 and 5

Solution: Since they have the same sign:
           Add the number parts: 3 + 5 = 8
           Sum has the same sign: +
           Therefore, the sum is +8

Example 2. Add: -7 and -11

----

*In algebra, what we call the number part is called the absolute value of the number.

Solution: Since they have the same sign:
Add the number parts: 7 + 11 = 18
Sum has the same sign: -
Therefore, the sum is -18.

Example 3. Add: -24 and 17

Solution: Since they have different signs:
Subtract the number parts: 24 - 17 = 7
Sum has sign of larger number part: -
(since -24 has the larger number part)
Therefore, the sum is -7.

Example 4. Add: +18 and -32

Solution: Since they have different signs:
Subtract the number parts: 32 - 18 = 14
Sum has sign of larger number part: -
(since -32 has the larger number part)
Therefore, the sum is -14.

Example 5. Add: -9 and 23

Solution: The signs are
different ──────────────→ $\begin{array}{r} -\,9 \\ +\,23 \\ \hline \end{array}$ 23 - 9 = (14)
Sign of the larger
number part ──────────→ + 14 ←─────────────────┘

Example 6. Add: -29 and -35

Solution: The signs are
the same ──────────────→ $\begin{array}{r} -\,29 \\ -\,35 \\ \hline \end{array}$ 29 + 35 = (64)
Same sign as
both numbers ─────────→ - 64 ←─────────────────┘

EXERCISES 1102, SET I. Add the signed numbers.

1. 7 and 6   2. 8 and 9   3. -5 and -9

4. -10 and -7   5. 12 and -7   6. 17 and -8

7. -10 and 25   8. -11 and 34   9. 15 and -23

10. 18 and -47   11. -74 and 35   12. -105 and 71

13. 14   14. 88   15.   35   16.   72
    27   19   -17   -49

17. -156   18. -284   19. -286   20. -756
     29   167   -354   -378

EXERCISES 1102, SET II. Add the signed numbers.

1. 12 and 6   2. -13 and -9   3. 18 and -5

4. -14 and 29   5. 17 and -33   6. -92 and 48

7. 36   8. 53   9. -162   10. -503
   29   -38   88   -279

Subtracting Signed Numbers

Now that we have learned how to *add* signed numbers, we can *subtract* them by using addition.

```
┌─────────────────────────────────────────────────────┐
│                                                       │
│   To Subtract One Signed Number from Another          │
│                                                       │
│   1. Change the sign of the number being subtracted.  │
│   2. Add the resulting signed numbers as shown in     │
│      Section 1102.                                    │
│                                                       │
└─────────────────────────────────────────────────────┘
```

Example 1. Subtract: 9 from 6

Solution: Change the sign of the 9 (since it is being
            subtracted), making it -9. Then add 6 and -9.

$$\begin{array}{r} 6 \\ -9 \\ \hline -3 \end{array}$$ These numbers are added
                                                          as shown in Section 1102.

Check: Subtraction is checked very easily by addition.

subtrahend + difference = minuend
$$(9) \quad + \quad (-3) \quad = \quad 6$$

Example 2. Subtract: -11 from -7

Solution: Change the sign of the -11, making it 11. Then
            Then add -7 and 11.

$$\begin{array}{r} -7 \\ 11 \\ \hline 4 \end{array}$$

Check: *subtrahend + difference = minuend*
$$(-11) \quad + \quad (4) \quad = \quad -7$$

Example 3. Subtract: +13
                     -14

Solution:
$$\begin{array}{r} +13 \\ -14 \\ + \\ \hline +27 \end{array}$$ We suggest that you write
                                                                  the changed sign below the
                                                                  given sign, then add the
                                                                  numbers using the bottom
                                                                  sign.

Example 4. Subtract: -147   Solution: -147
                     +59   +59
                                             -
                                           ──────
                                           -206

Example 5. Subtract: $-3\frac{1}{2}$ Solution: $-3\frac{1}{2}$ = $-3\frac{2}{4}$

$+2\frac{1}{4}$ $+2\frac{1}{4}$ $-2\frac{1}{4}$

$-$

$-5\frac{3}{4}$

Example 6. Subtract: -4.56 Solution: -4.56

-7.48 $\underset{+}{-7.48}$

$+2.92$

EXERCISES 1103, SET I. Subtract the lower number from the upper number.

1. 10
   4
	2. 12
	   5
		3. 8
		   -2
			4. 10
			   -3

5. -10
   -4
	6. -12
	   -5
		7. -15
		   11
			8. -24
			   13

9. 86
   96
	10. 72
	   89
		11. 156
		   -97
			12. 284
			   -89

13. -354
   -286
	14. -484
	   -375
		15. 780
		   840
			16. 579
			   700

17. 1,786
   -295
	18. 3,544
	   $-1,297$
		19. $-16,780$
		   3,915
			20. $-27,451$
			   28,762

21. $-3,005$ 22. $-7,000$ 23. $-5\frac{3}{4}$ 24. $-17\frac{5}{6}$

   $-5,001$ $-2,009$ $2\frac{1}{2}$ $8\frac{1}{3}$

25. -16.71 26. -7.45 27. $+3\frac{1}{5}$ 28. $4\frac{3}{16}$

   -18.39 -14.91 $+\frac{7}{10}$ $\frac{7}{8}$

29. 7.015   30. -0.875

   -2.94 -1.25

EXERCISES 1103, SET II. Subtract the lower number from the upper number.

1. 15
   9
	2. 16
	   -7
		3. -20
		   -7
			4. -28
			   16

5. 73
   59
	6. 314
	   -88
		7. -592
		   -346
			8. 670
			   830

9. 2,677   10. $-32,018$ 11. $-5,000$ 12. $-13\frac{3}{4}$

   $-1,982$ 29,402 $-3,008$ $8\frac{1}{3}$

13. -14.06 <u>-18.32</u>	14. $7\frac{2}{3}$ $\frac{5}{6}$	15. -0.614 <u>-1.38</u>

Multiplying Signed Numbers

> To Multiply One Signed Number by Another
>
> 1. Multiply the number parts of the two signed numbers.
> 2. The product is *positive* when the signed numbers have the same sign.
> The product is *negative* when the signed numbers have different signs.

Example 1. Multiply: $(-7)(4)$

Solution: $(-7)(4) = \ominus \, \boxed{28}$

Product of number parts:
$7 \times 4 = 28$
Product negative because numbers have different signs

Example 2. Multiply: $(23)(-11)$

Solution: $(23)(-11) = \ominus \, \boxed{253}$

$23 \times 11 = 253$
Because the numbers have different signs

Example 3. Multiply: $(-14)(-10)$

Solution: $(-14)(-10) = \oplus \, \boxed{140}$

$14 \times 10 = 140$
Because the numbers have the same sign

In the first three examples, we multiplied only *two* signed numbers. When *three* signed numbers are to be multiplied, multiply any two of them, then multiply the resulting product by the remaining number. This method can be extended to products with any number of factors.

Example 4. Multiply: $(-7)(12)(-8)$

Solution: $(-7)(12) = -84$

then, $(-84)(-8) = +672$

Therefore, $(-7)(12)(-8) = 672$

Example 5. Multiply: $\left(4\frac{1}{2}\right)\left(-1\frac{1}{3}\right)$

Solution: $\left(4\frac{1}{2}\right)\left(-1\frac{1}{3}\right) = \left(\frac{9}{2}\right)\left(-\frac{4}{3}\right) = -\left(\frac{\overset{3}{\cancel{9}}}{\underset{1}{\cancel{2}}} \cdot \frac{\overset{2}{\cancel{4}}}{\underset{1}{\cancel{3}}}\right) = -\frac{6}{1} = -6$

Example 6. Multiply: $(-2.7)(-4.6)$

Solution: $(-2.7)(-4.6) = +(2.7 \times 4.6) = 12.42$

EXERCISES 1104, SET I. Find the following products.

1. $8(-4)$ 2. $9(-5)$ 3. $(-7)(9)$

4. $(-6)(9)$ 5. $(-10)(-10)$ 6. $(-9)(-9)$

7. $75(-15)$ 8. $86(-13)$ 9. $(17.5)(-150)$

10. $(12.5)(-16)$ 11. $(-2)(3)(-4)$ 12. $3(-5)(-6)$

13. $(-7)(-2)(-5)$ 14. $(-2)(-13)(-5)$ 15. $(-700)(500)$

16. $(300)(-5,000)$ 17. $10^2(-47)$ 18. $10^2(-15.6)$

19. $10^4(-2)(-5)$ 20. $2(-3)(-3)(-2)$ 21. $3(-2)(2)(-5)$

22. $(-4)(-5)(-3)(2)$ 23. $\left(3\frac{1}{2}\right)\left(-\frac{8}{21}\right)$ 24. $\left(-\frac{6}{7}\right)\left(4\frac{2}{3}\right)$

25. $(-5.6)(-3.8)$ 26. $(-0.28)(-3.05)$ 27. $\left(-5\frac{1}{5}\right)\left(+1\frac{2}{13}\right)$

28. $\left(6\frac{3}{4}\right)\left(-5\frac{1}{3}\right)$ 29. $(0.06)(-875)$ 30. $(-880)(0.0075)$

EXERCISES 1104, SET II. Find the following products.

1. $6(-9)$ 2. $(-8)(7)$ 3. $(-10)(-8)$

4. $45(-25)$ 5. $(18.5)(-14)$ 6. $(-5)(4)(-6)$

7. $(-4)(-8)(-5)$ 8. $(-7000)(300)$ 9. $10^3(-9.7)$

10. $(-6)(-5)10^2$ 11. $(-2)(-7)(3)(-5)$ 12. $\left(-\frac{4}{5}\right)\left(2\frac{1}{7}\right)$

13. $(-6.5)(-8.4)$ 14. $\left(-3\frac{3}{5}\right)\left(2\frac{2}{9}\right)$ 15. $(0.055)(-625)$

Division of Signed Numbers

> To Divide One Signed Number by Another
>
> 1. Divide their number parts.
> 2. The quotient is *positive* when the signed numbers have the same sign.
>   The quotient is *negative* when the signed numbers have different signs.

Example 1. $(-30) \div (5)$

Solution: $(-30) \div (5) = \ominus \; ⑥$

             Quotient of number parts:
             $30 \div 5 = 6$
             Quotient negative because
             numbers have different
             signs

Example 2. $(-64) \div (-8)$

Solution: $(-64) \div (-8) = \oplus \; ⑧$

             $64 \div 8 = 8$
             Because numbers have
             same sign

Example 3. $\dfrac{35}{-7}$

Solution: $\dfrac{35}{-7} = \ominus \; ⑤$

             $35 \div 7 = 5$
             Because numbers have
             different signs

Example 4. $\dfrac{-42}{8}$

Solution: $\dfrac{\overset{21}{\cancel{-42}}}{\underset{4}{\cancel{8}}} = -\dfrac{21}{4} = -5\dfrac{1}{4}$

EXERCISES 1105, SET I. Find the following quotients.

1. $(-40) \div (8)$ 2. $(-60) \div (10)$ 3. $16 \div (-4)$

4. $25 \div (-5)$ 5. $(-15) \div (-5)$ 6. $(-27) \div (-9)$

7. $\dfrac{12}{-4}$ 8. $\dfrac{24}{-6}$ 9. $\dfrac{-18}{-2}$

10. $\dfrac{-49}{-7}$ 11. $\dfrac{-150}{10}$ 12. $\dfrac{-250}{100}$

13. $\dfrac{-15}{-6}$ 14. $\dfrac{-27}{-12}$ 15. $25.5 \div (-3)$

16. $45 \div (-7.5)$ 17. $\dfrac{(-367)}{10^2}$ 18. $\dfrac{-4,860}{10^3}$

19. $\dfrac{78.5}{-10^3}$ 20. $\dfrac{98.5}{-10^2}$

EXERCISES 1105, SET II. Find the following quotients.

1. $(-50) \div (5)$ 2. $35 \div (-7)$ 3. $(-42) \div (-6)$

4. $\dfrac{54}{-9}$ 5. $\dfrac{-44}{-11}$ 6. $\dfrac{-350}{10}$

7. $\dfrac{-56}{-8}$

8. $12.6 \div (-9)$

9. $\dfrac{-473}{10^2}$

10. $\dfrac{84.5}{-10^3}$

Finding the Value of an Expression Having Letters and Numbers

We have already used letters to represent numbers in several parts of this book. In this section we will use what we have already learned about signed numbers to help us find the value of expressions having letters as well as numbers.

To Find the Value of an Expression Having Letters and Numbers

1. Replace each letter by its number value.
2. Carry out all arithmetic operations as we have done before.

Example 1. Find the value of $a + 2b$ if $a = 5$ and $b = 3$.

Solution:
$$a + 2b$$
$$= (5) + 2(3)$$
$$= 5 + 6 = 11$$

You notice that we simply replace each letter by its number value, then carry out the arithmetic operations as we have before.

Example 2. Find the value of $3x - 5y$ if $x = 10$ and $y = 4$.

Solution:
$$3x - 5y$$
$$= 3(10) - 5(4)$$
$$= 30 - 20 = 10$$

Example 3. Find the value of $a + 2b$ if $a = -3$ and $b = 4$.

Solution:
$$a + 2b$$
$$= (-3) + 2(4)$$
$$= -3 + 8 = 5$$

Here you are using what you learned about adding signed numbers in Section 1102.

Example 4. Find the value of $3x - 5y$ if $x = -4$ and $y = -6$.

Solution:
$$3x - 5y$$
$$= 3(-4) - 5(-6)$$
$$= -12 + 30 = 18$$

Example 5. Find the value of $\dfrac{2a - b}{10c}$ if $a = -1$, $b = 3$, and $c = -2$.

Solution: $\dfrac{2a - b}{10c} = \dfrac{2(-1) - (3)}{10(-2)} = \dfrac{-2 - 3}{-20} = \dfrac{-5}{-20} = \dfrac{1}{4} = 0.25$

Example 6. Find the value of $\dfrac{5hgk}{2m}$ if $h = -2$, $g = 3$, $k = -4$, and $m = 6$.

Solution: $\dfrac{5hgk}{2m} = \dfrac{5(-2)(3)(-4)}{2(6)} = \dfrac{120}{12} = 10$

Example 7. Find the area of a circle with a radius of 5 feet (use $\pi \doteq 3.14$).

Solution:
$$A = \pi r^2 \text{ (see Section 903)}$$
$$= 3.14(5)^2$$
$$= 3.14(25) = 78.5 \text{ sq ft}$$

Example 8. Find the total surface area of a right circular cylinder having $r = 2$ centimeters and $h = 4$ centimeters.

Solution: Total surface area $= 2\pi rh + 2\pi r^2$ (see Section 905)
$$= 2(3.14)(2)(4) + 2(3.14)(2)^2$$
$$= 50.24 + 25.12 \doteq 75.4 \text{ sq cm}$$

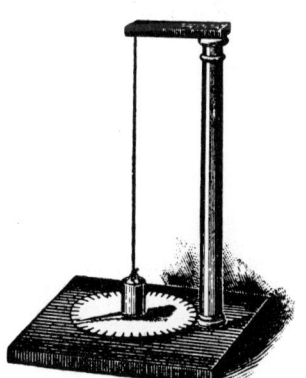

EXERCISES 1106, SET I

1. Find the value of $2x + 3y$
    (a) when $x = 10$ and $y = -5$
    (b) when $x = -10$ and $y = 5$
    (c) when $x = -10$ and $y = -5$

2. Find the value of $5x + 2y$
    (a) when $x = 10$ and $y = -5$
    (b) when $x = -10$ and $y = 5$
    (c) when $x = -10$ and $y = -5$

3. Find the value of $2x - 7y$
    (a) when $x = 5$ and $y = 4$
    (b) when $x = -8$ and $y = 3$
    (c) when $x = -9$ and $y = -8$

4. Find the value of $3x - 8y$

    (a) when $x = 3$ and $y = 5$

    (b) when $x = -7$ and $y = 4$

    (c) when $x = 9$ and $y = 6$

5. Find the value of $\dfrac{h(a + b)}{2}$

    (a) when $h = 6$, $a = 4$, and $b = 5$

    (b) when $h = 7$, $a = 4$, and $b = 9$

    (c) when $h = 2\frac{1}{2}$, $a = 3\frac{1}{2}$, and $b = 9\frac{1}{2}$

6. If $I = \dfrac{E}{R}$, find the value of I

    (a) when $E = 110$ and $R = 55$

    (b) when $E = 110$ and $R = 220$

    (c) when $E = 220$ and $R = 66$

7. If $C = \dfrac{5}{9}(F - 32)$, find C when $F = 77$.

8. Find the area of a circle having a radius of 2.5 meters (see Example 7).

9. Find the total surface area of a right circular cylinder having a radius of 3 feet and a height of 15 feet.

10. If $I = Prt$, find the value of I

    (a) when $P = 250$, $r = 0.08$, and $t = 3$

    (b) when $P = 400$, $r = 0.09$, and $t = \dfrac{3}{4}$

    (c) when $P = 1,000$, $r = 0.045$, and $t = 2\frac{1}{2}$

11. If $A = P(1 + rt)$, find the value of A

    (a) when $P = 100$, $r = 0.08$, and $t = 1.5$

    (b) when $P = 500$, $r = 0.09$, and $t = 2.5$

    (c) when $P = 250$, $r = 0.075$, and $t = 24$

12. If $I = \dfrac{nE}{R + nr}$, find the value of I when $n = 5$, $E = 110$, $R = 200$, and $r = 4$.

EXERCISES 1106, SET II

1. Find the value of $4x + 5y$

    (a) when $x = 8$ and $y = -3$

    (b) when $x = -6$ and $y = 4$

    (c) when $x = -3$ and $y = -7$

2. Find the value of $2x - 7y$

    (a) when $x = 6$ and $y = 5$

    (b) when $x = -4$ and $y = 7$

    (c) when $x = -9$ and $y = -3$

3. Find the value of $\dfrac{h(e - f)}{2}$

    (a) when $h = 8$, $e = -3$, and $f = 5$

    (b) when $h = -7$, $e = 3$, and $f = -9$

    (c) when $h = -5$, $e = -11$, and $f = -7$

4. Find the area of a circle having a radius of 3.5 inches (see Example 7).

5. If $F = \dfrac{9}{5}C + 32$, find F when $C = -10$.

6. Find the value of $b^2 - 4ac$ when $a = 2$, $b = -5$, and $c = -9$.

Equations

An *equation* looks like this:

$$5x - 8 = 3x + 2$$

left side ⟶   ⟵ right side

        equal sign

An equation is made up of three parts:

1. The equal sign (=).
2. The expression to the left of the = sign, called the *left side* (or left member) of the equation.
3. The expression to the right of the = sign, called the *right side* (or right member) of the equation.

Example 1. $2x - 3 = 7$

        The left side of the equation is $2x - 3$.
        The right side of the equation is 7.

Example 2. $5x - 2 = 10x + 3$

        The left side of the equation is $5x - 2$.
        The right side of the equation is $10x + 3$.

The Meaning of the Equal Sign in an Equation. The equal sign (=) in an equation means that the number represented by the left side *is the same* as the number represented by the right side.

Example 3. $x = -2$
and $-2 = x$
⎫ both say that x and -2 are the same number

Solving Equations by Adding the Same Signed Number to Both Sides

If we add the same signed number to both sides of an equation, the resulting sums must be equal.

Example 1.
$$
\begin{array}{ccc}
7 & = & 7 \\
+5 & & +5 \\
\hline
12 & = & 12
\end{array}
$$
We are adding +5 to both sides.

Example 2.
$$
\begin{array}{ccc}
7 & = & 7 \\
-5 & & -5 \\
\hline
2 & = & 2
\end{array}
$$
We are adding -5 to both sides.

Example 3.
$$
\begin{array}{ccc}
x - 3 & = & 6 \\
+ 3 & & +3 \\
\hline
x & = & 9
\end{array}
$$
We are adding +3 to both sides.

We call 9 the *solution* of the equation $x - 3 = 6$. A solution of an equation is a number which when put in place of the letter makes the two sides of the equation equal. An equation is solved when we have nothing but the unknown letter on one side of the = sign.

To Solve an Equation Using Addition
(Equations in which the unknown letter is only multiplied by 1)

1. Look at the side of the equation where the unknown letter appears.
2. Add the signed number(s) necessary to remove any number(s) from that side of the equation.
3. Add the same signed number(s) to the other side of the equation.
4. The resulting equation is the solution.

Example 4. Solve the equation: $2 + m = 7$

Solution:
$$
\begin{array}{rcl}
2 + m & = & 7 \\
-2 & & -2 \\
\hline
m & = & 5
\end{array}
$$
Adding -2 removes everything except m from the left side.

↑—— 5 is the solution

Example 5. Solve the equation: $8 = H - 4$

Solution:
$$
\begin{array}{rcl}
8 & = & H - 4 \\
+4 & & +4 \\
\hline
12 & = & H
\end{array}
$$
Adding +4 removes everything except H from the right side.

—— Since this means that 12 and H are symbols for the same number, we can also say $H = 12$.

> To Check the Solution of an Equation
>
> 1. Replace the unknown letter in the given equation by the number found in the solution.
> 2. Perform the indicated operations on both sides of the = sign.
> 3. If the resulting number on each side of the = sign is the same, the solution is correct.

Example 6. Solve and check: $x - 5 = 3$

Solution:

$$\begin{array}{rl} x - 5 &= 3 \\ + 5 & +5 \\ \hline x &= 8 \end{array}$$

Check:

$x - 5 = 3$

$8 - 5 = 3$ x was replaced by the number in the solution, 8

$3 = 3$ This shows that the solution is correct.

EXERCISES 1108, SET I. Solve and check the following equations.

1. $x + 5 = 8$ 2. $x + 4 = 9$

3. $x - 3 = 4$ 4. $x - 7 = 2$

5. $3 + x = -4$ 6. $2 + x = -5$

7. $x + 4 = 21$ 8. $x + 15 = 24$

9. $x - 35 = 7$ 10. $x - 42 = 9$

11. $9 = x + 5$ 12. $11 = x + 8$

13. $12 = x - 11$ 14. $14 = x - 15$

15. $-17 + x = 28$ 16. $-14 + x = 33$

17. $-21 + x = -42$ 18. $-37 + x = -51$

19. $-28 = -15 + x$ 20. $-47 = -18 + x$

EXERCISES 1108, SET II. Solve and check the following equations.

1. $x + 7 = 12$ 2. $x - 6 = 5$

3. $4 + x = -8$ 4. $x + 17 = 33$

5. $x - 29 = 13$ 6. $14 = x + 18$

7. $16 = x - 9$ 8. $-19 + x = 13$

9. $-31 + x = -47$ 10. $-38 = -21 + x$

Solving Equations by Dividing Both Sides by the Same Signed Number

If we divide both sides of an equation by the same signed number, the resulting quotients must be equal.

Example 1. $8 = 8$

$$\frac{8}{4} = \frac{8}{4}$$ We are dividing both sides by 4.

$$2 = 2$$ The resulting quotients are equal.

Example 2. $6 = 6$

$$\frac{6}{-2} = \frac{6}{-2}$$ We are dividing both sides by -2.

$$-3 = -3$$ The resulting quotients are equal.

Example 3. $2x = 10$

$$\frac{2x}{2} = \frac{10}{2}$$ Dividing both sides by 2.

$$x = 5$$ Solution

Example 4. $9x = -27$

$$\frac{9x}{9} = \frac{-27}{9}$$

$$x = -3$$ Solution

Example 5. $12x = 8$

$$\frac{12x}{12} = \frac{8}{12}$$

$$x = \frac{2}{3}$$ Solution

Example 6. $-6x = 30$

$$\frac{-6x}{-6} = \frac{30}{-6}$$

$$x = -5$$ Solution

Example 7. $16 = 2x$

$$\frac{16}{2} = \frac{2x}{2}$$

$$8 = x$$ Solution

Example 8. $8 = -24x$

$$\frac{8}{-24} = \frac{-24x}{-24}$$

$$-\frac{1}{3} = x$$ Solution

Example 9. $3x - 2 = 10$

$$\underline{ + 2 \quad +2} \qquad \text{Adding 2 to both sides}$$
$$3x \quad\;\; = 12$$

$$\frac{\cancel{3}x}{\cancel{3}} = \frac{12}{3} \qquad \text{Dividing both sides by 3}$$

$$x = 4 \qquad \text{Solution}$$

An equation is solved when we have nothing but the unknown letter on one side of the = sign. Therefore, in solving an equation, we must remove all the numbers from the side that contains the unknown letter. In the last example, two numbers had to be removed: 3 and -2. When more than one number must be removed, use the following procedure.

To Solve an Equation Using Addition and Division

All numbers on the same side as the unknown letter must be removed.

1. Remove those numbers being added or sub-tracted using the method of the last section.
2. Remove the number multiplied by the unknown letter by dividing both sides by that signed number.
3. Check your solution in the *original* equation.

Example 10. Solve the equation: $2x + 3 = 11$

Solution: The numbers 2 and 3 must be removed from the side with the x.

1st. Since the 3 is added, it is removed first.

$$2x + 3 = 11$$
$$\underline{ - 3 \quad - 3} \qquad \text{Adding -3 to both sides.}$$
$$2x \quad\;\; = \;\; 8$$

2nd. Since the 2 is multiplied by the x, it is removed by dividing both sides by 2.

$$2x = 8$$

$$\frac{\cancel{2}x}{\cancel{2}} = \frac{8}{2} \qquad \text{Dividing both sides by 2.}$$

$$x = 4$$

Check: $2x + 3 = 11$

$$2(4) + 3 = 11$$
$$8 + 3 = 11$$
$$11 = 11$$

Example 11. Solve the equation: $-12 = 3x + 15$

Solution: 15 and 3 must be removed from the side with the x.

1st.
$$-12 = 3x + 15$$
$$\underline{-15 \quad\quad\ \ -15}$$
$$-27 = 3x$$
Adding -15 to both sides.

2nd.
$$\frac{-27}{3} = \frac{3x}{3}$$
Dividing both sides by 3.

$$-9 = x$$

Check:
$$-12 = 3x + 15$$
$$-12 = 3(-9) + 15$$
$$-12 = -27 + 15$$
$$-12 = -12$$

EXERCISES 1109, SET I. Solve and check the following equations.

1. $2x = 8$ 2. $3x = 15$

3. $21 = 7x$ 4. $42 = 6x$

5. $11x = 33$ 6. $12x = 48$

7. $-9x = 6$ 8. $-20x = 16$

9. $36 = -3x$ 10. $15 = -5x$

11. $-24 = 4x$ 12. $-32 = 8x$

13. $4x + 1 = 9$ 14. $5x + 2 = 12$

15. $6x - 2 = 10$ 16. $7x - 3 = 4$

17. $2x - 15 = 11$ 18. $3x - 4 = 14$

19. $4x + 2 = -14$ 20. $5x + 5 = -10$

21. $14 = 9x - 13$ 22. $25 = 8x - 15$

23. $12x + 17 = 65$ 24. $11x + 19 = 41$

25. $8x - 23 = 31$ 26. $6x - 33 = 29$

27. $-14 + 4x = 28$ 28. $-18 + 6x = 44$

29. $-8 = 3x - 25$ 30. $-10 = 2x - 27$

31. $-73 = 24x + 31$ 32. $-48 = 36x + 42$

EXERCISES 1109, SET II. Solve and check the following equations.

1. $5x = 15$ 2. $12 = 3x$

3. $11x = 44$ 4. $-15x = 60$

5. $20 = -4x$ 6. $-28 = 7x$

7. $5x + 7 = 37$ 8. $6x - 4 = 32$

9. $6x - 5 = 37$ 10. $3x + 7 = 28$

11. $55 = 10x - 45$ 12. $9x + 5 = 23$

13. $8x - 21 = 19$ 14. $-13 + 3x = 14$

15. $-12 = 2x - 16$ 16. $-34 = 23x + 12$

 **Solving Equations by Multiplying Both Sides by the Same Signed Number**

If we multiply both sides of an equation by the same signed number, the resulting products must be equal.

Example 1. $4 = 4$

$\quad\quad\quad 2(4) = 2(4)$ Multiplying both sides by 2.

$\quad\quad\quad\quad 8 = 8$ The resulting products are equal.

Example 2. $\dfrac{x}{3} = 5$

$\quad\quad\quad 3\left(\dfrac{x}{3}\right) = 3(5)$ Multiplying both sides by 3.

$\quad\quad\quad\quad x = 15$ Solution

Example 3. $\dfrac{x}{6} = -8$

$\quad\quad\quad 6\left(\dfrac{x}{6}\right) = 6(-8)$ Multiplying both sides by 6.

$\quad\quad\quad\quad x = -48$ Solution

Example 4. $\dfrac{2x}{3} = 4$

$\quad\quad\quad 3\left(\dfrac{2x}{3}\right) = 3(4)$ Multiplying both sides by 3.

$\quad\quad\quad\quad 2x = 12$

$\quad\quad\quad \dfrac{2x}{2} = \dfrac{12}{2}$ Dividing both sides by 2.

$\quad\quad\quad\quad x = 6$ Solution

Example 5. Solve: $\dfrac{2x}{5} - 6 = 4$

$$\begin{array}{ll} \dfrac{2x}{5} - 6 = 4 & \rightarrow\; 5\left(\dfrac{2x}{5}\right) = 5(10) \\ \underline{\quad\; +6 \;\; +6\quad} & \qquad\quad 2x = 50 \\ \dfrac{2x}{5} \qquad = 10 & \qquad\quad \dfrac{2x}{2} = \dfrac{50}{2} \\ & \qquad\quad\; x = 25 \qquad \text{Solution} \end{array}$$

In the last example, three numbers had to be removed: 6, 5, and 2. When more than one number must be removed, use the following procedure.

```
┌─────────────────────────────────────────────┐
│  To Solve an Equation Using Addition, Division,│
│  and Multiplication                           │
│                                               │
│  All numbers on the same side as the unknown  │
│  letter must be removed.                       │
│                                               │
│  1. First remove those numbers being added or │
│     subtracted.                               │
│  2. Multiply both sides by the signed number the│
│     letter is divided by.                      │
│  3. Divide both sides by the signed number the │
│     letter is multiplied by.                   │
│  4. Check your solution in the *original* equation.│
└─────────────────────────────────────────────┘
```

Example 6. Solve: $\dfrac{3x}{4} + 2 = 11$

Solution: The numbers 2, 4, and 3 must be removed from the side with the x.

1st. Since the 2 is added, it is removed first.

$$\dfrac{3x}{4} + 2 = 11$$
$$\underline{\quad -2 \quad -2\quad}$$
$$\dfrac{3x}{4} = 9$$

2nd. Since the letter is divided by 4, we multiply both sides by 4.

$$\dfrac{3x}{4} = 9$$
$$4\left(\dfrac{3x}{4}\right) = 4(9)$$
$$3x = 36$$

3rd. Since the letter is multiplied by 3, we divide both sides by 3.

$$\dfrac{3x}{3} = \dfrac{36}{3}$$
$$x = 12 \qquad \text{Solution}$$

Check: $\qquad\qquad \dfrac{3x}{4} + 2 = 11$

$$\dfrac{3(\overset{3}{\cancel{12}})}{\underset{1}{\cancel{4}}} + 2 = 11$$

$$9 + 2 = 11$$

$$11 = 11$$

Sometimes we have equations in which the unknown letter is multiplied by a negative number.

Example 7. Solve: $3 - 2x = 9$

Solution:

$$
\begin{array}{r}
3 - 2x = 9 \\
-3 \qquad -3 \\
\hline
-2x = 6
\end{array}
$$

Since x is multiplied by (-2), we divide both sides by (-2).

$$\frac{-2x}{-2} = \frac{6}{-2}$$

$$x = -3 \qquad \text{Solution}$$

Check:

$$3 - 2x = 9$$
$$3 - 2(-3) = 9$$
$$3 + 6 = 9$$
$$9 = 9$$

Example 8. Solve:

$$
\begin{array}{r}
7 = 5 - x \\
-5 \quad -5 \\
\hline
2 = \quad -x
\end{array}
$$

$$(-1)(2) = (-1)(-x)$$

Multiplying both sides by (-1) changes $(-x)$ to (x).

$$-2 = x$$

or

$$x = -2$$

EXERCISES 1110, SET I. Solve and check the following equations.

1. $\frac{x}{3} = 4$ 2. $\frac{x}{5} = 2$ 3. $\frac{x}{5} = -2$

4. $\frac{x}{6} = -4$ 5. $\frac{x}{10} = 3.14$ 6. $\frac{x}{5} = 7.8$

7. $-4 = \frac{2x}{7}$ 8. $-1.5 = \frac{3x}{4}$ 9. $\frac{20x}{5} = 12$

10. $\frac{35x}{14} = 10$ 11. $\frac{x}{4} + 6 = 9$ 12. $\frac{x}{5} + 3 = 8$

13. $\frac{x}{10} - 5 = 13$ 14. $\frac{x}{20} - 4 = 12$ 15. $7 = \frac{2x}{5} + 3$

16. $-8 = \frac{3x}{4} - 5$ 17. $4 - \frac{7x}{5} = 11$ 18. $3 - \frac{2x}{9} = 12$

19. $10^3 = 10^2 - \frac{10x}{7}$ 20. $10^2 = 10^3 - \frac{10x}{9}$

EXERCISES 1110, SET II. Solve and check the following equations.

1. $\frac{x}{4} = 5$ 2. $\frac{x}{7} = -2$ 3. $\frac{x}{10} = 3.4$

4. $-2.5 = \dfrac{5x}{4}$ 5. $\dfrac{18x}{15} = 12$ 6. $\dfrac{x}{8} + 4 = 20$

7. $\dfrac{x}{10} - 9 = 11$ 8. $-7 = \dfrac{4x}{3} - 3$ 9. $5 - \dfrac{3x}{7} = 14$

10. $10^2 = 10^3 - \dfrac{10x}{11}$

Chapter Summary

Negative numbers are used to name the points to the left of 0 on the number line. (Sec. 1101)

Positive numbers are used to name the points to the right of 0 on the number line. (Sec. 1101)

Signed numbers include both negative and positive numbers. (Sec. 1101)

Integers are the numbers in the set

$$\{ \ldots , -3, -2, -1, 0, 1, 2, 3, \ldots \} \quad \text{(Sec. 1101)}$$

To add signed numbers: (Sec. 1102)

1. When the numbers have the same sign ⎯ Add the number parts.
   The sum has the same sign.
2. When the numbers have different signs ⎯ Subtract the number parts.
   The sum has sign of larger number part.

To subtract one signed number from another: (Sec. 1103)

1. Change the sign of the number being subtracted.
2. Add the resulting signed numbers as shown above.

To multiply one signed number by another: (Sec. 1104)

1. Multiply the number parts of the two signed numbers.
2. The product is positive when the signed numbers have the same sign.
   The product is negative when the signed numbers have different signs.

To divide one signed number by another: (Sec. 1105)

1. Divide their number parts.
2. The quotient is positive when the signed numbers have the same sign.
   The quotient is negative when the signed numbers have different signs.

To find the value of an expression having letters and numbers: (Sec. 1106)

1. Replace each letter by its number value.
2. Carry out all arithmetic operations as we have done before.

To solve an equation: All numbers on the same side as the unknown letter must be removed. (Sec. 1110)

1. First remove those numbers being added or subtracted.
2. If the unknown letter is divided by a signed number, multiply both sides of the equation by that signed number.
3. If the unknown letter is multiplied by a signed number, divide both sides of the equation by that signed number.
4. Check your solution by substituting it in the original equation.

REVIEW EXERCISES 1111, SET I

1. Add the signed numbers.

   (a)   274   (b) −706   (c) $-3\frac{1}{4}$ (d)   7.64
        −345   −315 $2\frac{1}{2}$ −9.01

2. Add the signed numbers.

   (a)   586   (b) −35.4   (c) $-5\frac{1}{5}$ (d)   0.49
        −794   −20.9 $2\frac{3}{10}$ −1.64

3. Subtract the lower number from the upper number.

   (a)   354   (b) −507   (c) $-2\frac{7}{8}$ (d) 2.99
      −286   −314 $3\frac{3}{4}$ 3.08

4. Subtract the lower number from the upper number.

   (a)   209   (b) −788   (c) $-5\frac{1}{6}$ (d) 3.07
      −156   −254 $3\frac{5}{12}$ 3.55

5. Find the following products.

   (a) (−21)(−50)   (b) 8.4(−55)

   (c) (−2)(3)(−5)   (d) (4)(−6)(−25)(−3)

6. Find the following products.

   (a) (−54)(−37)   (b) (−75)(5.6)

   (c) (7)(−2)(5)   (d) (−7)(8)(−125)(−2)

7. Find the following quotients.

(a) $(-32) \div (4)$

(b) $\dfrac{-125}{-5}$

(c) $3.75 \div (-12.5)$

(d) $\left(-2\dfrac{3}{4}\right) \div 1\dfrac{3}{8}$

8. Find the following quotients.

(a) $54 \div (-9)$

(b) $\dfrac{-1,000}{-125}$

(c) $(-8.75) \div 7.5$

(d) $1\dfrac{1}{6} \div \left(-4\dfrac{2}{3}\right)$

In Exercises 9–14, perform the indicated operations.

9. $\dfrac{(4)(-5)}{-2}$

10. $\dfrac{(-8)(9)}{-6}$

11. $(-5)(10^2)$

12. $(-3.6)(10^3)$

13. $-2 + \dfrac{12}{-3}$

14. $-6 + \dfrac{-15}{3}$

15. If $I = Prt$, find I when $P = 500$, $r = 0.08$, and $t = 1\dfrac{1}{2}$.

16. If $C = \dfrac{5}{9}(F - 32)$, find C when $F = 14$.

In Exercises 17–30, solve and check the given equations.

17. $x - 5 = 7$ 18. $x - 12 = -4$ 19. $-5 = x + 9$

20. $-6 = x + 7$ 21. $5x = 55$ 22. $7x = 56$

23. $18 = \dfrac{x}{7}$ 24. $14 = \dfrac{x}{8}$ 25. $2x + 7 = -15$

26. $3x + 4 = -11$ 27. $\dfrac{5x}{4} + 5 = 20$ 28. $\dfrac{6x}{5} - 7 = 11$

29. $17 = 9 - x$ 30. $14 = 2 - 3x$

31. At 10 A.M. the temperature was 45°. By 2 P.M. it had risen 17°. Between 2 P.M. and midnight, the temperature dropped 33°. What was the temperature at midnight?

32. At 9 A.M. a stock opened at $42\dfrac{3}{8}$. By noon it had dropped $2\dfrac{3}{4}$ points. Between noon and closing time, it rose $5\dfrac{7}{8}$ points. What was the price at closing time?

REVIEW EXERCISES 1111, SET II

1. Add the signed numbers.

(a) $\begin{array}{r} 586 \\ -249 \\ \hline \end{array}$

(b) $\begin{array}{r} -204 \\ -617 \\ \hline \end{array}$

(c) $\begin{array}{r} -5\dfrac{1}{6} \\ 2\dfrac{2}{3} \\ \hline \end{array}$

(d) $\begin{array}{r} 0.72 \\ -2.86 \\ \hline \end{array}$

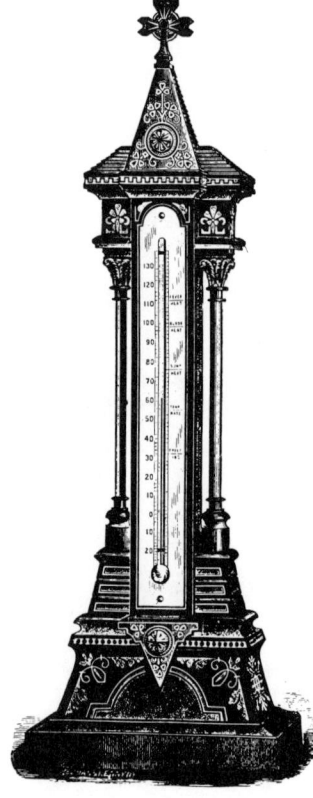

2. Subtract the lower number from the upper number.

(a) 294
   -506

(b) 208
   -712

(c) $-2\frac{7}{10}$
    $5\frac{1}{5}$

(d) 5.09
    7.90

3. Find the following products.

(a) $(-31)(-60)$

(b) $7.5(-48)$

(c) $(-4)(-2)(-3)$

(d) $(6)(-2)(-25)(-5)$

4. Find the following quotients.

(a) $(-28) \div (7)$

(b) $\frac{-128}{-8}$

(c) $75.0 \div (-12.5)$

(d) $\left(-3\frac{3}{4}\right) \div 2\frac{1}{2}$

In Exercises 5-7, perform the indicated operations.

5. $\frac{(-12)(15)}{-10}$

6. $(-3.7)(10^3)$

7. $-8 + \frac{-18}{6}$

8. If $I = Prt$, find I when $P = 600$, $r = 0.09$, and $t = 2\frac{1}{3}$.

In Exercises 9-15, solve and check the given equations.

9. $x - 13 = -5$

10. $-9 = x + 3$

11. $6x = 42$

12. $11 = \frac{x}{5}$

13. $4x + 3 = -17$

14. $\frac{7x}{2} - 5 = 9$

15. $12 = 2 - 5x$

16. At 8 A.M. the temperature was 68°. By 2 P.M. it had risen 22°. Between 2 P.M. and midnight, the temperature dropped 33°. What was the temperature at midnight?

Chapter 11: Diagnostic Test

Name _____

_____

The purpose of this test is to see how well you understand the basic ideas of beginning algebra. We recommend that you work this diagnostic test *before* your instructor tests you on this chapter. Allow yourself about 40 minutes to do this test.

Complete solutions for all the problems on this test, together with section references, are given in the Answer Section. You should study the sections referred to for the problems you do incorrectly.

---

1. Add the following signed numbers:

   (a) 7 and -3   (1a) _____

   (b) -10 and -6   (1b) _____

   (c) -22 and 17   (1c) _____

   (d) 25 and 16   (1d) _____

   (e) -16
        -35   (1e) _____

   (f)   31
        -47   (1f) _____

   (g) 73
        18   (1g) _____

   (h) -59
        84   (1h) _____

---

2. Subtract the lower number from the upper number.

   (a)   57   (b) -14
        -32   -22   (2a) _____

                                    (2b) _____

   (c) -93   (d) 138
         27   481   (2c) _____

                                    (2d) _____

3. Find the following products.

    (a) $5(-9)$ (3a) _____

    (b) $(-6)(-7)$ (3b) _____

    (c) $(-4)(12)$ (3c) _____

    (d) $(8)(9)$ (3d) _____

    (e) $(18)(-2)(-5)$ (3e) _____

    (f) $(-5)(7)(-2)(-4)$ (3f) _____

---

4. Find the following quotients.

    (a) $(126) \div (9)$ (4a) _____

    (b) $(39) \div (-13)$ (4b) _____

    (c) $(-64) \div (-16)$ (4c) _____

    (d) $\dfrac{-75}{15}$ (4d) _____

    (e) $\dfrac{84}{-12}$ (4e) _____

    (f) $\dfrac{-144}{-9}$ (4f) _____

---

5. Find the value of $3x + 5y$ when $x = -10$ and $y = 6$.

                                                  (5) _____

6. If $A = \dfrac{1}{2}bh$, find A when $b = 7$ and $h = 18$.

                                                  (6) _____

7. If $A = P(1 + rt)$, find A when $P = 600$, $r = 0.07$, and $t = 1.5$.

                                                  (7) _____

8. If $C = \frac{5}{9}(F - 32)$, find C when $F = 77$.

(8) _____

---

9. Solve the following equations.

(a) $x - 3 = 7$ (9a) _____

(b) $5x + 8 = -22$ (9b) _____

(c) $\frac{x}{6} = -2$ (9c) _____

(d) $3 = \frac{3x}{7} + 15$ (9d) _____

---

10. Check to see if $x = \frac{3}{5}$ is a solution of the equation $7 + 5x = 12$.

(10) _____

TWELVE

Scientific Notation

In science we often work with very large and very small numbers. To write and calculate with such numbers we need a different system from the one we have used so far. In this chapter we study a number system used in science. It is called *Scientific Notation* or *powers of 10*.

At this time, you should review Sections 119 and 120 dealing with powers of numbers.

$$10 \cdot 10 \cdot 10 = 10^3 = 1,000$$

exponent $\downarrow$

base $\rightarrow$ 3rd power of 10

Powers of Ten

Before we can write numbers in Scientific Notation we must understand positive and negative powers of 10. In Section 120 we showed positive powers of 10. We now extend this system of powers of 10 to include the negative powers of 10. See Figure 1201.

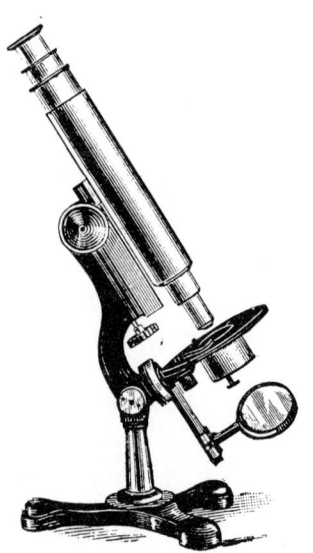

						decimal point						
10^6	10^5	10^4	10^3	10^2	10^1	10^0	10^{-1}	10^{-2}	10^{-3}	10^{-4}	10^{-5}	10^{-6}
= 1,000,000	= 100,000	= 10,000	= 1,000	= 100	= 10	= 1	= 0.1	= 0.01	= 0.001	= 0.0001	= 0.00001	= 0.000001
							= 1/10	= 1/100	= 1/1,000	= 1/10,000	= 1/100,000	= 1/1,000,000
millions	hundred-thousands	ten-thousands	thousands	hundreds	tens	ones	tenths	hundredths	thousandths	ten-thousandths	hundred-thousandths	millionths

Figure 1201 Powers of 10

MULTIPLYING POWERS OF TEN

Example 1. $10^2 \cdot 10^4$

Since $\begin{cases} 10^2 = 10 \cdot 10 \\ 10^4 = 10 \cdot 10 \cdot 10 \cdot 10 \end{cases}$

Therefore, $10^2 \cdot 10^4 = \overbrace{(10 \cdot 10)}^{\substack{2 \\ \text{factors}}} \overbrace{(10 \cdot 10 \cdot 10 \cdot 10)}^{\substack{4 \\ \text{factors}}}$

$$= \overbrace{10 \cdot 10 \cdot 10 \cdot 10 \cdot 10 \cdot 10}^{2 + 4 \text{ factors}} = 10^6$$

This means $10^2 \cdot 10^4 = 10^{2+4} = 10^6$

This example leads to the rule:

To Multiply Two Powers of 10

Add their exponents as shown:

$$10^a \cdot 10^b = 10^{a+b}$$

Example 2. $10^3 \cdot 10^5 = 10^{3+5} = 10^8$

This method of multiplying powers of 10 can be extended to products of more than two factors.

Example 3. $10^7 \cdot 10^2 \cdot 10^1 \cdot 10^5 = 10^{7+2+1+5} = 10^{15}$

DIVIDING POWERS OF TEN

Example 4

$$\frac{10^5}{10^3} = \frac{\overbrace{10 \cdot 10 \cdot \cancel{10} \cdot \cancel{10} \cdot \cancel{10}}^{\substack{5 \\ \text{factors}}}}{\underbrace{\cancel{10} \cdot \cancel{10} \cdot \cancel{10}}_{\substack{3 \\ \text{factors}}}} = \overbrace{10 \cdot 10}^{\substack{5 - 3 \\ \text{factors}}} = 10^2$$

This means $\dfrac{10^5}{10^3} = 10^{5-3} = 10^2$

This example leads to the following rule:

To Divide One Power of 10 by Another

Subtract their exponents as shown:

$$\frac{10^a}{10^b} = 10^{a-b}$$

Note: The exponent in the denominator is subtracted *from* the exponent in the numerator.

Example 5.

$$\frac{10^2}{10^6} = \frac{\overbrace{10 \cdot 10}^{2 \text{ factors}}}{\underbrace{10 \cdot 10 \cdot 10 \cdot 10 \cdot 10 \cdot 10}_{6 \text{ factors}}} = \frac{1}{\underbrace{10 \cdot 10 \cdot 10 \cdot 10}_{4 \text{ factors}}} = \frac{1}{10^4}$$

If we use the rule $\frac{10^a}{10^b} = 10^{a-b}$, then $\frac{10^2}{10^6} = 10^{2-6} = 10^{-4}$

This means $\frac{10^2}{10^6} = 10^{-4} = \frac{1}{10^4}$. This leads to the following

definitions:

The Meaning of a Negative Power of 10

$$10^{-a} = \frac{1}{10^a} \qquad \text{and} \qquad \frac{1}{10^{-b}} = 10^b$$

Example 6

   (a) $10^{-2} = \frac{1}{10^2} = \frac{1}{100} = 0.01$

   (b) $10^{-1} = \frac{1}{10^1} = \frac{1}{10} = 0.1$

   (c) $\frac{1}{10^{-3}} = 10^3 = 1,000$

   (d) $\frac{1}{10^{-4}} = 10^4 = 10,000$

Example 7. $\frac{10^3}{10^3} = \frac{\overset{1}{10} \cdot \overset{1}{10} \cdot \overset{1}{10}}{\underset{1}{10} \cdot \underset{1}{10} \cdot \underset{1}{10}} = 1$

Also $\dfrac{10^3}{10^3} = 10^{3-3} = 10^0$, using the rule $\dfrac{10^a}{10^b} = 10^{a-b}$.

Therefore, $10^0 = 1$, because they both equal $\dfrac{10^3}{10^3}$.

This leads to the definition:

$$10^0 = 1$$

Example 8

    (a) $10^4 \cdot 10^{-3} = 10^{4+(-3)} = 10^1 = 10$

    (b) $10^{-4} \cdot 10^{-3} = 10^{-4+(-3)} = 10^{-7}$

    (c) $10^{-2} \cdot 10^0 \cdot 10^5 = 10^{-2+0+5} = 10^3$

    (d) $\dfrac{10^3 \cdot 10^{-2}}{10^4} = \dfrac{10^{3+(-2)}}{10^4} = \dfrac{10^1}{10^4} = 10^{1-4} = 10^{-3}$

    (e) $\dfrac{10^5}{10^{-3}} = 10^{5-(-3)} = 10^{5+(3)} = 10^8$

    (f) $\dfrac{10^2 \cdot 10^{-3}}{10^{-4} \cdot 10^0 \cdot 10^5} = \dfrac{10^{2+(-3)}}{10^{-4+0+5}} = \dfrac{10^{-1}}{10^1} = 10^{-1-1} = 10^{-2}$

Since multiplying and dividing powers of 10 involve adding and subtracting signed numbers, you may need to review Sections 1102 and 1103 before working the following exercises.

EXERCISES 1201, SET I. Perform the indicated operations.

1. $10^2 \cdot 10^3$ 2. $10^4 \cdot 10^6$ 3. $10^8 \cdot 10^5$

4. $10^7 \cdot 10^{10}$ 5. $\dfrac{10^6}{10^2}$ 6. $\dfrac{10^5}{10^3}$

7. $\dfrac{10^2}{10^5}$ 8. $\dfrac{10^3}{10^8}$ 9. $10 \cdot 10^2 \cdot 10^5$

10. $10^0 \cdot 10^4 \cdot 10^7$ 11. $10^4 \cdot 10^{-2}$ 12. $10^6 \cdot 10^{-3}$

13. $\dfrac{10^5 \cdot 10^7}{10^{-2}}$ 14. $\dfrac{10^4 \cdot 10^{12}}{10^{-3}}$ 15. $\dfrac{10^{-5} \cdot 10^{-6}}{10^{-8}}$

16. $\dfrac{10^{-4} \cdot 10^{-3}}{10^{-7}}$ 17. $\dfrac{10^8 \cdot 10^{-5}}{10^3}$ 18. $\dfrac{10 \cdot 10^5}{10^0 \cdot 10^{-3}}$

19. $10^4 \cdot 10^{-6} \cdot 10^0 \cdot 10^{-1}$ 20. $10^{10} \cdot 10^{-6} \cdot 10^3 \cdot 10^0$

1. $10^5 \cdot 10^3$ 2. $10^4 \cdot 10^7$ 3. $\dfrac{10^8}{10^5}$

4. $\dfrac{10^3}{10^9}$ 5. $10^2 \cdot 10 \cdot 10^4$ 6. $10^3 \cdot 10^0 \cdot 10^5$

7. $10^{-4} \cdot 10^7$ 8. $\dfrac{10^{-8} \cdot 10^4}{10^{-6}}$ 9. $\dfrac{10^2 \cdot 10}{10^0 \cdot 10^{-3}}$

10. $10^3 \cdot 10^0 \cdot 10^{-2} \cdot 10^4$

1202   Multiplying a Number by a Power of 10

At this time you will find it helpful to review Section 410 dealing with multiplying and dividing decimals by powers of 10.

Example 1

(a) $2 \cdot 10^3$ = $2 \cdot 1{,}000$ = $2{\wedge}0\ 0\ 0.$ = $2{,}000.$

(b) $2 \cdot 10^2$ = $2 \cdot 100$ = $2{\wedge}0\ 0.$ = $200.$

(c) $2 \cdot 10^1$ = $2 \cdot 10$ = $2{\wedge}0.$ = $20.$

(d) $2 \cdot 10^0$ = $2 \cdot 1$ = $2.$ = $2.$

(e) $2 \cdot 10^{-1}$ = $2 \cdot \dfrac{1}{10}$ = $0.2{\wedge}$ = 0.2

(f) $2 \cdot 10^{-2}$ = $2 \cdot \dfrac{1}{100}$ = $0.0\ 2{\wedge}$ = 0.02

(g) $2 \cdot 10^{-3}$ = $2 \cdot \dfrac{1}{1{,}000}$ = $0.0\ 0\ 2{\wedge}$ = 0.002

The preceding examples lead us to the following rule:

When Multiplying a Number by a Power of 10

1. The decimal point moves $\begin{cases} \text{right if the exponent of 10} \\ \text{is positive.} \\ \text{left if the exponent of 10} \\ \text{is negative.} \end{cases}$

2. The decimal point moves as many places as the number part of the exponent of 10.

Example 2

(a) 24.6×10^2 = 2 4∧6 0. = 2,460.

(b) 0.00075×10^4 = ∧0 0 0 7.5 = 7.5

(c) $5,680. \times 10^{-3}$ = 5.6 8 0∧ = 5.680

(d) 0.0739×10^{-2} = .0 0∧0 7 3 9 = 0.000739

EXERCISES 1202, SET I. Find the following products.

1. 75.4×10^2 2. 3.86×10^3 3. 88×10^{-1}

4. 49×10^{-2} 5. 0.075×10^3 6. 0.055×10^4

7. $6,840 \times 10^{-4}$ 8. 378.5×10^{-3} 9. 0.00086×10^6

10. 0.00063×10^{-2} 11. 15×10^{12} 12. 5.8×10^{12}

EXERCISES 1202, SET II. Find the following products.

1. 8.35×10^3 2. 6.3×10^{-2} 3. 0.027×10^4

4. 90.26×10^{-3} 5. 0.00051×10^6 6. 4.935×10^7

Writing Numbers Using Powers of 10 (Scientific Notation)

To write a number in Scientific Notation, place one nonzero digit to the left of the decimal point and then multiply this number by the appropriate power of 10.

Example 1

	Common Notation		*Scientific Notation*

(a)   245 = 2∧4 5. = 2.45×10^2

(b)   24.5 = 2∧4.5 = 2.45×10^1

(c)   2.45 = 2.4 5 = 2.45×10^0
                     ∧

(d)   0.245 = 0.2∧4 5 = 2.45×10^{-1}

(e)   0.0245 = 0.0 2∧4 5 = 2.45×10^{-2}

(f)   0.00245 = 0.0 0 2∧4 5 = 2.45×10^{-3}

The preceding examples lead us to the following rule:

To Write a Number in Scientific Notation

1. Place a caret (∧) to the right of the first nonzero digit.
2. Draw an arrow *from* the caret to the actual decimal point.
3. The exponent of 10 is $\begin{cases} \underrightarrow{(+)} & \text{if arrow points right.} \\ \underleftarrow{(-)} & \text{if arrow points left.} \end{cases}$
4. The number part of the exponent of 10 is equal to the number of places between the caret and the actual decimal point.
5. Write the number with decimal point after the first nonzero digit and multiply by the power of 10 found in steps (3) and (4).

Note: The convention of signs used for the exponent of 10 is the same as that used on the number line.

$\underrightarrow{(+)}$

Positive Direction

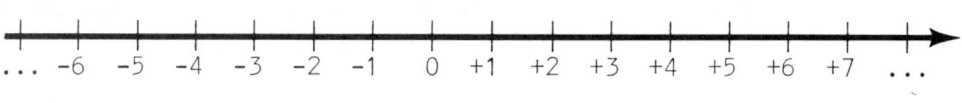

... -6 -5 -4 -3 -2 -1   0 +1 +2 +3 +4 +5 +6 +7 ...

$\underleftarrow{(-)}$

Negative Direction

Example 2. Changing numbers from Common to Scientific Notation.

(a) $92{,}900{,}000. = 9\!\wedge\!2\ 9\ 0\ 0\ 0\ 0\ 0. = 9.29 \times 10^7$

(b) $0.0056 = 0.0\ 0\ 5\!\wedge\!6 \qquad = 5.6 \times 10^{-3}$

(c) $0.01745 = 0.0\ 1\!\wedge\!7\ 4\ 5 \qquad = 1.745 \times 10^{-2}$

(d) $684.5 = 6\!\wedge\!8\ 4.5 \qquad = 6.845 \times 10^2$

Example 3. Changing numbers from Scientific Notation to Common Notation.

(a) $2.54 \times 10^{-2} = 0.0\ 2\!\wedge\!5\ 4 \qquad = 0.0254$

(b) $1.609 \times 10^3 = 1\!\wedge\!6\ 0\ 9. \qquad = 1609.$

(c) $4.67 \times 10^{-5} = 0.0\ 0\ 0\ 0\ 4\!\wedge\!6\ 7 = 0.0000467$

(d) $3.57 \times 10^5 = 3\!\wedge\!5\ 7\ 0\ 0\ 0. \qquad = 357{,}000.$

EXERCISES 1203, SET I. Complete the following table.

	Common Notation	Scientific Notation
1.	748	
2.	25,000	
3.	0.063	
4.		6.7×10^3
5.	0.001732	
6.		2.81×10^{-2}
7.		3.47×10^6
8.	86.48	
9.		1.91×10^{-6}
10.	588,000	
11.	0.0000563	
12.	27,800	
13.	0.000058	
14.	1,761,000	
15.		6.547×10^{-27}
16.	0.00000078	
17.		4.77×10^{-10}
18.	5,780,000,000,000	
19.	(50)(2,000)	
20.		$(2 \times 10^2)(3 \times 10^3)$

EXERCISES 1203, SET II. Complete the following table.

	Common Notation	Scientific Notation
1.	6,356	
2.	0.000752	
3.		9.32×10^3

Writing Numbers Using Powers of 10 (Scientific Notation)   403

	Common Notation	Scientific Notation
4.		4.875×10^{-2}
5.	0.0015625	
6.	92,426,000	
7.		6.03×10^{5}
8.		1.67339×10^{-24}
9.	(6000)(400)	
10.		$(5 \times 10^{2})(3 \times 10^{4})$

Multiplying and Dividing Numbers Using Scientific Notation

Example 1. Multiply: $200 \times 3,000$

Solution:

$$200 \times 3,000$$
$$= (2 \times 10^{2}) \times (3 \times 10^{3})$$
$$= 2 \times 10^{2} \times 3 \times 10^{3}$$
$$= 2 \times 3 \times 10^{2} \times 10^{3}$$
$$= 6 \times 10^{5}$$
$$= 600,000$$

Example 2. Multiply: $250 \times 4,000 \times 0.0012$

Solution:

$$250 \times 4,000 \times 0.0012$$
$$= (2.5 \times 10^{2}) \times (4 \times 10^{3}) \times (1.2 \times 10^{-3})$$
$$= 2.5 \times 4 \times 1.2 \times 10^{2} \times 10^{3} \times 10^{-3}$$
$$= 12 \times 10^{2} = 1,200$$

Example 3. $\dfrac{6,000}{150,000} = \dfrac{6 \times 10^{3}}{1.5 \times 10^{5}} = \dfrac{6}{1.5} \times \dfrac{10^{3}}{10^{5}} = 4 \times 10^{-2} = 0.04$

Example 4. $\dfrac{5,400 \times 0.0016}{0.036 \times 60}$

Solution: $\dfrac{(5.4 \times 10^{3}) \times (1.6 \times 10^{-3})}{(3.6 \times 10^{-2}) \times (6 \times 10^{1})} = \dfrac{5.4 \times 1.6}{3.6 \times 6} \times \dfrac{10^{3} \times 10^{-3}}{10^{-2} \times 10^{1}}$

$$= \dfrac{\overset{1}{\cancel{5.4}} \times \overset{.4}{\cancel{1.6}}}{\underset{1}{\cancel{3.6}} \times \underset{1}{\cancel{6}}} \times 10^{1}$$

$$= 0.4 \times 10 = 4$$

EXERCISES 1204, SET I. Perform the indicated operations using Scientific Notation.

1. $400 \times 7{,}000$ 2. $300 \times 8{,}000$

3. $(0.05)(0.007)$ 4. $(0.04)(0.009)$

5. 50×0.006 6. 0.007×90

7. $(20)(3{,}000)(0.08)$ 8. $(40)(500)(0.002)$

9. $(0.9)(0.07)(4{,}000)$ 10. $(0.6)(0.001)(3{,}000)$

11. $\dfrac{9{,}000}{0.03}$ 12. $\dfrac{4{,}000}{0.02}$

13. $\dfrac{0.0008}{400}$ 14. $\dfrac{0.006}{300}$

15. $\dfrac{(80)(0.005)}{4{,}000}$ 16. $\dfrac{(0.009)(200)}{0.06}$

17. $\dfrac{(0.03)(200)(40)}{(0.008)(30)}$ 18. $\dfrac{(0.5)(80)(0.07)}{(0.2)(300)}$

19. $\dfrac{(5{,}000)(25)(0.08)}{(40)(0.002)}$ 20. $\dfrac{(12.5)(400)(0.06)}{(30)(1{,}000)}$

EXERCISES 1204, SET II. Perform the indicated operations using Scientific Notation.

1. $(5{,}000)(400)$ 2. $(0.009)(0.06)$

3. 0.008×50 4. $(30)(600)(0.005)$

5. $(7{,}000)(0.08)(0.4)$ 6. $\dfrac{8{,}000}{0.04}$

7. $\dfrac{0.0012}{300}$ 8. $\dfrac{(40)(0.008)}{2{,}000}$

9. $\dfrac{(60)(300)(0.02)}{(0.009)(40)}$ 10. $\dfrac{(2{,}500)(0.04)(80)}{(0.002)(50)}$

Approximations Using Scientific Notation

Before starting this section, you will find it helpful to review Section 407 on rounding off decimals.

> To Approximate a Calculation
>
> 1. Round off each number to one nonzero digit.
> 2. Write each resulting number in Scientific Notation.
> 3. Perform the indicated operations using the method shown in Section 1204.

Example 1. 417 × 7,810

Solution:

1. 400 × 8,000   Rounding off to one nonzero digit

2. $4 \times 10^2 \times 8 \times 10^3$ Writing numbers in Scientific Notation

3. $4 \times 8 \times 10^2 \times 10^3 = 32 \times 10^5$ ⎫
                            $= 3,200,000$ ⎬ Method shown in
                                          ⎭ Section 1204

(Actual answer = 3,256,770)

Example 2. 0.00628 × 0.0289

Solution:

1. 0.006 × 0.03

2. $(6 \times 10^{-3}) \times (3 \times 10^{-2})$

3. $18 \times 10^{-5} = 0.00018$ (Actual answer = 0.000181492)

Example 3. 191 × 3,250 × 0.00796

Solution:

1. 200 × 3,000 × 0.008

2. $(2 \times 10^2) \times (3 \times 10^3) \times (8 \times 10^{-3})$

3. $48 \times 10^2 = 4,800$ (Actual answer = 4,941.17)

Example 4. $\dfrac{3,964}{0.01897}$

Solution:

1. $\dfrac{4,000}{0.02}$

2. $\dfrac{4 \times 10^3}{2 \times 10^{-2}}$

3. $2 \times 10^5 = 200,000$ (Answer $\doteq$ 208,962)

Example 5. $\dfrac{758 \times 0.0276}{5,550}$

Solution:

1. $\dfrac{800 \times 0.03}{6,000}$

2. $\dfrac{(8 \times 10^2) \times (3 \times 10^{-2})}{6 \times 10^3}$

3. $4 \times 10^{-3} = 0.004$ (Answer $\doteq$ 0.0037695)

EXERCISES 1205, SET I. Approximate the value of each of the
following expressions using Scientific Notation.

1. 7.86×204

2. 15.5×34.6

3. 0.714×0.286

4. 38.9×0.0076

5. $\dfrac{75.5}{0.894}$

6. $\dfrac{0.0666}{404}$

7. $\dfrac{17.1}{57,000}$

8. $\dfrac{37,000}{0.511}$

9. $21.9 \times 0.0071 \times 0.864$

10. $86.4 \times 0.866 \times 217$

11. $328 \times 0.91 \times 0.0088$

12. $565 \times 19.5 \times 0.000384$

13. $\dfrac{45,000 \times 0.0159}{0.788}$

14. $\dfrac{186,000 \times 0.0605}{5,280}$

15. $\dfrac{2,800 \times 0.416}{70.6}$

16. $\dfrac{0.0381 \times 41,600}{0.328}$

17. $\dfrac{25.6 \times 2,100}{0.064 \times 550}$

18. $\dfrac{27.5 \times 4,710}{0.304 \times 546}$

19. $\dfrac{285 \times 1.99 \times 72,000}{0.691 \times 566 \times 34.2}$

20. $\dfrac{81.5 \times 11.8 \times 0.744}{291 \times 0.558 \times 0.0324}$

EXERCISES 1205, SET II. Approximate the value of each of the
following expressions using Scientific Notation.

1. 3.15×867

2. 0.0052×781

3. $\dfrac{58.2}{0.0329}$

4. $\dfrac{0.283}{6,050}$

5. $39.2 \times 0.0051 \times 0.683$

6. $0.000527 \times 63.4 \times 175$

7. $\dfrac{0.0639 \times 4,892}{21.5}$

8. $\dfrac{0.729 \times 76.4}{6,800}$

9. $\dfrac{6.34 \times 1,900}{0.0084 \times 450}$

10. $\dfrac{8,450 \times 17.9 \times 0.0624}{0.432 \times 561 \times 2.2}$

Chapter 12: Diagnostic Test

Name_____

_____

The purpose of this test is to see how well you understand
Scientific Notation. We recommend that you work this diagnostic
test *before* your instructor tests you on this chapter. Allow
yourself about 40 minutes to do this test.

Complete solutions for all the problems on this test, together
with section references, are given in the Answer Section. You
should study the sections referred to for the problems you do
incorrectly.

---

1. Perform the indicated operations. Express your answer as
   a power of 10.

   (a) $10^2 \cdot 10^4$ (1a)_____

   (b) $\dfrac{10^5}{10^3}$ (1b)_____

   (c) $\dfrac{10^4}{10^7}$ (1c)_____

   (d) $10 \cdot 10^{-3}$ (1d)_____

   (e) $\dfrac{10^4}{10^{-2}}$ (1e)_____

   (f) $\dfrac{10^6 \cdot 10^{-2}}{10}$ (1f)_____

   (g) $10^5 \cdot 10^{-3} \cdot 10^0$ (1g)_____

---

2. Find the following products.

   (a) 3.46×10^3 (2a)_____

   (b) 0.068×10^2 (2b)_____

   (c) 742×10^{-2} (2c)_____

   (d) 54.9×10^0 (2d)_____

---

3. Write the following numbers in Scientific Notation.

   (a) 4,500   (3a)_____

   (b) 79.64   (3b)_____

   (c) 0.00684   (3c)_____

---

4. Change the following numbers from Scientific Notation to Common Notation.

   (a) 4.75×10^{-2} (4a)_____

   (b) 9.3×10^6 (4b)_____

   (c) 8.654×10^2 (4c)_____

---

5. Perform the indicated operations using Scientific Notation. Express your answer in Scientific Notation.

(a) $(300)(0.04)(50)$

(5a) _____

(b) $\dfrac{(0.008)(1,200)}{0.048}$

(5b) _____

---

6. Round off each number to one nonzero digit, then approximate the value of each of the following expressions using Scientific Notation. Write answers in Scientific Notation.

(a) $186,300 \times 427$

(6a) _____

(b) $\dfrac{0.0754}{17.5}$

(6b) _____

(c) $\dfrac{(387.8)(0.05136)}{4.19}$

(6c) _____

THIRTEEN
Sets

Ideas in all branches of mathematics such as arithmetic, algebra, geometry, calculus, statistics, etc., can be explained in terms of sets. For this reason, you will find it helpful to have a basic understanding of sets.

Basic Definitions

<u>SET</u>. A set is a collection of objects or things.

<u>Example 1</u>

    (a) A 48-piece set of dishes

    (b) A basket of Christmas presents

    (c) A basket containing an apple, a pillow, and a cat

    (d) The number of students attending college in the United States

    (e) The set of natural numbers

In this course, we are mainly concerned with sets of numbers, because numbers are the tools of arithmetic.

<u>ELEMENT OF A SET</u>. The objects or things that make up a set are called its *elements* (or *members*). Sets are usually represented by listing their elements within braces { } as we did in Section 101 when we described the set of natural numbers and the set of whole numbers.

<u>Example 2</u>

    (a) Set $\{1, 2, 3, \dots \}$ has elements 1, 2, 3, etc.

    (b) Set $\{5, 7, 9\}$ has elements 5, 7, and 9.

    (c) Set $\{a, d, f, h, k\}$ has elements $a, d, f, h,$ and k.

    (d) Set $\{$Ben, Kay, Frank, Albert$\}$ has elements Ben, Kay, Frank, and Albert.

    (e) Set $\{\square, \Sigma, \Delta, \odot\}$ has elements $\square, \Sigma, \Delta,$ and $\odot$.

It is sometimes helpful to think of the braces as a basket or box which contains the elements of the set. This basket, { }, may contain just a few elements, many elements, or no elements at all. A set is usually named by a capital letter such as $A, N, W,$ etc. The expression "$A = \{1, 5, 7\}$" is read "A is the set whose elements are 1, 5, and 7."

<u>ROSTER METHOD OF REPRESENTING A SET</u>. A class roster is a list of the members of the class. When we represent a set by $\{3, 8, 9, 11\}$, we are representing the set by a roster (or list) of its members. This method of representing a set is called the *roster method*.

<u>THE SYMBOL $\in$</u>. $2 \in A$ is read "2 is an element of set A" (or "2 is a member of set A"). If $A = \{2, 3, 4\}$, we can say: $2 \in A$, $3 \in A$, and $4 \in A$. If we wish to show that a number or

object is *not* a member of a given set, we use the symbol $\notin$, which is read "is not an element of" or "is not a member of." If $A = \{2, 3, 4\}$, then $5 \notin A$, which is read "5 is not an element of set A." To help you remember this symbol, notice that $\in$ looks like the first letter of the word "element." Actually, $\in$ is the Greek letter *epsilon*.

Example 3

    (a) If $B = \{5, 9\}$, then $5 \in B$, $9 \in B$, $3 \notin B$, and $1 \notin B$.

    (b) If $C = \{b, e, g, m\}$, then $g \in C$, $e \in C$, and $k \notin C$.

    (c) If $D = \{$Abe, Helen, John$\}$, then Helen $\in D$, but Mary $\notin D$.

CARDINAL NUMBER OF A SET. The cardinal number of a set is the number of elements in that set. The symbol $n(A)$ is read "the cardinal number of set A" and means the number of elements in set A. When the set is represented by the roster method, its cardinal number is found by counting its elements.

Example 4

    (a) If $A = \{5, 8, 6, 9\}$, then $n(A) = 4$.

    (b) If $H = \{$Ed, Mabel$\}$, then $n(H) = 2$.

    (c) If $Q = \{a, h, l, s, t, v\}$, then $n(Q) = 6$.

    (d) If $E = \{2, 4, 3, 2\}$, then $n(E) = 3$. The cardinal number is 3 instead of 4 because the set has only three *different* elements.

EQUAL SETS. Two sets are equal if they both have exactly the same elements.

Example 5

    (a) $\{1, 5, 7\} = \{5, 1, 7\}$. Notice that both sets have exactly the same elements, even though they are not listed in the same order.

    (b) $\{1, 5, 5, 5\} = \{5, 1\}$. Notice that both sets have exactly the same elements. It is not necessary to write the same element more than once when writing the roster of a set.

    (c) $\{7, 8, 11\} \neq \{7, 11\}$. These sets are not equal because they both do not have exactly the same elements.

THE EMPTY SET (OR NULL SET). If we think of the braces $\{\ \}$ as a basket, then set $A = \{1, 5, 7\}$ has three elements in its basket—namely, 1, 5, and 7. Set $B = \{1, 5\}$ has two elements in its basket, and set $C = \{5\}$ has only one element. Set $D = \{\ \}$ is empty, having no elements in its basket. A set having no elements is called the *empty set* (or *null set*). We use symbols $\{\ \}$ or $\emptyset$ to represent the empty set. Whenever either symbol appears, read it as "the empty set." The symbol $\emptyset$ is a Norwegian letter having a sound somewhat like the u in *hurt* (also like $\ddot{o}$ in German $\ddot{o}st$). The symbol ϕ, the Greek letter *phi*, is also often used to represent the empty set.

A word of caution:

$\phi = \{ \ \}$ has *no* elements

$\{\phi\}$ has *one* element, ϕ; therefore $\{\phi\} \neq \{ \ \}$.

$\{0\}$ has *one* element, 0; therefore $\{0\} \neq \{ \ \}$.

also, $\phi \neq \{\phi\}$

and $\phi \neq \{0\}$, for the same reason.

Example 6

(a) The set of all people in this classroom that are 10 feet tall = $\emptyset$.

(b) The set of all digits that are greater than 10 = $\{ \ \}$.

UNIVERSAL SET. A *universal set* is a set containing all the elements being considered in a particular problem. We use the symbol U to represent the universal set under consideration.

Example 7

(a) Suppose we are going to consider only digits. Then $U = \{0, 1, 2, 3, 4, 5, 6, 7, 8, 9\}$, since U contains *all* digits.

(b) Suppose we are going to consider whole numbers. Then $U = \{0, 1, 2, 3, \ldots \}$, since U contains *all* whole numbers.

(c) Suppose we are going to consider sets of football players at ELAC in 1974. Then U is the set of *all* football players at ELAC in 1974.

(d) Suppose we are going to consider different sets of cats. Then U is the set of *all* cats.

Notice that there can be different universal sets. We might want to consider the set of all students in this class as the universal set if we were going to deal only with class members. On the other hand, we might use the set of all 50 of the United States as our universal set if we were going to consider only sets of states.

FINITE SET. If in counting the elements of a set the counting comes to an end, the set is called a *finite set*. This means that the number of elements in a finite set must be a particular whole number (which we call its cardinal number).

Example 8. Examples of finite sets.

(a) $A = \{5, 9, 10, 13\}$. $n(A) = 4$.

(b) $D =$ the set of digits. $n(D) = 10$.

(c) $C =$ the set of whole numbers less than 100. $n(C) = 100$.

(d) $\emptyset = \{ \ \}$. $n(\emptyset) = 0$.

INFINITE SET. If in counting the elements of a set the counting never comes to an end, the set is called an *infinite set*. A set is infinite if it is not finite. The cardinal number of an infinite set is not a whole number.

Example 9. Examples of infinite sets.

   (a) $N = \{1, 2, 3, \ldots \}$

   (b) $W = \{0, 1, 2, 3, \ldots \}$

   (c) The set of all fractions

EXERCISES 1301, SET I

1. Can the collection □, X, =, and 5 be called a set?
2. Are $\{2, 7\}$ and $\{7, 2\}$ equal sets?
3. Are $\{2, 2, 7, 7\}$ and $\{7, 2\}$ equal sets?
4. Write the set of digits < 3. See Section 101 for the meaning of < and >.
5. Write the set of digits > 9.
6. Write the set of whole numbers < 3.
7. Write the set of whole numbers > 9.
8. Write the set of whole numbers > 4 and < 6.
9. Write the set of whole numbers > 4 and < 5.
10. Write the set of whole numbers < 4 and < 5.
11. Write all the elements of the set $\{2, a, 3\}$.
12. Write all the elements in the set of digits.
13. Write the cardinal number of each of the following sets:
    (a) $\{1, 1, 3, 5, 5, 5\}$
    (b) $\{0\}$
    (c) $\{a, b, g, x\}$
    (d) The set of whole numbers < 8
    (e) $\emptyset$
14. The empty set has no elements. Therefore, the statement $\emptyset = \{ \ \}$ is true. However, the statement $\emptyset = \{0\}$ is not true. Explain.
15. State which of the following sets are finite and which are infinite:
    (a) The set of digits
    (b) The set of whole numbers
    (c) The set of days in the week
    (d) The set of books in the ELAC library
    (e) The set of fish in all the seas of the earth at this instant
16. State which of the following statements are true and which are false:
    (a) If $A = \{5, 11, 19\}$, then $11 \in A$.
    (b) If $B = \{x, y, z, w\}$, then $a \notin B$.
    (c) If $C = \{\text{Ann, Bill, Charles}\}$, then $\text{Dan} \in C$.
    (d) $0 \in \emptyset$.
    (e) $\emptyset \notin \{ \ \}$.
17. Given the universal set $U = \{7, 12, 15, 20, 23\}$:
    (a) Write the set of numbers < 15.
    (b) Write the set of numbers > 20.
18. Given the universal set $U = \{5, 7, 11, 14, 20\}$:
    (a) Write the set of numbers > 7 and < 14.
    (b) Write the set of numbers < 7 and < 20.

1. Can the collection *, 5, A, and $\neq$ be called a set?
2. Are {5, 8, 8, 5} and {8, 5} equal sets?
3. Write the set of digits < 2. See Section 101 for the meaning of < and >.
4. Write the set of whole numbers > 3 and < 7.
5. Write the set of whole numbers > 7 and < 8.
6. Write the set of even digits.
7. Write the cardinal number of each of the following sets:
   (a) {ϕ}
   (b) {2, 3, 2}
   (c) {a, b, d}
   (d) ϕ
   (e) The set of whole numbers < 5
8. State which of the following sets are finite and which are infinite:
   (a) The set of letters in the English alphabet
   (b) The set of counting numbers
   (c) The set of men taller than seven feet at this instant
9. State which of the following statements are true and which are false:
   (a) If $P = \{h, s, t, k\}$, then $t \notin P$.
   (b) If $Q = \{9, 13, 7\}$, then $12 \in Q$.
   (c) $0 \notin \phi$
10. Given the universal set $U = \{22, 14, 9, 4, 17\}$:
    (a) Write the set of numbers > 9 and < 17.
    (b) Write the set of numbers < 4 and < 22.

Subsets

DEFINITION. A set A is called a *subset* of set B if every member of A is also a member of B. "A is a subset of B" is written "$A \subseteq B$."

Example 1

(a) $A = \{3, 5\}$ is a subset of $B = \{3, 5, 7\}$ because every member of A is also a member of B. Therefore, $A \subseteq B$.

(b) $P = \{a, c, g, f\}$ is a subset of $Q = \{d, f, a, g, h, c\}$ because every member of P is also a member of Q. Therefore, $P \subseteq Q$.

(c) $X = \{$Joe, Betty$\}$ is a subset of $Y = \{$Mary, Betty, Jack, Joe$\}$ because every member of X is also a member of Y. Therefore, $X \subseteq Y$.

(d) $D = \{4, 7\}$ is *not* a subset of $E = \{7, 8, 5\}$ because $4 \in D$, but $4 \notin E$. Therefore, $D \not\subseteq E$, read "D is not a subset of E."

(e) $K = \{d, f, m\}$ is *not* a subset of $L = \{f, m\}$ because $d \in K$, but $d \notin L$. Therefore, $K \not\subseteq L$, read "K is not a subset of L."

PROPER SUBSET. Subset A is called a *proper subset* of B if there is at least one member of B that is not a member of subset A. "A is a proper subset of B" is written "$A \subset B$."

Example 2

(a) Subset $A = \{3, 5\}$ is a proper subset of $B = \{3, 5, 7\}$ because $7 \in B$, but $7 \notin A$. Therefore, $A \subset B$.

(b) Subset $X = \{$Joe, Betty$\}$ is a proper subset of $Y = \{$Mary, Betty, Jack, Joe$\}$ because Mary $\in Y$, but Mary $\notin X$. Therefore, $X \subset Y$.

IMPROPER SUBSET. Subset A is called an *improper subset* of B if there is no member of B that is not also a member of A. This means that A is an improper subset of B if $A = B$.

Example 3

(a) Subset $A = \{10, 12\}$ is an improper subset of $B = \{12, 10\}$, because no member of B is not also a member of A. Note that $A = B$. This means that A is an improper subset of A. In other words, *any set is an improper subset of itself*. Therefore:

(b) D is an improper subset of D.

The empty set is a proper subset of every set except itself.

ALL THE SUBSETS OF A SET. By considering set $A = \{3, 5, 7\}$ we find all its subsets are $\emptyset$, $\{3\}$, $\{5\}$, $\{7\}$, $\{3, 5\}$, $\{3, 7\}$, $\{5, 7\}$, $\{3, 5, 7\}$. Of these eight subsets, only $A = \{3, 5, 7\}$ is an improper subset of A; the seven others are proper subsets of A.

Example 4

(a) All the subsets of $\{6, 8\}$ are $\emptyset$, $\{6\}$, $\{8\}$, $\{6, 8\}$. Of these four subsets, only $\{6, 8\}$ is an improper subset.

(b) All the subsets of $\{a\}$ are $\emptyset$, $\{a\}$. Of these two subsets, only $\{a\}$ is an improper subset.

EXERCISES 1302, SET I

1. $M = \{1, 2, 3, 4, 5\}$. State whether each of the following sets is a proper subset, an improper subset, or not a subset of M.

(a) $A = \{3, 5\}$ (b) $B = \{0, 1, 7\}$

(c) $\emptyset$ (d) $C = \{2, 4, 1, 3, 5\}$

2. $P = \{x, z, w, r, t, y\}$. State whether each of the following sets is a proper subset, an improper subset, or not a subset of P.

(a) $D = \{x, y, r\}$ (b) P

(c) $\emptyset$ (d) $E = \{x, y, z, s, t\}$

3. Write all the subsets of $\{R, G, Y\}$.

4. Write all the subsets of { □, △}.

EXERCISES 1302, SET II

1. $S = \{9, 0, 1, 3, 6\}$. State whether each of the following
   sets is a proper subset, an improper subset, or not a subset of S.

   (a) $D = \{0, 1, 7\}$ (b) { }

   (c) $\{1, 3, 6, 9, 0\}$ (d) $\{3, 0\}$

2. Write all the subsets of $\{A, B, C\}$.

1303   Union and Intersection of Sets

<u>VENN DIAGRAMS</u>. A useful tool for helping you understand set
concepts is the Venn Diagram. A simple Venn diagram is shown
in Figure 1303A.

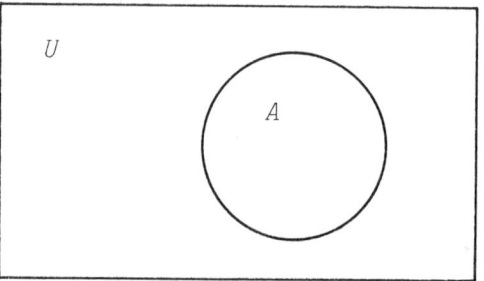

Venn Diagram

Figure 1303A

It is customary to represent the universal set U by a rectangle
and the other sets by circles enclosed in the U-rectangle.
Just as we enclose all members of a particular set within
braces in conventional set notation, in Venn diagrams we think
of all members of a set A as being enclosed in a circle marked
by the A. Using the same reasoning, we see that all elements in
the universe under consideration are enclosed in the U-rectangle.

<u>UNION OF SETS</u>. The union of sets A and B, written $A \cup B$, is
the set that contains all the elements of A as well as all the
elements of B.

In Figure 1303B, $A \cup B$ is represented by the shaded area. In
terms of set notation, suppose $A = \{b, c, g\}$ and $B = \{1, 2, 5, 7\}$. Then

$$A \cup B = \{b, c, g, 1, 2, 5, 7\}$$

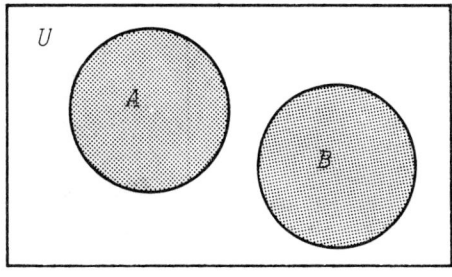

$A \cup B$

Figure 1303B

You will notice in the example just chosen that no member of A or B is in both sets.

Example 1

   (a) If $C = \{x,\ y\}$ and $D = \{4\}$, then $C \cup D = \{x,\ y,\ 4\}$.

   (b) If $F = \{$Mary, Helen$\}$ and $G = \{$high, low$\}$, then
       $F \cup G = \{$Mary, Helen, high, low$\}$.

   (c) If $H = \{4,\ 8,\ 26\}$ and $K = \{15,\ 17\}$, then
       $H \cup K = \{4,\ 8,\ 26,\ 15,\ 17\}$.

   (d) If $S = \{2,\ 5,\ 9\}$ and $T = \{9,\ 2,\ 7\}$, then
       $S \cup T = \{2,\ 5,\ 7,\ 9\}$.

   (e) If $P = \{a,\ 4,\ 3,\ c\}$ and $Q = \{3,\ k,\ a\}$, then
       $P \cup Q = \{a,\ 4,\ 3,\ c,\ k\}$.

INTERSECTION OF SETS. The intersection of sets C and D, written $C \cap D$, is the set that contains only elements in *both* C and D. Consider $C = \{g,\ f,\ m\}$ and $D = \{g,\ m,\ t,\ z\}$. Then $C \cap D = \{g,\ m\}$, because g and m are the only elements in both C and D. In the Venn diagram (Figure 1303C) the shaded area represents $C \cap D$ because that lies in both circles.

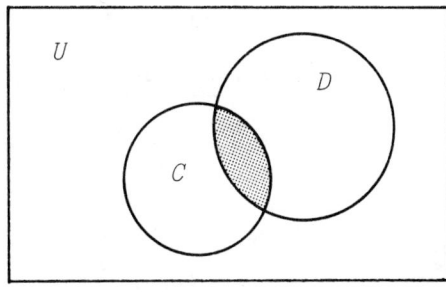

$C \cap D$

Figure 1303C

To be certain you can distinguish between the union and intersection of sets, two Venn diagrams (Figures 1303D and 1303E) are shown for the same sets, P and Q.

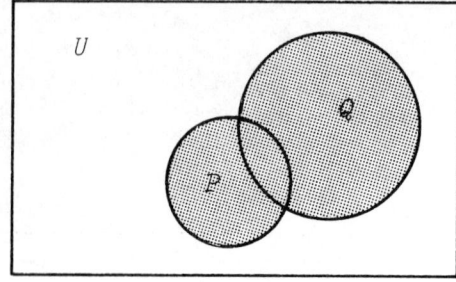

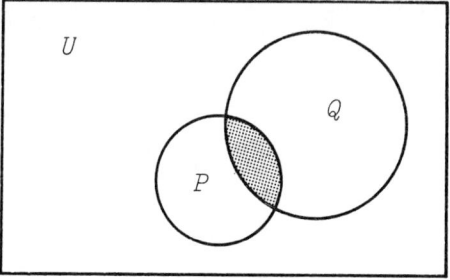

Shaded Area Is $P \cup Q$ Shaded Area Is $P \cap Q$

Figure 1303D   *Figure 1303E*

Example 2

    (a) If A = {John, Henry} and B = {Bill, Henry, Tom}, then $A \cap B$ = {Henry} and $A \cup B$ = {John, Henry, Bill, Tom}.

    (b) If G = {1, 5, 7} and H = {2, 5, 6, 7, 8}, then $G \cap H$ = {5, 7} and $G \cup H$ = {1, 5, 7, 2, 6, 8}.

    (c) If K = {a, 6, b} and M = {c, 5, 4}, then $K \cap M$ = { } and $K \cup M$ = {a, 6, b, c, 5, 4}.

DISJOINT SETS. Disjoint sets are sets whose intersection is the empty set. A Venn diagram showing disjoint sets is Figure 1303B, in which sets A and B are disjoint.

Example 3

    (a) If A = {1, 2} and B = {3, 4}, then $A \cap B$ = $\emptyset$. Therefore, A and B are disjoint sets.

    (b) If P = {a, b, c} and Q = {2, 8}, then $P \cap Q$ = $\emptyset$. Therefore, P and Q are disjoint sets.

    (c) If R = {5, 7, 9} and T = {9, 10, 12}, then $R \cap T$ = {9} $\neq \emptyset$. Therefore, R and T are *not* disjoint sets.

EXERCISES 1303, SET I

1. Write the union and intersection of each pair of given sets:

    (a) {1, 5, 7}, {2, 4}

    (b) {a, b}, {x, y, z, a}

    (c) { }, {k, 2}

    (d) {river, boat}, {boat, streams, down}

2. Given that A = {1, 3, 5, 7}, B = {2, 4, 6}, C = {1, 2, 3, 4} and D = {5, 6, 7}. Find the following:

    (a) $A \cup B$ (b) $C \cup D$

    (c) $A \cap B$ (d) $B \cap D$

3. Given that A = {Bob, John}, B = {Charles, Tom, Bob}, C = {Tom, John, Dick}, and D = {Ray, Bob}. Find any two sets that are disjoint.

4. Given the Venn diagram below, write each of the following sets in roster notation. (The small letters within a particular circle in the diagram are elements of that set.)

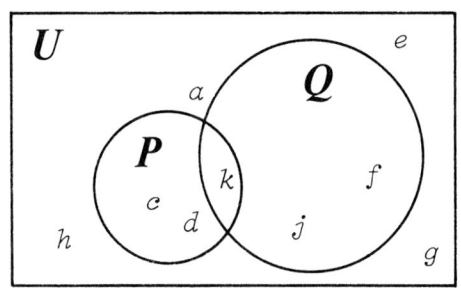

(a) P (b) Q (c) $P \cup Q$ (d) $P \cap Q$ (e) U

5. Given X = {2, 5, 6, 11} and Y = {7, 5, 11, 13}:

(a) Find $X \cap Y$.   (b) Find $Y \cap X$.

(c) Is $X \cap Y = Y \cap X$?

6. Given sets K = {a, 4, 7, b}, L = {m, 4, 6, b}, and M = {n, 4, 7, t}.

(a) Write $K \cap L$ in roster notation.

(b) Find $n(K \cap L)$.

(c) Write $L \cup M$ in roster notation.

(d) Find $n(L \cup M)$.

EXERCISES 1303, SET II

1. Write the union and intersection of each pair of given sets:

(a) {5, 0, 3, 1}, {3, 7}

(b) {able, can, do}, {do, well}

(c) {3, 8}, {1, 2, 5}

(d) {13, m}, ϕ

2. Given that K = {a, c, e}, L = {b, d, f, g}, M = {e, b, h}, and T = {f, i, k, e}. Find the following:

(a) $K \cup L$ (b) $T \cap L$ (c) $T \cup M$ (d) $L \cap K$

(e) Name a pair of disjoint sets.

(f) $n(T \cup M)$

(g) $n(L \cap K)$

3. Given the Venn diagram below, write each of the following sets in roster notation. (The small letters within a particular circle in the diagram are elements of that set.)

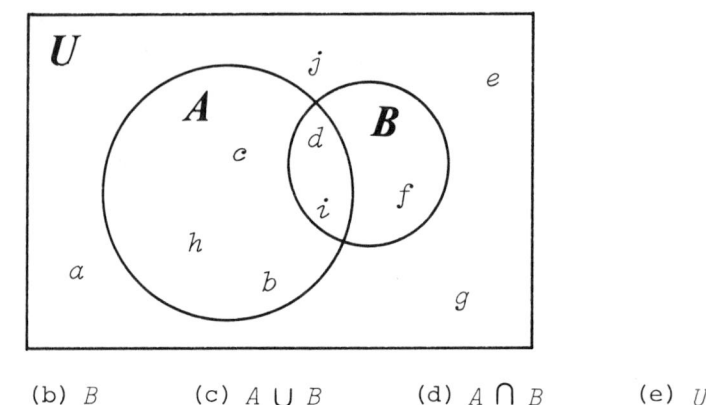

(a) A (b) B (c) $A \cup B$ (d) $A \cap B$ (e) U

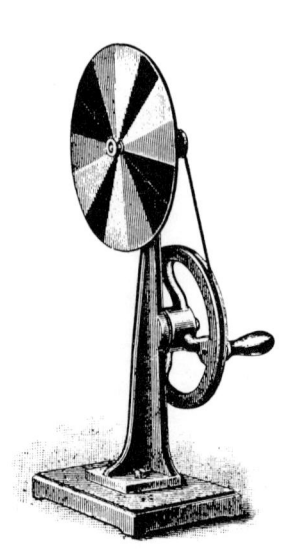

Chapter Summary

A *set* is a collection of objects or things. (Sec. 1301)

The *elements of a set* are the objects that make up the set. (Sec. 1301)

The *roster method* of representing a set is a list of its members enclosed in braces. (Sec. 1301)

The *cardinal number of a set* is the number of elements in the set. (Sec. 1301)

Equal sets are those having exactly the same members. (Sec. 1301)

The *empty set* is a set having no elements. (Sec. 1301)

The *universal set* is a set containing all the elements being considered in a particular problem. (Sec. 1301)

A *finite set* is one in which the counting of all its elements comes to an end. (Sec. 1301)

An *infinite set* is one in which the counting of all its elements never comes to an end. (Sec. 1301)

Set A is a *subset* of set B if every member of A is also a member of B. (Sec. 1302)

Subset A is a *proper subset* of B if there is at least one member of B that is not also a member of A. (Sec. 1302)

Subset A is an *improper subset* of B if no member of B is not also a member of A. A is an improper subset of B if $A = B$. (Sec. 1302)

Union of Sets. The union of sets A and B is the set containing all the elements of A as well as all the elements of B. (Sec. 1303)

Intersection of Sets. The intersection of sets C and D is the set containing only those elements which are elements of C as well as elements of D. (Sec. 1303)

Disjoint sets are sets whose intersection is the empty set. (Disjoint sets have no elements in common.) (Sec. 1303)

Symbols Used with Sets

$5 \in A$	is read "5 is an element of set A."
$9 \notin B$	is read "9 is not an element of set B."
$P \subseteq Q$	is read "P is a subset of Q."
$P \nsubseteq Q$	is read "P is not a subset of Q."
$P \subset Q$	is read "P is a proper subset of Q."
$P \not\subset Q$	is read "P is not a proper subset of Q."
$n(A)$	is read "the cardinal number of set A."
$\emptyset$, ϕ, or { }	are read "the empty set."
U	is read "the universal set."

REVIEW EXERCISES 1304, SET I

1. Given: $U = \{1, 2, 3, 4, 5, 6, 7, 8\}$, $A = \{5, 7, 8\}$,
   $B = \{2, 5, 7\}$, and $C = \{1, 3, 4, 6\}$.
   Find:

   (a) $A \cup B$ (b) $A \cap B$ (c) $C \cap B$ (d) $n(U)$

   (e) $n(A \cup B)$ (f) Is $\{2\}$ a subset of A?

   (g) A and C are said to be _?_ sets.

   (h) Is $U = \{1, 2, 3, 4, 5, 6, 7, 8\}$ a finite or infinite set?

   (i) Is A a proper subset of U?

2. Given: $U = W = \{0, 1, 2, 3, \ldots \}$, $D = \{2, 3\}$
   $E = \{3, 5, 7\}$, and $F = \{6, 7, 8\}$.
   Find:

   (a) $D \cap E$ (b) $D \cup F$ (c) $D \cap F$ (d) $n(D \cup E)$

   (e) $n(U)$

   (f) D and F are said to be _?_ sets.

   (g) Is $W = \{0, 1, 2, 3, \ldots \}$ a finite or infinite set?

   (h) Is E a proper subset of W?

3. Are the sets $\{1, 2, 5\}$ and $\{5, 1, 2\}$ equal sets?
4. Write the set of whole numbers > 15 and < 19.
5. Write the set of whole numbers < 5 and < 4.
6. Are the sets $\{a, b, c, b\}$ and $\{a, b, c\}$ equal sets?
7. What is the cardinal number of the set of letters in the word "Mississippi"?
8. Write all the subsets of the set $\{ \square, \triangle \}$.

9. Shade in $A \cup B$ in the Venn diagram given here.

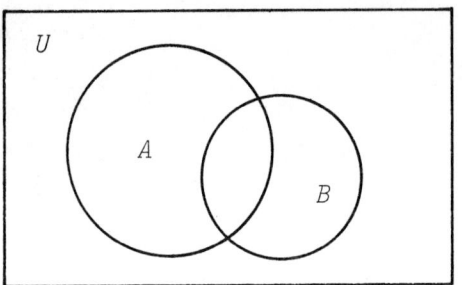

10. Shade in $P \cap Q$.

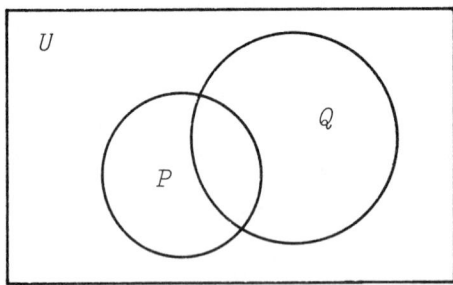

11. Write the name of the set representing the shaded area.

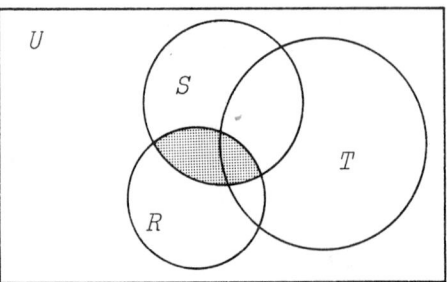

12. Write the name of the set representing the shaded area.

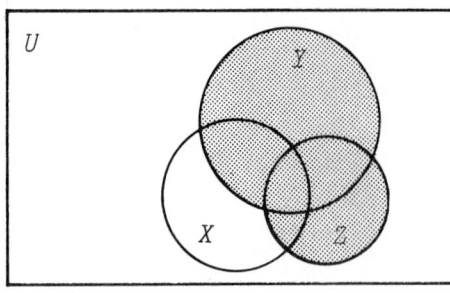

REVIEW EXERCISES 1304, SET II

1. Given: $U = \{a, b, c, d, e, f, g\}$, $E = \{a, c, d\}$, $F = \{d, a\}$, and $G = \{b, e, f, g\}$.
   Find:

   (a) $E \cup F$ (b) $E \cap F$ (c) $G \cap F$ (d) $n(U)$

   (e) $n(E \cup F)$ (f) Is $\{a\}$ a subset of F?

   (g) F and G are called __?__ sets.

   (h) Is U finite or infinite?

   (i) Is F a proper subset of U?

2. Are the sets {11, 5, 7, 5} and {5, 7, 11} equal sets?
3. Write the set of whole numbers < 5.
4. What is the cardinal number of the set of letters in the word "Alabama"?
5. Write all the subsets of the set {5, 8}.

6. Shade in $H \cap K$.

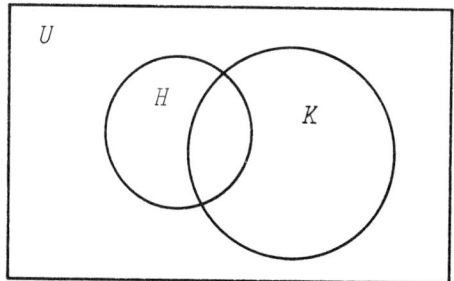

7. <u>Brain Teaser</u>. Express the set represented by the shaded area in terms of the given sets.

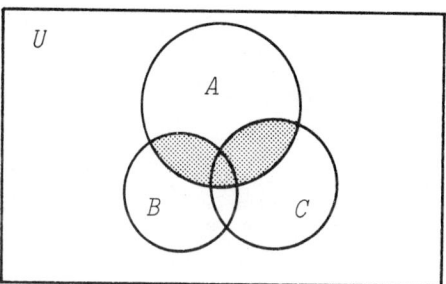

Chapter 13: Diagnostic Test

Name_____

_____

The purpose of this test is to see how well you understand
the basic ideas of sets. We recommend that you work this
diagnostic test *before* your instructor tests you on this chapter. Allow yourself about 40 minutes to do this test.

Complete solutions for all the problems on this test, together
with section references, are given in the Answer Section. You
should study the section referred to for the problems you do
incorrectly.

---

1. Can the collection $\{\theta, 5, \Delta, \div\}$ be called a set?

    (1)_____

2. Are the sets $\{5, 4, 4\}$ and $\{5, 4\}$ equal sets?

    (2)_____

3. Write the set of whole numbers > 4 and < 8.

    (3)_____

4. Write the cardinal number of the set $\{1, 2, 3, 3, 4\}$.

    (4)_____

5. Write the set of digits > 8.

    (5)_____

6. State which of the following sets is finite and which is
   infinite.

    (a) The set of all the grains of sand on the earth today

    (b) The set of natural numbers

    (6a)_____

    (6b)_____

7. Which of the following statements are true and which are false? Circle T if true or F if false.

(a) If $D = \{7, 10, 13, 15\}$, then $10 \in D$.   (7a)   T   F

(b) If $P = \{a, b, c\}$, then $a \notin P$.   (7b)   T   F

(c) If the universal set U is the set of digits, then $11 \in U$.   (7c)   T   F

(d) $0 \notin \emptyset$.   (7d)   T   F

(e) $\emptyset \in \{\ \}$.   (7e)   T   F

_____

8. $R = \{4, 0, 7, 3\}$. State whether each of the following sets is a proper subset, an improper subset, or not a subset of R.

(a) $A = \{4, 3\}$.   (8a) _____

(b) $B = \{7, 1, 3\}$.   (8b) _____

(c) $\emptyset$ (8c) _____

(d) $C = \{0, 3, 4, 7\}$.   (8d) _____

_____

9. Write all the subsets of $\{a, b\}$.

(9) _____

_____

10. Given: $A = \{1, 3, 5\}$, $B = \{2, 6, 8\}$, and $C = \{1, 2, 3\}$. Find the following:

(a) $A \cup C$ (10a) _____

(b) $B \cap C$ (10b) _____

(c) $A \cap B$ (10c) _____

(d) $B \cup C$ (10d) _____

_____

11. Given the Venn diagram, write each of the following sets in roster notation. (The small letters within a particular circle are the elements of that set.)

(a) A (11a) _____

(b) $A \cap B$ (11b) _____

(c) B (11c) _____

(d) U (11d) _____

(e) $A \cup B$ (11e) _____

12. Write the name of the set representing the shaded area in the Venn diagram below.

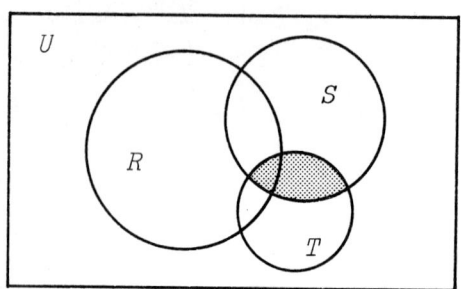

(12)_____

Answers

EXERCISES 101, SET I (page 5)

1. 5   2. 1
3. 0   The digits are: 0, 1, 2, 3, 4, 5, 6, 7, 8, 9.
4. 0   5. 5
6. 10
7. 100   The next smallest whole number, 99,
         has two digits.
8. 9   9. 99   10. 7, seven
11. No   12. Yes   13. Yes
14. There is no largest natural number.
15. 9
16. 0, 1, 2, 3   17. 6, 7, 8, 9   18. 15, 16
19. 0, 1, 2, 3, 4
20. >   21. <   22. >   23. <   24. >   25. <

EXERCISES 102A, SET I (page 7)

1. (a) 6 units   (b) 7 tens = 70 units
    (c) 5 hundreds = 500 units
2. (a) 4 units   (b) 0 units
    (c) 900 units
3. (a) 4 tens   (b) 4 tens = 40 units
    (c) 3 hundreds   (d) 30 tens
4. (a) 6 tens   (b) 60 units
    (c) 8 hundreds   (d) 80 tens

EXERCISES 102B, SET I (page 9)

1. Seven thousand, nine hundred twenty-six
2. One million, eight hundred ninety thousand
3. Twenty-four thousand, nine hundred two
4. Seventy-six million, one hundred sixteen thousand,
   two hundred four
5. (a) Seven hundred fifty million
   (b) Five hundred forty-seven million
   (c) Two hundred forty-one million, seven hundred
       forty-eight thousand
   (d) Two hundred four million, seven hundred sixty-
       five thousand, seven hundred seventy
6. One trillion, one hundred fifty-five billion,
   one hundred fifty-five million
7. One hundred ninety-six million, nine hundred
   forty thousand
   Fifty-seven million, five hundred six thousand
   One hundred thirty-nine million, four hundred
   thirty-four thousand
8. Two million, fifty thousand
   Fifty-four thousand, eight hundred
9. Five trillion, eight hundred seventy-nine billion,
   two hundred million
10. Seven hundred twenty-three thousand
11. Fifty trillion, five hundred sixty-one billion
12. (a) 710,405. Seven hundred ten thousand, four
       hundred five.
    (b) 655,430,186. Six hundred fifty-five million,
        four hundred thirty thousand, one hundred
        eighty-six
    (c) 700,005,009. Seven hundred million, five
        thousand, nine
    (d) 1,002,003,004,005. One trillion, two billion,
        three million, four thousand, five
    (e) 10,020,030,040,050. Ten trillion, twenty
        billion, thirty million, forty thousand,
        fifty
    (f) 100,200,300,400,500. One hundred trillion,
        two hundred billion, three hundred million,
        four hundred thousand, five hundred

13. (a) 8,008,808   (b) 7,000,007
    (c) 10,000,000,010,010   (d) 107,035,000,075
14.   75   15.   56
         1,005   3,080
        10,004   721,049,008
       300,156   5,000,235,000,796
    34,186,075

EXERCISES 103A, SET I (page 13)

1. 13   2. 14   3. 14   4. 15
5. 17   6. 16   7. 15   8. 13
9. 13   10. 12   11. 16   12. 17

EXERCISES 103B, SET I (page 16)

1. 15   2. 30   3. 39   4. 53
5. 80   6. 103   7. 28   8. 17
9. 88   10. 99   11. 32   12. 58
13. 93   14. 83   15. 99   16. 73
17. 82   18. 94   19. 53   20. 41
21. 97   22. 61   23. 63   24. 94

EXERCISES 104A, SET I (page 18)

1. 3 + 5 + 4 = (3 + 5) + 4 = 3 + (5 + 4)
               8   + 4 = 3 +   9
                  12   =   12
2. 14
3. 8 + 2 + 5 = (8 + 2) + 5 = 8 + (2 + 5)
              10   + 5 = 8 +   7
               15   =   15
4. 16
5. 9 + 8 + 7 = (9 + 8) + 7 = 9 + (8 + 7)
              17   + 7 = 9 +   15
               24   =   24
6. 14
7.   5 }11   5 }18   8. 24   9.   6 }15   6 — }20
     6 }18   6 }13   9 }20   9 }14
     7 — }18   7 }18   5 — }20   5 }14
      18   18   20   20
10. 25   11. 10 }14   10 — }20   12. 22
                4 }20   4 }10
                6 — }20   6 }10
                 20   20

EXERCISES 104B, SET I (page 18)

1.
   3 ⌐ = 7 ⌐
   4 ⌐   = 9 ⌐
   2 ⌐   = 16 ⌐
   7 ⌐   = 17
   1 ⌐
   ——
   17
                                      2. 20

3.
   7 ⌐ = 11 ⌐
   4 ⌐   = 17 ⌐
   6 ⌐   = 22 ⌐
   5 ⌐   = 25
   3 ⌐
   ——
   25
                                     4. 27
                                     5. 29
                                     6. 32
                                     7. 35
                                     8. 35

9. 7 + 5 + 6 + 3 + 2 + 8 + 4 = 35   10. 44

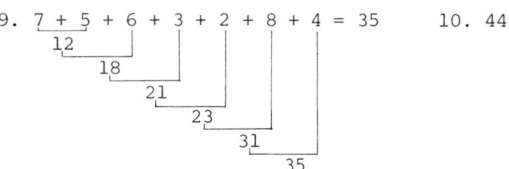

429

11. 3 + 9 + 7 + 6 + 5 + 2 + 8 = 40   12. 41
        12
          19
            25
              30
                32
                  40

13. 10 + 8 + 12 + 25 + 13 + 41 = 109   14. 123
        18
          30
            55
              68
                109

EXERCISES 105, SET I (page 20)

1. (a) 10   (b) 32   (c)   1   (d) 60
        15   25   54   37
        ──   ──   74   ──
        25   57   ───   97
                                 128

   (e)   1   (f) 11   (g) 11   (h) 11
         56   78   85   55
         29   87   58   97
         ──   ───   ───   ───
         85   165   143   152

   (i) 11   (j) 11
         79   98
         92   77
        ───   ───
        171   175

2. (a) 109   (b) 52   (c) 97   (d) 244
   (e) 154   (f) 154   (g) 224   (h) 106
   (i) 168   (j) 254

3. (a) 102   (b) 243   (c)   1   (d) 678
       213   746   354   54
       ───   ───   83   ───
       315   989   ───   732
                                 437

   (e) 1 11   (f) 1 1   (g) 1 1   (h) 1 1
        987   555   817   987
        715   894   209   807
      ─────   ─────   ─────   ─────
      1,702   1,449   1,026   1,794

4. (a) 1,024   (b) 1,071   (c) 2,253   (d) 1,806
   (e) 867   (f) 787   (g) 1,011   (h) 965

5. (a) 1 1   (b)   (c) 1 21   (d) 111 11
       5,840   75   61,304   8
         218   3,594   7,105   78,481
          21   315   70,009   4,154
       3,009   2,171   801   51,816
      ──────   ──────   ───────   ───────
       9,088   6,155   139,219   134,459

   (e) 1 12   (f) 11 12   (g) 11 12
         85   1,009   716
        319   54,311   75
      4,667   31,555   48,315
          8   9,999   8,888
      ──────   ──────   ──────
      5,079   96,874   57,994

6. (a) 1,014,924,913   (b) 1,049,125,745

7.   1
          30,006   8. 45
      75,000,100
   2,000,000,500
      50,100,010
   ─────────────
   2,125,130,616

9. 8 5   10. 2,200
     876
     877   11.   1 11
     878   36,198
     879   + 29,028
     880   ─────────
     881   65,226 ft
     882
     883   12. 1,635,748,852
     884
     885
     886
   ─────
   9,691

13.   John $75
          Jim $18
                 ───
                 $93   Amount John and Jim have together
               + 12
          Marie $105
   Then, John   75
          Jim   18
                ────
                $198 Total amount the three have
                      together

14. $125
15. 27 + 15 + 10 + 36 + 71 + 44 = 203
    18 + 30 + 26 + 82 + 28 + 37 = 221
     9 + 58 + 69 + 77 + 39 + 8 = 260
    ──   ───   ───   ───   ───   ──
    54 + 103 + 105 + 195 + 138 + 89 = 684

16. 7,706   17. 15 + 6 + 15 + 6 = 42 ft
18. 168 in

19. 1 22   20. 842 ft
    2,435
    1,742
    5,078
    4,659
   ──────
   13,914 yd

EXERCISES 106, SET I (page 25)

1. 1 + 1 = 2   2. 6
3. 2 + 7 = 9 → 0   4. 0
5. 4 + 7 = 11 → 2   6. 3
7. 2 + 3 = 5   8. 2

9. 945   10. 0
    ⁹5̸ → 0

11. 4 + 0 + 2 + 6 = 12 → 3   12. 6
13. 2 + 8 + 3 + 7   14. 2
      10   10
       1 +   1 = 2

15. 7 + 0 + 3 + 5 + 8   16. 2
         10
       1 + 5 + 8 = 14 → 5

17.   9   18. 4
    1 + 6 + 4 + 8 + 7
          10
        1 + 7 = 8

19. 20736   20. 8
     5̸ 5̸ → 0

EXERCISES 107, SET I (page 26)

1.   1   2. 114   3.   1
    74   2   85   4
    38   2   55   1
    46   1   94   4
   ───   ─   ───   ─
   158   ⑤   234   ⑨
    ⑤ ←── check   ⓪ ←── ⓪ check

4. 81   5.   1 2   6. 5,515
            726   6
             49   4
            657   0
          ─────   ──
          1,432   10
                  ①
             ① ←── check

7.   11   8. 12,591
    6,123   3
      365   5   9.   1 12
    7,216   7   7,856   8
   ────── ──   2,094   6
   13,704 15   5,436   0
           ⑥   309   3
       ⑥ ←── check   ──────   ──
                       15,695   17
                               ⑧
                           ⑧ ←── check

10. 20,123

11.
```
  1 22
  7,238    2
  6,272    8
  7,854    6
  4,259    2
 -------  ---
 25,623   18
                0
           0 ← check
```

12. 204,570

EXERCISES 108, SET I (page 28)

1. 42 2. 48 3. 45 4. 54
5. 72 6. 63 7. 56 8. 40
9. 36 10. 35 11. 63 12. 72
13. 64 14. 28 15. 30 16. 6
17. 9 18. 9

EXERCISES 109, SET I (page 30)

1. 12, 18, 24   2. 1, 2, 3, 6   3. 1, 2, 4, 8
4. 9, 18, 27, 36, 45   5. 1, 3, 5, 9, 15, 45

EXERCISES 111, SET I (page 32)

1. 0 2. 0 3. 0 4. 0
5. 3 6. 6 7. 0 8. 8

EXERCISES 112, SET I (page 32)

1. 9 · 8   2. 5 · 6
   9(8)   (5)6
   (9)(8)   5(6)
   (9)8   (5)(6)

3. 3 · 8   4. 7 · 4
   3(8)   (7)(4)
   (3)(8)   7(4)
   (3)8   (7)4

EXERCISES 113, SET I (page 33)

1. (a)
```
 34
  2
---
 68
```
(b)
```
 1
 74
  3
---
222
```
(c)
```
 2
 56
  4
---
224
```
(d)
```
 1
 83
  6
---
498
```

(e)
```
 4
 48
  5
---
240
```
· (f)
```
 3
 29
  4
---
116
```
(g)
```
 5
 68
  7
---
476
```
(h)
```
 4
 95
  8
---
760
```

2. (a) 156 (b) 390 (c) 252 (d) 234
   (e) 368 (f) 174 (g) 651 (h) 234

3. (a)
```
 11
135
  3
---
405
```
(b)
```
 31
283
  4
-----
1,132
```
(c)
```
  4
506
  7
-----
3,542
```
(d)
```
  5
209
  6
-----
1,254
```

(e)
```
310
  8
-----
2,480
```
(f)
```
400
  9
-----
3,600
```
(g)
```
 43
687
  5
-----
3,435
```
(h)
```
 76
898
  8
-----
7,184
```

4. (a) 468 (b) 1,251 (c) 5,000 (d) 1,436
   (e) 4,746 (f) 3,024 (g) 4,945 (h) 5,400

5. (a)
```
 1 1
2,453
    3
-----
7,359
```
(b)
```
 4 43
6,987
    5
------
34,935
```
(c)
```
 23
1,069
    4
-----
4,276
```

(d)
```
 3 77
6,499
    8
------
51,992
```
(e)
```
 65
7,088
    7
------
49,616
```
(f)
```
 6 86
8,697
    9
------
78,273
```

6. (a) 69,468 (b) 280,205 (c) 364,716
   (d) 490,378 (e) 160,072 (f) 620,802

---

7. (a)
```
 700
   6
-----
4,200
```
(b)
```
     5
30,080
     7
-------
210,560
```
(c)
```
    24
12,500
     8
-------
100,000
```

(d)
```
     24
526,000
      8
---------
4,208,000
```
(e)
```
     2
25,000
     4
-------
100,000
```
(f)
```
     8
70,900
     9
-------
638,100
```

EXERCISES 114, SET I (page 35)

1. (a)
```
 35
 21
---
 35
 70
---
735
```
(b)
```
 28
 32
---
 56
 84
---
896
```
(c)
```
  56
  43
-----
 168
2 24
-----
2,408
```

(d)
```
  71
  28
-----
 568
1 42
-----
1,988
```
(e)
```
  78
  36
-----
 468
2 34
-----
2,808
```
(f)
```
  95
  42
-----
 190
3 80
-----
3,990
```

2. (a) 7,310 (b) 16,848 (c) 26,085
   (d) 29,904 (e) 24,966 (f) 87,906

3. (a)
```
2,314
   52
------
4 628
115 70
-------
120,328
```
(b)
```
4,536
   34
-------
18 144
136 08
-------
154,224
```
(c)
```
5,076
   45
-------
25 380
203 04
-------
228,420
```

(d)
```
5,070
   64
-------
20 280
304 20
-------
324,480
```
(e)
```
7,836
   58
-------
62 688
391 80
-------
454,488
```
(f)
```
9,267
   49
-------
83 403
370 68
-------
454,083
```

4. (a) 4,872,744 (b) 6,365,112 (c) 677,820
   (d) 2,983,228 (e) 6,866,370

5. (a)
```
  235
  415
------
1 175
2 35
94 0
------
97,525
```
(b)
```
  426
  351
------
  426
21 30
127 8
-------
149,526
```
(c)
```
7,023
  542
-------
14 046
280 92
3 511 5
---------
3,806,466
```

(d)
```
 8,107
   623
-------
24 321
162 14
4 864 2
---------
5,050,661
```
(e)
```
23,016
   524
-------
92 064
460 32
11 508 0
----------
12,060,384
```

6. (a) 22,594,496   (b) 4,256,392
   (c) 254,216,655   (d) 841,043,216

EXERCISES 115, SET I (page 36)

1.
```
  356
  204
------
1 424
71 2
------
72,624
```
2. 539,372
3.
```
  705
  206
------
4 230
141 0
-------
145,230
```

4. 2,685,624
5.
```
9,067
  504
------
36 268
4 533 5
--------
4,569,768
```
6. 21,569,156

7.
```
 95,046
  3,007
--------
665 322
285 138
---------
285,803,322
```
8. 540,822,290
9.
```
 7,802
 5,008
-------
62 416
39 010
--------
39,072,416
```

10. 22,505,520,063

EXERCISES 116, SET I (page 37)

1.
```
   376
  2,500
 1880
  752
 940,000
```
2. 6,084,000

3.
```
  3,154
  9,200
  6308
 28386
 2,901,6800
```

4. 2,356,500

5.
```
  3,751
    500
 1,875,500
```
6. 1,598,000

7.
```
   449
  6600
 2694
 2694
 2,963,400
```
8. 35,040,000

9.
```
  7893
  9500
 39465
 71037
 74,983,500
```

10. 26,250,000

11.
```
   8960
   5600
  5376
  4480
 50,176000
```
12. 296,400,000

13.
```
  1427
   500
 713,500  sheets
```
14. $3,780,000

15.
```
  16000
     8
 128000¢
 = $1,280
```

EXERCISES 117, SET I (page 39)

1.
```
   785  →    2
    37     × 1
  5 495
  23 55      ②
  29,045 →   ②    check
```
2. 20,790

3.
```
    834  →    6
     68     × 5
  6 672     30  → ③
 50 04
 56,712 →    ③    check
```
4. 106,658

5.
```
   5,048  →    8
      56     × 2
  30 288     16  → ⑦
 252 40
 282,688 →    ⑦    check
```
6. 3,098,100

7.
```
   18,075  →    3
     284      × 5
  72 300      15  → ⑥
 1 446 00
 3 615 0            check
 5,133,300 →   ⑥
```
8. 7,019,980

9.
```
    89,765  →    8
       789     × 6
   807 885      48  → ③
 7 181 20
 62 835 5            check
 70,824,585 →   ③
```
10. 132,829,968

EXERCISES 118, SET I (page 41)

1. $A = L \times W$
   $= 18 \times 11 = 198$ sq in

2. 3,196 sq ft

3. (a) $A = L \times W$
       $= 7 \times 5 = 35$ sq yd
   (b) $P = 2L + 2W$
       $= 2(7) + 2(5)$
       $= 14 + 10$
       $= 24$ yd

4. $A = 72$ sq yd
   $P = 34$ yd

5. $A = L \times W$
   $= 7 \times 4$
   $= 28$ sq ft

6. $42

7. $A = L \times W = 5 \times 3 = 15$ sq yd
   Cost $= \$7 \times$ number of sq yd
        $= \$7 \times 15 = \$105$

8. 54 sq ft

9. Area of front $= 8 \times 9 = 72$ sq in
   Area of back $= 8 \times 9 = 72$ sq in
   Area of end $= 8 \times 6 = 48$ sq in
   Area of
     other end $= 8 \times 6 = 48$ sq in
   Total area of sides $= 240$ sq in

10. (a) 30 sq yd
    (b) $240

EXERCISES 119, SET I (page 43)

1. $3^3 = 3 \times 3 \times 3 = 27$
2. 16
3. $5^2 = 5 \times 5 = 25$
4. 216
5. $7^2 = 7 \times 7 = 49$
6. 81
7. $0^3 = 0 \times 0 \times 0 = 0$
8. 1
9. 10
10. 100
11. $10^3 = 10 \times 10 \times 10 = 1,000$
12. 10,000
13. $10^5 = 10 \times 10 \times 10 \times 10 \times 10 = 100,000$
14. $10^6 = 1,000,000$
15. $5^0 = 1$
16. 32
17. $2^6 = 2 \times 2 \times 2 \times 2 \times 2 \times 2 = 64$
18. 512
19. $2^8 = 2 \times 2 \times 2 \times 2 \times 2 \times 2 \times 2 \times 2 = 256$
20. 625
21. $40^3 = 40 \times 40 \times 40 = 64,000$
22. 0
23. $12^3 = 12 \times 12 \times 12 = 1,728$
24. 225
25. $1^5 = 1 \times 1 \times 1 \times 1 \times 1 = 1$

EXERCISES 120, SET I (page 45)

1. 20
2. 20
3. 350
4. 3,500
5. 27,000
6. 50,000
7. 400
8. 7,000
9. 1,000
10. 10,000
11. 1,000,000
12. 100,000
13. 700,000
14. 8
15. 800
16. 5,000
17. 70
18. 9
19. 84,000
20. 750,000

EXERCISES 121, SET I (page 46)

1. $7 \times 5 + 45$
   $= 35 + 45 = 80$
2. 82

3. $20 + 2^3$
   $= 20 + 8 = 28$
4. 76

5. $8 \cdot 7 + 9 \cdot 6 + 45$
   $= 56 + 54 + 45 = 155$
6. 80

7. $(10^2)10 + 100$
   $= (100)10 + 100$
   $= 1,000 + 100 = 1,100$
8. 2,000

9. $2 \cdot 5^2 + 3 \cdot 2^2 + 4$
   $= 2 \cdot 25 + 3 \cdot 4 + 4$
   $= 50 + 12 + 4 = 66$
10. 45

11. $6^0 + 10^0 + 10 + 10^4$
    $= 1 + 1 + 10 + 10,000 = 10,012$
12. 100,001

1. (thousands)

30,050
30,000

Since the first digit to be dropped is less than 5, the part retained is unchanged.

2. 14,000

3. (hundreds)

1,299
1,300

Since the first digit to be dropped is greater than 5, the part retained is increased by 1.

4. 300

5. (millions)

204,765,770
205,000,000

Since the first digit to be dropped is greater than 5, the part retained is increased by 1.

6. 242,000,000

7. (thousands)

188,380
188,000

Since the first digit to be dropped is less than 5, the part retained is unchanged.

8. 12,000,000

9. (hundred-thousands)

9,490,000
9,400,000

Since the first digit to be dropped is more than 5, the part retained is increased by 1.

10. 4,100

11. (hundreds)

2,348
2,300

Since the first digit to be dropped is less than 5, the part retained is unchanged.

12. 1,000

13. (tens)

859
860

Since the first digit to be dropped is greater than 5, the part retained is increased by 1.

14. 260

15. (ten-thousands)

245,429,444
245,430,000

Since the first digit to be dropped is greater than 5, the part retained is increased by 1.

16. 200,000,000

1. Three billion, seventy-five million, six hundred thousand, eight.

2. 5,072,006

3.
```
    1 2 23
    7,825
       84
      900
   45,788
        9
2,000,085
─────────
2,054,691
```

4. (a) 1,140
   (b) 1,941
   (c) 1,210
   (d) 86,314

5. (a)
```
    786
     35
  3 930
 23 58
──────
 27,510
```
   (b)
```
    342
     28
  2 736
  6 84
──────
  9,576
```
   (c)
```
  2,847
      9
──────
 25,623
```
   (d)
```
    3,967
      867
   27 769
  238 02
  3 173 6
────────
3,439,389
```
   (e)
```
    9,207
      704
   36 828
  6 444 9
────────
6,481,728
```
   (f) 586,000

   (g)
```
176 000
 35 00
   880
   528
────────
616 000 000
```

6. (a) 12,390   (b) 62,512   (c) 7,665
   (d) 5,967,996   (e) 5,400   (f) 490,000
   (g) 156,762,930

7. $8 + 7 \cdot 9$
   $= 8 + 63 = 71$

8. 698

9. $5 \times 0 + 8 \cdot 9 + 0$
   $= 0 + 72 + 0 = 72$

10. 11

11. $25 \times 4 + 10^2$
    $= 100 + 100 = 200$

12. 1,000

13. $A = L \times W$
    $= 3 \times 2 = 6$ sq ft

14. $1,792

15. $A = L \times W$
    $= 12 \times 9 = 108$ sq ft

16. $210

17.
```
  503,024
      207
  3 521 168
100 604 8
─────────────
104,125,968 gal
```

18. 250,268

19.
```
  7,568
      6
──────
$45,408
```

20. 1,100

21.
```
    168
     52
    336
  8 40
──────
$8,736
```

22. 352 mi

23. (a) $2^{10} = 1,024$¢

    (c) $2^{15} = 32,768$¢

    (d) $2^{20} = 1,048,576$¢

    (b)
```
 2^1  =     2
 2^2  =     4
 2^3  =     8
 2^4  =    16
 2^5  =    32
 2^6  =    64
 2^7  =   128
 2^8  =   256
 2^9  =   512
 2^10 = 1,024
       ──────
       2,046¢
```

24. First, both boys cross the river in the boat. Boy one returns with the boat. Man one then crosses with the boat. Boy two returns with the boat; then both boys cross the river with the boat. Boy one returns in the boat and gives it to man two, who crosses with the boat. Then boy two returns with the boat to get boy one. Then both boys cross the river in the boat.

25. Yes, if you don't mind walking around with your socks on over your shoes.

SOLUTIONS FOR CHAPTER 1 DIAGNOSTIC TEST (page 57)

Following each problem number is the textbook section reference (in parentheses) where that kind of problem is discussed.

1. (101) 8, 9   2. (101) 0, 1, 2, 3
3. (101) 99   4. (101) (a) 6 < 14
                      (b) 12 > 0
5. (102B) Five trillion, eight hundred seventy-nine billion, two hundred million
6. (102B) 54,007,506,080   7. (104) 242

8. (107)

```
7,856    8
3,942    0
8,674    7
3,597    6
24,069   21
```
$(3) \longleftrightarrow (3)$
check

9. (105)

```
         75
      3,086
 70,500,006
        108
      8,009
 70,511,284
```

10. (a) (114)
```
7,546
   89
67914
60368
671,594
```
(b) (114)
```
3,084
  706
18504
21588
2,177,304
```

11. (116)
```
75,000
 8,600
   450
   600
645,000,000
```
12. (117)
```
836    8
 74    2
3 344   16
58 52   (7)
61,864
```
$(7) \longleftarrow$ check

13. (120) (a) 7,500   (b) 8,000   (c) 300,000
          (d) 5,080,000   (e) 100,000
14. (119) 29 × 29 = 841
15. (121) (a) 10 + 8 × 9
              = 10 + 72 = 82
         (b) $2^3 \cdot 3 + 5$
              = 8 · 3 + 5
              = 24 + 5 = 29
         (c) $5^2 + 6^1 + 10^0 + 10^3$
              = 25 + 6 + 1 + 1,000
              = 1,032
16. (118) (a) P = 65 + 48 + 65 + 48   (b) 226
              = 226 ft   4
                                          $904
17. (123) John $45   John $45
          Harry $23   Harry $23
              $68   Bill $80
           + $12 $148   Total for all
          Bill $80   three
18. (123) (a) Living room 7 × 5 = 35 sq yd
              Bedroom   4 × 4 = 16 sq yd
              Total area   = 51 sq yd
          (b) Cost = 6 × 51
                   = $306

EXERCISES 202A, SET I (page 64)

1. (a)   7,564   (b)   38,921
       - 2,341   - 17,211
         5,223   21,710

   (c) 77,806   (d)   91,105
       - 35,002   - 1,102
        42,804   90,003

2. (a) 6,108   (b) 5,005
   (c) 87,504   (d) 63,322

3. (a)   9,673   (b)   2,745
       - 3,542   - 324
         6,131   2,421

   (c) 84,273   (d)   14,863
       - 61,022   - 3,521
        23,251   11,342

4. (a) 3,392   (b) 61,151
   (c) 51,122   (d) 313,332

5.   785   6. 8,244   7. 92,889,000
   - 281   - 239,000
     504   92,650,000 miles

8. Fill the 5-gallon bucket. Pour just enough from the 5-gallon bucket to fill the 3-gallon bucket. This leaves 2 gallons in the 5-gallon bucket. Empty the 3-gallon bucket. Pour the 2 gallons remaining in the 5-gallon bucket into the empty 3-gallon bucket. Fill the 5-gallon bucket. Pour just enough from the 5-gallon bucket into the 3-gallon bucket to fill it. Since there were 2 gallons in the 3-gallon bucket to begin with, only 1 gallon is needed to fill it, leaving *4 gallons* in the 5-gallon bucket.

EXERCISES 202B, SET I (page 66)

1. check [→ 37; − 24; 13; → 37] add   2. 33
3. check [→ 79; − 51; 28; → 79] add   4. 53
5. check [→ 7,864; − 2,033; 5,831; → 7,864] add   6. 8,205
7. check [→ 2,806; − 1,502; 1,304; → 2,806] add   8. 3,005
9. check [→ 7,186; − 7,136; 50; → 7,186] add   10. 94,200

1. (a) $2^1 5$
    $- \quad 8$
    $\quad\; 1$
    $\overline{\quad 1\; 7}$

    (b) $7^1 4$
    $- \quad 9$
    $\quad\; 1$
    $\overline{\quad 6\; 5}$

    (c) $3^1 6$
    $- \quad 7$
    $\quad\; 1$
    $\overline{\quad 2\; 9}$

    (d) $5^1 5$
    $- \quad 8$
    $\quad\; 1$
    $\overline{\quad 4\; 7}$

    (e) $9^1 2$
    $- \quad 5$
    $\quad\; 1$
    $\overline{\quad 8\; 7}$

    (f) $8^1 0$
    $- \quad 6$
    $\quad\; 1$
    $\overline{\quad 7\; 4}$

    (g) $8^1 8$
    $- \quad 9$
    $\quad\; 1$
    $\overline{\quad 7\; 9}$

2. (a) 45   (b) 38   (c) 58
    (d) 16   (e) 54   (f) 87
    (g) 69

3. (a) $4^1 2$
    $- \; 2\; 8$
    $\quad\; 1$
    $\overline{\quad 1\; 4}$

    (b) $5^1 4$
    $- \; 3\; 7$
    $\quad\; 1$
    $\overline{\quad 1\; 7}$

    (c) $2^1 5$
    $- \; 1\; 9$
    $\quad\; 1$
    $\overline{\qquad 6}$

    (d) $7^1 3$
    $- \; 4\; 6$
    $\quad\; 1$
    $\overline{\quad 2\; 7}$

    (e) $6^1 0$
    $- \; 5\; 4$
    $\quad\; 1$
    $\overline{\qquad 6}$

    (f) $8^1 1$
    $- \; 6\; 2$
    $\quad\; 1$
    $\overline{\quad 1\; 9}$

    (g) $9^1 7$
    $- \; 7\; 9$
    $\quad\; 1$
    $\overline{\quad 1\; 8}$

4. (a) 28   (b) 34   (c) 27
    (d) 22   (e) 56   (f) 32
    (g) 6

5. (a) $7\; 2^1 3$
    $- \; 2\; 1\; 8$
    $\qquad 1$
    $\overline{\; 5\; 0\; 5}$

    (b) $4^1 2^1 5$
    $- \; 1\; 5\; 7$
    $\quad\; 1\; 1$
    $\overline{\; 2\; 6\; 8}$

    (c) $6^1 8^1 4$
    $- \; 2\; 9\; 5$
    $\quad\; 1\; 1$
    $\overline{\; 3\; 8\; 9}$

    (d) $5\; 4^1 1$
    $- \; 2\; 0\; 7$
    $\qquad 1$
    $\overline{\; 3\; 3\; 4}$

    (e) $2^1 5^1 6$
    $- \; 1\; 7\; 7$
    $\quad\; 1\; 1$
    $\overline{\qquad 7\; 9}$

    (f) $3^1 0^1 0$
    $- \; 1\; 5\; 7$
    $\quad\; 1\; 1$
    $\overline{\; 1\; 4\; 3}$

    (g) $8^1 0^1 2$
    $- \; 5\; 0\; 9$
    $\quad\; 1\; 1$
    $\overline{\; 2\; 9\; 3}$

6. (a) 208   (b) 434   (c) 229
    (d) 634   (e) 328   (f) 538
    (g) 39

7. (a) $2^1 1^1 0\; 8$
    $- 1,8\; 9\; 6$
    $\quad\; 1\; 1$
    $\overline{\qquad 2\; 1\; 2}$

    (b) $6,9^1 1^1 4$
    $- 6,0\; 5\; 7$
    $\qquad 1\; 1$
    $\overline{\qquad 8\; 5\; 7}$

    (c) $3,0^1 0^1 5$
    $- \quad 6\; 8\; 4$
    $\quad\; 1\; 1$
    $\overline{\; 2,3\; 2\; 1}$

    (d) $1,8^1 0^1 4$
    $- \quad 3\; 0\; 8$
    $\qquad 1\; 1$
    $\overline{\; 1,4\; 9\; 6}$

    (e) $6^1 0,7^1 0^1 1$
    $- 1\; 0,8\; 0\; 8$
    $\quad\; 1\; 1\; 1\; 1$
    $\overline{\; 4\; 9,8\; 9\; 3}$

    (f) $2\; 8,9^1 6\; 4$
    $- \quad 4,9\; 7\; 2$
    $\qquad\; 1\; 1$
    $\overline{\; 2\; 3,9\; 9\; 2}$

8. (a) 6,578   (b) 1,799   (c) 49,555
    (d) 337,866   (e) 84,691   (f) 230,792
    (g) 465,088

9. $7\; 5^1 0^1 0^1 0^1 0^1 0^1 0\; 0$
    $- 2\; 0\; 4,7\; 7\; 6,7\; 7\; 0$
    $\qquad 1\; 1\; 1\; 1\; 1$
    $\overline{\; 5\; 4\; 5,2\; 2\; 3,2\; 3\; 0}$

10. 3,448 mi

11. $2\; 9,0^1 2\; 8$
    $- 1\; 4,4\; 9\; 5$
    $\qquad 1\; 1$
    $\overline{\; 1\; 4,5\; 3\; 3 \text{ ft}}$

12. $3,665

13. $74
    $\times\; 27$
    $\overline{\;518}$
    $\;1\;48$
    $\overline{\$1,998}$

    $2,6^1 6^1 4$
    $- 1,9\; 9\; 8$
    $\quad 1\; 1\; 1$
    $\overline{\$\;\; 6\; 6\; 6}$ Unpaid balance

14. $210

15. Jim $75
    Joe   57
    Jack   92 = 75 + 17
    Mike   119 = (75 + 57) − 13
    $\overline{\$343}$

16. A 50¢ piece and a dime

17. 5 sheep

    4 ahead   1 in the middle

    ⟵   X   X   (X)   X   X

    4 behind

1. 8
    − 2 ← 1st subtraction
    $\overline{\;6}$
    − 2 ← 2nd subtraction
    $\overline{\;4}$
    − 2 ← 3rd subtraction
    $\overline{\;2}$
    − 2 ← ④th subtraction
    $\overline{\;0}$
    Therefore, 8 ÷ 2 = ④

2. 2

3. 10
    − 5 ← 1st subtraction
    $\overline{\;5}$
    − 5 ← ②nd subtraction
    $\overline{\;0}$
    Therefore, 10 ÷ 5 = ②

4. 5

5. 16
    − 4 ← 1st subtraction
    $\overline{12}$
    − 4 ← 2nd subtraction
    $\overline{\;8}$
    − 4 ← 3rd subtraction
    $\overline{\;4}$
    − 4 ← ④th subtraction
    $\overline{\;0}$
    Therefore, 16 ÷ 4 = ④

6. 3

7. 20
    − 4 ← 1st subtraction
    $\overline{16}$
    − 4 ← 2nd subtraction
    $\overline{12}$
    − 4 ← 3rd subtraction
    $\overline{\;8}$
    − 4 ← 4th subtraction
    $\overline{\;4}$
    − 4 ← ⑤th subtraction
    $\overline{\;0}$
    Therefore, 20 ÷ 4 = ⑤

8. 7

9. 35
    − 7 ← 1st subtraction
    $\overline{28}$
    − 7 ← 2nd subtraction
    $\overline{21}$
    − 7 ← 3rd subtraction
    $\overline{14}$
    − 7 ← 4th subtraction
    $\overline{\;7}$
    − 7 ← ⑤th subtraction
    $\overline{\;0}$
    Therefore, 35 ÷ 7 = ⑤

10. 4

11. 45
```
    - 5 ←1st subtraction
    ──
    40
    - 5 ←2nd subtraction
    ──
    35
    - 5 ←3rd subtraction
    ──
    30
    - 5 ←4th subtraction
    ──
    25
    - 5 ←5th subtraction
    ──
    20
    - 5 ←6th subtraction
    ──
    15
    - 5 ←7th subtraction
    ──
    10
    - 5 ←8th subtraction
    ──
    5
    - 5 ←⑨th subtraction
    ──
    0
```
Therefore, 45 ÷ 5 = ⑨

12. 8

13. 36
```
    - 9 ←1st subtraction
    ──
    27
    - 9 ←2nd subtraction
    ──
    18
    - 9 ←3rd subtraction
    ──
    9
    - 9 ←④th subtraction
    ──
    0
```
Therefore, 36 ÷ 9 = ④

14. 3

15. 48
```
    - 12 ←1st subtraction
    ──
    36
    - 12 ←2nd subtraction
    ──
    24
    - 12 ←3rd subtraction
    ──
    12
    - 12 ←④th subtraction
    ──
    0
```
Therefore, 48 ÷ 12 = ④

16. 4

17. 15
```
    - 3 ←1st subtraction
    ──
    12
    - 3 ←2nd subtraction
    ──
    9
    - 3 ←3rd subtraction
    ──
    6
    - 3 ←4th subtraction
    ──
    3
    - 3 ←⑤th subtraction
    ──
    0
```
Therefore, 15 ÷ 3 = ⑤ tickets

18. 5 bushels

EXERCISES 207B, SET I (page 74)

1. $18 = 1 \cdot 18 = 2 \cdot 9 = 3 \cdot 6$
   Therefore, divisors = 1, 2, 3, 6, 9, 18
2. 1, 2, 3, 4, 6, 12
3. $35 = 1 \cdot 35 = 5 \cdot 7$
   Therefore, divisors = 1, 5, 7, 35
4. 1, 2, 3, 4, 6, 8, 12, 16, 24, 48
5. $39 = 1 \cdot 39 = 3 \cdot 13$
   Therefore, divisors = 1, 3, 13, 39
6. 1, 2, 3, 6, 9, 18, 27, 54
7. $7 = 1 \cdot 7$
   Therefore, divisors = 1, 7
8. 1, 2, 4, 8, 16, 32, 64
9. $13 = 1 \cdot 13$
   Therefore, divisors = 1, 13
10. 1, 3, 9, 27, 81

EXERCISES 207C, SET I (page 76)

1. $7 \div 2 = 2\overline{)7}\ \ ^{3\ R1}$ because $7 = 2 \cdot 3 + 1$
   $= 6 + 1$
   $= 7$

   2 can be subtracted 3 times, leaving a remainder of 1.

2. 2 R2

3. $9 \div 5 = 5\overline{)9}\ \ ^{1\ R4}$ because $9 = 5 \cdot 1 + 4$
   $= 5 + 4$
   $= 9$

   5 can be subtracted 1 time, leaving a remainder of 4.

4. 5 R1

5. $15 \div 4 = 4\overline{)15}\ \ ^{3\ R3}$ because $15 = 4 \cdot 3 + 3$
   $= 12 + 3$
   $= 15$

   4 can be subtracted 3 times, leaving a remainder of 3.

6. 2 R5

7. $12 \div 5 = 5\overline{)12}\ \ ^{2\ R2}$ because $12 = 5 \cdot 2 + 2$
   $= 10 + 2$
   $= 12$

   5 can be subtracted 2 times, leaving a remainder of 2.

8. 2 R7

9. $15 \div 7 = 7\overline{)15}\ \ ^{2\ R1}$ because $15 = 7 \cdot 2 + 1$
   $= 14 + 1$
   $= 15$

   7 can be subtracted 2 times, leaving a remainder of 1.

10. 7 R2

11. $33 \div 10 = 10\overline{)33}\ \ ^{3\ R3}$ because $33 = 10 \cdot 3 + 3$
    $= 30 + 3$
    $= 33$

    10 can be subtracted 3 times, leaving a remainder of 3.

12. 4 R6

13. $49 \div 6 = 6\overline{)49}\ \ ^{8\ R1}$ because $49 = 6 \cdot 8 + 1$
    $= 48 + 1$
    $= 49$

    6 can be subtracted 8 times, leaving a remainder of 1.

14. 7 R3

15. $64 \div 11 = 11\overline{)64}\ \ ^{5\ R9}$ because $64 = 11 \cdot 5 + 9$
    $= 55 + 9$
    $= 64$

    11 can be subtracted 5 times, leaving a remainder of 9.

16. 4 R9

17. $45 \div 9 = 9\overline{)45}\ \ ^{5\ rows}$ because $45 = 9 \cdot 5$
    $= 45$

18. 5 months

19. 15 ft = 15 ÷ 3 = 5 yd
    27 ft = 27 ÷ 3 = 9 yd
    $A = LW = 9 \cdot 5 = 45$ sq yd

20. 5 hamburgers,
    25¢ change

21. $1,250; No

EXERCISES 208, SET I (page 79)

1. $\dfrac{9\quad \text{R1}}{3\,\overline{)28}}$ 2. 14 R1

3. $\dfrac{1\ 5\quad \text{R4}}{5\,\overline{)7\,^29}}$ 4. 47

5. $\dfrac{4\ 9}{3\,\overline{)1\ 4\,^27}}$ 6. 66 R1

7. $\dfrac{1\ 8\ 9}{2\,\overline{)3\,^17\,^18}}$ 8. 171

9. $\dfrac{1\ 4\ 8\quad \text{R5}}{6\,\overline{)8\,^29\,^53}}$ 10. 129 R1

11. $\dfrac{1\ 3\ 7}{5\,\overline{)6\,^18\,^35}}$ 12. 354

13. $\dfrac{2\ 8\ 6}{4\,\overline{)1\ 1\,^34\,^24}}$ 14. 851

15. $\dfrac{2\ 1\ 6}{7\,\overline{)1\ 5\,^11\,^42}}$ 16. 154 R3

17. $\dfrac{7\ 9\ 6\quad \text{R2}}{9\,\overline{)7\ 1\,^86\,^56}}$ 18. 1,050 R4

19. $\dfrac{3\ 4\ 0\ 6\quad \text{R5}}{6\,\overline{)2\ 0\,^24\ 4\,^41}}$ 20. 6,796 R7

21. $\dfrac{14¢\ \text{per ounce}}{7\,\overline{)9\,^28}}$ 22. $358

23. $\dfrac{1\ 2\quad \text{R4}}{8\,\overline{)1\ 0\,^20}}$ Therefore, leader carries 12 + 4 = 16 boxes.

24. No matter what order the nine digits are arranged in, the resulting number is exactly divisible by 9.

EXERCISES 209, SET I (page 81)

1.
```
    17
14│238
    14
    98
    98
```
2. 25

3.
```
      47
35│1645
   140
    245
    245
```

4. 51

5.
```
    35  R10
21│745
   63
   115
   105
    10
```
6. 27 R10

7.
```
   43  R3
12│519
   48
   39
   36
    3
```
8. 17 R3

9.
```
    48  R5
13│629
   52
   109
   104
     5
```

10. 20 R25

11.
```
      27
52│1404
   104
    364
    364
```
12. 35

13.
```
    67  R4
78│5230
   468
   550
   546
     4
```
14. 286

15.
```
      53  R66
87│4677
   435
    327
    261
     66
```

16. 845 17.
```
      944
85│80240
   765
    374
    340
    340
    340
```
18. 1,071 R53

19.
```
      2084  R27
89│185503
   178
    750
    712
    383
    356
     27
```
20. 351 R123

21.
```
       715  R89
206│147379
    1442
     317
     206
    1119
    1030
      89
```
22. 1,414 R129

23.
```
       790  R240
715│565090
    5005
     6459
     6435
      240
```
24. 16,058 R783

25.
```
        28035  R378
784│21979818
    1568
     6299
     6272
      2781
      2352
      4298
      3920
       378
```
26. 34,671 R679

27.
```
   62 hr
38│2356
   228
    76
    76
```

28. (a) 7 bars
    (b) 3¢

29.
```
            4 tires per wheel; 16 tires in all
24,000│96,000
       96,000
```

30. 9 min

EXERCISES 210, SET I (page 82)

1. Average $=\dfrac{7+5}{2}=\dfrac{12}{2}=6$ 2. 7

3. Average $=\dfrac{3+6+9}{3}=\dfrac{18}{3}=6$ 4. 5

5. Average $=\dfrac{6+8+9+5}{4}=\dfrac{28}{4}=7$ 6. 6

7. Average $=\dfrac{21+24+33}{3}=\dfrac{78}{3}=26$ 8. 8

9. Average $=\dfrac{74+88+85+69}{4}=\dfrac{316}{4}=79$ 10. 92

11. $\dfrac{75+83+74+86+95+61}{6}=\dfrac{474}{6}=79$ 12. 149 lb

13. $\dfrac{76+78+84+72+75}{5}=\dfrac{385}{5}=77$ in   14. $19

15. $\dfrac{73+74+75+76+77+78+79}{7}=76$ 16. $17

17. $\dfrac{17+14+19+15+7+2+1+2+4+7+11+9}{12}$

$=\dfrac{108}{12}=9$ in

18. Group A: 84; Group B: 79; Group A average is 5 higher.

1.
$$\begin{array}{r} 17 \nearrow^{8} \; \text{R0} \\ 14\,\overline{)238} \\ {}_{5}\swarrow\; \underline{14}\,\searrow_{4} \\ 98 \\ \underline{98} \\ 0 \end{array}$$

Check: Using 9-rems
$$\begin{aligned} 4 &= 5 \cdot 8 + 0 \\ &= \;40\; + 0 \\ &= \;\;4\;\; + 0 \\ 4 &= 4 \end{aligned}$$

2. Using 9-rems: $4 = 7 \cdot 7 + 0$
$$\qquad\qquad\qquad 4 = 4$$

3.
$$\begin{array}{r} 47 \nearrow^{2} \; \text{R0} \\ 35\,\overline{)1645} \\ {}_{8}\swarrow\; \underline{140}\,\searrow_{7} \\ 245 \\ \underline{245} \\ 0 \end{array}$$

Check: Using 9-rems
$$\begin{aligned} 7 &= 8 \cdot 2 + 0 \\ &= \;16\; + 0 \\ &= \;\;7\;\; + 0 \\ 7 &= 7 \end{aligned}$$

4. Using 9-rems: $6 = 1 \cdot 6 + 0$
$$\qquad\qquad\qquad 6 = 6$$

5.
$$\begin{array}{r} 35 \nearrow^{8} \; \text{R10} \to 1 \\ 21\,\overline{)745} \\ {}_{3}\swarrow\; \underline{63}\,\searrow_{7} \\ 115 \\ \underline{105} \\ 10 \end{array}$$

Check: Using 9-rems
$$\begin{aligned} 7 &= 3 \cdot 8 + 1 \\ &= \;24\; + 1 \\ &= \;\;6\;\; + 1 \\ 7 &= 7 \end{aligned}$$

6. Using 9-rems: $1 = 4 \cdot 0 + 1$
$$\qquad\qquad\qquad 1 = 1$$

7.
$$\begin{array}{r} 43 \nearrow^{7} \; \text{R3} \\ 12\,\overline{)519} \\ {}_{3}\swarrow\; \underline{48}\,\searrow_{6} \\ 39 \\ \underline{36} \\ 3 \end{array}$$

Check: Using 9-rems
$$\begin{aligned} 6 &= 3 \cdot 7 + 3 \\ &= \;21\; + 3 \\ &= \;\;3\;\; + 3 \\ 6 &= 6 \end{aligned}$$

8. Using 9-rems: $5 = 7 \cdot 8 + 3$
$$\qquad\qquad\qquad = \;\;2\;\; + 3$$
$$\qquad\qquad\qquad 5 = 5$$

9.
$$\begin{array}{r} 48 \nearrow^{3} \; \text{R5} \\ 13\,\overline{)629} \\ {}_{4}\swarrow\; \underline{52}\,\searrow_{8} \\ 109 \\ \underline{104} \\ 5 \end{array}$$

Check: Using 9-rems
$$\begin{aligned} 8 &= 4 \cdot 3 + 5 \\ &= \;12\; + 5 \\ &= \;\;3\;\; + 5 \\ 8 &= 8 \end{aligned}$$

10. Using 9-rems: $3 = 7 \cdot 2 + 7$
$$\qquad\qquad\qquad = \;\;5\;\; + 7$$
$$\qquad\qquad\qquad 3 = 3$$

11.
$$\begin{array}{r} 27 \nearrow^{0} \; \text{R0} \\ 52\,\overline{)1404} \\ {}_{7}\swarrow\; \underline{104}\,\searrow_{0} \\ 364 \\ \underline{364} \\ 0 \end{array}$$

Check: Using 9-rems
$$\begin{aligned} 0 &= 7 \cdot 0 + 0 \\ &= \;\;0\;\; + 0 \\ 0 &= 0 \end{aligned}$$

12. Using 9-rems: $3 = 6 \cdot 8 + 0$
$$\qquad\qquad\qquad = \;\;3\;\; + 0$$
$$\qquad\qquad\qquad 3 = 3$$

13.
$$\begin{array}{r} 67 \nearrow^{4} \; \text{R4} \\ 78\,\overline{)5230} \\ {}_{6}\swarrow\; \underline{468}\,\searrow_{1} \\ 550 \\ \underline{546} \\ 4 \end{array}$$

Check: Using 9-rems
$$\begin{aligned} 1 &= 6 \cdot 4 + 4 \\ &= \;24\; + 4 \\ &= \;\;6\;\; + 4 \\ 1 &= 1 \end{aligned}$$

14. Using 9-rems: $1 = 4 \cdot 7 + 0$
$$\qquad\qquad\qquad = \;\;1\;\; + 0$$
$$\qquad\qquad\qquad 1 = 1$$

15.
$$\begin{array}{r} 53 \nearrow^{8} \; \text{R66} \to 3 \\ 87\,\overline{)4677} \\ {}_{6}\swarrow\; \underline{435}\,\searrow_{6} \\ 327 \\ \underline{261} \\ 66 \end{array}$$

Check: Using 9-rems
$$\begin{aligned} 6 &= 6 \cdot 3 + 3 \\ &= \;48\; + 3 \\ &= \;12\; + 3 \\ &= \;\;3\;\; + 3 \\ 6 &= 6 \end{aligned}$$

1.

Number	Square of number
0	$0^2 = 0 \cdot 0 = 0$
1	$1^2 = 1 \cdot 1 = 1$
2	$2^2 = 2 \cdot 2 = 4$
3	$3^2 = 3 \cdot 3 = 9$
4	$4^2 = 4 \cdot 4 = 16$
5	$5^2 = 5 \cdot 5 = 25$
6	$6^2 = 6 \cdot 6 = 36$
7	$7^2 = 7 \cdot 7 = 49$
8	$8^2 = 8 \cdot 8 = 64$
9	$9^2 = 9 \cdot 9 = 81$
10	$10^2 = 10 \cdot 10 = 100$
11	$11^2 = 11 \cdot 11 = 121$
12	$12^2 = 12 \cdot 12 = 144$
13	$13^2 = 13 \cdot 13 = 169$
14	$14^2 = 14 \cdot 14 = 196$
15	$15^2 = 15 \cdot 15 = 225$
16	$16^2 = 16 \cdot 16 = 256$

2. 5

3. $\sqrt{81} = 9$ because $9^2 = 81$

4. 7

5. $\sqrt{100} = 10$ because $10^2 = 100$

6. 12

7. $\sqrt{196} = 14$ because $14^2 = 196$

8. 16

9. $11^2 = 11 \cdot 11 = 121$

10. 64

1.

Number	Square	Cube	Fourth power
0	$0 \cdot 0 = 0$	$0 \cdot 0 \cdot 0 = 0$	$0 \cdot 0 \cdot 0 \cdot 0 = 0$
1	$1 \cdot 1 = 1$	$1 \cdot 1 \cdot 1 = 1$	$1 \cdot 1 \cdot 1 \cdot 1 = 1$
2	$2 \cdot 2 = 4$	$2 \cdot 2 \cdot 2 = 8$	$2 \cdot 2 \cdot 2 \cdot 2 = 16$
3	$3 \cdot 3 = 9$	$3 \cdot 3 \cdot 3 = 27$	$3 \cdot 3 \cdot 3 \cdot 3 = 81$
4	$4 \cdot 4 = 16$	$4 \cdot 4 \cdot 4 = 64$	$4 \cdot 4 \cdot 4 \cdot 4 = 256$
5	$5 \cdot 5 = 25$	$5 \cdot 5 \cdot 5 = 125$	$5 \cdot 5 \cdot 5 \cdot 5 = 625$
6	$6 \cdot 6 = 36$	$6 \cdot 6 \cdot 6 = 216$	$6 \cdot 6 \cdot 6 \cdot 6 = 1296$
10	$10 \cdot 10 = 100$	$10 \cdot 10 \cdot 10 = 1000$	$10 \cdot 10 \cdot 10 \cdot 10 = 10,000$

2. 3

3. $\sqrt[3]{27} = 3$ because $3^3 = 27$

4. 5

5. $\sqrt[4]{1} = 1$ because $1^4 = 1$

6. 0

7. $\sqrt[3]{1,000} = 10$ because $10^3 = 1,000$

8. 2

9. $7^4 = 7 \cdot 7 \cdot 7 \cdot 7 = 2,401$

10. 512

1. $10 \div 2 \times 5$
    $= 5 \times 5$
    $= 25$

2. 60

3. $3(2^4)$
    $= 3(16)$
    $= 48$

4. 15

5. $10 \cdot 15^2 - 4^3$
    $= 10 \cdot 225 - 64$
    $= 2,250 - 64$
    $= 2,186$

6. 1

7. $(10^2)\sqrt{16} \cdot 5$
    $= (100) \cdot 4 \cdot 5$
    $= 400 \cdot 5$
    $= 2,000$

8. 43

9. $2 \cdot 3 + 3^2 - 4 \cdot 2$
    $= 6 + 9 - 8$
    $= 15 - 8$
    $= 7$

10. 624

11. $4 + 77 \div 11 \cdot 2 - 5$
    $= 4 + 7 \cdot 2 - 5$
    $= 4 + 14 - 5$
    $= 18 - 5$
    $= 13$

12. 161,500

13. $(10^2)\sqrt{4} - 5 + 36$
    $= (100) \cdot 2 - 5 + 36$
    $= 200 - 5 + 36$
    $= 195 + 36$
    $= 231$

14. 14

15. $28 \div 4 \cdot 2(6)$
    $= 7 \cdot 2(6)$
    $= 14(6)$
    $= 84$

1. Inverse

2. Subtraction

3. Dividend

4. Not possible

5. Divisors (or factors)

6. 0

7. Short

8. Trial divisor

9. Not possible

10. Subtraction and division

11. (a)
$$\begin{array}{r} 7{,}5^1 0^1 6 \\ -\ 2{,}7\ 8\ 9 \\ \underline{\ \ \ 1\ 1\ 1\ } \\ 4{,}7\ 1\ 7 \\ \hline 7{,}5\ 0\ 6 \end{array} \text{ add}$$

(b)
$$\begin{array}{r} 8{,}2^1 4^1 7 \\ -\ \ \ 3\ 5\ 8 \\ \underline{\ \ \ 1\ 1\ 1\ } \\ 7{,}8\ 8\ 9 \\ \hline 8{,}2\ 4\ 7 \end{array} \text{ add}$$

(c)
$$\begin{array}{r} 3\ 1{,}5^1 6^1 2 \\ -\ 2\ 0{,}0\ 9\ 9 \\ \underline{\ \ \ \ \ 1\ 1\ } \\ 1\ 1{,}4\ 6\ 3 \\ \hline 3\ 1{,}5\ 6\ 2 \end{array}$$

(d)
$$\begin{array}{r} 7^1 0^1 0^1 0^1 5^1 1 \\ -\ \ \ 2\ 0{,}8\ 9\ 3 \\ \underline{\ \ \ 1\ 1\ 1\ 1\ 1\ } \\ 6\ 7\ 9{,}1\ 5\ 8 \\ \hline 7\ 0\ 0{,}0\ 5\ 1 \end{array}$$

12. (a) 2,632   (b) 854   (c) 35,104 R2

13. (a)
$$\begin{array}{r} 24 \ \ \text{R12} \to 3 \\ 31\overline{)756} \\ \underline{62} \ \ \\ 136 \\ \underline{124} \\ 12 \end{array}$$

Check: Using 9-rems
$0 = 4 \cdot 6 + 3$
$\ \ \ = 24 + 3$
$\ \ \ = 6 + 3$
$0 = 0$

(b)
$$\begin{array}{r} 3 \ \ \text{R45} \to 0 \\ 51\overline{)198} \\ \underline{153} \\ 45 \end{array}$$

Check: Using 9-rems
$0 = 6 \cdot 3 + 0$
$\ \ \ = 0 + 0$
$0 = 0$

(c)
$$\begin{array}{r} 9 \ \ \text{R11} \to 2 \\ 77\overline{)704} \\ \underline{693} \\ 11 \end{array}$$

Check: Using 9-rems
$2 = 5 \cdot 0 + 2$
$\ \ \ = 0 + 2$
$2 = 2$

(d)
$$\begin{array}{r} 37 \ \ \text{R48} \to 3 \\ 84\overline{)3156} \\ \underline{252} \\ 636 \\ \underline{588} \\ 48 \end{array}$$

Check: Using 9-rems
$6 = 3 \cdot 1 + 3$
$\ \ \ = 3 + 3$
$6 = 6$

(e)
$$\begin{array}{r} 476 \ \ \text{R98} \to 8 \\ 126\overline{)60074} \\ \underline{504} \\ 967 \\ \underline{882} \\ 854 \\ \underline{756} \\ 98 \end{array}$$

Check: Using 9-rems
$8 = 0 \cdot 8 + 8$
$\ \ \ = 0 + 8$
$8 = 8$

(f)
$$\begin{array}{r} 4 \ \ \text{R342} \to 0 \\ 708\overline{)3174} \\ \underline{2832} \\ 342 \end{array}$$

Check: Using 9-rems
$6 = 6 \cdot 4 + 0$
$\ \ \ = 6 + 0$
$6 = 6$

14. 5

15.
$$\begin{array}{r} \$758 \text{ per month} \\ 12\overline{)9096} \\ \underline{84} \\ 69 \\ \underline{60} \\ 96 \\ \underline{96} \\ 0 \end{array}$$

16. $48 a month

17. 3 years = 36 months
$$\begin{array}{r} \$56 \text{ a month} \\ 36\overline{)2016} \\ \underline{180} \\ 216 \\ \underline{216} \end{array}$$

18. $20

19.
$$\begin{array}{r} \$58 \\ \times\ 23 \\ \hline 174 \\ 116 \\ \hline \$1334\ \text{Paid} \end{array}$$
$$\begin{array}{r} \$1,350 \\ -\ 1,334 \\ \hline \$16\ \text{Final} \\ \text{payment} \end{array}$$

20. $127

21. (a)
$$\begin{array}{r} \$19\ \text{a month} \\ 24\,\overline{)456} \\ 24 \\ \hline 216 \\ 216 \\ \hline \end{array}$$
(b)
$$\begin{array}{r} \$19 \\ \times\ 17 \\ \hline 133 \\ 19 \\ \hline \$323\ \text{Paid} \end{array}$$
$$\begin{array}{r} \$456 \\ -\ 323 \\ \hline \$133\ \text{Still} \\ \text{owed} \end{array}$$

22. 83

23. $\dfrac{77 + 89 + 95 + 61 + 81 + 92 + 86}{7} = \dfrac{581}{7} = 83$

24. 871,200 times

25.
$$\begin{array}{r} 53,731 \\ -\ 53,408 \\ \hline 323\ \text{mi} \end{array}$$
$$\begin{array}{r} 17\ \text{mi per gal} \\ 19\,\overline{)323} \\ 19 \\ \hline 133 \\ 133 \\ \hline 0 \end{array}$$

26. 7 on upper limb, 5 on lower limb

SOLUTIONS FOR CHAPTER 2 DIAGNOSTIC TEST (page 94)

Following each problem number is the textbook section reference (in parentheses) where that kind of problem is discussed.

1. (201)
$$\begin{array}{r} 8\ \ \text{minuend} \\ -\ 2\ \ \text{subtrahend} \\ \hline 6\ \ \text{difference} \end{array}$$

2. (202) (a)
$$\begin{array}{r} 4,768 \\ -\ 3,205 \\ \hline 1,563 \end{array}$$

3. (202)
$$\begin{array}{r} \overset{1\ 1\ 11}{71,304} \\ -\ 67,856 \\ \hline {}^{11\ 11}\ \ \\ 3,448 \end{array}$$

(b)
$$\begin{array}{r} \overset{1\ 1}{3,564} \\ -\ \ \ 782 \\ \hline {}^{1\ 1}\ \ \\ 2,782 \end{array}$$

4. (207B)
$$\begin{array}{r} 14 \leftarrow \text{quotient} \\ \text{divisor} \rightarrow 17\,\overline{)243} \leftarrow \text{dividend} \\ 17 \\ \hline 73 \\ 68 \\ \hline 5 \leftarrow \text{remainder} \end{array}$$

(c)
$$\begin{array}{r} \overset{1\ 11}{50,406} \\ -\ 35,008 \\ \hline {}^{1\ 11}\ \ \\ 15,398 \end{array}$$

5. (207C) (a) $\dfrac{0}{5} = 0$ (b) $\dfrac{4}{4} = 1$ (c) $\dfrac{7}{0} = ?$ Not possible

6. (207B) 1, 2, 3, 4, 6, 12

7. (a) (208)
$$\begin{array}{r} 9,800\ \ \text{R7} \\ 8\,\overline{)78{,}407} \end{array}$$
(b) (209)
$$\begin{array}{r} 34 \\ 65\,\overline{)2,210} \\ 1\ 95 \\ \hline 260 \\ 260 \\ \hline 0 \end{array}$$

(c) (209)
$$\begin{array}{r} 706\ \ \text{R126} \\ 495\,\overline{)349,596} \\ 346\ 5 \\ \hline 3\ 096 \\ 2\ 970 \\ \hline 126 \end{array}$$

8. (209)
$$\begin{array}{r} 62\ \text{hr} \\ 38\,\overline{)2,356} \\ 2\ 28 \\ \hline 76 \\ 76 \\ \hline 0 \end{array}$$

9. (209) 3 yr = 36 mo
$$\begin{array}{r} \$56\ \text{per mo} \\ 36\,\overline{)2,016} \\ 1\ 80 \\ \hline 216 \\ 216 \\ \hline 0 \end{array}$$

10. (210) Average $= \dfrac{73 + 84 + 88 + 92 + 68}{5} = \dfrac{405}{5} = 81$

11. (215) (a) 5, because $5^2 = 5 \times 5 = 25$

(b) 2, because $2^3 = 2 \cdot 2 \cdot 2 = 8$

(c) 10, because $10^2 = 10 \cdot 10 = 100$

(d) 2, because $2^4 = 2 \cdot 2 \cdot 2 \cdot 2 = 16$

12. (216) (a) $100 \div 4 \cdot 5 - 3$
$= 25 \cdot 5 - 3 = 125 - 3 = 122$

(b) $2(5^2) - 3\sqrt{16}$
$= 2(25) - 3 \cdot 4 = 50 - 12 = 38$

(c) $2^4 + 10^2 + \sqrt[3]{8}$
$= 16 + 100 + 2 = 118$

13. (217)
$$\begin{array}{r} \$58 \\ \times\ 23 \\ \hline 174 \\ 116 \\ \hline \$1334 \end{array}$$
$$\begin{array}{r} \$1,350 \\ -\ 1,334 \\ \hline \$16\ \text{Final payment} \end{array}$$

EXERCISES 301, SET I (page 102)

1. $\dfrac{3}{8}$ 2. $\dfrac{3}{5}$ 3. $\dfrac{1}{6}$

4. $\dfrac{3}{11}$ 5. $\dfrac{3}{7}$ 6. $\dfrac{5}{8}$

7. $\dfrac{5}{11},\ \dfrac{17}{22},\ \dfrac{1}{2},\ \dfrac{98}{107},\ \dfrac{1}{31}$ 8. $\dfrac{28}{13},\ \dfrac{8}{8},\ \dfrac{316}{219},\ \dfrac{4}{4}$

9. 5   10. 13

11. 17 and 22   12. $\dfrac{8}{8} = \dfrac{4}{4}$

EXERCISES 302, SET I (page 103)

1. $\dfrac{2}{3} \cdot \dfrac{5}{7} = \dfrac{2 \cdot 5}{3 \cdot 7} = \dfrac{10}{21}$ 2. $\dfrac{5}{6}$

3. $\dfrac{3}{4} \cdot \dfrac{7}{8} = \dfrac{3 \cdot 7}{4 \cdot 8} = \dfrac{21}{32}$ 4. $\dfrac{15}{16}$

5. $\dfrac{4}{5} \cdot \dfrac{3}{5} = \dfrac{4 \cdot 3}{5 \cdot 5} = \dfrac{12}{25}$ 6. $\dfrac{35}{48}$

7. $\dfrac{4}{9} \cdot \dfrac{2}{3} = \dfrac{4 \cdot 2}{9 \cdot 3} = \dfrac{8}{27}$ 8. $\dfrac{35}{27}$

9. $\dfrac{3}{4} \cdot \dfrac{3}{4} = \dfrac{3 \cdot 3}{4 \cdot 4} = \dfrac{9}{16}$ 10. $\dfrac{36}{49}$

11. $\dfrac{5}{12} \cdot \dfrac{5}{8} = \dfrac{5 \cdot 5}{12 \cdot 8} = \dfrac{25}{96}$ 12. $\dfrac{15}{128}$

13. $\dfrac{11}{32} \cdot \dfrac{3}{2} = \dfrac{11 \cdot 3}{32 \cdot 2} = \dfrac{33}{64}$ 14. $\dfrac{91}{180}$

15. $\dfrac{7}{8} \cdot \dfrac{11}{13} = \dfrac{7 \cdot 11}{8 \cdot 13} = \dfrac{77}{104}$ 16. $\dfrac{555}{688}$

17. $\dfrac{24}{23} \cdot \dfrac{6}{17} = \dfrac{24 \cdot 6}{23 \cdot 17} = \dfrac{144}{391}$ 18. $\dfrac{779}{986}$

19. $\dfrac{81}{28} \cdot \dfrac{13}{55} = \dfrac{81 \cdot 13}{28 \cdot 55} = \dfrac{1,053}{1,540}$ 20. $\dfrac{19,071}{24,276}$

21. $A = L \times W = \dfrac{11}{15} \times \dfrac{2}{5} = \dfrac{11 \cdot 2}{15 \cdot 5} = \dfrac{22}{75}$ sq mi

22. $\dfrac{1}{2}$ sq in

EXERCISES 303A, SET I (page 105)

1. $2\overline{)8}\;^4$ 2. 3   3. $6\overline{)24}\;^4$ 4. 6

5. $35\overline{)35}\;^1$ 6. 12   7. $16\overline{)144}\;^9$ $\underline{144}$ 0 8. 1

9. $97\overline{)2231}\;^{23}$
$\underline{194}$
291
$\underline{291}$
0

10. 234

EXERCISES 303B, SET I (page 106)

1. $2 \cdot \frac{1}{3} = \frac{2}{1} \cdot \frac{1}{3} = \frac{2 \cdot 1}{1 \cdot 3} = \frac{2}{3}$ 2. $\frac{6}{7}$

3. $\frac{1}{5} \cdot 4 = \frac{1}{5} \cdot \frac{4}{1} = \frac{1 \cdot 4}{5 \cdot 1} = \frac{4}{5}$ 4. $\frac{5}{8}$

5. $3 \cdot \frac{5}{2} = \frac{3}{1} \cdot \frac{5}{2} = \frac{3 \cdot 5}{1 \cdot 2} = \frac{15}{2}$ 6. $\frac{18}{5}$

7. $\frac{7}{12} \cdot 7 = \frac{7}{12} \cdot \frac{7}{1} = \frac{7 \cdot 7}{12 \cdot 1} = \frac{49}{12}$ 8. $\frac{15}{16}$

9. $5 \cdot \frac{1}{12} = \frac{5}{1} \cdot \frac{1}{12} = \frac{5 \cdot 1}{1 \cdot 12} = \frac{5}{12}$ 10. $\frac{27}{32}$

11. $\frac{2}{3} \cdot 420 = \frac{2}{3} \cdot \frac{420}{1} = \frac{840}{3} = 3\overline{)840}\;^{280}$ 12. $\frac{4}{5}$ sq mi

13. $A = L \times W$
$A = 120 \times 60 = 7,200$ sq yd

$\cdot \frac{2}{3} \cdot \frac{7200}{1} = \frac{14400}{3} = 3\overline{)14400}\;^{4800}$ sq yd reseeded

EXERCISES 305, SET I (page 108)

1. $3\overline{)5}\;^{1\;R2} = 1\frac{2}{3}$ 2. $1\frac{3}{4}$

3. $5\overline{)9}\;^{1\;R4} = 1\frac{4}{5}$ 4. $2\frac{3}{4}$

5. $5\overline{)13}\;^{2\;R3} = 2\frac{3}{5}$ 6. $3\frac{2}{3}$

7. $4\overline{)15}\;^{3\;R3} = 3\frac{3}{4}$ 8. $17\frac{1}{2}$

9. $6\overline{)23}\;^{3\;R5} = 3\frac{5}{6}$ 10. $9\frac{5}{9}$

11. $13\overline{)16}\;^{1\;R3} = 1\frac{3}{13}$ 12. $1\frac{9}{23}$

13. $7\overline{)20}\;^{2\;R6} = 2\frac{6}{7}$ 14. $3\frac{11}{15}$

15. $19\overline{)207}\;^{10\;R17} = 10\frac{17}{19}$
$\underline{19}$
17 16. $1\frac{16}{21}$

17. $17\overline{)54}\;^{3\;R3} = 3\frac{3}{17}$
$\underline{51}$
3 18. $2\frac{25}{31}$

19. $45\overline{)136}\;^{3\;R1} = 3\frac{1}{45}$
$\underline{135}$
1 20. $3\frac{45}{64}$

EXERCISES 306, SET I (page 109)

1. $1\frac{1}{2} = \frac{1 \cdot 2 + 1}{2} = \frac{2 + 1}{2} = \frac{3}{2}$ 2. $\frac{13}{5}$

3. $3\frac{1}{4} = \frac{3 \cdot 4 + 1}{4} = \frac{12 + 1}{4} = \frac{13}{4}$ 4. $\frac{21}{8}$

5. $4\frac{5}{6} = \frac{4 \cdot 6 + 5}{6} = \frac{24 + 5}{6} = \frac{29}{6}$ 6. $\frac{9}{2}$

7. $3\frac{7}{10} = \frac{3 \cdot 10 + 7}{10} = \frac{30 + 7}{10} = \frac{37}{10}$ 8. $\frac{53}{16}$

9. $5\frac{7}{12} = \frac{5 \cdot 12 + 7}{12} = \frac{60 + 7}{12} = \frac{67}{12}$ 10. $\frac{44}{7}$

11. $12\frac{2}{3} = \frac{12 \cdot 3 + 2}{3} = \frac{36 + 2}{3} = \frac{38}{3}$ 12. $\frac{48}{13}$

13. $6\frac{3}{4} = \frac{6 \cdot 4 + 3}{4} = \frac{24 + 3}{4} = \frac{27}{4}$ 14. $\frac{49}{11}$

15. $3\frac{7}{15} = \frac{3 \cdot 15 + 7}{15} = \frac{45 + 7}{15} = \frac{52}{15}$ 16. $\frac{25}{17}$

17. $15\frac{23}{44} = \frac{15 \cdot 44 + 23}{44} = \frac{660 + 23}{44}$
$= \frac{683}{44}$ 18. $\frac{710}{33}$

19. $2\frac{8}{63} = \frac{2 \cdot 63 + 8}{63} = \frac{126 + 8}{63} = \frac{134}{63}$ 20. $\frac{959}{117}$

EXERCISES 307, SET I (page 111)

1. Yes, because $2 \cdot 12 = 3 \cdot 8$ 2. Yes
3. No, because $7 \cdot 4 \neq 27 \cdot 1$ 4. No
   $7 \cdot 4 > 27 \cdot 1$, therefore $\frac{7}{27} > \frac{1}{4}$

5. Yes, because $25 \cdot 6 = 30 \cdot 5$ 6. Yes
   $150 = 150$
7. Yes, because $30 \cdot 28 = 35 \cdot 24$ 8. No
   $840 = 840$
9. No, because $8 \cdot 18 \neq 12 \cdot 10$ 10. No
   $144 \neq 120$
   $8 \cdot 18 > 12 \cdot 10$; therefore $\frac{8}{12} > \frac{10}{18}$
11. Yes, because $8 \cdot 15 = 10 \cdot 12$
   $120 = 120$
12. Yes
13. No, because $28 \cdot 30 \neq 40 \cdot 24$
   $840 \neq 960$
   $28 \cdot 30 < 40 \cdot 24$; therefore $\frac{28}{40} < \frac{24}{30}$
14. Yes
15. No, because $54 \cdot 56 \neq 70 \cdot 48$
   $3,024 \neq 3,360$
   $54 \cdot 56 < 70 \cdot 48$; therefore $\frac{54}{70} < \frac{48}{56}$

1. $\frac{1}{2} = \frac{2}{4}$ Since the denominator must be multiplied by 2 to give 4, we must multiply the numerator by 2, giving 2.

2. $\frac{1}{2} = \frac{3}{6}$

3. $\frac{1}{2} = \frac{5}{10}$ Since the denominator must be multiplied by 5 to give 10, we must multiply the numerator by 5, giving 5.

4. $\frac{1}{3} = \frac{2}{6}$

5. $\frac{2}{3} = \frac{4}{6}$ Since the denominator was multiplied by 2 to give 6, we must also multiply the numerator by 2, giving 4.

6. $\frac{2}{3} = \frac{4}{6}$

7. $\frac{1}{4} = \frac{3}{12}$ Since the numerator was multiplied by 3 to give 3, we must also multiply the denominator by 3, giving 12.

8. $\frac{3}{4} = \frac{30}{40}$

9. $\frac{2}{5} = \frac{8}{20}$ Since the numerator was multiplied by 4 to give 8, we must also multiply the denominator by 4, giving 20.

10. $\frac{5}{6} = \frac{10}{12}$

11. $\frac{3}{5} = \frac{9}{15}$ Numerator and denominator multiplied by 3.

12. $\frac{4}{7} = \frac{24}{42}$

13. $\frac{9}{13} = \frac{36}{52}$ By dividing $13\overline{)52}$ ^{4}, we find that the numerator must be multiplied by 4.

14. $\frac{12}{16} = \frac{60}{80}$

15. $\frac{15}{18} = \frac{30}{36}$ The denominator was multiplied by 2; therefore, the numerator must be multiplied by 2.

16. $\frac{5}{9} = \frac{35}{63}$

17. $\frac{8}{11} = \frac{40}{55}$ Numerator and denominator multiplied by 5.

18. $\frac{14}{25} = \frac{42}{75}$

19. $\frac{20}{26} = \frac{40}{52}$ Numerator and denominator multiplied by 2.

20. $\frac{6}{15} = \frac{48}{120}$

1. $\frac{3}{6} = \frac{1}{2}$ Since the denominator 6 was divided by 3 to give 2, we must divide the numerator by 3.

2. $\frac{2}{4} = \frac{1}{2}$

3. $\frac{6}{8} = \frac{3}{4}$ Since the denominator 8 was divided by 2 to give 4, we must divide the numerator by 2.

4. $\frac{2}{10} = \frac{1}{5}$

5. $\frac{6}{10} = \frac{3}{5}$ Since the denominator 10 was divided by 2 to give 5, we must divide the numerator by 2.

6. $\frac{9}{12} = \frac{3}{4}$

7. $\frac{6}{16} = \frac{3}{8}$ Since the denominator 16 was divided by 2 to give 8, we must divide the numerator by 2.

8. $\frac{15}{20} = \frac{3}{4}$

9. $\frac{14}{20} = \frac{7}{10}$ Both numerator and denominator were divided by 2.

10. $\frac{10}{16} = \frac{5}{8}$

11. $\frac{18}{45} = \frac{6}{15}$ Both numerator and denominator were divided by 3.

12. $\frac{63}{35} = \frac{9}{5}$

13. $\frac{36}{54} = \frac{6}{9}$ Both numerator and denominator were divided by 6.

14. $\frac{6}{14} = \frac{3}{7}$

15. $\frac{72}{24} = \frac{9}{3}$ Divide numerator and denominator by 8.

16. $\frac{80}{100} = \frac{20}{25}$

17. $\frac{85}{34} = \frac{5}{2}$ Divide numerator and denominator by 17.

18. $\frac{27}{78} = \frac{9}{26}$

19. $\frac{33}{77} = \frac{3}{7}$ Divide numerator and denominator by 11.

20. $\frac{72}{60} = \frac{6}{5}$

EXERCISES 310, SET I (page 116)

1. $2\overline{)5} \;\; 2\text{ R1} = 2\frac{1}{2}$

2. $1\frac{3}{4}$

3. $3\overline{)8} \;\; 2\text{ R2} = 2\frac{2}{3}$

4. $3\frac{3}{5}$

5. $16\overline{)25} \;\; 1\text{ R9},\; \underline{16},\; 9 = 1\frac{9}{16}$

6. $4\frac{7}{19}$

7. $1\frac{2}{3} = \frac{1 \cdot 3 + 2}{3} = \frac{3 + 2}{3} = \frac{5}{3}$

8. $\frac{9}{4}$

9. $4\frac{1}{2} = \frac{4 \cdot 2 + 1}{2} = \frac{8 + 1}{2} = \frac{9}{2}$

10. $\frac{48}{5}$

11. $3\frac{7}{12} = \frac{3 \cdot 12 + 7}{12} = \frac{36 + 7}{12} = \frac{43}{12}$

12. $\frac{403}{16}$

13. $\frac{7}{8} \cdot \frac{3}{4} = \frac{7 \cdot 3}{8 \cdot 4} = \frac{21}{32}$

14. $\frac{16}{27}$

15. $\frac{7}{3} \cdot \frac{5}{13} = \frac{7 \cdot 5}{3 \cdot 13} = \frac{35}{39}$

16. $2\frac{2}{35}$

17. $\frac{17}{32} \cdot \frac{21}{13} = \frac{17 \cdot 21}{32 \cdot 13} = \frac{357}{416}$

18. $\frac{2{,}438}{3{,}071}$

19. $\frac{7}{9} = \frac{56}{72}$ Since 72 is 8 times 9, the numerator must be multiplied by 8.

20. $\frac{24}{16} = \frac{6}{4}$

21. $\frac{3}{5} = \frac{12}{20}$ Since 12 is 4 times 3, the denominator must be multiplied by 4.

22. $\frac{18}{27} = \frac{2}{3}$

23. $\frac{2}{7} = \frac{12}{42}$ Since 42 is 6 times 7, the numerator must be multiplied by 6.

24. $\frac{21}{36} = \frac{7}{12}$

25. $\frac{16}{27} = \frac{32}{54}$ Since 32 is 2 times 16, the denominator must be multiplied by 2.

26. $\frac{33}{15} = \frac{11}{5}$

27. $\frac{4}{3} = \frac{72}{54}$ Since 54 is 18 times 3, the numerator must be multiplied by 18.

28. $\frac{45}{80} = \frac{9}{16}$

29. $\frac{7}{9} = \frac{56}{72}$ Since the numerator 7 is multiplied by 8 to give 56, the denominator must be multiplied by 8.

30. $\frac{56}{40} = \frac{7}{5}$

31. $\frac{15}{11} = \frac{135}{99}$ Since the denominator is multiplied by 9 to give 99, the numerator must also be multiplied by 9.

32. $\frac{35}{84} = \frac{5}{12}$

33. $\frac{23}{19}$, $\frac{19}{18}$, and $\frac{12}{12}$

34. Not equivalent

35. Equivalent because $24 \cdot 30 = 40 \cdot 18$
$$720 = 720$$

36. All are equivalent

37. $\frac{27}{18}$ and $\frac{45}{30}$ are equivalent because $27 \cdot 30 = 18 \cdot 45$
$$810 = 810$$

38. $\frac{15}{32}$ sq yd

39. $\frac{2}{5}$

40. $\frac{5}{12}$

41. $\frac{1}{6}$

42. $\frac{17}{52}$

43. $\frac{2}{7}$

EXERCISES 311A, SET I (page 119)

1. $\frac{\overset{2}{\cancel{6}}}{\underset{3}{\cancel{9}}} = \frac{2}{3}$ Both 6 and 9 were divided by 3.

2. $\frac{3}{4}$

3. $\frac{\overset{3}{\cancel{12}}}{\underset{4}{\cancel{16}}} = \frac{3}{4}$ Both 12 and 16 were divided by 4.

4. $\frac{2}{3}$

5. $\frac{\overset{3}{\cancel{30}}}{\underset{4}{\cancel{40}}} = \frac{3}{4}$ Both 30 and 40 were divided by 10.

6. $\frac{4}{5}$

7. $\frac{\overset{3}{\cancel{24}}}{\underset{4}{\cancel{32}}} = \frac{3}{4}$ Both 24 and 32 were divided by 8.

8. $\frac{2}{5}$

9. $\frac{\overset{5}{\cancel{10}}}{\underset{4}{\cancel{8}}} = \frac{5}{4} = 1\frac{1}{4}$

10. $\frac{2}{3}$

11. $\frac{\overset{4}{\cancel{32}}}{\underset{5}{\cancel{40}}} = \frac{4}{5}$

12. $\frac{3}{4}$

13. $\frac{\cancel{54}}{\cancel{90}} = \frac{3}{5}$

14. $\frac{3}{7}$

15. $\frac{\cancel{84}}{\cancel{105}} = \frac{4}{5}$

16. $\frac{1}{2}$

17. $\frac{\overset{3}{\cancel{33}}}{\underset{5}{\cancel{55}}} = \frac{3}{5}$

18. $\frac{2}{3}$

19. $\frac{\cancel{210}}{\cancel{270}} = \frac{7}{9}$

20. $\frac{5}{7}$

EXERCISES 311B, SET I (page 120)

1. $\frac{2}{\underset{1}{\cancel{3}}} \cdot \frac{\overset{2}{\cancel{6}}}{7} = \frac{4}{7}$

2. $\frac{1}{4}$

3. $\frac{\overset{1}{\cancel{6}}}{\underset{2}{\cancel{8}}} \cdot \frac{\overset{1}{\cancel{4}}}{\underset{3}{\cancel{9}}} = \frac{1}{3}$

4. $1\frac{1}{2}$

5. $\frac{\overset{1}{\cancel{10}}}{\underset{2}{\cancel{18}}} \cdot \frac{\overset{1}{\cancel{9}}}{\underset{1}{\cancel{5}}} = 1$

6. $3\frac{3}{4}$

7. $\frac{\overset{5}{\cancel{35}}}{\underset{8}{\cancel{72}}} \cdot \frac{\overset{1}{\cancel{18}}}{\underset{8}{\cancel{56}}} = \frac{5}{32}$

8. $\frac{4}{5}$

9. $\frac{\overset{2}{\cancel{22}}}{\underset{15}{\cancel{75}}} \cdot \frac{\overset{5}{\cancel{20}}}{\underset{11}{\cancel{44}}} = \frac{2}{15}$

10. $\frac{1}{16}$

11. $\frac{\overset{1}{\cancel{5}}}{\underset{1}{\cancel{7}}} \cdot \frac{\cancel{3}}{\underset{2}{\cancel{10}}} \cdot \frac{\overset{1}{\cancel{14}}}{\underset{5}{\cancel{15}}} = \frac{1}{5}$

12. $\frac{4}{9}$

13. $\frac{\cancel{16}}{\underset{7}{\cancel{49}}} \cdot \frac{\cancel{14}}{\cancel{8}} \cdot \frac{\overset{3}{\cancel{9}}}{\cancel{12}} = \frac{3}{7}$

14. $\frac{8}{9}$

15. $\dfrac{\overset{1}{\cancel{4}}}{\underset{5}{\cancel{15}}} \cdot \dfrac{\overset{1}{\cancel{10}}}{\cancel{28}} \cdot \dfrac{\overset{\cancel{3}/1}{\cancel{21}}}{\underset{1}{\cancel{10}}} = \dfrac{1}{5}$ 16. $\dfrac{1}{9}$

EXERCISES 311C, SET I (page 121)

1. $\dfrac{\overset{13}{\cancel{39}}}{\underset{17}{\cancel{51}}}$ 39 is divisible by 3.
51 is divisible by 3, because $5 + 1 = 6$, which is divisible by 3.

2. $\dfrac{31}{37}$ 3. $\dfrac{\overset{7}{\cancel{35}}}{\underset{11}{\cancel{55}}} = \dfrac{7}{11}$ 4. $\dfrac{3}{4}$

5. $\dfrac{\overset{\overset{3}{\cancel{9}}}{\cancel{45}}}{\underset{\underset{5}{\cancel{15}}}{\cancel{75}}} = \dfrac{3}{5}$ 6. $\dfrac{21}{130}$ 7. $\dfrac{\overset{\overset{\overset{1}{\cancel{11}}}{\cancel{22}}}{\cancel{44}}}{\underset{\underset{\underset{6}{\cancel{66}}}{\cancel{132}}}{\cancel{264}}} = \dfrac{1}{6}$

8. $\dfrac{17}{43}$ 9. $\dfrac{\overset{29}{\cancel{290}}}{\underset{61}{\cancel{610}}} = \dfrac{29}{61}$ 10. $\dfrac{17}{23}$

11. $\dfrac{\overset{\overset{47}{\cancel{141}}}{\cancel{423}}}{\underset{\underset{59}{\cancel{177}}}{\cancel{531}}} = \dfrac{47}{59}$ 12. $\dfrac{37}{47}$

13. $\dfrac{\overset{\overset{7}{\cancel{35}}}{\cancel{175}}}{\underset{\underset{13}{\cancel{65}}}{\cancel{325}}} = \dfrac{7}{13}$ 14. $\dfrac{8}{15}$

15. $\dfrac{583}{689}$ The rules of divisibility will not help here because the smallest number that will divide into each of these numbers is 53. To reduce this fraction, we will use Euclid's Algorithm, which is discussed next.

EXERCISES 311D, SET I (page 123)

1.
```
        1              7             9
  182│234       26│182       26│234
     182    3       182           234
      52│182         0             0
        156   2
      →26│52
           52
            0
```
└─largest number that divides both 182 and 234

Therefore, $\dfrac{182}{234} = \dfrac{182 \div 26}{234 \div 26} = \dfrac{7}{9}$

2. $\dfrac{11}{13}$

3. $\dfrac{\overset{111}{\cancel{555}}}{\underset{164}{\cancel{820}}} = \dfrac{111}{164}$

```
        1
  111│164
     111   2
      53│111
         106  10
          5│53
            50  1
             3│5
              3   1
              2│3
                2   2
                1│2
                  2
                  0
```
This shows that $\dfrac{111}{164}$ is in lowest terms because 1 is the largest number that divides both 111 and 164.

4. $\dfrac{2}{3}$

5. $\dfrac{\overset{\overset{72}{\cancel{144}}}{\cancel{576}}}{\underset{\underset{82}{\cancel{164}}}{\underset{\cancel{328}}{\cancel{656}}}} = \dfrac{\overset{36}{\cancel{72}}}{\underset{41}{\cancel{82}}} = \dfrac{36}{41}$ 6. $\dfrac{236}{265}$

7.
```
        1              4             5
  188│235       47│188       47│235
     188    4       188           235
      →47│188        0             0
          188
            0
```
└─largest number that divides both 188 and 235

Therefore, $\dfrac{188}{235} = \dfrac{188 \div 47}{235 \div 47} = \dfrac{4}{5}$

8. $\dfrac{2}{3}$

9. $\dfrac{\overset{118}{\cancel{354}}}{\underset{295}{\cancel{885}}} = \dfrac{118}{295}$

```
        2              2             5
  118│295       59│118       59│295
     236   2       118           295
      →59│118        0             0
          118
            0
```
└─largest number that divides both 118 and 295

Therefore, $\dfrac{354}{885} = \dfrac{118}{295} = \dfrac{118 \div 59}{295 \div 59} = \dfrac{2}{5}$

10. $\dfrac{9}{14}$

11. Divide smaller term into larger term.

```
         1              19            23
  1577│1909      83│1577      83│1909
      1577   4       83           166
       332│1577      747          249
         1328   1     747          249
          249│332     0             0
            249   3
            →83│249
               249
                 0
```
└─largest number that divides both 1,577 and 1,909

Therefore, $\dfrac{1,909}{1,577} = \dfrac{1,909 \div 83}{1,577 \div 83} = \dfrac{23}{19}$

12. $\dfrac{73}{91}$

EXERCISES 312, SET I (page 125)

1. $\dfrac{1}{6} + \dfrac{3}{6} = \dfrac{1+3}{6} = \dfrac{\overset{2}{\cancel{4}}}{\underset{3}{\cancel{6}}} = \dfrac{2}{3}$ 2. $\dfrac{1}{2}$

3. $\dfrac{3}{5} + \dfrac{2}{5} = \dfrac{3+2}{5} = \dfrac{5}{5} = 1$ 4. 1

5. $\dfrac{5}{6} + \dfrac{5}{6} = \dfrac{5+5}{6} = \dfrac{\overset{5}{\cancel{10}}}{\underset{3}{\cancel{6}}} = \dfrac{5}{3} = 1\dfrac{2}{3}$ 6. $1\dfrac{2}{5}$

7. $\dfrac{1}{2} + \dfrac{3}{2} + \dfrac{5}{2} = \dfrac{1+3+5}{2} = \dfrac{9}{2} = 4\dfrac{1}{2}$ 8. $2\dfrac{2}{3}$

9. $\dfrac{5}{8} + \dfrac{4}{8} + \dfrac{7}{8} = \dfrac{5+4+7}{8} = \dfrac{16}{8} = 2$ 10. $1\dfrac{1}{2}$

11. $\dfrac{3}{15} + \dfrac{1}{15} + \dfrac{6}{15} = \dfrac{3+1+6}{15} = \dfrac{\overset{2}{\cancel{10}}}{\underset{3}{\cancel{15}}} = \dfrac{2}{3}$ 12. $\dfrac{5}{8}$

13. $\dfrac{1}{12} + \dfrac{5}{12} + \dfrac{2}{12} + \dfrac{3}{12} = \dfrac{1+5+2+3}{12} = \dfrac{11}{12}$ 14. $\dfrac{3}{4}$

15. $\dfrac{35}{80} + \dfrac{27}{80} = \dfrac{35+27}{80} = \dfrac{\overset{31}{\cancel{62}}}{\underset{40}{\cancel{80}}} = \dfrac{31}{40}$ 16. $1\dfrac{1}{15}$

EXERCISES 313, SET I (page 126)

1. $\dfrac{3}{4} - \dfrac{1}{4} = \dfrac{3-1}{4} = \dfrac{2}{4} = \dfrac{1}{2}$

2. $\dfrac{2}{3}$

3. $\dfrac{7}{8} - \dfrac{3}{8} = \dfrac{7-3}{8} = \dfrac{4}{8} = \dfrac{1}{2}$

4. 0

5. $\dfrac{5}{3} - \dfrac{2}{3} = \dfrac{5-2}{3} = \dfrac{3}{3} = 1$

6. $1\dfrac{1}{5}$

7. $\dfrac{7}{10} - \dfrac{5}{10} = \dfrac{7-5}{10} = \dfrac{2}{10} = \dfrac{1}{5}$

8. $\dfrac{2}{3}$

9. $\dfrac{6}{7} - \dfrac{6}{7} = \dfrac{6-6}{7} = \dfrac{0}{7} = 0$

10. $\dfrac{1}{3}$

11. $\dfrac{9}{14} - \dfrac{5}{14} = \dfrac{9-5}{14} = \dfrac{4}{14} = \dfrac{2}{7}$

12. $\dfrac{3}{5}$

13. $\dfrac{45}{52} - \dfrac{6}{52} = \dfrac{45-6}{52} = \dfrac{39}{52} = \dfrac{39 \div 13}{52 \div 13} = \dfrac{3}{4}$

14. $\dfrac{17}{27}$

$$39\overline{)52} \qquad 13\overline{)39} \qquad 13\overline{)52}$$

$$\dfrac{39}{13}\overline{)39} \qquad \dfrac{39}{0} \qquad \dfrac{52}{0}$$

$$\dfrac{39}{}$$

└─ largest number that divides both 39 and 52

15. $\dfrac{123}{144} - \dfrac{43}{144} = \dfrac{123-43}{144} = \dfrac{80}{144} = \dfrac{20}{36} = \dfrac{5}{9}$

16. $\dfrac{9}{47}$

EXERCISES 314, SET I (page 127)

1. Prime
2. Composite
3. Prime
4. Composite
5. Composite, because $12 = 3 \cdot 2 \cdot 2$
6. Prime
7. Composite, because $18 = 2 \cdot 3 \cdot 3$
8. Prime
9. Prime
10. Composite
11. Composite, because $55 = 5 \cdot 11$
12. Prime
13. Composite, because $49 = 7 \cdot 7$
14. Prime
15. Composite, because $51 = 3 \cdot 17$
16. Composite
17. Composite, because $111 = 3 \cdot 37$
18. Prime

EXERCISES 315A, SET I (page 130)

1.

2	2	3	4
	1	3	2

LCD = $2 \cdot 1 \cdot 3 \cdot 2 = 12$

2. 20

3.

2	2	8	4
2	1	4	2
	1	2	1

LCD = $2 \cdot 2 \cdot 1 \cdot 2 \cdot 1 = 8$

4. 18

5.

5	3	5	15
3	3	1	3
	1	1	1

LCD = $5 \cdot 3 \cdot 1 \cdot 1 \cdot 1 = 15$

6. 14

7.

2	14	10
	7	5

LCD = $2 \cdot 7 \cdot 5 = 70$

8. 48

9.

7, 5 — Since no prime number divides into both 7 and 5, the LCD = $7 \cdot 5 = 35$.

10. 36

11.

2	6	8	9
3	3	4	9
	1	4	3

LCD = $2 \cdot 3 \cdot 1 \cdot 4 \cdot 3 = 72$

12. 180

13.

5	40	15	25
	8	3	5

LCD = $5 \cdot 8 \cdot 3 \cdot 5 = 600$

14. 105

15.

4, 5, 21 — Since no prime number divides into any two of these numbers, the LCD = $4 \cdot 5 \cdot 21 = 420$.

16. 36

17.

2	6	13	26
13	3	13	13
	3	1	1

LCD = $2 \cdot 13 \cdot 3 = 78$

18. 4,410

19.

3	66	33	132
2	22	11	44
11	11	11	22
	1	1	2

LCD = $3 \cdot 2 \cdot 11 \cdot 1 \cdot 1 \cdot 2 = 132$

20. 1,680

EXERCISES 315B, SET I (page 132)

1.

$$\dfrac{1}{2} = \dfrac{2}{4}$$
$$+ \dfrac{3}{4} = \dfrac{3}{4}$$
$$\dfrac{5}{4} = 1\dfrac{1}{4}$$

2. $1\dfrac{1}{6}$

3.

$$\dfrac{3}{5} = \dfrac{6}{10}$$
$$+ \dfrac{3}{10} = \dfrac{3}{10}$$
$$\dfrac{9}{10}$$

4. $1\dfrac{1}{8}$

5.

$$\dfrac{2}{3} = \dfrac{8}{12}$$
$$+ \dfrac{1}{4} = \dfrac{3}{12}$$
$$\dfrac{11}{12}$$

6. $\dfrac{11}{15}$

7.

$$\dfrac{3}{4} = \dfrac{9}{12}$$
$$+ \dfrac{1}{12} = \dfrac{1}{12}$$
$$\dfrac{10}{12} = \dfrac{5}{6}$$

8. $1\dfrac{1}{12}$

9.

$$\dfrac{3}{6} = \dfrac{1}{2} = \dfrac{3}{6}$$
$$+ \dfrac{5}{15} = \dfrac{1}{3} = \dfrac{2}{6}$$
$$\dfrac{5}{6}$$

10. $\dfrac{1}{2}$

11. $\dfrac{5}{16} = \dfrac{25}{80}$

$+\ \dfrac{3}{5} = \dfrac{48}{80}$

$\overline{\qquad\quad \dfrac{73}{80}}$

Since 16 and 5 have no common factor, the LCD = 5 · 16 = 80.

12. $1\dfrac{11}{42}$

13. $\dfrac{3}{20} = \dfrac{6}{40}$

$+\ \dfrac{1}{8} = \dfrac{5}{40}$

$\overline{\qquad\quad \dfrac{11}{40}}$

$\begin{array}{r|rr} 2 & 20 & 8 \\ 2 & 10 & 4 \\ \hline & 5 & 2 \end{array}$

LCD = 2 · 2 · 5 · 2 = 40

14. $1\dfrac{11}{12}$

15. $\dfrac{3}{5} = \dfrac{6}{10}$

$\dfrac{1}{2} = \dfrac{5}{10}$

$+\ \dfrac{3}{10} = \dfrac{3}{10}$

$\overline{\qquad\quad \dfrac{14}{10}} = 1\dfrac{4}{10} = 1\dfrac{2}{5}$

$\begin{array}{r|rrr} 2 & 5 & 2 & 10 \\ 5 & 5 & 1 & 5 \\ \hline & 1 & 1 & 1 \end{array}$

LCD = 2 · 5 = 10

16. $1\dfrac{29}{30}$

17. $\dfrac{1}{4} = \dfrac{6}{24}$

$\dfrac{7}{12} = \dfrac{14}{24}$

$+\ \dfrac{5}{8} = \dfrac{15}{24}$

$\overline{\qquad\quad \dfrac{35}{24}} = 1\dfrac{11}{24}$

$\begin{array}{r|rrr} 2 & 4 & 12 & 8 \\ 2 & 2 & 6 & 4 \\ \hline & 1 & 3 & 2 \end{array}$

LCD = 2 · 2 · 3 · 2 = 24

18. $1\dfrac{13}{24}$

19. $\dfrac{4}{6} = \dfrac{2}{3} = \dfrac{14}{21}$

$\dfrac{6}{14} = \dfrac{3}{7} = \dfrac{9}{21}$

$+\ \dfrac{2}{3} = \dfrac{2}{3} = \dfrac{14}{21}$

$\overline{\qquad\qquad\qquad \dfrac{37}{21}} = 1\dfrac{16}{21}$

3, 7

LCD = 3 · 7 = 21

20. $2\dfrac{7}{30}$

21. $\dfrac{5}{12} = \dfrac{5}{12} = \dfrac{10}{24}$

$\dfrac{6}{16} = \dfrac{3}{8} = \dfrac{9}{24}$

$+\ \dfrac{12}{32} = \dfrac{3}{8} = \dfrac{9}{24}$

$\overline{\qquad\qquad\qquad \dfrac{28}{24}} = \dfrac{7}{6} = 1\dfrac{1}{6}$

$\begin{array}{r|rr} 2 & 12 & 8 \\ 2 & 6 & 4 \\ \hline & 3 & 2 \end{array}$

LCD = 2 · 2 · 3 · 2 = 24

22. $1\dfrac{19}{165}$

23. $\dfrac{3}{4} = \dfrac{3}{4} = \dfrac{21}{28}$

$\dfrac{2}{14} = \dfrac{1}{7} = \dfrac{4}{28}$

$+\ \dfrac{1}{2} = \dfrac{1}{2} = \dfrac{14}{28}$

$\overline{\qquad\qquad\qquad \dfrac{39}{28}} = 1\dfrac{11}{28}$

$\begin{array}{r|rrr} 2 & 4 & 7 & 2 \\ \hline & 2 & 7 & 1 \end{array}$

LCD = 2 · 2 · 7 = 28

24. $\dfrac{161}{240}$

25. $\dfrac{5}{6} = \dfrac{20}{24}$

$\dfrac{3}{8} = \dfrac{9}{24}$

$+\ \dfrac{1}{12} = \dfrac{2}{24}$

$\overline{\qquad\quad \dfrac{31}{24}} = 1\dfrac{7}{24}$

$\begin{array}{r|rrr} 2 & 6 & 8 & 12 \\ 2 & 3 & 4 & 6 \\ 3 & 3 & 2 & 3 \\ \hline & 1 & 2 & 1 \end{array}$

LCD = 2 · 2 · 3 · 2 = 24

26. $1\dfrac{41}{112}$

27. $\dfrac{2}{28} = \dfrac{1}{14} = \dfrac{4}{56}$

$\dfrac{2}{16} = \dfrac{1}{8} = \dfrac{7}{56}$

$+\ \dfrac{3}{21} = \dfrac{1}{7} = \dfrac{8}{56}$

$\overline{\qquad\qquad\qquad \dfrac{19}{56}}$

$\begin{array}{r|rrr} 2 & 14 & 8 & 7 \\ 7 & 7 & 4 & 7 \\ \hline & 1 & 4 & 1 \end{array}$

LCD = 2 · 7 · 1 · 4 · 1 = 56

28. $\dfrac{13}{15}$

29. $\dfrac{13}{28} = \dfrac{39}{84}$

$+\ \dfrac{5}{42} = \dfrac{10}{84}$

$\overline{\qquad\quad \dfrac{49}{84}} = \dfrac{7}{12}$

$\begin{array}{r|rr} 2 & 28 & 42 \\ 7 & 14 & 21 \\ \hline & 2 & 3 \end{array}$

LCD = 2 · 7 · 2 · 3 = 84

30. $\dfrac{3}{4}$

EXERCISES 316, SET I (page 134)

1. $\dfrac{3}{4} = \dfrac{3}{4}$

$-\ \dfrac{1}{2} = -\ \dfrac{2}{4}$

$\overline{\qquad\quad \dfrac{1}{4}}$

2. $\dfrac{1}{2}$

3. $\dfrac{5}{8} = \dfrac{5}{8}$

$-\ \dfrac{2}{4} = -\ \dfrac{4}{8}$

$\overline{\qquad\quad \dfrac{1}{8}}$

4. $\dfrac{1}{10}$

5. $\dfrac{6}{12} = \dfrac{1}{2} = \dfrac{3}{6}$

$-\ \dfrac{1}{3} = -\ \dfrac{1}{3} = -\ \dfrac{2}{6}$

$\overline{\qquad\qquad\qquad \dfrac{1}{6}}$

6. $\dfrac{7}{20}$

7. $\dfrac{6}{7} = \dfrac{6}{7} = \dfrac{18}{21}$

$-\ \dfrac{4}{12} = -\ \dfrac{1}{3} = -\ \dfrac{7}{21}$

$\overline{\qquad\qquad\qquad \dfrac{11}{21}}$

7, 3

LCD = 7 · 3 = 21

8. $\dfrac{5}{24}$

9. $\dfrac{12}{30} = \dfrac{2}{5}$

$-\ \dfrac{1}{5} = -\ \dfrac{1}{5}$

$\overline{\qquad\quad \dfrac{1}{5}}$

10. $\dfrac{1}{3}$

11. $\dfrac{25}{32} = \dfrac{25}{32}$

$-\ \dfrac{3}{4} = -\ \dfrac{24}{32}$

$\overline{\qquad\quad \dfrac{1}{32}}$

$\begin{array}{r|rr} 2 & 32 & 4 \\ 2 & 16 & 2 \\ \hline & 8 & 1 \end{array}$

LCD = 2 · 2 · 8 = 32

12. $\dfrac{3}{16}$

13. $\dfrac{56}{64} = \dfrac{7}{8} = \dfrac{21}{24}$

$-\ \dfrac{14}{24} = -\ \dfrac{7}{12} = -\ \dfrac{14}{24}$

$\overline{\qquad\qquad\qquad \dfrac{7}{24}}$

$\begin{array}{r|rr} 2 & 8 & 12 \\ 2 & 4 & 6 \\ \hline & 2 & 3 \end{array}$

LCD = 2 · 2 · 2 · 3 = 24

14. $\dfrac{3}{20}$

15. $\dfrac{11}{18} = \dfrac{55}{90}$

$-\ \dfrac{4}{15} = -\ \dfrac{24}{90}$

$\overline{\qquad\quad \dfrac{31}{90}}$

$\begin{array}{r|rr} 3 & 18 & 15 \\ \hline & 6 & 5 \end{array}$

LCD = 3 · 6 · 5 = 90

16. $\dfrac{2}{7}$

17. $\dfrac{5}{12} = \dfrac{25}{60}$

$-\ \dfrac{2}{15} = -\ \dfrac{8}{60}$

$\overline{\qquad\quad \dfrac{17}{60}}$

$\begin{array}{r|rr} 3 & 12 & 15 \\ \hline & 4 & 5 \end{array}$

LCD = 3 · 4 · 5 = 60

18. $\dfrac{35}{48}$

19. $\dfrac{19}{35} = \dfrac{38}{70}$

   $-\dfrac{5}{14} = -\dfrac{25}{70}$

          $\dfrac{13}{70}$

$7\,\big|\ 35\quad 14$

      $5\quad\ 2$

LCD $= 7 \cdot 5 \cdot 2$

     $= 70$

20. $\dfrac{19}{80}$

21. $\dfrac{26}{36} = \dfrac{13}{18} = \dfrac{130}{180}$

   $-\dfrac{11}{20} = -\dfrac{11}{20} = -\dfrac{99}{180}$

                     $\dfrac{31}{180}$

$2\,\big|\ 18\quad 20$

      $9\quad 10$

LCD $= 2 \cdot 9 \cdot 10$

     $= 180$

22. $\dfrac{121}{504}$

EXERCISES 317, SET I (page 135)

1. $\dfrac{3}{4} \div \dfrac{1}{2} = \dfrac{3}{\overset{}{\underset{2}{4}}} \cdot \dfrac{\overset{1}{2}}{1} = \dfrac{3}{2} = 1\dfrac{1}{2}$ 2. $1\dfrac{1}{5}$

3. $\dfrac{1}{2} \div \dfrac{2}{3} = \dfrac{1}{2} \cdot \dfrac{3}{2} = \dfrac{3}{4}$ 4. 10

5. $\dfrac{1}{2} \div \dfrac{5}{1} = \dfrac{1}{2} \cdot \dfrac{1}{5} = \dfrac{1}{10}$ 6. 1

7. $\dfrac{5}{2} \div \dfrac{5}{8} = \dfrac{\overset{1}{5}}{\underset{1}{2}} \cdot \dfrac{\overset{4}{8}}{\underset{1}{5}} = 4$ 8. $\dfrac{1}{8}$

9. $1 \div \dfrac{1}{2} = \dfrac{1}{1} \cdot \dfrac{2}{1} = 2$ 10. $1\dfrac{1}{2}$

11. $\dfrac{3}{5} \div \dfrac{3}{10} = \dfrac{\overset{1}{3}}{\underset{1}{5}} \cdot \dfrac{\overset{2}{10}}{\underset{1}{3}} = 2$ 12. $1\dfrac{3}{4}$

13. $\dfrac{3}{16} \div \dfrac{9}{20} = \dfrac{\overset{1}{3}}{\underset{4}{16}} \cdot \dfrac{\overset{5}{20}}{\underset{3}{9}} = \dfrac{5}{12}$ 14. $\dfrac{1}{84}$

15. $34 \div \dfrac{17}{56} = \dfrac{\overset{2}{34}}{1} \cdot \dfrac{56}{\underset{1}{17}} = 112$ 16. $\dfrac{2}{9}$

17. $\dfrac{35}{16} \div \dfrac{42}{22} = \dfrac{\overset{5}{35}}{16} \cdot \dfrac{\overset{11}{22}}{\underset{\underset{3}{21}}{42}} = \dfrac{55}{48} = 1\dfrac{7}{48}$ 18. $\dfrac{5}{18}$

19. $36 \div \dfrac{4}{5} = \dfrac{\overset{9}{36}}{1} \cdot \dfrac{5}{\underset{1}{4}} = 45$ 20. $\dfrac{1}{81}$

21. $\dfrac{6}{35} \div \dfrac{8}{15} = \dfrac{\overset{3}{6}}{\underset{7}{35}} \cdot \dfrac{\overset{3}{15}}{\underset{4}{8}} = \dfrac{9}{28}$ 22. $\dfrac{5}{14}$

23. $\dfrac{14}{24} \div 210 = \dfrac{\overset{1}{14}}{\underset{12}{24}} \cdot \dfrac{1}{\underset{30}{210}} = \dfrac{1}{360}$ 24. 10

25. $\dfrac{56}{15} \div \dfrac{28}{90} = \dfrac{\overset{2}{56}}{\underset{1}{15}} \cdot \dfrac{\overset{6}{90}}{\underset{1}{28}} = 12$ 26. 3

EXERCISES 318, SET I (page 137)

1. $\dfrac{\frac{3}{4}}{\frac{1}{6}} = \dfrac{3}{4} \div \dfrac{1}{6} = \dfrac{3}{\underset{2}{4}} \cdot \dfrac{\overset{3}{6}}{1} = \dfrac{9}{2} = 4\dfrac{1}{2}$ 2. $\dfrac{25}{64}$

3. $\dfrac{\frac{2}{3}}{\frac{1}{2}} = \dfrac{2}{3} \div \dfrac{1}{2} = \dfrac{2}{3} \cdot \dfrac{2}{1} = \dfrac{4}{3} = 1\dfrac{1}{3}$ 4. $\dfrac{6}{7}$

5. $\dfrac{\frac{3}{5}}{\frac{3}{10}} = \dfrac{3}{5} \div \dfrac{3}{10} = \dfrac{\overset{1}{3}}{\underset{1}{5}} \cdot \dfrac{\overset{2}{10}}{\underset{1}{3}} = 2$ 6. $1\dfrac{1}{2}$

7. $\dfrac{\frac{3}{8}}{\frac{5}{12}} = \dfrac{3}{8} \div \dfrac{5}{12} = \dfrac{3}{\underset{2}{8}} \cdot \dfrac{\overset{3}{12}}{5} = \dfrac{9}{10}$ 8. $1\dfrac{1}{2}$

9. $\dfrac{6}{\frac{2}{3}} = 6 \div \dfrac{2}{3} = \dfrac{\overset{3}{6}}{1} \cdot \dfrac{3}{\underset{1}{2}} = 9$ 10. $\dfrac{5}{18}$

11. $\dfrac{14}{\frac{8}{5}} = 14 \div \dfrac{8}{5} = \dfrac{\overset{7}{14}}{1} \cdot \dfrac{5}{\underset{4}{8}} = \dfrac{35}{4} = 8\dfrac{3}{4}$ 12. $\dfrac{3}{32}$

13. $\dfrac{\frac{1}{4} + \frac{2}{5}}{\frac{1}{6}} = \dfrac{\frac{5+8}{20}}{\frac{1}{6}} = \dfrac{\frac{13}{20}}{\frac{1}{6}} = \dfrac{13}{20} \div \dfrac{1}{6} = \dfrac{13}{\underset{10}{20}} \cdot \dfrac{\overset{3}{6}}{1} = \dfrac{39}{10} = 3\dfrac{9}{10}$

14. $5\dfrac{1}{4}$

15. $\dfrac{4 + \frac{1}{4}}{2 - \frac{1}{2}} = \dfrac{\frac{17}{4}}{\frac{3}{2}} = \dfrac{17}{4} \div \dfrac{3}{2} = \dfrac{17}{\underset{2}{4}} \cdot \dfrac{\overset{1}{2}}{3} = \dfrac{17}{6} = 2\dfrac{5}{6}$

16. $1\dfrac{1}{82}$

17. $\dfrac{\frac{11}{4} - \frac{5}{9}}{\frac{7}{18} + \frac{13}{36}} = \dfrac{\frac{99 - 20}{36}}{\frac{14 + 13}{36}} = \dfrac{\frac{79}{36}}{\frac{27}{36}} = \dfrac{79}{36} \div \dfrac{27}{36} = \dfrac{79}{36} \cdot \dfrac{36}{27}$

                     $= \dfrac{79}{27} = 2\dfrac{25}{27}$

18. $\dfrac{13}{14}$

19. $\dfrac{\frac{16}{5} - \frac{7}{15}}{\frac{9}{30} + \frac{3}{10}} = \dfrac{\frac{48 - 7}{15}}{\frac{3}{10} + \frac{3}{10}} = \dfrac{\frac{41}{15}}{\frac{3}{5}} = \dfrac{41}{\underset{3}{15}} \cdot \dfrac{\overset{1}{3}}{3} = \dfrac{41}{9} = 4\dfrac{5}{9}$

20. $1\dfrac{3}{16}$

EXERCISES 319, SET I (page 138)

1. $\dfrac{\overset{6}{36}}{\underset{10}{60}} = \dfrac{\overset{3}{6}}{\underset{5}{10}} = \dfrac{3}{5}$ 2. $1\dfrac{2}{5}$

3. $\dfrac{99}{143} = \dfrac{99 \div 11}{143 \div 11} = \dfrac{9}{13}$ Using Euclid's Algorithm.

$99\,\overline{\big)\,143}$ $11\,\overline{\big)\,99}$ $11\,\overline{\big)\,143}$

   $\underline{99}\ \ 2$ $\underline{99}$ $\underline{11}$

   $44\,\overline{\big)\,99}$ 33

    $\underline{88}\ \ 4$ $\underline{33}$

    $\longrightarrow 11\,\overline{\big)\,44}$

             $\underline{44}$

largest number that divides both 99 and 143

4. $\frac{97}{53}$ is in lowest terms.

5. $\frac{\overset{104}{\cancel{208}}}{\underset{143}{\cancel{286}}} = \frac{104}{143} = \frac{104 \div 13}{143 \div 13} = \frac{8}{11}$

$104\overline{\smash{)}143}$ $13\overline{\smash{)}104}$ $13\overline{\smash{)}143}$
$\underline{104}$ $\underline{104}$ $\underline{13}$
$39\overline{\smash{)}104}$ 13
$\underline{78}$ $\underline{13}$
$26\overline{\smash{)}39}$
$\underline{26}$
$13\overline{\smash{)}26}$
$\underline{26}$
largest number that divides both 104 and 143

6. $\frac{6}{7}$ 7. $\frac{\overset{105}{\cancel{210}}}{\underset{126}{\cancel{252}}} = \frac{\overset{35}{\cancel{105}}}{\underset{42}{\cancel{126}}} = \frac{\overset{5}{\cancel{35}}}{\underset{6}{\cancel{42}}} = \frac{5}{6}$ 8. $\frac{11}{17}$

9. $\frac{\overset{507}{\cancel{2535}}}{\underset{624}{\cancel{3120}}} = \frac{\overset{169}{\cancel{507}}}{\underset{208}{\cancel{624}}} = \frac{169}{208} = \frac{169 \div 13}{208 \div 13} = \frac{13}{16}$

$169\overline{\smash{)}208}$ $13\overline{\smash{)}169}$ $13\overline{\smash{)}208}$
$\underline{169}$ $\underline{13}$ $\underline{13}$
$39\overline{\smash{)}169}$ 39 78
$\underline{156}$ $\underline{39}$ $\underline{78}$
$13\overline{\smash{)}39}$
$\underline{39}$
largest number that divides both 169 and 208

10. $\frac{15}{22}$ 11. $\frac{6}{7} + \frac{3}{7} = \frac{9}{7} = 1\frac{2}{7}$ 12. 10

13. $\frac{5}{8} - \frac{3}{8} = \frac{2}{8} = \frac{1}{4}$ 14. 14

15. $\frac{\frac{6}{9}}{\frac{12}{1}} = \frac{6}{9} \div \frac{12}{1} = \frac{\overset{1}{\cancel{6}}}{9} \cdot \frac{1}{\underset{2}{\cancel{12}}} = \frac{1}{18}$ 16. $1\frac{3}{8}$

17. $\frac{3}{5} \div \frac{7}{10} = \frac{3}{\underset{1}{\cancel{5}}} \cdot \frac{\overset{2}{\cancel{10}}}{7} = \frac{6}{7}$ 18. $\frac{3}{8}$

19. $\frac{2 + \frac{2}{3}}{\frac{1}{2} - \frac{1}{6}} = \frac{\frac{8}{3}}{\frac{1}{3}} = \frac{8}{3} \div \frac{1}{3} = \frac{8}{\cancel{3}} \cdot \frac{\overset{1}{\cancel{3}}}{1} = 8$ 20. $\frac{5}{24}$

21.
$\frac{1}{3} = \frac{4}{12}$
$\frac{1}{4} = \frac{3}{12}$
$+ \frac{1}{12} = \frac{1}{12}$
$\frac{8}{12} = \frac{2}{3}$ of day used

Part of day left $= 1 - \frac{2}{3} = \frac{1}{3}$

22. $2\frac{3}{16}$

23. $5 \div 3 = \frac{5}{3} = 1\frac{2}{3}$ lb   24. 8

EXERCISES 320A, SET I (page 140)

1. $2\frac{1}{4} + 1\frac{3}{4}$ 2. $4\frac{3}{5}$

$= \frac{9}{4} + \frac{7}{4} = \frac{16}{4} = 4$

3. $1\frac{1}{3} + 2\frac{1}{6}$ 4. $5\frac{3}{4}$

$= \frac{4}{3} + \frac{13}{6} = \frac{8}{6} + \frac{13}{6} = \frac{\overset{7}{\cancel{21}}}{\underset{2}{\cancel{6}}} = \frac{7}{2} = 3\frac{1}{2}$

5. $2\frac{3}{5} + 1\frac{1}{10}$ 6. $3\frac{5}{8}$

$= \frac{13}{5} + \frac{11}{10} = \frac{26}{10} + \frac{11}{10} = \frac{37}{10} = 3\frac{7}{10}$

7. $3\frac{1}{3} + 2\frac{1}{2}$ 8. 7

$= \frac{10}{3} + \frac{5}{2} = \frac{20}{6} + \frac{15}{6} = \frac{35}{6} = 5\frac{5}{6}$

9. $4\frac{1}{6} + 3\frac{2}{3}$ 10. $6\frac{1}{4}$

$= \frac{25}{6} + \frac{11}{3} = \frac{25}{6} + \frac{22}{6} = \frac{47}{6} = 7\frac{5}{6}$

11. $1\frac{5}{8} + 2\frac{1}{2}$ 12. $5\frac{13}{40}$

$= \frac{13}{8} + \frac{5}{2} = \frac{13}{8} + \frac{20}{8} = \frac{33}{8} = 4\frac{1}{8}$

13. $7\frac{5}{6} + 3\frac{2}{3}$ 14. $11\frac{3}{40}$

$= \frac{47}{6} + \frac{11}{3} = \frac{47}{6} + \frac{22}{6} = \frac{\overset{23}{\cancel{69}}}{\underset{2}{\cancel{6}}} = \frac{23}{2} = 11\frac{1}{2}$

15. $2\frac{5}{8} + 3 = 5\frac{5}{8}$ 16. $6\frac{3}{5}$

17. $1\frac{1}{2} + 2\frac{1}{3} + 3\frac{1}{4}$ 18. $7\frac{11}{12}$

$= \frac{3}{2} + \frac{7}{3} + \frac{13}{4} = \frac{18}{12} + \frac{28}{12} + \frac{39}{12} = \frac{85}{12} = 7\frac{1}{12}$

19. $5\frac{1}{4} + 3 + 2\frac{3}{8}$ 20. $10\frac{3}{10}$

$= \frac{21}{4} + \frac{3}{1} + \frac{19}{8} = \frac{42}{8} + \frac{24}{8} + \frac{19}{8} = \frac{85}{8} = 10\frac{5}{8}$

21. $9\frac{9}{9}$

EXERCISES 320B, SET I (page 141)

1. $3\frac{4}{5} - 1\frac{1}{5}$ 2. $3\frac{1}{3}$

$= \frac{19}{5} - \frac{6}{5} = \frac{13}{5} = 2\frac{3}{5}$

3. $2\frac{1}{3} - 1\frac{3}{5}$ 4. $1\frac{1}{8}$

$= \frac{7}{3} - \frac{8}{5} = \frac{35}{15} - \frac{24}{15} = \frac{11}{15}$

5. $5 - 2\frac{3}{8}$ 6. $2\frac{1}{6}$

$= \frac{5}{1} - \frac{19}{8} = \frac{40}{8} - \frac{19}{8} = \frac{21}{8} = 2\frac{5}{8}$

7. $3\frac{3}{4} - 2 = 1\frac{3}{4}$ 8. $4\frac{1}{5}$

9. $5\frac{8}{9} - 1\frac{2}{3}$ 10. $4\frac{1}{24}$

$= \frac{53}{9} - \frac{5}{3} = \frac{53}{9} - \frac{15}{9} = \frac{38}{9} = 4\frac{2}{9}$

11. $3\frac{3}{4} - 2\frac{1}{3}$ 12. $3\frac{1}{14}$

$= \frac{15}{4} - \frac{7}{3} = \frac{45}{12} - \frac{28}{12} = \frac{17}{12} = 1\frac{5}{12}$

13. $7 - 5\frac{3}{4} = \frac{7}{1} - \frac{23}{4} = \frac{28}{4} - \frac{23}{4} = \frac{5}{4} = 1\frac{1}{4}$ sq yd

14. Yes, and have 1 inch left over.

1. $1\frac{2}{3} \times 2\frac{1}{2}$

   $= \frac{5}{3} \times \frac{5}{2} = \frac{25}{6} = 4\frac{1}{6}$

2. 3

3. $1\frac{3}{7} \div 1\frac{1}{4}$

   $= \frac{10}{7} \div \frac{5}{4} = \frac{10}{7} \cdot \frac{4}{5} = \frac{8}{7} = 1\frac{1}{7}$

4. $\frac{2}{3}$

5. $2\frac{2}{3} \times 2\frac{1}{4}$

   $= \frac{8}{3} \times \frac{9}{4} = 6$

6. 6

7. $2\frac{3}{5} \div 1\frac{4}{35}$

   $= \frac{13}{5} \div \frac{39}{35} = \frac{13}{5} \cdot \frac{35}{39} = \frac{7}{3} = 2\frac{1}{3}$

8. $2\frac{1}{2}$

9. $8 \times 3\frac{3}{4}$

   $= \frac{8}{1} \cdot \frac{15}{4} = 30$

10. 28

11. $7 \div 4\frac{2}{3}$

    $= 7 \div \frac{14}{3} = \frac{7}{1} \cdot \frac{3}{14} = \frac{3}{2} = 1\frac{1}{2}$

12. $\frac{1}{5}$

13. $2\frac{5}{8} \times 4$

    $= \frac{21}{8} \times \frac{4}{1} = \frac{21}{2} = 10\frac{1}{2}$

14. $14\frac{1}{2}$

15. $3\frac{1}{3} \div 5$

    $= \frac{10}{3} \cdot \frac{1}{5} = \frac{2}{3}$

16. $3\frac{1}{2}$

17. $3\frac{3}{10} \times \frac{6}{11} \times 1\frac{2}{3}$

    $= \frac{33}{10} \times \frac{6}{11} \times \frac{5}{3} = 3$

18. $\frac{11}{12}$

19. $3\frac{1}{5} \times 75 \times \frac{7}{10}$

    $= \frac{16}{5} \times \frac{75}{1} \times \frac{7}{10} = 168$

20. 68

21. $8 \times 2\frac{7}{16}$

    $= \frac{8}{1} \times \frac{39}{16} = \frac{39}{2} = 19\frac{1}{2}$ lb

22. $1\frac{1}{2}$

1. $13\frac{1}{5} = 13\frac{3}{15}$

   $+ \ 4\frac{2}{3} = 4\frac{10}{15}$

   $17\frac{13}{15}$

2. $38\frac{11}{16}$

3. $12\frac{1}{2} = 12\frac{2}{4}$

   $+ 23\frac{3}{4} = 23\frac{3}{4}$

   ①

   $\frac{5}{4} = 1\frac{1}{4}$

   $36\frac{1}{4}$

4. $38\frac{11}{20}$

5. $37\frac{5}{6} = 37\frac{10}{12}$

   $+ 44\frac{1}{4} = 44\frac{3}{12}$

   ①

   $\frac{13}{12} = 1\frac{1}{12}$

   $82\frac{1}{12}$

6. $124\frac{7}{24}$

7. $125\frac{3}{7}$

   $+ 208$

   $333\frac{3}{7}$

8. $796\frac{7}{9}$

9. $72\frac{2}{3} = 72\frac{8}{12}$

   $81\frac{3}{4} = 81\frac{9}{12}$

   $+ 93\frac{1}{2} = 93\frac{6}{12}$

   ①

   $\frac{23}{12} = 1\frac{11}{12}$

   $247\frac{11}{12}$

10. $221\frac{7}{18}$

11. $17\frac{1}{3} = 17\frac{5}{15}$

    $28\frac{2}{5} = 28\frac{6}{15}$

    $+ 15\frac{4}{15} = 15\frac{4}{15}$

    ①

    $\frac{15}{15} = 1$

    61

12. $132\frac{17}{28}$

13. $56\frac{3}{4} = 56\frac{9}{12}$

    $72 \ \ = 72$

    $+ 48\frac{2}{3} = 48\frac{8}{12}$

    ①

    $\frac{17}{12} = 1\frac{5}{12}$

    $177\frac{5}{12}$

14. $200\frac{11}{30}$

15. $107\frac{5}{6} = 107\frac{15}{18}$

    $293\frac{1}{3} = 293\frac{6}{18}$

  $+ 480\frac{7}{9} = 480\frac{14}{18}$

                ①⟵

                $\frac{35}{18} =$ ①$\frac{17}{18}$

           $881\frac{17}{18}$⟵

16. $1{,}273\frac{7}{16}$

17. $156\frac{4}{5} = 156\frac{24}{30}$

     $93 \ \ = \ 93$

    $81\frac{7}{15} = \ 81\frac{14}{30}$

 $+ 204\frac{3}{10} = 204\frac{9}{30}$

                ①⟵

               $\frac{47}{30} =$ ①$\frac{17}{30}$

        $535\frac{17}{30}$⟵

18. $194\frac{17}{32}$

19. $7\frac{3}{4} = 7\frac{60}{80}$

   $5\frac{7}{8} = 5\frac{70}{80}$

   $8\frac{5}{16} = 8\frac{25}{80}$

 $+ 10\frac{2}{5} = 10\frac{32}{80}$

            ②⟵

          $\frac{187}{80} =$ ②$\frac{27}{80}$

       $32\frac{27}{80}$ oz

```
2 | 4   8   16   5
2 | 2   4    8   5
2 | 1   2    4   5
    1   1    2   5  →
```

LCD = 2 · 2 · 2 · 2 · 5

    = 80

20. $456\frac{1}{2}$ lb

EXERCISES 322, SET I (page 146)

1. $14\frac{3}{4}$

  $- 10\frac{1}{4}$

    $4\frac{2}{4} = 4\frac{1}{2}$

2. $3\frac{2}{3}$

3. $17\frac{3}{4} = 17\frac{6}{8}$

  $- 5\frac{1}{8} = -5\frac{1}{8}$

         $12\frac{5}{8}$

4. $13\frac{1}{2}$

5. $8 \ = 7 + \frac{2}{2} = 7\frac{2}{2}$

  $- 4\frac{1}{2} = \ \ \ \ \ -4\frac{1}{2}$

           $3\frac{1}{2}$

6. $3\frac{2}{3}$

7. $6 \ = 5 + \frac{5}{5} = 5\frac{5}{5}$

  $- 2\frac{3}{5} = \ \ \ \ \ -2\frac{3}{5}$

           $3\frac{2}{5}$

8. $2\frac{1}{6}$

9. $4\frac{1}{4} = 3 + 1 + \frac{1}{4} = 3 + \frac{4}{4} + \frac{1}{4} = 3\frac{5}{4}$

  $- 1\frac{3}{4} =$

                              $2\frac{2}{4} = 2\frac{1}{2}$

10. $3\frac{2}{3}$

11. $12\frac{2}{5} = \ 12\frac{4}{10}$

  $- 7\frac{3}{10} = -7\frac{3}{10}$

         $5\frac{1}{10}$

12. $15\frac{1}{2}$

13. $3\frac{1}{12} = 2 + 1 + \frac{1}{12} = 2 + \frac{12}{12} + \frac{1}{12} = 2\frac{13}{12}$

  $- 1\frac{1}{6} =$

                             $- 1\frac{2}{12}$

                              $1\frac{11}{12}$

14. $5\frac{7}{8}$

15. $45 \ = 44 + \frac{3}{3} = 44\frac{3}{3}$

  $- 38\frac{2}{3} = \ \ \ \ \ \ -38\frac{2}{3}$

                 $6\frac{1}{3}$

16. $3\frac{8}{15}$

17. $68\frac{5}{16} = 67 + 1 + \frac{5}{16} = 67 + \frac{16}{16} + \frac{5}{16} = 67\frac{21}{16}$

  $- 53\frac{3}{4} =$

                             $- 53\frac{12}{16}$

                              $14\frac{9}{16}$

18. $8\frac{1}{2}$

19. $234\frac{5}{14} = 233 + \frac{14}{14} + \frac{5}{14} = 233\frac{19}{24}$

  $- 157\frac{3}{7} =$

                     $- 157\frac{6}{14}$

                     $76\frac{13}{14}$

20. $4{,}137\frac{11}{15}$

21. $1\frac{1}{4} - \frac{7}{8}$

  $= \frac{5}{4} - \frac{7}{8} = \frac{10}{8} - \frac{7}{8} = \frac{3}{8}$ lb

22. $5\frac{1}{6}$ days

EXERCISES 324, SET I (page 150)

1. $10 - 3 + 2$

  $= \ \ \ \ 7 \ \ + 2 = 9$

2. 2

3. $8 \cdot 4 \div 2$

  $= \ \ \ 32 \ \ \div 2 = 16$

4. 4

5. $8 \div 4 \cdot 2$

  $= \ \ \ 2 \ \ \ \cdot 2 = 4$

6. 36

7. $5 \cdot 2 + 3 \div 6$

  $= \ \ 10 \ \ + \ \frac{3}{6} = 10 + \frac{1}{2} = 10\frac{1}{2}$

8. $24\frac{2}{3}$

9. $\frac{\overset{3}{\cancel{6}}}{1} \cdot \frac{1}{\underset{1}{\cancel{2}}} - 2 \div 8$

  $= \ \ \ 3 \ \ \ - \ \frac{2}{8} = 3 - \frac{1}{4} = 2\frac{3}{4}$

10. $5\frac{2}{3}$

11. $(5)^2 + 3 \cdot 1\frac{1}{3}$

  $= 5 \cdot 5 + \frac{\overset{1}{\cancel{3}}}{1} \cdot \frac{4}{\underset{1}{\cancel{3}}} = 25 + 4 = 29$

12. 34

13. $\left(\dfrac{3}{4}\right)^2 + \dfrac{1}{4} \cdot 1\dfrac{3}{4}$ 14. 1

$= \dfrac{3}{4} \cdot \dfrac{3}{4} + \dfrac{1}{4} \cdot \dfrac{7}{4} = \dfrac{9}{16} + \dfrac{7}{16} = \dfrac{16}{16} = 1$

15. $4 - \dfrac{2}{3} + 1\dfrac{1}{2} = \dfrac{12}{3} - \dfrac{2}{3} + \dfrac{3}{2}$ 16. $\dfrac{3}{4}$

$= \dfrac{10}{3} + \dfrac{3}{2} = \dfrac{20}{6} + \dfrac{9}{6} = \dfrac{29}{6} = 4\dfrac{5}{6}$

17. $\dfrac{3}{4} + 33 \div 4\dfrac{1}{8}$ 18. $3\dfrac{1}{4}$

$= \dfrac{3}{4} + 33 \div \dfrac{33}{8} = \dfrac{3}{4} + \overset{1}{\cancel{33}} \cdot \dfrac{8}{\underset{1}{\cancel{33}}} = 8\dfrac{3}{4}$

19. $\left(8\dfrac{1}{4} - 1\dfrac{2}{3}\right) \cdot 3$ 20. 4

$= \left(\dfrac{33}{4} - \dfrac{5}{3}\right) \cdot 3 = \left(\dfrac{99}{12} - \dfrac{20}{12}\right) \cdot 3$

$= \dfrac{79}{\underset{4}{\cancel{12}}} \cdot \dfrac{\overset{1}{\cancel{3}}}{1} = \dfrac{79}{4} = 19\dfrac{3}{4}$

21. $2\dfrac{2}{5} \div \left(3 - \dfrac{3}{10}\right) = \dfrac{12}{5} \div \left(\dfrac{30}{10} - \dfrac{3}{10}\right)$ 22. $\dfrac{1}{2}$

$= \dfrac{12}{5} \div \dfrac{27}{10} = \dfrac{\overset{4}{\cancel{12}}}{\underset{1}{\cancel{5}}} \cdot \dfrac{\overset{2}{\cancel{10}}}{\underset{9}{\cancel{27}}} = \dfrac{8}{9}$

23. $2\dfrac{2}{3} \cdot \left(\dfrac{3}{7} + 2\right) \div 4\dfrac{6}{7}$ 24. $\dfrac{1}{2}$

$= \dfrac{8}{3} \cdot \left(\dfrac{3}{7} + \dfrac{14}{7}\right) \div \dfrac{34}{7}$

$= \dfrac{\overset{4}{\cancel{8}}}{3} \cdot \dfrac{\overset{1}{\cancel{17}}}{\underset{1}{\cancel{7}}} \cdot \dfrac{\overset{1}{\cancel{7}}}{\underset{\underset{1}{2}}{\cancel{34}}} = 1\dfrac{1}{3}$

EXERCISES 325, SET I (page 151)

1. LCD $= 2 \cdot 3 \cdot 2 = 12$

2	4	6	3
3	2	3	3
	2	1	1

$\left.\begin{array}{l} \dfrac{3}{4} = \dfrac{9}{12} \\ \dfrac{5}{6} = \dfrac{10}{12} \\ \dfrac{2}{3} = \dfrac{8}{12} \end{array}\right\}$ since $\dfrac{10}{12} > \dfrac{9}{12} > \dfrac{8}{12}$

therefore $\dfrac{5}{6} > \dfrac{3}{4} > \dfrac{2}{3}$

2. $\dfrac{5}{12} > \dfrac{1}{3} > \dfrac{2}{9}$

3. LCD $= 2 \cdot 7 \cdot 2 = 28$

2	14	7	4
7	7	7	2
	1	1	2

$\left.\begin{array}{l} \dfrac{5}{14} = \dfrac{10}{28} \\ \dfrac{3}{7} = \dfrac{12}{28} \\ \dfrac{3}{4} = \dfrac{21}{28} \end{array}\right\}$ since $\dfrac{21}{28} > \dfrac{12}{28} > \dfrac{10}{28}$

therefore $\dfrac{3}{4} > \dfrac{3}{7} > \dfrac{5}{14}$

4. $\dfrac{4}{5} > \dfrac{3}{4} > \dfrac{7}{10}$

5. LCD $= 2 \cdot 3 \cdot 5 = 30$

2	15	6	10
3	15	3	5
5	5	1	5
	1	1	1

$\left.\begin{array}{l} \dfrac{2}{15} = \dfrac{4}{30} \\ \dfrac{1}{6} = \dfrac{5}{30} \\ \dfrac{3}{10} = \dfrac{9}{30} \end{array}\right\}$ since $\dfrac{9}{30} > \dfrac{5}{30} > \dfrac{4}{30}$

therefore $\dfrac{3}{10} > \dfrac{1}{6} > \dfrac{2}{15}$

6. $\dfrac{5}{8} > \dfrac{19}{32} > \dfrac{9}{16}$

7. $\left.\begin{array}{l} \dfrac{5}{8} = \dfrac{5}{8} \\ \dfrac{3}{4} = \dfrac{6}{8} \end{array}\right\}$ since $\dfrac{6}{8} > \dfrac{5}{8}$

therefore $\dfrac{3}{4} > \dfrac{5}{8}$

Therefore, $12\dfrac{3}{4} > 12\dfrac{5}{8}$ by $\dfrac{6}{8} - \dfrac{5}{8} = \dfrac{1}{8}$

Stock went up $\dfrac{1}{8}$.

8. Stock went down $\dfrac{1}{8}$.

REVIEW EXERCISES 327, SET I (page 153)

1. (a) $\dfrac{5}{2} = 2\dfrac{1}{2}$ (b) $\dfrac{11}{8} = 1\dfrac{3}{8}$

(c) $\dfrac{18}{13} = 1\dfrac{5}{13}$ (d) $\dfrac{26}{16} = 1\dfrac{5}{8}$

(e) $\dfrac{63}{32} = 1\dfrac{31}{32}$

2. (a) $2\dfrac{1}{3}$ (b) $1\dfrac{2}{19}$

(c) $1\dfrac{22}{25}$ (d) $2\dfrac{4}{11}$

(e) $2\dfrac{10}{57}$

3. (a) $2\dfrac{3}{5} = \dfrac{2 \cdot 5 + 3}{5} = \dfrac{10 + 3}{5} = \dfrac{13}{5}$

(b) $3\dfrac{7}{8} = \dfrac{3 \cdot 8 + 7}{8} = \dfrac{24 + 7}{8} = \dfrac{31}{8}$

(c) $9\dfrac{2}{11} = \dfrac{9 \cdot 11 + 2}{11} = \dfrac{99 + 2}{11} = \dfrac{101}{11}$

(d) $13\dfrac{5}{16} = \dfrac{13 \cdot 16 + 5}{16} = \dfrac{208 + 5}{16} = \dfrac{213}{16}$

(e) $23\dfrac{27}{31} = \dfrac{23 \cdot 31 + 27}{31} = \dfrac{713 + 27}{31} = \dfrac{740}{31}$

4. (a) $\dfrac{7}{4}$ (b) $\dfrac{25}{9}$

(c) $\dfrac{109}{13}$ (d) $\dfrac{295}{18}$

(e) $\dfrac{958}{35}$

5. (a) $\begin{array}{r} 2 \\ + 4\dfrac{2}{5} \\ \hline 6\dfrac{2}{5} \end{array}$

(b) $2\frac{2}{3} = 2\frac{10}{15}$

$+ 1\frac{3}{5} = 1\frac{9}{15}$

———

$①$

$\frac{19}{15} = ①\frac{4}{15}$

$4\frac{4}{15}$

(c) $4\frac{5}{16} = 4\frac{5}{16}$

$+ \frac{5}{8} = \frac{10}{16}$

———

$4\frac{15}{16}$

(d) $153\frac{2}{5} = 153\frac{8}{20}$

$+ 135\frac{3}{4} = 135\frac{15}{20}$

———

$①$

$\frac{23}{20} = ①\frac{3}{20}$

$289\frac{3}{20}$

6. (a) $7\frac{1}{3}$ (b) $7\frac{3}{4}$ (c) $8\frac{1}{6}$ (d) $212\frac{17}{24}$

7. (a) $5\frac{4}{5} = 5\frac{8}{10}$

$- 3\frac{7}{10} = - 3\frac{7}{10}$

———

$2\frac{1}{10}$

(b) $5\frac{1}{3} = 4\frac{4}{3} = 4\frac{16}{12}$

$- 2\frac{3}{4} = - 2\frac{3}{4} = - 2\frac{9}{12}$

———

$2\frac{7}{12}$

(c) $16 = 15\frac{8}{8}$

$- 5\frac{7}{8} = - 5\frac{7}{8}$

———

$10\frac{1}{8}$

(d) $351\frac{2}{3} = 351\frac{6}{9} = 350\frac{15}{9}$

$- 272\frac{7}{9} = - 272\frac{7}{9} = - 272\frac{7}{9}$

———

$78\frac{8}{9}$

8. (a) $2\frac{1}{6}$ (b) $1\frac{9}{14}$ (c) $12\frac{11}{16}$ (d) $89\frac{7}{8}$

9. (a) $4\frac{1}{5} \cdot 2\frac{3}{7}$

$= \frac{21}{5} \cdot \frac{17}{7} = \frac{51}{5} = 10\frac{1}{5}$

(b) $3\frac{2}{3} \cdot 6\frac{3}{5}$

$= \frac{11}{3} \cdot \frac{33}{5} = \frac{121}{5} = 24\frac{1}{5}$

(c) $\frac{3}{5} \times \frac{105}{1} = 63$

(d) $\frac{12}{13} \times 8\frac{1}{3} = \frac{12}{13} \times \frac{25}{3} = \frac{100}{13} = 13\overline{)100} \quad {}^{7} \ {}^{R9} = 7\frac{9}{13}$

$\underline{91}$

9

10. (a) $10\frac{1}{2}$ (b) 8 (c) 35 (d) 4

11. (a) $2\frac{2}{5} \div 1\frac{1}{15}$

$= \frac{12}{5} \div \frac{16}{15} = \frac{12}{5} \cdot \frac{15}{16} = \frac{9}{4} = 2\frac{1}{4}$

(b) $1\frac{1}{6} \div 4\frac{2}{3}$

$= \frac{7}{6} \div \frac{14}{3} = \frac{7}{6} \cdot \frac{3}{14} = \frac{1}{4}$

(c) $16 \div \frac{8}{13} = \frac{16}{1} \cdot \frac{13}{8} = 26$

(d) $1\frac{9}{16} \div \frac{10}{1}$

$= \frac{25}{16} \cdot \frac{1}{10} = \frac{5}{32}$

12. (a) $\frac{2}{3}$ (b) $\frac{1}{8}$ (c) $13\frac{1}{3}$ (d) $\frac{2}{5}$

13. (a) $\frac{42}{54} = \frac{7}{9}$ (b) $\frac{45}{105} = \frac{9}{21} = \frac{3}{7}$

(c) $\frac{28}{57}$ in lowest terms

(d) $\frac{207}{253} = \frac{207 \div 23}{253 \div 23} = \frac{9}{11}$

$207\overline{)253}$ $23\overline{)207}$ $23\overline{)253}$

$\underline{207}$ $\underline{207}$ $\underline{23}$

$46\overline{)207}$ 23

$\underline{184}$ 23

$23\overline{)46}$

$\underline{46}$

└—largest number that divides both 207 and 253

14. (a) $\frac{5}{6}$ (b) $\frac{5}{7}$ (c) $\frac{5}{8}$ (d) $\frac{15}{13}$

15. (a) Equivalent because $10 \cdot 21 = 35 \cdot 6$

$210 = 210$

(b) Equivalent because $12 \cdot 55 = 15 \cdot 44$

$660 = 660$

(c) Not equivalent because $14 \cdot 45 \neq 30 \cdot 18$

$630 \neq 540$

$14 \cdot 45 > 30 \cdot 18$

Therefore, $\frac{14}{30} > \frac{18}{45}$

(d) Equivalent because $12 \cdot 35 = 21 \cdot 20$

$420 = 420$

16. (a) Equivalent

(b) Not equivalent; $\frac{27}{18} > \frac{18}{24}$

(c) Equivalent

(d) Equivalent

17. LCD = $3 \cdot 3 \cdot 2 \cdot 2 = 36$

$$\begin{array}{r|ccc} 3 & 9 & 12 & 18 \\ 3 & 3 & 4 & 6 \\ 2 & 1 & 4 & 2 \\ \hline & 1 & 2 & 1 \end{array}$$

$\dfrac{4}{9} = \dfrac{16}{36}$

$\dfrac{5}{12} = \dfrac{15}{36}$ Since $\dfrac{16}{36} > \dfrac{15}{36} > \dfrac{14}{36}$

$\dfrac{7}{18} = \dfrac{14}{36}$ Therefore, $\dfrac{4}{9} > \dfrac{5}{12} > \dfrac{7}{18}$

18. $\dfrac{7}{8} > \dfrac{5}{6} > \dfrac{2}{3}$

19. (a) $\dfrac{7}{45} + \dfrac{25}{36}$

$$\begin{array}{r|cc} 3 & 45 & 36 \\ 3 & 15 & 12 \\ \hline & 5 & 4 \end{array}$$

$= \dfrac{28}{180} + \dfrac{125}{180}$

LCD = $3 \cdot 3 \cdot 5 \cdot 4$

$= \dfrac{\overset{51}{\cancel{153}}}{\underset{60}{\cancel{180}}} = \dfrac{\overset{17}{\cancel{51}}}{\underset{20}{\cancel{60}}} = \dfrac{17}{20}$

$= 180$

(b) $\dfrac{11}{35} - \dfrac{7}{30}$

$$\begin{array}{r|cc} 5 & 35 & 30 \\ \hline & 7 & 6 \end{array}$$

$= \dfrac{66}{210} - \dfrac{49}{210} = \dfrac{17}{210}$

LCD = $5 \cdot 7 \cdot 6$

$= 210$

20. (a) $\dfrac{37}{42}$ (b) $\dfrac{17}{45}$

21. (a) $\dfrac{\frac{5}{8}}{\frac{5}{6}} = \dfrac{5}{8} \div \dfrac{5}{6} = \dfrac{\cancel{5}}{\underset{4}{\cancel{8}}} \cdot \dfrac{\overset{3}{\cancel{6}}}{\cancel{5}} = \dfrac{3}{4}$

(b) $\dfrac{\frac{12}{1}}{\frac{6}{11}} = \dfrac{12}{1} \div \dfrac{6}{11} = \dfrac{\overset{2}{\cancel{12}}}{1} \cdot \dfrac{11}{\underset{1}{\cancel{6}}} = 22$

(c) $\dfrac{\frac{3}{4}}{6} = \dfrac{3}{4} \div 6 = \dfrac{\cancel{3}}{4} \cdot \dfrac{1}{\underset{2}{\cancel{6}}} = \dfrac{1}{8}$

(d) $\dfrac{\frac{2}{3} + \frac{1}{2}}{\frac{5}{6}} = \dfrac{\frac{7}{6}}{\frac{5}{6}} = \dfrac{7}{6} \div \dfrac{5}{6} = \dfrac{7}{\cancel{6}} \cdot \dfrac{\cancel{6}}{5} = \dfrac{7}{5} = 1\dfrac{2}{5}$

22. (a) $\dfrac{1}{2}$ (b) 9   (c) $\dfrac{1}{16}$ (d) $\dfrac{3}{8}$

23. (a) $\left(3\dfrac{1}{2} - 2\dfrac{3}{4}\right) \div 6 \cdot \dfrac{4}{7}$

$= \left(\dfrac{7}{2} - \dfrac{11}{4}\right) \div 6 \cdot \dfrac{4}{7}$

$= \left(\dfrac{14}{4} - \dfrac{11}{4}\right) \div 6 \cdot \dfrac{4}{7}$

$= \dfrac{3}{4} \div 6 \cdot \dfrac{4}{7}$

$= \dfrac{\cancel{3}}{\underset{1}{\cancel{4}}} \cdot \dfrac{1}{\underset{2}{\cancel{6}}} \cdot \dfrac{\cancel{4}}{7} = \dfrac{1}{14}$

(b) $\left(\dfrac{3}{4}\right)^2 + 2\dfrac{2}{3} \cdot \dfrac{3}{16}$

$= \dfrac{3}{4} \cdot \dfrac{3}{4} + \dfrac{\overset{1}{\cancel{8}}}{\underset{1}{\cancel{3}}} \cdot \dfrac{\cancel{3}}{\underset{2}{\cancel{16}}} = \dfrac{9}{16} + \dfrac{1}{2} = \dfrac{9}{16} + \dfrac{8}{16} = \dfrac{17}{16} = 1\dfrac{1}{16}$

24. (a) $\dfrac{2}{3}$ (b) $3\dfrac{1}{2}$

25. $\dfrac{8}{24} = \dfrac{1}{3}$

26. (a) $19\dfrac{1}{4}$ (b) $70

27. $4\dfrac{1}{2} = 4\dfrac{6}{12}$

$12\dfrac{3}{4} = 12\dfrac{9}{12}$

$8\dfrac{1}{3} = 8\dfrac{4}{12}$

$+ 15\dfrac{1}{6} = 15\dfrac{2}{12}$

$\dfrac{21}{12} = 1\dfrac{9}{12} = 1\dfrac{3}{4}$

$40\dfrac{3}{4}$

28. 56

29. $24 \times 4\dfrac{3}{4} = \dfrac{\overset{6}{\cancel{24}}}{1} \times \dfrac{19}{\underset{1}{\cancel{4}}} = \$114$

30. $24\dfrac{1}{4}$ in

31. (a) $16 \div 3\dfrac{1}{2} = 16 \div \dfrac{7}{2} = 16 \cdot \dfrac{2}{7} = \dfrac{32}{7} = 4\dfrac{4}{7}$

Therefore, four shelves can be cut from the 16-foot board.

(b) $4 \times 3\dfrac{1}{2} = \dfrac{\overset{2}{\cancel{4}}}{1} \times \dfrac{7}{\underset{1}{\cancel{2}}} = 14$ ft for shelves

Four cuts $= 4 \times \dfrac{1}{8} = \dfrac{4}{8} = \dfrac{1}{2}$ inches for saw cuts

Part left $= 16$ ft $- 14$ ft $- \dfrac{1}{2}$ in

$= 2$ ft $- \dfrac{1}{2}$ in $= 24$ in $- \dfrac{1}{2}$ in

$= 23\dfrac{1}{2}$ in

32. Because $\dfrac{1}{2} + \dfrac{1}{3} + \dfrac{1}{9} = \dfrac{9}{18} + \dfrac{6}{18} + \dfrac{2}{18} = \dfrac{17}{18}$.
According to the will, the sons were only to receive $\dfrac{17}{18}$ of the 17 camels. They actually got all 17 camels, which was slightly more than their proper shares, in order that no camels would have to be cut up.

SOLUTIONS FOR CHAPTER 3 DIAGNOSTIC TEST (page 158)

Following each problem number is the textbook section reference (in parentheses) where that kind of problem is discussed.

1. (305) (a) $\dfrac{7}{4} = 4\overline{)7}^{\,1\ R3} = 1\dfrac{3}{4}$

(b) $\dfrac{63}{29} = 29\overline{)63}^{\,2\ R5}\ \dfrac{58}{5} = 2\dfrac{5}{29}$

2. (306) (a) $2\dfrac{3}{4} = \dfrac{2 \cdot 4 + 3}{4} = \dfrac{8 + 3}{4} = \dfrac{11}{4}$

(b) $15\dfrac{7}{12} = \dfrac{15 \cdot 12 + 7}{12} = \dfrac{180 + 7}{12} = \dfrac{187}{12}$

3. (307) (a) Equivalent because $7 \times 27 = 9 \times 21$

$189 = 189$

(b) Not equivalent because $5 \times 73 \neq 9 \times 41$

$365 \neq 369$

Since $5 \times 73 < 9 \times 41$, then $\dfrac{5}{9} < \dfrac{41}{73}$

4. (314) (a) Prime
(b) Composite because $51 = 3 \times 17$
(c) Prime

5. (a) (321) $2 + 3\frac{1}{5} = 5\frac{1}{5}$

(b) (321) $4\frac{2}{3} = 4\frac{4}{6}$

$+ 3\frac{1}{2} = 3\frac{3}{6}$

─────

$\boxed{1}$

$\frac{7}{6} = \boxed{1}\ \frac{1}{6}$

$8\frac{1}{6}$

(c) (302) $\overset{1}{\cancel{\frac{5}{8}}}_2 \times \overset{1}{\cancel{\frac{2}{20}}}_4 = \frac{1}{8}$

(d) (302) $\cancel{\frac{2}{3}}_1 \cdot \overset{37}{\cancel{\frac{111}{1}}} = 74$

(e) (317) $\frac{3}{8} \div \frac{9}{16} = \frac{3}{\cancel{8}_1} \cdot \frac{\cancel{16}^2}{\cancel{9}_3} = \frac{2}{3}$

(f) (320C) $4\frac{1}{5} \cdot 2\frac{1}{7} = \frac{\cancel{21}^3}{\cancel{5}_1} \cdot \frac{\cancel{15}^3}{\cancel{7}_1} = 9$

(g) (320C) $2\frac{2}{9} \div 3\frac{1}{3} = \frac{20}{9} \div \frac{10}{3} = \frac{\cancel{20}^2}{\cancel{9}_3} \cdot \frac{\cancel{3}}{\cancel{10}_1} = \frac{2}{3}$

(h) (318) $\frac{\frac{5}{8}}{\frac{5}{6}} = \frac{5}{8} \div \frac{5}{6} = \frac{\cancel{5}}{\cancel{8}_4} \cdot \frac{\cancel{6}^3}{\cancel{5}} = \frac{3}{4}$

(i) (318) $\frac{\frac{3}{8} + \frac{5}{8}}{4\frac{1}{3}} = \frac{\frac{8}{8}}{\frac{13}{3}} = 1 \div \frac{13}{3}$

$= \frac{1}{1} \cdot \frac{3}{13} = \frac{3}{13}$

6. (311) (a) $\frac{\cancel{30}^5}{\cancel{42}_7} = \frac{5}{7}$

(b) $\frac{\cancel{180}}{\cancel{540}} = \frac{\cancel{18}^2}{\cancel{54}_6} = \frac{2}{6} = \frac{1}{3}$

7. (311D)

$217\overline{)248}$ quotient 1

$\underline{217}$

$31\overline{)217}$ quotient 7

$\underline{217}$

$31\overline{)217}$ quotient 7 $\qquad$ $31\overline{)248}$ quotient 8

$\underline{217}$ $\qquad\qquad$ $\underline{248}$

largest number that divides both 217 and 248

Therefore, $\frac{217}{248} = \frac{217 \div 31}{248 \div 31} = \frac{7}{8}$

8. (321) $27\frac{2}{9} = 27\frac{16}{72}$

$18\frac{1}{2} = 18\frac{36}{72}$

$4\frac{3}{4} = 4\frac{54}{72}$

$+ 10\frac{7}{8} = 10\frac{63}{72}$

─────

$\boxed{2}$

$\frac{169}{72} = \boxed{2}\ \frac{25}{72}$

$61\frac{25}{72}$

$\begin{array}{r|llll} 2 & 9 & 2 & 4 & 8 \\ 2 & 9 & 1 & 2 & 4 \\ \hline & 9 & 1 & 1 & 2 \end{array}$

LCD $= 2 \cdot 2 \cdot 9 \cdot 2$

$= 72$

9. (322) $124\frac{2}{3} = 124\frac{10}{15} = 123 + \frac{15}{15} + \frac{10}{15} = 123\frac{25}{15}$

$- 17\frac{4}{5} = - 17\frac{12}{15} \qquad\qquad = - 17\frac{12}{15}$

─────

$106\frac{13}{15}$

10. (324) (a) $6 + \frac{\cancel{2}^1}{1} \times \frac{3}{\cancel{4}_2} = 6 + \frac{3}{2} = 6 + 1\frac{1}{2} = 7\frac{1}{2}$

(b) $4\left(\frac{3}{4}\right)^2 - \frac{1}{2} = \frac{\cancel{4}}{1} \cdot \frac{3}{\cancel{4}} \cdot \frac{3}{4} - \frac{1}{2}$

$= \frac{9}{4} - \frac{2}{4} = \frac{7}{4} = 1\frac{3}{4}$

(324) (c) $\frac{2}{3} \div \frac{4}{3} \cdot \frac{2}{5}$

$= \frac{\cancel{2}^1}{\cancel{3}_1} \cdot \frac{\cancel{3}^1}{\cancel{4}_2} \cdot \frac{\cancel{2}^1}{5} = \frac{1}{5}$

11. (322) $1\frac{1}{8} = \frac{9}{8}$

$- \frac{3}{4} = - \frac{6}{8}$

─────

$\frac{3}{8}$

12. (320C) $7\frac{1}{2} \div 3 = \frac{15}{2} \div \frac{3}{1} = \frac{\cancel{15}^5}{2} \cdot \frac{1}{\cancel{3}_1} = \frac{5}{2} = 2\frac{1}{2}$

13. (320C) Area = Length × Width

$= 7\frac{1}{2} \times 4\frac{2}{3} = \frac{\cancel{15}^5}{\cancel{2}_1} \times \frac{\cancel{14}^7}{\cancel{3}_1} = 35$ sq yd

EXERCISES 402, SET I (page 167)

1. 6.4 $\qquad$ 2. 7.003
3. 12.02 $\qquad$ 4. 7.21
5. 0.35 $\qquad$ 6. 122.6
7. 5,086.07 $\qquad$ 8. 0.135
9. 700,052.0009 $\qquad$ 10. 8,040,005.2746
11. Eight tenths $\qquad$ 12. Ninety-five hundredths
13. Four and three hundred seventy-five thousandths
14. Twenty and six tenths·
15. Fifteen and sixty-five hundredths
16. One hundred thirty-seven and ninety-five hundredths
17. One thousand, one hundred fifteen
18. Six and forty-five ten-thousandths
19. Five and three thousand, seven hundred fifty-six ten-thousandths
20. Forty-seven thousand, twenty-eight and five thousand, three hundred sixty-one hundred-thousandths

EXERCISES 403, SET I (page 169)

1.
```
  111 1
   6.5
   0.66
  80.75
 287.
   0.078
 374.988
```

2. 127.698

3.
```
  122 1
 $ 0.35
  24.79
 127.50
  18.84
  96.
 $267.48
```

4. $361.52

5.
```
   1
  75.5
   3.45
 180.
   0.0056
 258.9556
```

6. 220.2467

7.
```
  3 222 32
    987.46
     35.778
  1,750.46
    706.188
  7,556.189
 11,036.075
```

8. 261,818.0489

9.
```
  23 3
 $ 5.33
   7.47
   3.89
   6.28
   4.96
  11.24
 $39.17
```

10. 52.2 gal

11.
```
 3,050.37
     5.00002
    70.0150
 3,125.38502
```

12. 13.375

EXERCISES 404, SET I (page 172)

1.
```
   7.85
 - 3.44
   4.41
```

2. 83.41

3.
```
         1
 208.50
 -  7.16
        1
 201.34
```

4. 687.56

5.
```
  11 111
 300.000
 -  0.145
 111 11
 299.855
```

6. 6,996.32

7.
```
 1 111 1
 81,284.56
 - 2,784.80
 11 111
 78,499.76
```

8. 1,970,030.2

9.
```
   1 11
   5.7850
 - 0.9665
   1 11
   4.8185
```

10. 5.1574

11.

Beginning balance plus deposits	Checks written	Computing balance
254.39	27.15	750.50
183.50	86.94	- 566.09
233.75	123.47	184.41 Ending balance
78.86	167.66	
750.50	122.20	
	38.67	
	566.09	

12. $389.42

EXERCISES 405, SET I (page 173)

1. 1   2. 2   3. 3   4. 0   5. 3
6. 4   7. 3   8. 3   9. 3   10. 4

EXERCISES 406, SET I (page 174)

2. 11.248

3.
```
    2 7.9
      1.5 4
    1 1 1 6
  1 3 9 5
  2 7 9
  4 2.9 6 6
```

4. 39,916.8

5.
```
      8.4 1 2
        0.2 5
    4 2 0 6 0
  1 6 8 2 4
  2.1 0 3 0 0
```

6. 129.1545

7.
```
   3 8 6.4 5
       0.0 0 5 6
   2 3 1 8 7 0
 1 9 3 2 2 5
 2.1 6 4 1 2 0
```

8. 76.8576

9.
```
   0.0 1 2 8
         3.2
       2 5 6
     3 8 4
   0.0 4 0 9 6
```

10. 0.38010

11.
```
   5.6 0 7
       8.7
   3 9 2 4 9
 4 4 8 5 6
 4 8.7 8 0 9
```

12. 6.71230

13.
```
   0.0 0 2 5 6 8
           0.8 5
       1 2 8 4 0
     2 0 5 4 4
   0.0 0 2 1 8 2 8 0
```

14. 141,480.00

15.
```
      2.5 6 |
        9 3|0 0 0
        7 6 8
      2 3 0 4
  2 3 8,0 8|0.0 0
```

16. 17,250.000

17.
```
 $ 1 7.6 3
       1 8
   1 4 1 0 4
   1 7 6 3
 $3 1 7.3 4
```

18. (a) $107.20
    (b) $147.40

19.
```
      2 1 5|0
        1 4.7|
      1 5 0 5
        8 6 0
        2 1 5
  3 1,6 0 5.|0  lb
```

20. (a) 31.128 lb per sq in
    (b) 66,925.200 lb
    (c) 146,157 lb
    (d) 2,316,472.128 lb

EXERCISES 407, SET I (page 179)

1. 7.2   2. 3.2   3. 6.2   4. 3.2
5. 0.1   6. 0.1   7. 13.1   8. 5.0
9. 3.1   10. 18.0   11. 8   12. 8
13. 10   14. 10   15. 11   16. 3
17. 1   18. 0   19. 5   20. 140
21. 1.24   22. 0.04   23. 0.04   24. 1.376
25. 5.007   26. 0.0568   27. 90   28. 10
29. 6.744   30. 0.500

EXERCISES 408A, SET I (page 180)

1.
```
    1 0.8 7
 8|8 6.9⁶9⁵6
```

2. 35.6

3.
```
    1 5.6
 6|9³3.³6
```

4. 74.8

5.
```
       1.6 5
 32|5 2.8 0
    3 2
    2 0 8
    1 9 2
      1 6 0
      1 6 0
```

6. 2.46

7.
```
       0.1 7 5
39 | 6.8 2 5
     3 9
     ─────
     2 9 2
     2 7 3
     ─────
       1 9 5
       1 9 5
```

8. 0.485

9.
```
        7.5
63 | 4 7 2.5
     4 4 1
     ─────
       3 1 5
       3 1 5
```

10. 3.66

11. $1.2 2 2 \doteq 1.22$
```
7 | 8.⁵5¹6²0
```

12. 50.7

13. $4 7.0 3 7 5 \doteq 47.038$
```
8 | 3 7⁵6.3³0⁶0⁴0
```

14. 85.783

15. $1.3 9 5 \doteq 1.40$
```
42 | 5 8.6 0 0
     4 2
     ─────
     1 6 6
     1 2 6
     ─────
       4 0 0
       3 7 8
       ─────
         2 2 0
         2 1 0
         ─────
           1 0
```

16. 1.48

17. $0.0 5 0 7 \doteq 0.051$
```
76 | 3.8 6 0 0
     3 8 0
     ─────
       6 0 0
       5 3 2
       ─────
         6 8
```

18. 0.0687

19. $0.3 6 7 \doteq 0.37$
```
208 | 7 6.5 0 0
      6 2 4
      ─────
      1 4 1 0
      1 2 4 8
      ───────
        1 6 2 0
        1 4 5 6
        ───────
          1 6 4
```

20. 0.163

EXERCISES 408B, SET I (page 183)

1.
```
          3.5 6
2.7∧ | 9.6∧1 2
       8 1
       ─────
       1 5 1
       1 3 5
       ─────
         1 6 2
         1 6 2
```

2. 4.71

3.
```
           7 8.4
6.1∧ | 4 7 8.2∧4
       4 2 7
       ─────
         5 1 2
         4 8 8
         ─────
           2 4 4
           2 4 4
```

4. 1.35

5.
```
        0.2 6 1
3∧0 0 | ∧7¹8.3
```

6. 0.081

7.
```
          0.1 4
1.4∧0 | 0.1∧9 6
         1 4
         ─────
          5 6
          5 6
```

8. 0.22

9.
```
          2 8 6.9
0.4 5∧ | 1 2 9.1 0∧5
         9 0
         ─────
         3 9 1
         3 6 0
         ─────
           3 1 0
           2 7 0
           ─────
             4 0 5
             4 0 5
```

10. 4.816

11.
```
                3.5 6
0.0 0 5 7∧ | 0.0 2 0 2∧9 2
             1 7 1
             ─────
             3 1 9
             2 8 5
             ─────
               3 4 2
               3 4 2
```

12. 0.00443

13.
```
              2.8 5
0.3 6 7∧ | 1.0 4 5∧9 5
           7 3 4
           ─────
           3 1 1 9
           2 9 3 6
           ───────
             1 8 3 5
             1 8 3 5
```

14. 3.76

15.
```
           0.2 2 7 0 ≐ 0.227
1 7∧0. | 3∧8.6 0 0
         3 4
         ─────
         4 6
         3 4
         ─────
         1 2 0
         1 1 9
         ─────
           1 0
```

16. 2.64

17.
```
           2 2.4 0 ≐ 22.4
3 5.7∧ | 8 0 0.0∧0 0
         7 1 4
         ─────
         8 6 0
         7 1 4
         ─────
         1 4 6 0
         1 4 2 8
         ───────
           3 2 0
```

18. 8.76

19.
```
            0.1 2 7 2 ≐ 0.127
7.5 5∧ | 0.9 6∧1 0 0 0
         7 5 5
         ─────
         2 0 6 0
         1 5 1 0
         ───────
           5 5 0 0
           5 2 8 5
           ───────
             2 1 5 0
             1 5 1 0
             ───────
               6 4 0
```

20. 0.01006

21.
```
           6 8.5 3 1 ≐ $68.53
36 | 2 4 6 7.1 4 0
     2 1 6
     ─────
       3 0 7
       2 8 8
       ─────
         1 9 1
         1 8 0
         ─────
           1 1 4
           1 0 8
           ─────
             6 0
             3 6
             ───
             2 4
```

22. 15.7

23.
```
            2 4. mo
16.27∧ | 3 9 0.4 8∧
         3 2 5 4
         ───────
         6 5 0 8
         6 5 0 8
```

24. 18 mo

EXERCISES 409, SET I (page 185)

1.
```
   2 1
  26.4      3
   0.072    0
 138.       3
   5.6      2
 ───────   ──
 170.072   (8)
```
(8) ────→ (8)

2. 48.80

3.
```
  3.7 4 6     2
  0.5 2       7
  ───────    ──
  7 4 9 2    14  ────→ (5)
1 8 7 3 0
─────────
1.9 4 7 9 2  → (5)
```

4. 4.15188

5.
```
         0.8 1 5        Check:
  5.8∧│4.7∧2 9 0        4.7290 = 5.8 × 0.815 + 20
       4 6 4          (4)      4 × 5    + 2
       ─────                   2        + 2  ───→ (4)
         8 9
         5 8
         ───
         3 1 0
         2 9 0
         ─────
           2 0
```

6. 3.72

EXERCISES 410, SET I (page 187)

1. Move decimal point 1 place to left in 9∧5.6, making it 9.56.
2. 0.798
3. Move decimal point 2 places to right in 27.80∧, making it 2,780.
4. 89.5
5. Move decimal point 2 places to left in 7∧5 0.2, making it 7.502.
6. 0.00614
7. Move decimal point 2 places to right in 9.8 4∧6, making it 984.6.
8. 83.7
9. Move decimal point 3 places to left in ∧100., making it 0.100.
10. 200,000
11. Move decimal point 4 places to right in 27.4000∧, making it 274,000.
12. 4,570
13. $\dfrac{146.35}{10} = 14.635 \doteq \$14.64$
14. $15,000
15. 537 × 100 = $53,700
16. $237.58

EXERCISES 411, SET I (page 188)

1.
```
   18│27
 = 2│3    Both dividend and divisor were
          divided by 9.
      1  R1
 = 2│3      = 1½
```

2. $33\overline{)714} = 11\overline{)238} = 21\dfrac{7}{11}$

3.
```
   25│2.45
 = 5│0.49    Both divided by 5
     0.0 9 8
 = 5│0.4⁴9⁴0
```

4. $24\overline{)14.4} = 6\overline{)3.6} = 0.6$

5.
```
   1.2│4.26
 = 12│42.6      Both multiplied by 10
 =  2│7.1       Both divided by 6
      3.5 5
 =  2│7.¹1¹0
```

6. $3.5\overline{)1.75} = 35\overline{)17.5} = 7\overline{)3.5} = 0.5$

7.
```
   30│71.1
 =  3│7.11      Both divided by 10
      2.3 7
 =  3│7.¹1 1
```

8. $200\overline{)785.6} = 2\overline{)7.856} = 3.928$

9.
```
   3,000│729.3
 =     3│0.7293    Both divided by 1,000
        0.2 4 3 1
 =     3│0.7¹2 9 3
```

10. $500\overline{)8,500} = 5\overline{)85} = 17$

11.
```
   200│0.0248
 =   2│0.000248    Dividing both by 100
      0.000124
 =   2│0.000248
```

12. $130\overline{)0.52} = 13\overline{)0.052} = 0.004$

13.
```
   4.44│88.80
 =  444│8,880     Multiplying both by 100
 =  111│2,220     Dividing both by 4
          20
 =  111│2,220
```

14. $10.5∧\overline{)3.9∧0} = 35\overline{)13} \doteq 0.37$

15.
```
   4.80│270.0
 =   48│2,700     Multiplying both by 10
 =   12│675       Dividing both by 4
 =    4│225       Dividing both by 3
       5 6.2 5  ≐ 56.2
 =    4│2 2²5⋮0²0
```

EXERCISES 412, SET I (page 190)

1. $\dfrac{3}{4} = 4\overline{)3.0²0}\;\;^{0.7\,5}$

2. 0.625

3. $2\dfrac{1}{2} = 2 + \dfrac{1}{2} = 2 + 0.5 = 2.5$

4. 5.75

5. $\dfrac{3}{8} = 8\overline{)3.0⁶0⁴0}\;\;^{0.3\,7\,5}$

6. 4.6

7. $\dfrac{5}{16} = 16\overline{)5.0000}\;\;^{0.3125}$
```
      4 8
      ──
       20
       16
       ──
        40
        32
        ──
         80
```

8. 1.25

9.
```
     0.28125
32 │ 9.00000
     6 4
     ───
     2 60
     2 56
     ────
       40
       32
       ──
       80
       64
       ──
      160
      160
```

10. 4.025

11.
```
      0.636  ≐ 0.64
11 │ 7.000
     6 6
     ───
      40
      33
      ──
      70
      66
      ──
       4
```

12. 4.714

13.
```
     0.2 2 2  ≐ 0.22
9 │ 2.0²0²0
```

14. 2.583

15. $7\frac{7}{8} = 7 + \frac{7}{8}$

$$\frac{7}{8} = 8 \overline{\smash{)}7.0^60^40} \quad \frac{0.8\ 7\ 5}{} \doteq 0.88$$

$7\frac{7}{8} \doteq 7 + 0.88$

$7\frac{7}{8} \doteq 7.88$

16. 5.156

17. No. The 0.315 pin is too large for the 0.3125 hole.

```
       .3125  hole
16 │ 5.0000
     4 8
     ───
      20
      16
      ──
      40
      32
      ──
      80
      80
```

18. $0.43

EXERCISES 413A, SET I (page 192)

1. $0.6 = \frac{6}{10} = \frac{3}{5}$

2. $\frac{4}{5}$

3. $0.05 = \frac{5}{100} = \frac{1}{20}$

4. $\frac{13}{20}$

5. $\begin{matrix}0.075\\1,000\end{matrix} \to \frac{75}{1,000} = \frac{3}{40}$

6. $\frac{3}{4}$

7. $\begin{matrix}0.875\\1,000\end{matrix} \to \frac{875}{1,000} = \frac{175}{200} = \frac{7}{8}$

8. $1\frac{4}{5}$

9. $\begin{matrix}2.5\\1\ 0\end{matrix} \to \frac{25}{10} = \frac{5}{2} = 2\frac{1}{2}$

10. $3\frac{7}{10}$

11. $\begin{matrix}5.9\\1\ 0\end{matrix} \to \frac{59}{10} = 5\frac{9}{10}$

12. $4\frac{3}{10}$

13. $\begin{matrix}0.0625\\1\ 0000\end{matrix} \to \frac{625}{10,000} = \frac{25}{400}$

14. $2\frac{1}{8}$

$$= \frac{5}{80} = \frac{1}{16}$$

15. $37.5 = 37\frac{5}{10} = 37\frac{1}{2}$

16. $2\frac{3}{16}$

17. $65.625 = 65 + 0.625$

$$\begin{matrix}0.625\\1,000\end{matrix} \to \frac{625}{1,000} = \frac{25}{40} = \frac{5}{8}$$

$$65.625 = 65 + \frac{5}{8} = 65\frac{5}{8}$$

18. $\frac{7}{8,000}$

19. $\begin{matrix}0.875\\1,000\end{matrix} \to \frac{875}{1,000} = \frac{7}{8}$ in

20. $\frac{5}{8}$ in

EXERCISES 413B, SET I (page 193)

1. $\begin{matrix}0.37\frac{1}{2}\\1\ 0\ 0\end{matrix} \to \frac{37\frac{1}{2}}{100} = \frac{\frac{75}{2}}{100} = \frac{75}{2} \div \frac{100}{1}$

$$= \frac{\overset{3}{\cancel{75}}}{2} \cdot \frac{1}{\underset{4}{\cancel{100}}} = \frac{3}{8}$$

2. $\frac{5}{8}$

3. $\begin{matrix}0.33\frac{1}{3}\\1\ 0\ 0\end{matrix} \to \frac{33\frac{1}{3}}{100} = \frac{\frac{100}{3}}{100} = \frac{100}{3} \div \frac{100}{1}$

$$= \frac{\cancel{100}}{3} \cdot \frac{1}{\cancel{100}} = \frac{1}{3}$$

4. $\frac{1}{8}$

5. $\begin{matrix}0.5\frac{3}{4}\\1\ 0\end{matrix} \to \frac{5\frac{3}{4}}{10} = \frac{\frac{23}{4}}{10} = \frac{23}{4} \div \frac{10}{1}$

$$= \frac{23}{4} \cdot \frac{1}{10} = \frac{23}{40}$$

6. $2\frac{1}{16}$

7. $\begin{matrix}1.0\frac{1}{5}\\1\ 0\end{matrix} \to \frac{10\frac{1}{5}}{10} = \frac{\frac{51}{5}}{10} = \frac{51}{5} \div \frac{10}{1}$

$$= \frac{51}{5} \cdot \frac{1}{10} = \frac{51}{50} = 1\frac{1}{50}$$

8. $\frac{1}{15}$

9. $\begin{matrix}0.00\frac{5}{12}\\1\ 0\ 0\end{matrix} \to \frac{\frac{5}{12}}{100} = \frac{5}{12} \div \frac{100}{1}$

$$= \frac{\overset{1}{\cancel{5}}}{12} \cdot \frac{1}{\underset{20}{\cancel{100}}} = \frac{1}{240}$$

10. $2\frac{1}{400}$

11. $\begin{matrix}0.001\frac{1}{6}\\1,00\ 0\end{matrix} \to \frac{1\frac{1}{6}}{1,000} = \frac{\frac{7}{6}}{1,000} = \frac{7}{6} \div \frac{1,000}{1}$

$$= \frac{7}{6} \cdot \frac{1}{1,000} = \frac{7}{6,000}$$

12. $1\frac{43}{800}$

1. $\frac{3}{4} \times 5.24 = 0.75 \times 5.24$

$$
\begin{array}{r}
5.2\ 4 \\
0.7\ 5 \\
\hline
2\ 6\ 2\ 0 \\
3\ 6\ 6\ 8 \\
\hline
3.9\ 3\ 0\ 0
\end{array}
$$

2. 19.5

3. $2\frac{1}{2} \times 7.4 = 2.5 \times 7.4$

$$
\begin{array}{r}
7.4 \\
2.5 \\
\hline
3\ 7\ 0 \\
1\ 4\ 8 \\
\hline
1\ 8.5\ 0
\end{array}
$$

4. 7.93

5. $2\frac{3}{8} \times 13.6 = 2.375 \times 13.6$
   $= 32.3$

6. 5.325

7. $\frac{7}{10} \times 56.5 = 0.7 \times 56.5$
   $= 39.55$

8. 18.72

9. $\frac{7}{\overset{\cancel{12}}{\underset{3}{}}} \times \overset{14 \cdot 1}{\cancel{56.4}} = \frac{98.7}{3} = 32.90$

10. 69.833

11. $1\frac{9}{14} \times 7.72 = \frac{23}{\underset{7}{\cancel{14}}} \times \frac{\overset{3 \cdot 86}{\cancel{7.72}}}{1}$

$= \frac{88.78}{7} \doteq 12.683$

12. 14.566

13. $7\frac{1}{2} \times 2.45 = 7.5 \times 2.45$
    $= \$18.375 \doteq \$18.38$

14. $\doteq \$9.22$

EXERCISES 415, SET I (page 196)

1. 0.409, 0.490, 0.410

   Since   0.490 > 0.410 > 0.409
   therefore, 0.49 > 0.41 > 0.409
2. 0.35 > 0.305 > 0.3
3. 3.075, 3.100, 3.050, 3.009

   Since   3.100 > 3.075 > 3.050 > 3.009
   therefore, 3.1 > 3.075 > 3.05 > 3.009
4. 7.1 > 7.099 > 7.08 > 7.0
5. 0.07500, 0.07501, 0.74900, 0.700000

   Since   0.70000 > 0.07501 > 0.07500 > 0.07490
   therefore, 0.7 > 0.07501 > 0.075 > 0.0749
6. 0.6 > 0.1998 > 0.06 > 0.059
7. 5.0500, 5.5000, 5.0501, 5.0000, 5.0496

   Since   5.5000 > 5.0501 > 5.0500 > 5.0496 > 5.0000
   therefore, 5.5 > 5.0501 > 5.05 > 5.0496 > 5
8. 3.199 > 3.0695 > 3.051 > 3.0505 > 3

REVIEW EXERCISES 416, SET I (page 197)

1.
$$
\begin{array}{r}
\overset{1\ \ 331\ \ 12}{} \\
75.23 \\
186.56 \\
7,896,448. \\
8.007 \\
386.759 \\
.0058 \\
\hline
7,897,104.5618
\end{array}
$$

2. 4,847.1326

3.
$$
\begin{array}{r}
509.\overset{1}{6}\overset{1}{0} \\
-\ 81.34 \\
\hline
\overset{1}{\ }\ \overset{1}{\ }\ 428.26
\end{array}
$$

4. 19.43

5. $100 \times 7.78_\wedge = 778.$
6. 65.
7. $7_\wedge 89. \div 100 = 7.89$
8. 0.47
9. $1,000 \times 75.000_\wedge = 75,000.$
10. 425.
11. $_\wedge 00.064 \div 10^2 = 0.00064$
12. 0.0425
13. $10^4 \times 0.0056_\wedge = 56.$
14. 7.5
15. $7\ 8_\wedge 6,0\ 0\ 0. \div 10,000 = 78.6$
16. 85.6
17. $2 \times 5 \times 78 = 10 \times 78.0_\wedge = 780.$
18. 45,000
19. $5 \times 85.4 \times 2 = 10 \times 85.4_\wedge$
    $= 854.$
20. 16.7
21. $4 \times 9.6 \times 25 = 100 \times 9.60_\wedge$
    $= 960.$
22. 350

23.
$$
\require{enclose}
\begin{array}{r}
1.4\ 4\ 2 \\
6_\wedge 0\ 0\ \enclose{longdiv}{8_\wedge 6^2 6^2 5^1 2}
\end{array}
$$

24. 0.2034

25.
$$
\begin{array}{r}
0.2\ 6 \\
2\ 6_\wedge 0\ \enclose{longdiv}{6_\wedge 7.6} \\
5\ 2 \\
\hline
1\ 5\ 6 \\
1\ 5\ 6 \\
\hline
\end{array}
$$

26. 0.51

27.
$$
\frac{7}{8} = \begin{array}{r}
0.8\ 7\ 5 \\
8\ \enclose{longdiv}{7.0^6 0^4 0}
\end{array}
$$

28. 0.3125

29. $2\frac{3}{4} = 2 + \frac{3}{4} = 2 + 0.75 = 2.75$

$$
\frac{3}{4} = \begin{array}{r}
0.7\ 5 \\
4\ \enclose{longdiv}{3.0^2 0}
\end{array}
$$

30. 5.375

31.
$$
\begin{array}{r}
3.4 \\
5\ \enclose{longdiv}{1\ 7.^2 0}
\end{array}
$$

32. 1.55

33.
$$
\begin{array}{r}
0.8\ 3\ 3 \doteq 0.83 \\
6\ \enclose{longdiv}{5.0^2 0^2 0}
\end{array}
$$

34. 0.58

35.
$$
3\frac{2}{3} = \frac{11}{3} = \begin{array}{r}
3.6\ 6\ 6\ 6 \doteq 3.667 \\
3\ \enclose{longdiv}{1\ 1.^2 0^2 0^2 0^2 0}
\end{array}
$$

36. 4.562

37. $0.\overset{|}{6}\overset{|}{8} \to \frac{68}{100} = \frac{34}{50} = \frac{17}{25}$

38. $\frac{27}{50}$

39. $3.\overset{|}{8}\overset{|}{5} \to \frac{385}{100} = \frac{77}{20} = 3\frac{17}{20}$

40. $5\frac{13}{50}$

41. $8.7\overset{|}{\frac{1}{2}} \to \frac{87\frac{1}{2}}{10} = \frac{\frac{175}{2}}{10} = \frac{175}{2} \div \frac{10}{1} = \frac{\overset{35}{\cancel{175}}}{2} \cdot \frac{1}{\underset{2}{\cancel{10}}} = \frac{35}{4} = 8\frac{3}{4}$

42. $\frac{1}{16}$

43. 0.1075, 0.0900, 0.1100, 0.2000

    Since   0.2000 > 0.1100 > 0.1075 > 0.0900
    therefore, 0.2 > 0.11 > 0.1075 > 0.09

44. 3.1 > 3.0501 > 3.05 > 3 > 2.99

45.
```
      9 4.7 8
        7 0.9
      ─────────
      8 5 3 0 2
    6 6 3 4 6 0
    ─────────────
    6,7 1 9.9 0 2 ≐ 6,719.9
```

46. 2,798.5

47.
```
                0.8 2 8 ≐ 0.83
    7.2 5∧│6.0 0∧7 0 0
            5 8 0 0
          ─────────
            2 0 7 0
            1 4 5 0
          ─────────
              6 2 0 0
              5 8 0 0
            ─────────
                4 0 0
```

48. 1.71

49.
```
    121 1
    56.75
    186.3
      8.388
    ─────────
    251.438 ≐ 251.44
```

50. 825.97

51.
```
    3 8 7.6¹1¹0
  - 2 4 6.5 9 3
        1 1
  ───────────────
    1 4 1.0 1 7 ≐ 141.02
```

52. 17.59

53. 100 · 5 ÷ 4 + 7.25
  =   500   ÷ 4 + 7.25
  =   125   + 7.25
  = 132.25

54. 107.25

55. $3(2^3) + 2.3 \times 5.7 - 7.55$
  = 3(8)   + 2.3 × 5.7 - 7.55
  = 24   + 13.11   - 7.55
  =   37.11   - 7.55
  = 29.56

56. 72.25

57. $\dfrac{2}{\cancel{3}_1} \times \dfrac{\overset{7.7}{\cancel{23.1}}}{1} = 2 \times 7.7 = 15.4$

58. 1.77

59. $2\dfrac{1}{4} \times 3.48 = \dfrac{9}{\cancel{4}_1} \times \dfrac{\overset{0.87}{\cancel{3.48}}}{1}$

             = 9 × 0.87 = 7.83

60. 29.7

61. $\dfrac{11}{\cancel{12}_6} \times \dfrac{\overset{4.37}{\cancel{8.74}}}{1} = \dfrac{48.07}{6} = 6\overline{)48.07}\;\;\overset{8.01}{}\;≐ 8.0$

62. 23.5

63. shoes   = $19.95
    2 neckties   2 × 3.95 =   7.90
    3 pair socks 3 × 1.25 =   3.75
    suit   =   89.95
                      $121.55

64. $267.11

65. $4\dfrac{2}{3} \times 3.49 = \dfrac{14}{3} \times \dfrac{3.49}{1} = \dfrac{48.86}{3} = 3\overline{)4^1 8.8^2 6^2 0}\;\;\overset{1 6.2 8 6}{}$

                                     ≐ $16.29

66. $19.72

67.
```
      2 3 4 5
        3 6.9
    ───────────
    2 1 1 0 5
  1 4 0 7 0
  7 0 3 5
  ─────────────
  8 6,5 3 0.5 lb
```

68. ≐ $0.065 per mi

    This is $6\dfrac{1}{2}$¢

    per mi.

69.
```
       14 gal used        $0.429
   14│196                  × 14
       14                  1 716
     ─────                 4 29
       56                  ────────
       56                  6.006 ≐ $6.01 spent
```

70. Pin: $0.05
    Tie: $1.05

71. Yes; 1,924 pennies = $19.24   72. 1 dozen

SOLUTIONS FOR CHAPTER 4 DIAGNOSTIC TEST (page 202)

Following each problem number is the textbook section reference (in parentheses) where that kind of problem is discussed.

1. (402) (a) 0.125
        (b) 8,050.07
        (c) 16.0973

2. (402) (a) Sixty-seven hundredths
        (b) Eighty-one and twelve thousandths

3. (403)
```
    ¹¹7.8
     56.
      .017
    ─────────
    500.94
    564.757
```

4. (404)
```
    4¹0.6¹0
  -   8.5 4
      1 1
  ─────────
    3 2.0 6
```

5. (404)
```
     ¹
     9.0 7 3
   - 0.8 7
     1
   ─────────
     8.2 0 3
```

6. (406)
```
          3.7 5
          0.0 5 8
        ─────────
          3 0 0 0
        1 8 7 5
        ───────────
        0.2 1 7 5 0
```

7. (408)
```
                7.8 6
    3.5∧│2 7.5∧1 0
            2 4 5
          ─────────
            3 0 1
            2 8 0
          ─────────
              2 1 0
              2 1 0
```

8. (407) (a) 2.817   Since the first digit to be
          2.82   dropped is greater than 5,
                    increase the preceding digit
                    by one.
        (b) 54.749 Since the first digit to be
          54.7   dropped is less than 5, the
                    part retained is unchanged.
        (c) 0.0465 Since the first digit to be
          0.046   dropped is 5 and the digit in
                    the place we are rounding off
                    to is even, it remains
                    unchanged.

9. (408)
```
                7.8 2 ≐ 7.8
    0.5 6∧│4.3 8∧0 0
            3 9 2
          ─────────
            4 6 0
            4 4 8
          ─────────
              1 2 0
              1 1 2
            ─────────
                  8
```

10. (412)
```
            0.8 3 3 ≐ 0.83
    5/6 = 6│5.0²0²0
```

460   Answers

11. (412) $5\frac{3}{16} = \frac{83}{16} = 5.1875$

$$\begin{array}{r} 5.1875 \\ 16\overline{\smash{)}83.0000} \\ \underline{80} \\ 30 \\ \underline{16} \\ 140 \\ \underline{128} \\ 120 \\ \underline{112} \\ 80 \\ \underline{80} \end{array}$$

12. (413) $0.78 = \frac{78}{100} = \frac{39}{50}$

13. (410) (a) $100 \times 5.81_\wedge6 = 581.6$

(b) $\frac{76_\wedge4.1}{10} = 76.41$

(c) $3.9 \times 10^3 = 3.900_\wedge \times 1,000 = 3,900.$

(d) $\frac{41.8}{10^2} = {_\wedge}\frac{41.8}{100} = 0.418$

14. (414) (a) $\frac{2}{\underset{1}{3}} \times \frac{\overset{22 \cdot 4}{67 \cdot 2}}{1} = 2 \times 22.4 = 44.8$

(414) (b) $2\frac{3}{5} \times 8.41 = 2.6 \times 8.41 = 21.866 \doteq 21.9$

(415) (c) $5(2^3) - 2.4 \div 0.8 = 5(8) - \frac{\overset{3}{2.4}}{\underset{1}{0.8}}$

$= 40 - 3 = 37$

15. (406)

$$\begin{array}{r} 2\ 3\ 2\ 5 \\ 3\ 0.2 \\ \hline 4\ 6\ 5\ 0 \\ 6\ 9\ 7\ 5 \\ \hline 7\ 0\ 2\ 1\ 5.0\ \text{lb} \end{array}$$

16. (408)

$$\begin{array}{r} 79.722 \doteq \$79.72 \\ 36\overline{\smash{)}2870.000} \\ \underline{252} \\ 350 \\ \underline{324} \\ 260 \\ \underline{252} \\ 80 \\ \underline{72} \\ 80 \\ \underline{72} \\ 8 \end{array}$$

17. (415)

Balance at beginning of month	$346.52
Deposit	325.00
	671.52
Less checks	495.53 ←
Balance at end of month	$175.99

Checks:	$ 17.75
	64.57
	91.35
	135.46
	186.40
	$495.53 ←

EXERCISES 501, SET I (page 206)

1. $\frac{40}{27}$ 2. $\frac{117}{38}$ 3. $\frac{53}{307}$ 4. $\frac{3}{11}$ 5. $\frac{240}{13}$

6. (a) $\frac{7}{9}$ (b) $\frac{7}{2}$ (c) $\frac{2}{7}$

EXERCISES 502, SET I (page 208)

1. $\frac{\overset{2}{28}}{\underset{1}{14}} = \frac{2}{1}$ 2. $\frac{29}{12}$

3. $\frac{52}{39} = \frac{52 \div 13}{39 \div 13} = \frac{4}{3}$ 4. $\frac{9}{4}$

$$39\overline{\smash{)}52}^{\ 1} \qquad 13\overline{\smash{)}52}^{\ 4} \qquad 13\overline{\smash{)}39}^{\ 3}$$
$$\underline{39}\ {}^{3}_{13\overline{)39}} \qquad \underline{52} \qquad \underline{39}$$
$$\underline{39}$$
└─largest number that divides both 52 and 39

5. $\frac{119}{153} = \frac{119 \div 17}{153 \div 17} = \frac{7}{9}$

$$119\overline{\smash{)}153}^{\ 1} \qquad 17\overline{\smash{)}119}^{\ 7} \qquad 17\overline{\smash{)}153}^{\ 9}$$
$$\underline{119}\ {}^{3}_{34\overline{)119}} \qquad \underline{119} \qquad \underline{153}$$
$$\underline{102}\ {}^{2}_{17\overline{)34}}$$
$$\underline{34}$$
└─largest number that divides both 119 and 153

6. $\frac{3}{4}$ 7. $\frac{\overset{17}{85}\text{ cents}}{\underset{3}{15}\text{ cans}} = \frac{17\text{ cents}}{3\text{ cans}}$

8. 5 qt to 11 children

9. $\frac{\overset{7}{42}\text{ yards}}{\underset{2}{12}\text{ dresses}} = \frac{7\text{ yards}}{2\text{ dresses}}$

10. 5 radios to 1 student

11. $\frac{6\text{ hours}}{40\text{ minutes}} = \frac{\overset{3}{6} \times \overset{3}{60}\text{ minutes}}{\underset{\underset{1}{2}}{40}\text{ minutes}} = \frac{9}{1}$ 12. $\frac{10}{9}$

13. $\frac{88\text{ cents}}{2\text{ dollars}} = \frac{\overset{11}{88}\text{ cents}}{\underset{25}{200}\text{ cents}} = \frac{11}{25}$ 14. $\frac{3}{20}$

15. $\frac{1\text{ hour}}{10\text{ minutes}} = \frac{\overset{6}{60}\text{ minutes}}{\underset{1}{10}\text{ minutes}} = \frac{6}{1}$ 16. $\frac{7}{1}$

17. $\frac{\overset{12}{24}\text{ hours}}{\underset{1}{2}\text{ hours}} = \frac{12}{1}$ 18. $\frac{1}{12}$

19. $\frac{3\text{ inches}}{1\text{ foot}} = \frac{\overset{1}{3}\text{ inches}}{\underset{4}{12}\text{ inches}} = \frac{1}{4}$ 20. $\frac{3}{4}$

21. $\frac{\overset{10}{50¢}}{\underset{1}{5¢}} = \frac{10}{1}$ 22. $\frac{13}{8}$

23. $\frac{\overset{7}{175}\text{ pounds}}{\underset{1}{25}\text{ pounds}} = \frac{7}{1}$ 24. $\frac{5}{3}$ 25. $\frac{\overset{11}{22}}{\underset{35}{70}} = \frac{11}{35}$

EXERCISES 503, SET I (page 210)

1. $\frac{\frac{1}{2}}{2} = \frac{1}{2} \div \frac{2}{1} = \frac{1}{2} \cdot \frac{1}{2} = \frac{1}{4}$ 2. $\frac{1}{32}$

3. $\frac{4}{\frac{3}{8}} = 4 \div \frac{3}{8} = \frac{4}{1} \cdot \frac{8}{3} = \frac{32}{3}$ 4. $\frac{27}{7}$

5. $\frac{\frac{3}{4}}{\frac{5}{8}} = \frac{3}{4} \div \frac{5}{8} = \frac{3}{\underset{1}{4}} \cdot \frac{\overset{2}{8}}{5} = \frac{6}{5}$ 6. $\frac{5}{2}$

7. $\dfrac{2\frac{1}{2}}{5} = 2\frac{1}{2} \div 5 = \dfrac{\cancel{5}^{1}}{2} \cdot \dfrac{1}{\cancel{5}_{1}} = \dfrac{1}{2}$ 8. $\dfrac{1}{3}$

9. $\dfrac{6}{1\frac{3}{5}} = 6 \div 1\frac{3}{5} = 6 \div \dfrac{8}{5} = \dfrac{\cancel{6}^{3}}{1} \cdot \dfrac{5}{\cancel{8}_{4}} = \dfrac{15}{4}$ 10. $\dfrac{5}{2}$

11. $\dfrac{2\frac{2}{3}}{1\frac{1}{15}} = 2\frac{2}{3} \div 1\frac{1}{15} = \dfrac{8}{3} \div \dfrac{16}{15} = \dfrac{\cancel{8}^{1}}{\cancel{3}_{1}} \cdot \dfrac{\cancel{15}^{5}}{\cancel{16}_{2}} = \dfrac{5}{2}$

12. $\dfrac{3}{4}$

13. (a) $\dfrac{6}{4\frac{1}{2}} = 6 \div 4\frac{1}{2} = 6 \div \dfrac{9}{2} = \dfrac{\cancel{6}^{2}}{1} \cdot \dfrac{2}{\cancel{9}_{3}} = \dfrac{4}{3}$

    (b) $\dfrac{4}{3} = 3\overline{)4}^{\,1\ \text{R}1} = 1\frac{1}{3}$ 14. $\dfrac{4}{1}$

15. $\dfrac{9\frac{3}{4}}{16\frac{1}{2}} = 9\frac{3}{4} \div 16\frac{1}{2} = \dfrac{\cancel{39}^{13}}{\cancel{4}_{2}} \cdot \dfrac{\cancel{2}^{1}}{\cancel{33}_{11}} = \dfrac{13}{22}$

16. (a) $\dfrac{16}{5}$ or $\dfrac{5}{16}$ (b) $3\frac{1}{5}$ times

EXERCISES 504, SET I (page 212)

1. (a) 12   (b) 5   (c) 24   (d) 10
   (e) 5 and 24   (f) 12 and 10
2. (a) 8   (b) 14   (c) 16   (d) x
   (e) 14 and 16   (f) 8 and x

EXERCISES 505, SET I (page 214)

1. Yes, because 3 × 10 = 5 × 6
            30   =   30   2. Yes
3. No, because 2 × 7 ≠ 3 × 5
           14   ≠   15   4. No
5. Yes, because 6 × 6 = 9 × 4
            36   =   36   6. Yes
7. No, because 21 × 15 ≠ 17 × 19
           315   ≠   323   8. No
9. Yes, because 36 × 26 = 39 × 24
            936   =   936   10. No
11. Yes, because 12 × 42 = 18 × 28
            504   =   504   12. Yes
13. No, because 114 × 130 ≠ 162 × 95
          14,820   ≠   15,390   14. Yes

EXERCISES 506, SET I (page 217)

1. $\dfrac{x}{4} = \dfrac{2}{3}$ 2. $7\frac{1}{2}$

   $3x = 8$

   $\dfrac{\cancel{3}x}{\cancel{3}} = \dfrac{8}{3}$

   $x = \dfrac{8}{3} = 2\frac{2}{3}$

3. $\dfrac{8}{x} = \dfrac{4}{5}$ 4. $2\frac{2}{3}$

   $4x = 40$

   $\dfrac{\cancel{4}x}{\cancel{4}} = \dfrac{\cancel{40}}{\cancel{4}}$

   $x = 10$

5. $\dfrac{4}{7} = \dfrac{x}{21}$ 6. $11\frac{1}{4}$

   $7x = 84$

   $\dfrac{\cancel{7}x}{\cancel{7}} = \dfrac{\cancel{84}}{\cancel{7}}$

   $x = 12$

7. $\dfrac{4}{13} = \dfrac{16}{x}$ 8. 9

   $4x = 13(16)$

   $\dfrac{\cancel{4}x}{\cancel{4}} = \dfrac{13(\cancel{16})}{\cancel{4}}$

   $x = 52$

9. $\dfrac{x}{18} = \dfrac{24}{30}$ 10. $9\frac{1}{3}$

   $5x = 4 \cdot 18$

   $\dfrac{\cancel{5}x}{\cancel{5}} = \dfrac{72}{5}$

   $x = 14\frac{2}{5}$

11. $\dfrac{15}{22} = \dfrac{x}{33}$ 12. 40

   $\dfrac{\cancel{22}x}{\cancel{22}} = \dfrac{15(\cancel{33})}{\cancel{22}}$

   $x = \dfrac{45}{2} = 22\frac{1}{2}$

13. $\dfrac{55}{x} = \dfrac{35}{28}$ 14. 8

   $5x = 4 \cdot 55$

   $\dfrac{\cancel{5}x}{\cancel{5}} = \dfrac{4 \cdot 55}{\cancel{5}}$

   $x = 44$

15. $\dfrac{100}{x} = \dfrac{40}{30}$ 16. 24

   $4x = 300$

   $\dfrac{\cancel{4}x}{\cancel{4}} = \dfrac{\cancel{300}}{\cancel{4}}$

   $x = 75$

17. $\dfrac{x}{100} = \dfrac{75}{125}$ 18. 36

   $5x = 300$

   $\dfrac{\cancel{5}x}{\cancel{5}} = \dfrac{\cancel{300}}{\cancel{5}}$

   $x = 60$

19. $\dfrac{39}{x} = \dfrac{\overset{13}{\cancel{104}}}{\underset{6}{\cancel{48}}}$ 20. 63

$13x = 6(39)$

$\dfrac{\cancel{13}x}{\cancel{13}} = \dfrac{6(\overset{3}{\cancel{39}})}{\underset{1}{\cancel{13}}}$

$x = 18$

EXERCISES 507, SET I (page 218)

1. $\dfrac{\frac{3}{4}}{6} = \dfrac{P}{16}$ 2. $2\dfrac{1}{2}$

$6P = \dfrac{3}{\cancel{4}} \cdot \dfrac{\overset{4}{\cancel{16}}}{1}$

$6P = 12$

$\dfrac{\cancel{6}P}{\cancel{6}} = \dfrac{\overset{2}{\cancel{12}}}{\underset{1}{\cancel{6}}}$

$P = 2$

3. $\dfrac{A}{9} = \dfrac{3\frac{1}{3}}{5}$ 4. 1

$5A = 3\dfrac{1}{3}(9) = \dfrac{10}{\cancel{3}} \cdot \dfrac{\overset{3}{\cancel{9}}}{1} = 30$

$\dfrac{\cancel{5}A}{\cancel{5}} = \dfrac{\overset{6}{\cancel{30}}}{\underset{1}{\cancel{5}}}$

$A = 6$

5. $\dfrac{7.7}{B} = \dfrac{3.5}{5}$ 6. 22.96

$3.5B = 5(7.7) = 38.5$

$\dfrac{\cancel{3.5}B}{\cancel{3.5}} = \dfrac{38.5}{3.5}$

$B = 11$

7. $\dfrac{P}{100} = \dfrac{\frac{3}{2}}{15}$ 8. 4

$15P = \dfrac{3}{\cancel{2}}\left(\dfrac{\overset{50}{\cancel{100}}}{1}\right) = 150$

$\dfrac{\cancel{15}P}{\cancel{15}} = \dfrac{\overset{10}{\cancel{150}}}{\underset{1}{\cancel{15}}}$

$P = 10$

9. $\dfrac{12\frac{1}{2}}{100} = \dfrac{A}{48}$ 10. 54

$100A = 12\dfrac{1}{2}(48) = \dfrac{25}{\cancel{2}} \cdot \dfrac{\overset{24}{\cancel{48}}}{1} = 600$

$\dfrac{\cancel{100}A}{\cancel{100}} = \dfrac{\overset{6}{\cancel{600}}}{\underset{1}{\cancel{100}}}$

$A = 6$

11. $\dfrac{2.54}{1} = \dfrac{x}{7.5}$ 12. 25.64

$x = 7.5(2.54) = 19.05$

EXERCISES 508, SET I (page 221)

1. $\dfrac{3 \text{ gallons}}{2 \text{ rooms}} = \dfrac{x \text{ gallons}}{20 \text{ rooms}}$ 2. $7\dfrac{1}{2}$ days

$2x = 60$

$\dfrac{\cancel{2}x}{\cancel{2}} = \dfrac{\overset{30}{\cancel{60}}}{\underset{1}{\cancel{2}}}$

$x = 30 \text{ gal}$

3. $\dfrac{6 \text{ height}}{4 \text{ shadow}} = \dfrac{H \text{ height}}{20 \text{ shadow}}$ 4. 50

$\dfrac{\cancel{4}H}{\cancel{4}} = \dfrac{\overset{30}{\cancel{120}}}{\underset{1}{\cancel{4}}}$

$H = 30 \text{ ft}$

5. $\dfrac{1 \text{ inch}}{8 \text{ feet}} = \dfrac{2\frac{1}{2} \text{ inches}}{W \text{ feet}}$ 6. $750

$W = 2\dfrac{1}{2}(8) = \dfrac{5}{\cancel{2}} \cdot \dfrac{\overset{4}{\cancel{8}}}{1} = 20 \text{ ft}$

$\dfrac{1 \text{ inch}}{8 \text{ feet}} = \dfrac{3 \text{ inches}}{L \text{ feet}}$

$L = 24 \text{ ft}$

7. $\dfrac{3{,}000 \text{ investment}}{180 \text{ income}} = \dfrac{P \text{ investment}}{540 \text{ income}}$ 8. 21

$\dfrac{\overset{100}{\cancel{3{,}000}}}{\underset{6}{\cancel{180}}} = \dfrac{P}{540}$

$6P = (100)(540)$

$\dfrac{\cancel{6}P}{\cancel{6}} = \dfrac{(100)(\overset{90}{\cancel{540}})}{\underset{1}{\cancel{6}}}$

$P = \$9{,}000$

9. $\dfrac{2\frac{1}{2} \text{ quarts}}{1{,}800 \text{ miles}} = \dfrac{x \text{ quarts}}{12{,}000 \text{ miles}}$

$1{,}800x = 2\dfrac{1}{2}(12{,}000) = \dfrac{5}{\cancel{2}} \cdot \dfrac{\overset{6{,}000}{\cancel{12{,}000}}}{1} = 30{,}000$

$\dfrac{\cancel{1{,}800}x}{\cancel{1{,}800}} = \dfrac{\overset{50}{\cancel{30{,}000}}}{\underset{3}{\cancel{1{,}800}}}$

$x = \dfrac{50}{3} = 16\dfrac{2}{3} \text{ quarts}$

10. 300

1. $\dfrac{\overset{9}{\cancel{27}}}{\underset{4}{\cancel{12}}} = \dfrac{9}{4}$

2. $\dfrac{9}{7}$

3. $\dfrac{2 \text{ yards}}{9 \text{ feet}} = \dfrac{\overset{2}{\cancel{6} \text{ feet}}}{\underset{3}{\cancel{9} \text{ feet}}} = \dfrac{2}{3}$

4. $\dfrac{25}{6}$

    (2 yd = 6 ft)

5. Yes, because $50 \times 300 = 120 \times 125$
                $15{,}000 = 15{,}000$

6. No

7. Yes, because $63 \times 36 = 81 \times 28$
                $2{,}268 = 2{,}268$

8. No

9. $\dfrac{\overset{9}{\cancel{45}}}{\underset{14}{\cancel{70}}} = \dfrac{x}{56}$

10. 63

    $14x = 504$

    $\dfrac{\cancel{14}x}{\cancel{14}} = \dfrac{504}{14}$

    $x = 36$

11. $\dfrac{\overset{8}{\cancel{24}}}{\underset{5}{\cancel{15}}} = \dfrac{18}{x}$

12. $\dfrac{9}{20}$

    $8x = 90$

    $\dfrac{\cancel{8}x}{\cancel{8}} = \dfrac{90}{8}$

    $x = 11\frac{1}{4}$

13. $\dfrac{\overset{7}{\cancel{2.8}}}{\underset{13}{\cancel{5.2}}} = \dfrac{A}{6.5}$

14. $\dfrac{3}{4}$

    $13A = 7(6.5)$

    $\dfrac{\cancel{13}A}{\cancel{13}} = \dfrac{7(6.5)}{13}$

    $A = 3.5$

15. $\dfrac{5 \text{ pounds}}{62 \text{ cents}} = \dfrac{2\frac{1}{2} \text{ pounds}}{x \text{ cents}}$

16. $1\frac{1}{7}$ pounds

    $5x = \dfrac{5}{\underset{1}{\cancel{2}}} \cdot \dfrac{\overset{31}{\cancel{62}}}{1}$

    $\dfrac{\cancel{5}x}{\cancel{5}} = \dfrac{\cancel{5}(31)}{\cancel{5}}$

    $x = 31$¢

17. $\dfrac{1 \text{ quart}}{1{,}500 \text{ miles}} = \dfrac{x \text{ quarts}}{12{,}000 \text{ miles}}$

18. 315

    $1{,}500x = 12{,}000$

    $\dfrac{\cancel{1{,}500}x}{\cancel{1{,}500}} = \dfrac{\overset{40}{\cancel{12{,}000}}}{\underset{5}{\cancel{1{,}500}}} = \dfrac{\overset{8}{\cancel{40}}}{\underset{1}{\cancel{5}}} = 8$

    $x = 8$

19. $\dfrac{\overset{3}{\cancel{6}} \text{ height}}{\underset{4}{\cancel{8}} \text{ shadow}} = \dfrac{H \text{ height}}{96 \text{ shadow}}$

20. 750 mi per hr

    $\dfrac{3}{4} = \dfrac{H}{96}$

    $4H = 3(96)$

    $\dfrac{\cancel{4}H}{\cancel{4}} = \dfrac{3(\overset{24}{\cancel{96}})}{\underset{1}{\cancel{4}}}$

    $H = 72$ ft

21. $\dfrac{2.8 \text{ miles}}{15 \text{ minutes}} = \dfrac{x \text{ miles}}{60 \text{ minutes}}$ (= 1 hr)

22. (a) $\dfrac{3}{8}$ oz

    $15x = 2.8(60)$

   (b) $\dfrac{5}{7}$ oz

    $\dfrac{\cancel{15}x}{\cancel{15}} = \dfrac{2.8(\overset{4}{\cancel{60}})}{\underset{1}{\cancel{15}}}$

    $x = 11.2$ mi per hr

23. 1 minute 58.1 seconds $= 1 + \dfrac{58.1}{60}$ minutes

                          $= \dfrac{118.1}{60}$ minutes.

    $\dfrac{x \text{ miles}}{60 \text{ minutes}} = \dfrac{1.25 \text{ miles}}{\dfrac{118.1}{60} \text{ minutes}}$

    $\dfrac{118.1}{60}x = 1.25(60) = 75$

    $\dfrac{\dfrac{\cancel{118.1}}{\cancel{60}}x}{\dfrac{118.1}{60}} = \dfrac{75}{\dfrac{118.1}{60}}$

    Then $x = 75 \div \dfrac{118.1}{60} = \dfrac{75}{1} \cdot \dfrac{60}{118.1}$

                $= \dfrac{4{,}500}{118.1} \doteq 38.1$ mi per hr

SOLUTIONS FOR CHAPTER 5 DIAGNOSTIC TEST (page 225)

Following each problem number is the textbook section reference (in parentheses) where that kind of problem is discussed.

1. (501) (a) $\dfrac{7}{10}$ (b) $\dfrac{7}{3}$

2. (502) (a) $\dfrac{\overset{5}{\cancel{15}}}{\underset{6}{\cancel{18}}} = \dfrac{5}{6}$

    (b) $\dfrac{\overset{3}{\cancel{12}} \text{ cars}}{\underset{2}{\cancel{8}} \text{ families}} = \dfrac{3 \text{ cars}}{2 \text{ families}}$

                           $= 3$ cars to 2 families

    (c) $\dfrac{9 \text{ inches}}{2 \text{ feet}} = \dfrac{\overset{3}{\cancel{9} \text{ inches}}}{\underset{8}{\cancel{24} \text{ inches}}} = \dfrac{3}{8}$

3. (505) (a) No, because $21 \times 23 \neq 37 \times 13$
                  $483 \neq 481$
    (b) Yes, because $24 \times 81 = 54 \times 36$
                    $1{,}944 = 1{,}944$

4. (506 and 507) (a) $\dfrac{x}{12} = \dfrac{\overset{2}{\cancel{40}}}{\underset{3}{\cancel{60}}}$ (b) $\dfrac{\overset{8}{\cancel{24}}}{\underset{25}{\cancel{75}}} = \dfrac{P}{100}$

$\dfrac{x}{12} = \dfrac{2}{3}$ $25P = 800$

$3x = 24$ $\dfrac{\cancel{25}P}{\cancel{25}} = \dfrac{800}{25}$

$\dfrac{\cancel{3}x}{\cancel{3}} = \dfrac{24}{3}$

$x = 8$ $P = 32$

(c) $\dfrac{3\frac{1}{2}}{B} = \dfrac{21}{40}$

$21B = (3\tfrac{1}{2})40 = \dfrac{7}{\cancel{2}} \cdot \dfrac{\overset{20}{\cancel{40}}}{1} = 140$

$\dfrac{\cancel{21}B}{\cancel{21}} = \dfrac{\overset{20}{\cancel{140}}}{\underset{3}{\cancel{21}}}$

$B = \dfrac{20}{3} = 6\tfrac{2}{3}$

5. (508) Let x = cost of $7\frac{1}{2}$ lb

$\dfrac{5 \text{ pounds}}{84 \text{ cents}} = \dfrac{7\frac{1}{2} \text{ pounds}}{x \text{ cents}}$

$5x = (7\tfrac{1}{2})84 = \left(\dfrac{15}{\cancel{2}}\right)\dfrac{\overset{42}{\cancel{84}}}{1} = (15)(42)$

$\dfrac{\cancel{5}x}{\cancel{5}} = \dfrac{(\overset{3}{\cancel{15}})(42)}{\underset{1}{\cancel{5}}}$

$x = 126 \text{ cents} = \$1.26$

6. (508) Let H = height of the flagpole.

$\dfrac{\overset{3}{\cancel{6}} \text{ height}}{\underset{2}{\cancel{4}} \text{ shadow}} = \dfrac{H \text{ height}}{18 \text{ shadow}}$

$2H = 3(18)$

$\dfrac{\cancel{2}H}{\cancel{2}} = \dfrac{3(\overset{9}{\cancel{18}})}{\underset{1}{\cancel{2}}}$

$H = 27 \text{ ft}$

7. (508) Let x = number of quarts needed for 6,000 miles.

$\dfrac{2 \text{ quarts}}{1,500 \text{ miles}} = \dfrac{x \text{ quarts}}{6,000 \text{ miles}}$

$1,500x = 2(6,000)$

$\dfrac{1,500x}{1,500} = \dfrac{12,000}{1,500} = \dfrac{\overset{24}{\cancel{120}}}{\underset{3}{\cancel{15}}} = 8$

$x = 8$

8. (508) Let x = number of ounces needed for a 175-pound man.

$\dfrac{\frac{1}{4} \text{ ounce}}{25 \text{ pounds}} = \dfrac{x \text{ ounces}}{175 \text{ pounds}}$

$25x = \dfrac{1}{4}(175) = \dfrac{175}{4}$

$\dfrac{\cancel{25}x}{\cancel{25}} = \dfrac{\frac{175}{4}}{25} = \dfrac{175}{4} \div 25$

$x = \dfrac{\overset{7}{\cancel{175}}}{4} \cdot \dfrac{1}{\underset{1}{\cancel{25}}} = \dfrac{7}{4} = 1\tfrac{3}{4}$

EXERCISES 602, SET I (page 229)

1. $\dfrac{1}{4} = 4\overline{\smash{\big)}1.0^20}$ $0.2\,5$ 2. 0.5

3. $\dfrac{3}{8} = 8\overline{\smash{\big)}3.0^60^40}$ $0.3\,7\,5$ 4. 0.2

5. $\dfrac{5}{6} = 6\overline{\smash{\big)}5.0^20^20}$ $0.8\,3\,3 \doteq 0.83$ 6. 0.8

7. $16\overline{\smash{\big)}5.0000}$ 0.3125
     4 8
     ——
     20
     16
     ——
     40
     32
     ——
     80
     80   8. 0.1875

9. $\dfrac{5}{7} = 7\overline{\smash{\big)}5.0^10^30}$ $0.7\,1\,4 \doteq 0.71$ 10. 1.67

11. $\dfrac{4}{25} = 25\overline{\smash{\big)}4.00}$ 0.16
     2 5
     ——
     1 50
     1 50   12. 0.14

13. $1\dfrac{2}{5} = \dfrac{7}{5} = 5\overline{\smash{\big)}7.^20}$ 1.4   14. 2.625

15. 2.

EXERCISES 603, SET I (page 230)

1. $0.27_\wedge = 27.\%$ 2. 35.%

3. $0.06_\wedge = 6.\%$ 4. 12.5%

5. $1.40_\wedge = 140.\%$ 6. 205.%

7. $0.18_\wedge6 = 18.6\%$ 8. 1.5%

9. $0.07_\wedge5 = 7.5\%$ 10. 17.5%

11. $2.90_\wedge = 290.\%$ 12. 380.%

13. $2.00_\wedge5 = 200.5\%$ 14. 301.5%

15. $1.36_\wedge = 136.\%$ 16. 211.%

17. $4.00_\wedge = 400.\%$ 18. 300.%

19. $5.74_\wedge = 574.\%$ 20. 715.%

EXERCISES 604, SET I (page 231)

1. $_\wedge 45.\% = 0.45$ 2. 0.78

3. $1_\wedge 25.\% = 1.25$ 4. 1.50

5. $_\wedge 06.5\% = 0.065$ 6. 0.086

7. $_\wedge 02.35\% = 0.0235$ 8. 0.0385

9. $2\frac{1}{2}\% = {}_\wedge 02.5\% = 0.025$ 10. 0.0475

11. $3\frac{1}{4}\% = {}_\wedge 03.25\% = 0.0325$ 12. 0.054

13. $_\wedge 10.05\% = 0.1005$ 14. 0.0208

15. $\frac{3}{4}\% = {}_\wedge 00.75\% = 0.0075$ 16. 0.005

17. $66\frac{2}{3}\% = {}_\wedge 66.67\% = 0.6667$ 18. 0.3333

19. $12\frac{1}{2}\% = {}_\wedge 12.5\% = 0.125$ 20. 0.375

EXERCISES 605, SET I (page 232)

1. $\underset{1\ \ 0}{0.7} \to \frac{7}{10}$ 2. $\frac{4}{5}$

3. $\underset{1\ \ 00}{0.64} \to \frac{64}{100} = \frac{16}{25}$ 4. $\frac{22}{25}$

5. $\underset{1\ \ 000}{0.125} \to \frac{125}{1,000} = \frac{5}{40} = \frac{1}{8}$ 6. $\frac{3}{8}$

7. $\underset{1\ \ 0}{3.5} \to \frac{35}{10} = \frac{7}{2} = 3\frac{1}{2}$ 8. $2\frac{4}{5}$

9. $\underset{1\ \ 00}{1.75} \to \frac{175}{100} = \frac{7}{4} = 1\frac{3}{4}$ 10. $1\frac{1}{4}$

11. $\underset{1\ \ 0000}{0.1875} \to \frac{1,875}{10,000} = \frac{75}{400} = \frac{3}{16}$ 12. $\frac{9}{16}$

13. $\frac{1}{2} = 2\overline{\smash{\big)}\,1.0}^{\,0.5} = 0.50_\wedge = 50.\%$ 14. $0.25;\ 25.\%$

15. $0.6 = \frac{6}{10} = \frac{3}{5}$ 16. $\frac{2}{5};\ 40.\%$

$\qquad 0.60_\wedge = 60.\%$

17. $_\wedge 10\% = 0.10 = \frac{1}{10}$ 18. $\frac{1}{5};\ 0.2$

19. $\frac{3}{4} = 4\overline{\smash{\big)}\,3.0^2 0}^{\,0.7\ 5} = 0.75_\wedge = 75.\%$ 20. $\frac{9}{25};\ 36.\%$

21. $_\wedge 06\% = 0.06 = \frac{6}{100} = \frac{3}{50}$ 22. $0.8;\ 80.\%$

23. $0.48_\wedge = 48.\%$ 24. $0.08;\ \frac{2}{25}$

$\qquad 0.48 = \frac{48}{100} = \frac{12}{25}$

25. $1\frac{1}{8} = \frac{9}{8} = 8\overline{\smash{\big)}\,9.^{1}0^2 0^4 0}^{\,1.1\ 2\ 5} = 1.12_\wedge 5 = 112.5\%$

26. $7.5\%;\ \frac{3}{40}$

27. $_\wedge 44\% = 0.44 = \frac{44}{100} = \frac{11}{25}$

28. $2.375;\ 237.5\%$

29. $0.02_\wedge 5 = 2.5\%$ 30. $0.28;\ \frac{7}{25}$

$\qquad 0.025 = \frac{25}{1,000} = \frac{1}{40}$

31. $3_\wedge 50\% = 3.50 = \frac{350}{100} = \frac{7}{2} = 3\frac{1}{2}$ 32. $2.75;\ 2\frac{3}{4}$

33. $6.25_\wedge = 625.\%$ 34. $0.045;\ \frac{9}{200}$

$\quad 6.25 = \frac{625}{100} = \frac{25}{4} = 6\frac{1}{4}$

35. $5\frac{1}{4}\% = {}_\wedge 05.25\% = 0.0525 = \frac{525}{10,000} = \frac{21}{400}$

36. $875.\%;\ 8\frac{3}{4}$

37. $\frac{3}{4}\% = {}_\wedge 00.75\% = 0.0075 = \frac{75}{10,000} = \frac{3}{400}$

38. $0.005;\ \frac{1}{200}$

39. $\frac{2}{3} = 3\overline{\smash{\big)}\,2.0^2 0^2 0^2 0}^{\,0.6\ 6\ 6\ 6} \doteq 0.66_\wedge 7 = 66.7\%$

40. $0.833;\ 83.3\%$

41. $\underset{1\ \ 0}{0.5\frac{3}{8}} \to \frac{5\frac{3}{8}}{10} = 5\frac{3}{8} \div 10 = \frac{43}{8} \cdot \frac{1}{10} = \frac{43}{80}$

$\qquad = 8_\wedge 0\overline{\smash{\big)}\,4_\wedge 3.^3 0^6 0^4 0}^{\,0.5\ 3\ 7\ 5}$

$\qquad \doteq 0.53_\wedge 8 = 53.8\%$

42. $45.2\%;\ \frac{181}{400}$

43. $24\frac{1}{4}\% = {}_\wedge 24.25 = 0.2425 = \frac{2,425}{10,000} = \frac{97}{400}$

$\qquad\qquad\qquad \uparrow$ — round off to 0.242

44. $0.066;\ \frac{53}{800}$

45. (a) $\frac{10}{40} = \frac{1}{4}$ (b) $\frac{1}{4} = 4\overline{\smash{\big)}\,1.0^2 0}^{\,0.2\ 5}$

(c) $0.25_\wedge = 25.\%$

46. $66\frac{2}{3}\% \doteq 66.7\%$

47. $100\% - 20\% = {}_\wedge 80\% = 0.8 = \frac{8}{10} = \frac{4}{5}$

48. (a) $\frac{9}{12} = \frac{3}{4}$ (b) 0.75 (c) 75%

49. $_\wedge 20\frac{5}{6}\% = 0.20\frac{5}{6} = \frac{20\frac{5}{6}}{100} = 20\frac{5}{6} \div 100 = \frac{\overset{5}{\cancel{125}}}{6} \cdot \frac{1}{\underset{4}{\cancel{100}}} = \frac{5}{24}$

EXERCISES 606, SET I (page 236)

1. $\underset{1}{\frac{1}{\cancel{6}}} \times \frac{\overset{8}{\cancel{48}}}{1} = 8$ 2. 19

3. $\begin{array}{r} 36 \\ 0.25 \\ \overline{1\ 80} \\ 7\ 2 \\ \overline{9.00} \end{array}$ 4. 3

5. $_\wedge 15\%$ of $32 = 0.15 \times 32 = 4.80$ 6. 9

7. $\underset{1}{\frac{3}{\cancel{4}}} \times \frac{\overset{13}{\cancel{52}}}{1} = 39$ 8. 40

9. $\begin{array}{r} 0.2\ 2\ 5 \\ \underline{1\ 4\ 0\ \ } \\ 9\ 0\ 0 \\ 2\ 2\ 5 \\ \overline{3\ 1.5\ 0\ 0} \end{array}$ 10. 56.25

11. $2_\wedge 00\%$ of $56 = 2.00 \times 56 = 112$ 12. 216

13. $\frac{5}{\cancel{6}_2} \times \frac{\cancel{27}^9}{1} = \frac{45}{2} = 22\frac{1}{2}$ 14. $18\frac{2}{3}$

15.
```
 0.0 3 1 2 5 |
         9 6|0
 ─────────────
   1 8 7 5 0
   2 8 1 2 5
 ─────────────
   3 0.0 0 0 0|0
```
16. 30

17. $13\frac{1}{3}\% = \frac{13\frac{1}{3}}{100} = 13\frac{1}{3} \div 100 = \frac{\cancel{40}^2}{3} \times \frac{1}{\cancel{100}_5} = \frac{2}{15}$

     Therefore, $13\frac{1}{3}\%$ of $702 = \frac{2}{\cancel{15}_5} \times \frac{\cancel{702}^{234}}{1} = \frac{468}{5} = 93.6$

18. 43.68

19. $\frac{1}{2}\% = {}_\wedge00.5\% = 0.005$

     Therefore, $\frac{1}{2}\%$ of $300 = 0.005 \times 300 = 1.5$

20. $\frac{1}{2}$

21. ${}_\wedge00.35\%$ of $550 = 0.0035 \times 550 = 1.925$

22. 4.125

23. ${}_\wedge27\%$ of $375 = 0.27 \times 375 = \$101.25$

24. 17

25. $\frac{1}{\cancel{11}_1} \times \frac{\cancel{88}^8}{1} = 8$ days   26. \$132

27. ${}_\wedge15\%$ of $2,150 = 0.15 \times 2,150 = \$322.50$

28. \$23,250

29. $\frac{1}{\cancel{3}_1} \times \frac{\cancel{24}^8}{1} = 8$ hr per da   30. $937\frac{1}{2}$ lb

     Therefore, $8 \times 7 = 56$ hr per wk

31. Discount $= {}_\wedge15\%$ of $2,490 = 0.15 \times 2,490 = \$373.50$

     Selling price $= 2,490 - 373.50 = \$2,116.50$

32. \$13,062.50

33. $100\% - 22\% = 78\%$
     78% of $237.50 = 0.78 \times 237.50 = \$185.25$
     78% of $\ \ 94.50 = 0.78 \times \ \ 94.50 = \ \ \ \ 73.71$
     78% of $248.50 = 0.78 \times 248.50 = \ \ 193.83$
     78% of $\ \ 89.50 = 0.78 \times \ \ 89.50 = \ \ \ \ 69.81$

34.   57,112,600 land
     139,827,400 water

EXERCISES 607, SET I (page 239)

1. $\frac{\cancel{15}^1}{\cancel{75}_5} = \frac{P}{100}$ 2. 48

     $5 \cdot P = 1 \cdot 100$

     $\frac{\cancel{5}^1 \cdot P}{\cancel{5}_1} = \frac{\cancel{100}^{20}}{\cancel{5}_1}$

     $P = 20$

3. $\frac{A}{20} = \frac{\cancel{60}^3}{\cancel{100}_5}$ 4. 6

     $5 \cdot A = 3 \cdot 20$

     $\frac{\cancel{5}^1 \cdot A}{\cancel{5}_1} = \frac{\cancel{60}^{12}}{\cancel{5}_1}$

     $A = 12$

5. $\frac{48}{B} = \frac{\cancel{75}^3}{\cancel{100}_4}$ 6. 20

     $3 \cdot B = 4 \cdot 48$

     $\frac{\cancel{3}^1 \cdot B}{\cancel{3}_1} = \frac{4 \cdot \cancel{48}^{16}}{\cancel{3}_1}$

     $B = 64$

7. $\frac{2.84}{40} = \frac{P}{100}$ 8. 3.0

     $40 \cdot P = 100 \times 2.84$

     $\frac{\cancel{40}^1 \cdot P}{\cancel{40}_1} = \frac{284}{40} = 7.1$

     $P = 7.1$

9. $\frac{A}{78.6} = \frac{\cancel{125}^5}{\cancel{100}_4}$ 10. 2.8

     $4 \cdot A = 5 \times 78.6 = 393.0$

     $\frac{\cancel{4}^1 \cdot A}{\cancel{4}_1} = \frac{393.0}{4} = 98.25$

     $A = 98.25$

11. $\frac{7.4}{B} = \frac{37.5}{100}$ 12. 299.2

     $37.5 \times B = 740$

     $B = \frac{740}{37.5} \doteq 19.7$

13. $\frac{14.7}{37\frac{1}{2}} = \frac{P}{100}$ 14. 6.72

     $37.5 \times P = 1,470$

     $P = \frac{1,470}{37.5} = 39.2$

15. $\frac{A}{58.6} = \frac{7\frac{1}{2}}{100}$ 16. 16.0

     $100 \cdot A = 7.5 \times 58.6$

     $A = \frac{7.5 \times 58.6}{100} = 4.395$

     $A \doteq 4.40$

1. 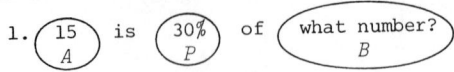 $\bigcirc$ 15 A is $\bigcirc$ 30% P of $\bigcirc$ what number? B

   B is the unknown.

2. $P = 20$
   $A = 16$
   B is the unknown.

3. $\bigcirc$ 115 A is $\bigcirc$ what percent P of $\bigcirc$ 250? B

   P is the unknown.

4. P is the unknown.
   $A = 330$
   $B = 225$

5. $\bigcirc$ What A is $\bigcirc$ 25% P of $\bigcirc$ 40? B

   A is the unknown.

6. $P = 45$
   $B = 65$
   A is the unknown.

7. $\bigcirc$ 15% P of $\bigcirc$ what number B is $\bigcirc$ 127.5? A

   B is the unknown.

8. $P = 32$
   B is the unknown.
   $A = 256$

9. $\bigcirc$ What % P of $\bigcirc$ 8 B is $\bigcirc$ 17? A

   P is the unknown.

10. P is the unknown.
    $B = 6$
    $A = 12$

11. $\bigcirc$ 63% P of $\bigcirc$ 48 B is $\bigcirc$ what number? A

    A is the unknown.

12. $P = 87$
    $B = 49$
    A is the unknown.

13. $\bigcirc$ 750 A is $\bigcirc$ 125% P of $\bigcirc$ what number? B

    B is the unknown.

14. $P = 130$
    B is the unknown.
    $A = 325$

15. $\bigcirc$ 23 A is $\bigcirc$ what % P of $\bigcirc$ 16? B

    P is the unknown.

16. P is the unknown.
    $B = 23$
    $A = 57$

17. $\bigcirc$ What A is $\bigcirc$ 200% P of $\bigcirc$ 12? B

    A is the unknown.

18. $P = 300$
    $B = 9$
    A is the unknown.

19. P is the unknown.
    $B = 500$
    $A = 27$

20. $P = 15$
    B is the unknown.
    $A = 37.5$

21. $P = 80$
    $A = 68$
    B is the unknown.

22. $P = 27$
    $B = 135$
    A is the unknown.

23. P is the unknown.
    $B = 42$
    $A = 7$

24. $P = 23$
    $B = 110$
    A is the unknown.

1. 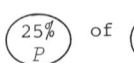 $\bigcirc$ 15 A is $\bigcirc$ 30% P of $\bigcirc$ what number? B

   $\dfrac{15}{B} = \dfrac{3\cancel{0}}{10\cancel{0}}$

   $3 \cdot B = 150$

   $B = \dfrac{150}{3} = 50$

2. 80

3. $\bigcirc$ 115 A is $\bigcirc$ what % P of $\bigcirc$ 250? B

   $\dfrac{115}{250} = \dfrac{P}{100}$

   $250 \cdot P = 11{,}500$

   $P = \dfrac{11{,}500}{250} = 46$

4. $146\frac{2}{3}\%$

5. $\bigcirc$ What A is $\bigcirc$ 25% P of $\bigcirc$ 40? B

   $\dfrac{A}{40} = \dfrac{\overset{1}{\cancel{25}}}{\underset{4}{\cancel{100}}}$

   $4 \cdot A = 40$

   $A = \dfrac{40}{4} = 10$

6. 29.25

7. $\bigcirc$ 15% P of $\bigcirc$ what number B is $\bigcirc$ 127.5? A

   $\dfrac{127.5}{B} = \dfrac{15}{100}$

   $15 \cdot B = 12{,}750$

   $B = \dfrac{12{,}750}{15} = 850$

8. 800

9. $\bigcirc$ What % P of $\bigcirc$ 8 B is $\bigcirc$ 17? A

   $\dfrac{17}{8} = \dfrac{P}{100}$

   $8 \cdot P = 1{,}700$

   $P = \dfrac{1{,}700}{8} = 212.5$

10. 200%

11. $\bigcirc$ 63% P of $\bigcirc$ 48 B is $\bigcirc$ what number? A

    $\dfrac{A}{48} = \dfrac{63}{100}$

    $100 \cdot A = 48 \times 63$

    $A = \dfrac{48 \times 63}{100} = 30.24$

12. 42.63

13. (750) is (125%) of (what number?)
    A   P   B

$$\frac{750}{B} = \frac{\overset{5}{\cancel{125}}}{\underset{4}{\cancel{100}}}$$

$$5 \cdot B = 4 \times 750$$

$$B = \frac{4 \times \overset{150}{\cancel{750}}}{\underset{1}{\cancel{5}}} = 600$$

14. 250

15. (23) is (what %) of (16?)
    A   P   B

$$\frac{23}{16} = \frac{P}{100}$$

$$16 \cdot P = 2,300$$

$$P = \frac{2,300}{16} = 143.75$$

16. $\doteq 247.8\%$

17. (What) is (200%) of (12?)
    A   P   B

$$\frac{A}{12} = \frac{\overset{2}{\cancel{200}}}{\underset{1}{\cancel{100}}}$$

$$A = 24$$

18. 27

19. $P = 66\frac{2}{3}$ ⎫ $\frac{42}{B} = \frac{66\frac{2}{3}}{100}$
    $A = 42$ ⎬
    B is unknown   ⎭ $66\frac{2}{3} \cdot B = 4,200$

$$B = \frac{4,200}{66\frac{2}{3}} = 4,200 \div 66\frac{2}{3}$$

$$= \frac{4,200}{1} \div \frac{200}{3}$$

$$B = \frac{\overset{21}{\cancel{4,200}}}{1} \cdot \frac{3}{\underset{1}{\cancel{200}}} = 63$$

20. 216

21. P is unknown ⎫ $\frac{27}{500} = \frac{P}{100}$
    $B = 500$ ⎬
    $A = 27$ ⎭ $500 \cdot P = 2,700$

$$P = \frac{2,700}{500} = 5.4$$

22. 250

23. $P = 80$ ⎫ $\frac{68}{B} = \frac{\cancel{80}}{\cancel{100}}$
    $A = 68$ ⎬
    B is unknown. ⎭ $8 \cdot B = 680$

$$B = \frac{680}{8} = 85$$

24. $36.45

25. $A = 7$ ⎫ $\frac{\overset{1}{\cancel{7}}}{\underset{6}{\cancel{42}}} = \frac{P}{100}$
    $B = 42$ ⎬
    P is unknown. ⎭ $6 \cdot P = 100$

$$P = \frac{100}{6} = 16\frac{2}{3}$$

26. $25.30

27. P is unknown. ⎫ $\frac{\overset{9}{\cancel{54}}}{\underset{35}{\cancel{210}}} = \frac{P}{100}$
    $A = 54$ ⎬
    $B = 210$ ⎭ $35 \cdot P = 900$

$$P = \frac{900}{35} \doteq 25.7$$

28. 11%

29. $\frac{A}{B} = \frac{P}{100}$ 30. $20,881.58

$$\frac{A}{460} = \frac{5.85}{100}$$

$$100A = 460(5.85)$$

$$A = \frac{460(5.85)}{100} = \$26.91$$

EXERCISES 610, SET I (page 248)

1. (a) Interest for 1 year = principal × rate × time
                          = 1,400 × 0.09 × 1
                          = $126
   (b) Interest for 3 years = 126 × 3 = $378
2. (a) $200
   (b) $400
3. (a) Interest for 1 month = 350 × 0.015 × 1
                            = $5.25
   (b) Interest for 2 months = 5.25 × 2 = $10.50
4. (a) $2.62
   (b) $7.88
5. (a) Interest for 1 day = 2,000 × 0.0004932 × 1
                          $\doteq$ $0.99
   (b) Interest for 30 days = 2,000 × 0.0004932 × 30
                            $\doteq$ $29.59
6. (a) $1.06
   (b) $21.21
7. Interest = principal × rate × time
            = 750 × 0.07 × 2.5 = $131.25
8. $150.00
9. Interest = 175 × 0.015 × 2 = $5.25
10. $5.80
11. Interest = 38.37 × 0.0004932 × 23 = $0.44
12. $0.53

EXERCISES 611, SET I (page 250)

1. Principal = cost − down payment = 375 − 40 = $335
   Interest = 335 × 0.07 × 2 = $46.90
   Total amount owed = principal + interest
                     = 335 + 46.90 = $381.90
   Monthly payment = $\frac{\text{total amount owed}}{\text{number of months}} = \frac{381.90}{24}$
                   $\doteq$ $15.91
2. $19.58
3. Principal = $2,300
   Interest = 2,300 × 0.065 × 3 = $448.50
   Total amount owed = 2,300 + 448.50 = $2,748.50
   Monthly payment = $\frac{2,748.50}{36} \doteq$ $76.35
4. $141.96
5. Principal = 250 − 25 = $225
   Interest = 225 × 0.0775 × 1 = $17.44
   Total amount owed = 225 + 17.44 = $242.44
   Monthly payment = $\frac{242.44}{12} \doteq$ $20.20
6. $13.84
7. Principal = 3,500 − 350 = $3,150
   Interest = 3,150 × 0.0625 × 3 = $590.62
   Total amount owed = 3,150 + 590.62 = $3,740.62
   Monthly payment = $\frac{3,740.62}{36} = $103.91
8. $77.50

1. $\dfrac{2}{5} = 5\overline{\smash{\big)}2.0}^{\,0.4} = 0.40_\wedge = 40.\%$

2. $\dfrac{9}{20}$; 45%

3. $_\wedge 30.\% = 0.3 = \dfrac{3}{10}$

4. 2.5; 250%

5. $1.75_\wedge = 175.\%;\ 1.75 = \dfrac{175}{100} = \dfrac{7}{4} = 1\dfrac{3}{4}$

6. $0.125;\ \dfrac{1}{8}$

7. $\overset{\text{What}}{A}$ is $\overset{35\%}{P}$ of $\overset{\$275?}{B}$

$\dfrac{A}{275} = \dfrac{\overset{7}{\cancel{35}}}{\underset{20}{\cancel{100}}}$

$20 \cdot A = 275 \times 7$

$A = \dfrac{275 \times 7}{20} = \$96.25$

8. 15%

9. $\overset{77.5}{A}$ is $\overset{31\%}{P}$ of $\overset{\text{what number?}}{B}$

$\dfrac{77.5}{B} = \dfrac{31}{100}$

$31 \cdot B = 7{,}750$

$B = \dfrac{7{,}750}{31} = 250$

10. $1,102.50

11. $\overset{7}{A}$ is $\overset{\text{what }\%}{P}$ of $\overset{8?}{B}$

$\dfrac{7}{8} = \dfrac{P}{100}$

$8 \cdot P = 700$

$P = \dfrac{700}{8} = 87.5$

12. 244

13. Find $\overset{2\frac{1}{2}\%}{P}$ of $\overset{\$400.}{B}$

$\dfrac{A}{400} = \dfrac{2.5}{100}$

$100 \cdot A = 400 \times 2.5$

$A = \dfrac{\overset{4}{\cancel{400}} \times 2.5}{\underset{1}{\cancel{100}}} = 10$

14. $9.50

15. $\overset{\text{Increase}}{A}$ is $\overset{5\%}{P}$ of $\overset{3,560.}{B}$

$\dfrac{A}{3{,}560} = \dfrac{\overset{1}{\cancel{5}}}{\underset{20}{\cancel{100}}}$

$20 \cdot A = 3{,}560$

$A = \dfrac{3{,}560}{20} = 178$

Fall enrollment = 3,560 + 178 = 3,738

16. 75%

17. $\overset{\text{Discount}}{A}$ is $\overset{20\%}{P}$ of $\overset{\$95.}{B}$

$\dfrac{A}{95} = \dfrac{\overset{1}{\cancel{20}}}{\underset{5}{\cancel{100}}}$

$5 \cdot A = 95$ Selling price = 95 − 19 = $76

$A = \dfrac{95}{5} = \$19$

18. $92.25

19. $\overset{75\text{ g}}{A}$ is $\overset{\text{what }\%}{P}$ of $\overset{500\text{ g?}}{B}$

$\dfrac{\overset{3}{\cancel{75}}}{\underset{20}{\cancel{500}}} = \dfrac{P}{100}$

$20 \cdot P = 300$

$P = \dfrac{300}{20} = 15$

20. $4.20

21. Interest = 1,150 × 0.096 × 1 = $110.40

22. $1,700

23. $\overset{\text{Raise}}{A}$ is $\overset{5.6\%}{P}$ of $\overset{\$9,500.}{B}$

$\dfrac{A}{9{,}500} = \dfrac{5.6}{100}$

$100 \cdot A = 5.6 \times 9{,}500$

$A = \dfrac{5.6 \times 9{,}500}{100} = \$532$

New salary = 9,500 + 532 = $10,032

24. $71.25

SOLUTIONS FOR CHAPTER 6 DIAGNOSTIC TEST (page 255)

Following each problem number is the textbook section reference (in parentheses) where that kind of problem is discussed.

1. (602) $\dfrac{4}{5} = 5\overline{\smash{\big)}4.0}^{\,0.8}$

2. (602) $4\dfrac{2}{3} = \dfrac{14}{3} = 3\overline{\smash{\big)}14.000}^{\,4.666\ \doteq\ 4.67}$

3. (603) $0.74 = \dfrac{74}{100} = 74\%$

4. (603) $2.05 = \dfrac{205}{100} = 205\%$

5. (604) $48\% = \dfrac{48}{100} = 0.48$

6. (604) $565\% = \dfrac{565}{100} = 5.65$

7. (605) $0.58 = \dfrac{58}{100} = \dfrac{29}{50}$

8. (605) $2.75 = \dfrac{275}{100} = \dfrac{55}{20} = \dfrac{11}{4} = 2\dfrac{3}{4}$

(or) $2.75 = 2\dfrac{75}{100} = 2\dfrac{3}{4}$

9. (606) $\dfrac{3}{8}$ of $56 = \dfrac{3}{\underset{1}{\cancel{8}}} \times \dfrac{\overset{7}{\cancel{56}}}{1} = 21$

10. (606) 0.32 of 480 = 0.32 × 480 = 153.6

11. (606) 78% = 0.78   Then 78% of 350 = 0.78 × 350 = 273

12. (609) $\dfrac{A}{B} = \dfrac{P}{100}$

$\dfrac{12}{30} = \dfrac{P}{100}$

$30P = 1,200$

$\dfrac{\cancel{30}P}{\cancel{30}} = \dfrac{1,20\cancel{0}}{\cancel{30}}$

$P = 40$

13. (609) $\dfrac{A}{B} = \dfrac{P}{100}$

$\dfrac{28}{B} = \dfrac{42}{100}$

$42B = 2,800$

$\dfrac{\cancel{42}B}{\cancel{42}} = \dfrac{2,800}{42}$

$B = 66\dfrac{2}{3} \doteq 67$

14. (609) $\dfrac{A}{B} = \dfrac{P}{100}$

$\dfrac{17}{20} = \dfrac{P}{100}$

$20P = 1,700$

$\dfrac{\cancel{20}P}{\cancel{20}} = \dfrac{1,70\cancel{0}}{\cancel{20}}$

$P = 85$

15. (610, 611) (a) $I = Prt$
$I = (2,000)(0.07)(3)$
$I = \$420$

       (b) Amount = principal + interest
$A = \$2,000 + \$420$
$A = \$2,420$

       (c)
$$36\overline{)2420.000} \quad 67.222 \doteq \$67.22$$
$$\begin{array}{r} 216 \\ \hline 260 \\ 252 \\ \hline 80 \\ 72 \\ \hline 80 \\ 72 \\ \hline 80 \\ 72 \end{array}$$

EXERCISES 702, SET I (page 264)

1. 5 yd = __?__ ft

$\dfrac{5 \text{ yd}}{1}\left(\dfrac{3 \text{ ft}}{1 \text{ yd}}\right) = 15 \text{ ft}$

Therefore, 5 yd = 15 ft

2. 21 ft

3. 3 ft = __?__ in

$\dfrac{3 \text{ ft}}{1}\left(\dfrac{12 \text{ in}}{1 \text{ ft}}\right) = 36 \text{ in}$

Therefore, 3 ft = 36 in

4. 60 in

5. $3\dfrac{1}{3}$ yd = __?__ ft

$3\dfrac{1}{3}\text{ yd}\left(\dfrac{3 \text{ ft}}{1 \text{ yd}}\right) = \dfrac{10}{\cancel{3}} \times \dfrac{\cancel{3} \text{ ft}}{1} = 10 \text{ ft}$

Therefore, $3\dfrac{1}{3}$ yd = 10 ft

6. 13 ft

7. $5\dfrac{2}{3}$ ft = __?__ in

$5\dfrac{2}{3}\text{ ft}\left(\dfrac{12 \text{ in}}{1 \text{ ft}}\right) = \dfrac{17}{\cancel{3}} \times \dfrac{\overset{4}{\cancel{12}} \text{ in}}{1} = 68 \text{ in}$

Therefore, $5\dfrac{2}{3}$ ft = 68 in

8. 38 in

9. 8 gal = __?__ qt

$\dfrac{8 \text{ gal}}{1}\left(\dfrac{4 \text{ qt}}{1 \text{ gal}}\right) = 32 \text{ qt}$

10. 28 qt

11. $2\dfrac{3}{4}$ gal = __?__ qt

$2\dfrac{3}{4}\text{ gal}\left(\dfrac{4 \text{ qt}}{1 \text{ gal}}\right) = \dfrac{11}{\cancel{4}} \times \dfrac{\cancel{4} \text{ qt}}{1} = 11 \text{ qt}$

12. 21 qt

13. 7 qt = __?__ pt

$\dfrac{7 \text{ qt}}{1}\left(\dfrac{2 \text{ pt}}{1 \text{ qt}}\right) = 14 \text{ pt}$

14. 22 pt

15. 2.5 qt = __?__ pt

$\dfrac{2.5 \text{ qt}}{1}\left(\dfrac{2 \text{ pt}}{1 \text{ qt}}\right) = 5 \text{ pt}$

16. 17 pt

17. $1\dfrac{1}{2}$ hr = __?__ min

$1\dfrac{1}{2}\text{ hr}\left(\dfrac{60 \text{ min}}{1 \text{ hr}}\right) = \dfrac{3}{\underset{1}{\cancel{2}}} \times \dfrac{\overset{30}{\cancel{60}} \text{ min}}{1} = 90 \text{ min}$

18. 200 min

19. 7.75 min = __?__ sec

$\dfrac{7.75 \text{ min}}{1}\left(\dfrac{60 \text{ sec}}{1 \text{ min}}\right) = 7.75 \times 60 \text{ sec}$
$= 465 \text{ sec}$

20. 195 sec

21. $1\dfrac{3}{4}$ mi = __?__ ft

$1\dfrac{3}{4}\text{ mi}\left(\dfrac{5,280 \text{ ft}}{1 \text{ mi}}\right) = \dfrac{7}{4}\left(\dfrac{5,280 \text{ ft}}{1}\right)$
$= 9,240 \text{ ft}$

22. 11,880 ft

23. $3\dfrac{1}{4}$ da = __?__ hr

$3\dfrac{1}{4}\text{ da}\left(\dfrac{24 \text{ hr}}{1 \text{ da}}\right) = \dfrac{13}{\underset{1}{\cancel{4}}}\left(\dfrac{\overset{6}{\cancel{24}} \text{ hr}}{1}\right) = 78 \text{ hr}$

24. 124 hr

25. 10 yd = __?__ ft

$\dfrac{10 \text{ yd}}{1}\left(\dfrac{3 \text{ ft}}{1 \text{ yd}}\right) = 30 \text{ ft}$

$\dfrac{30 \text{ ft}}{1}\left(\dfrac{12 \text{ in}}{1 \text{ ft}}\right) = 360 \text{ in}$

26. 21 ft
252 in

27. 8 gal = __?__ qt

$\dfrac{8 \text{ gal}}{1}\left(\dfrac{4 \text{ qt}}{1 \text{ gal}}\right) = 32 \text{ qt}$

$\dfrac{32 \text{ qt}}{1}\left(\dfrac{2 \text{ pt}}{1 \text{ qt}}\right) = 64 \text{ pt}$

28. 10 qt
20 pt

29. 24 in = __?__ ft

$\dfrac{24 \text{ in}}{1}\left(\dfrac{1 \text{ ft}}{12 \text{ in}}\right) = \dfrac{24}{12} \text{ ft} = 2 \text{ ft}$

30. 3 ft

31. 84 in = __?__ ft

$\dfrac{84 \text{ in}}{1}\left(\dfrac{1 \text{ ft}}{12 \text{ in}}\right) = \dfrac{84}{12} \text{ ft} = 7 \text{ ft}$

32. 8 ft

33. 90 sec = __?__ min

$\dfrac{90 \text{ sec}}{1}\left(\dfrac{1 \text{ min}}{60 \text{ sec}}\right) = \dfrac{90}{60} \text{ min} = 1\dfrac{1}{2} \text{ min}$

34. $1\dfrac{1}{4}$ min

35. 150 min = __?__ hr

$\dfrac{150 \text{ min}}{1}\left(\dfrac{1 \text{ hr}}{60 \text{ min}}\right) = \dfrac{150}{60} \text{ hr} = 2\dfrac{1}{2} \text{ hr}$

36. $1\dfrac{3}{4}$ hr

37. 5,280 ft = __?__ mi

$\dfrac{5,280 \text{ ft}}{1}\left(\dfrac{1 \text{ mi}}{5,280 \text{ ft}}\right) = \dfrac{5,280}{5,280} \text{ mi} = 1 \text{ mi}$

38. 2 mi

39. 730 da = __?__ yr

$\dfrac{730 \text{ da}}{1}\left(\dfrac{1 \text{ yr}}{365 \text{ da}}\right) = \dfrac{730}{365} \text{ yr} = 2 \text{ yr}$

40. 1.4 yr

41. 104 wk = __?__ yr

$\dfrac{104 \text{ wk}}{1}\left(\dfrac{1 \text{ yr}}{52 \text{ wk}}\right) = \dfrac{104}{52} \text{ yr} = 2 \text{ yr}$

42. 1.25 yr

43. 2 gal = _?_ cu in   44. 539 cu in

$$\frac{2 \text{ gal}}{1}\left(\frac{231 \text{ cu in}}{1 \text{ gal}}\right) = 2(231 \text{ cu in})$$
$$= 462 \text{ cu in}$$

45. $2\frac{1}{2}$ lb = _?_ oz   46. 52 oz

$$2\frac{1}{2}\text{ lb}\left(\frac{16 \text{ oz}}{1 \text{ lb}}\right) = \frac{5}{2}\left(\frac{16 \text{ oz}}{1}\right)^{8} = 40 \text{ oz}$$

47. 48 oz = _?_ lb   48. 2.25 lb

$$\frac{48 \text{ oz}}{1}\left(\frac{1 \text{ lb}}{16 \text{ oz}}\right) = \frac{48}{16}\text{ lb} = 3 \text{ lb}$$

49. $2\frac{1}{2}$ tons = _?_ lb   50. 6,500 lb

$$2\frac{1}{2}\text{ tons}\left(\frac{2,000 \text{ lb}}{1 \text{ ton}}\right) = \frac{5}{2}\left(\frac{2,000 \text{ lb}}{1}\right)^{1,000}$$
$$= 5,000 \text{ lb}$$

51. 2,200 lb = _?_ tons   52. 1.8 tons

$$\frac{2,200 \text{ lb}}{1}\left(\frac{1 \text{ ton}}{2,000 \text{ lb}}\right) = \frac{2,200}{2,000}\text{ tons}$$
$$= 1.1 \text{ tons}$$

53. Change 1.25 miles to feet.   54. $\doteq 1.67$ mi
$$\frac{1.25 \text{ mi}}{1}\left(\frac{5,280 \text{ ft}}{1 \text{ mi}}\right) = 1.25(5,280 \text{ ft})$$
$$= 6,600 \text{ ft}$$

55. Change 2 weeks to hours.   56. 8,800 yd
$$\frac{2 \text{ wk}}{1}\left(\frac{7 \text{ da}}{1 \text{ wk}}\right)\left(\frac{24 \text{ hr}}{1 \text{ da}}\right) = 2(7)(24 \text{ hr})$$
$$= 336 \text{ hr}$$

57. Change 2,160 minutes to days.
$$\frac{2,160 \text{ min}}{1}\left(\frac{1 \text{ hr}}{60 \text{ min}}\right)\left(\frac{1 \text{ da}}{24 \text{ hr}}\right) = \frac{2,160}{60 \times 24}\text{ da} = 1.5 \text{ da}$$

58. 3.75 lb

59. $\dfrac{8 \text{ mi}}{12 \text{ min}}\left(\dfrac{60 \text{ min}}{1 \text{ hr}}\right)^{5} = \dfrac{40 \text{ mi}}{\text{hr}} = 40 \text{ mph}$

60. 2.4 mph

61. $\dfrac{1,100 \text{ ft}}{1 \text{ sec}}\left(\dfrac{3,600 \text{ sec}}{1 \text{ hr}}\right)\left(\dfrac{1 \text{ mi}}{5,280 \text{ ft}}\right) = \dfrac{1,100(3,600) \text{ mi}}{5,280 \text{ hr}}$
$$= 750 \text{ mph}$$

62. 25 ft per month

EXERCISES 703, SET I (page 266)

1. $\frac{1}{4}$ cup = _?_ oz   2. 4 tbsp
$$\frac{1}{4}\text{ cup}\left(\frac{8 \text{ oz}}{1 \text{ cup}}\right) = \frac{1}{4}\left(\frac{8 \text{ oz}}{1}\right)^{2} = 2 \text{ oz}$$

3. $\frac{3}{8}$ cup = _?_ tbsp   4. 10 tbsp
$$\frac{3}{8}\text{ cup}\left(\frac{16 \text{ tbsp}}{1 \text{ cup}}\right) = \frac{3}{8}\left(\frac{16 \text{ tbsp}}{1}\right)^{2} = 6 \text{ tbsp}$$

5. 1 gal = _?_ oz   6. 2 pt
$$\frac{1 \text{ gal}}{1}\left(\frac{4 \text{ qt}}{1 \text{ gal}}\right)\left(\frac{2 \text{ pt}}{1 \text{ qt}}\right)\left(\frac{16 \text{ oz}}{1 \text{ pt}}\right) = 4(2)(16) \text{ oz}$$
$$= 128 \text{ oz}$$

7. $1\frac{1}{3}$ tbsp = _?_ tsp   8. 6 cups
$$1\frac{1}{3}\text{ tbsp}\left(\frac{3 \text{ tsp}}{1 \text{ tbsp}}\right) = \frac{4}{3}\left(\frac{3 \text{ tsp}}{1}\right) = 4 \text{ tsp}$$

9. 5 pt = _?_ cups   10. 8 tsp
$$\frac{5 \text{ pt}}{1}\left(\frac{2 \text{ cups}}{1 \text{ pt}}\right) = 5(2 \text{ cups}) = 10 \text{ cups}$$

11. $3 \times 1\frac{1}{4} = 3 \times \frac{5}{4} = \frac{15}{4} = 3\frac{3}{4}$ cups of sugar
$3 \times 2 \qquad\qquad\qquad\quad = 6$ tablespoons of flour
$3 \times \frac{1}{8} \qquad\qquad\qquad\quad = \frac{3}{8}$ teaspoon of salt
$3 \times \frac{1}{2} = \frac{3}{2} \qquad\qquad = 1\frac{1}{2}$ cups of water
$3 \times \frac{1}{4} \qquad\qquad\qquad\quad = \frac{3}{4}$ cup butter or margarine
$3 \times 3 \qquad\qquad\qquad\quad = 9$ eggs
$3 \times 1 \qquad\qquad\qquad\quad = 3$ teaspoons grated lemon peel
$3 \times 1 \qquad\qquad\qquad\quad = 3$ lemons

12. $3\frac{3}{4}$ cups butter   15 eggs
or margarine   5 cups flour
15 oz chocolate $2\frac{1}{2}$ teaspoons baking powder
5 cups sugar $7\frac{1}{2}$ teaspoons vanilla

EXERCISES 704, SET I (page 268)

1. {7 in, 6 in, 2 in}; {5 lb}; {4 gal, 3 gal}; {$1, $10}

2. {15, $\frac{5}{8}$, 6}; {8 mi, 7 gal, 4 lb}

3. Miles, pounds, dozen, ounces, grams, quarts, pint, dollars

4. {21, $\frac{3}{4}$, 6}; {$5, 7 in, 8 mi, 5 lb, 3 dozen}

5. {2 pt, 3 pt}; {5 gal, 2 gal}; {3 qt}; {$5, $4}; {75¢}

EXERCISES 705, SET I (page 270)

1.
4 ft	~~15~~ in
	⌐1 ft 3 in
1 ft	
5 ft	3 in

2. 8 yd 2 ft

3.
2 wk	~~9~~ da
	⌐1 wk 2 da
1 wk	
3 wk	2 da

4. 3 da 12 hr

5.
3 gal	~~15~~ qt
	⌐3 gal 3 qt
3 gal	
6 gal	3 qt

6. 12 qt 1 pt

7.
yd	ft	in
5	4	~~27~~
		3
	~~2~~	
	~~6~~	
2		
7 yd	0 ft	3 in

8. 11 yd 1 ft 10 in

9.
hr	min	sec	10. 4 hr 24 min 5 sec
2	73	~~110~~	
	1	50	
	~~74~~		
1	14		
3 hr	14 min	50 sec	

11.
gal	qt	pt	12. 6 gal
3	7	~~5~~	
	2	1	
	~~9~~		
2			
5 gal	1 qt	1 pt	

13.
```
mi      ft
2     6,000
1       720
3 mi    720 ft
```

14. 4 mi 1,120 ft

15.
```
yr     wk    da
2      48    75
       10     5
       58
1       6
3 yr    6 wk  5 da
```

16. 4 yr 6 wk 6 da

17.
```
tons     lb
2      3,500
1      1,500
3 tons 1,500 lb
```

18. 4 tons 250 lb

19.
```
lb      oz
4       20
1        4
5 lb     4 oz
```

20. 8 lb 8 oz

21.
```
da     hr     min
2      23      75
               1
               24
1
3 da   0 hr    15 min
```

22. 5 yd 3 in

EXERCISES 706, SET I (page 272)

1.
```
4 ft   5 in
3      6
5      2
1
       13 in
13 ft  1 in
```

2. 10 yd 2 ft

3.
```
3 hr   15 min
2      50
7      24
1
       89
13 hr  29 min
```

4. 1 hr 24 min 21 sec

5.
```
3 gal  2 qt  1 pt
5      3     1
8      1     1
1      1
       7     3
17 gal 3 qt  1 pt
```

6. 13 gal 3 qt 1 pt

7.
```
3 yd  2 ft  10 in
1     1     9
8     2     7
2     2
      7     26
14 yd 1 ft  2 in
```

8. 4 mi 2,520 ft

9.
```
1 da  12 hr  15 min
5     23     54
2     18     47
2     1
      54     116
10 da 6 hr   56 min
```

10. 7 yr 35 wk 5 da

11.
```
3 tons 1,500 lb
5        450
7      1,850
1
       3,800
16 tons 1,800 lb
```

12. 25 lb 13 oz

13.
```
7 lb  3 oz
15    9
22    17
1
      29
45 lb 13 oz
```

14. 11 hr 55 min 46 sec

EXERCISES 707, SET I (page 274)

1.
```
8 ft 10 in
3    4
5 ft  6 in
```

2. 2 ft 4 in

3.
```
13 yd 2 ft
7     1
6 yd  1 ft
```

4. 4 yd 1 ft

5.
```
  7     5/1    2
  8 gal 2 qt   0 pt
  3     3      1
  4 gal 2 qt   1 pt
```

6. 4 hr 25 min

7.
```
  4      19
  5 lb   3 oz
  2      8
  2 lb   11 oz
```

8. 1 lb 13 oz

9.
```
  2       2,700
  3 tons  700 lb
  1       1,200
  1 ton   1,500 lb
```

10. 1 ton 900 lb

11.
```
  2     29
  3 da  5 hr
  1     15
  1 da  14 hr
```

12. 1 gal 1 qt 1 pt

13.
```
  4    4/1    18
  5 yd 2 ft   6 in
  2    2      9
  2 yd 2 ft   9 in
```

14. 1 yd 1 ft 11 in

15.
```
  3      8,280
  4 mi  3,000 ft
  1     4,700
  2 mi  3,580 ft
```

16. 4,930 ft

17.
```
  34      100
  35 min  40 sec
  20      55
  14 min  45 sec
```

18. 11 min 48 sec

19.
```
  4    36     82
  5 da 12 hr  22 min
  2    18     45
  2 da 18 hr  37 min
```

20. 2 yr 44 wk 5 da

21. 2 lb 10 oz

EXERCISES 708A, SET I (page 275)

1.
```
  3 wk   5 da
×        4
12 wk   20 da
  2      6
14 wk    6 da
```

2. 7 yr 310 da

3.
```
  5 mi  2,850 ft
×        6
30 mi  17,100 ft
  3     1,260
33 mi   1,260 ft
```

4. 16 mi 2,470 ft

5.
```
  2 yd 1 ft  3 in
×          5
10 yd 5 ft  15 in
      1     3
      6 ft
  2   0
12 yd        3 in
```

6. 16 gal 3 qt 1 pt

7.
```
  1 hr 25 min  11 sec
×             4
  4 hr 100 min 44 sec
  1    40
  5 hr 40 min  44 sec
```

8. 19 yd 2 ft 6 in

9.
$$
\begin{array}{l}
\text{3 gal} \quad \text{3 qt} \quad \text{l pt}\\
\underline{\times \qquad\qquad\qquad 8}\\
\text{24 gal} \quad \text{24 qt} \quad \text{8 pt}\\
\qquad\qquad\quad 4 \qquad 0\\
\qquad\qquad\underline{\quad 28 \text{ qt}}\\
\ \ 7 \qquad\qquad 0\\
\overline{31 \text{ gal}}
\end{array}
$$

10. 13 hr 32 min 6 sec

11.
$$
\begin{array}{l}
\qquad\quad 0 \text{ lb} \quad\ 12\tfrac{4}{7}\text{ oz}\\
7\overline{\smash{)}5 \text{ lb} \qquad 8 \text{ oz}}\\
\ \underline{0}\\
\ 5 \text{ lb} = 80 \text{ oz}\\
\qquad\qquad 88 \text{ oz}\\
\qquad\qquad \underline{84}\\
\qquad\qquad \ 4 \text{ oz} \ \to \ \dfrac{4 \text{ oz}}{7} = \dfrac{4}{7} \text{ oz}
\end{array}
$$

12. $9\tfrac{1}{4}$ oz

13.
$$
\begin{array}{l}
\qquad\ \ 4 \text{ wk} \quad\ 4 \text{ da} \quad\ 5 \text{ hr}\\
3\overline{\smash{)}13 \text{ wk} \quad 5 \text{ da} \quad 15 \text{ hr}}\\
\ \underline{12}\\
\ 1 \text{ wk} = \underline{\ 7 \text{ da}}\\
\qquad\qquad 12 \text{ da}\\
\qquad\qquad \underline{12}\\
\qquad\qquad\ 0 \qquad\quad \underline{15 \text{ hr}}\\
\qquad\qquad\qquad\qquad\ \ 15
\end{array}
$$

14. 2 wk 3 da $16\tfrac{2}{5}$ hr

15.
$$
\begin{array}{l}
\qquad 1 \text{ mi} \quad\ 2{,}510 \text{ ft}\\
6\overline{\smash{)}8 \text{ mi} \quad 4{,}500 \text{ ft}}\\
\ \underline{6}\\
\ 2 \text{ mi} = \underline{10{,}560 \text{ ft}}\\
\qquad\qquad 15{,}060 \text{ ft}\\
\qquad\qquad \underline{15{,}060}
\end{array}
$$

16. $2{,}093\tfrac{1}{3}$ ft

EXERCISES 708B, SET I (page 276)

1. Area = $L \times W$
     = 176 ft × 48 ft
     = 8,448 sq ft

2. 476 sq in

3. Area = $L \times W$
     = 90 ft × 90 ft
     = 8,100 sq ft

4. 5,500 sq yd

5. Man hours = 12 × 17 × 8
              = 1,632
     Cost = $6.50 × 1,632
          = $10,608.00

6. $\doteq$ $13,609.57

EXERCISES 709A, SET I (page 279)

1.
$$
\begin{array}{l}
\quad\ 1 \text{ ft} \ 3 \text{ in}\\
2\overline{\smash{)}2 \text{ ft} \ 6 \text{ in}}\\
\ \underline{2} \qquad \underline{6}\\
\ 0 \qquad\ 0
\end{array}
$$

2. 1 qt $\tfrac{1}{3}$ pt

3.
$$
\begin{array}{l}
\qquad 0 \text{ qt} \ 1\tfrac{1}{4} \text{ pt}\\
4\overline{\smash{)}2 \text{ qt} \ 1 \text{ pt}}\\
\ \longrightarrow \underline{\ 4}\\
\qquad\quad 5 \text{ pt}\\
\qquad\quad \underline{4}\\
\qquad\quad 1 \text{ pt} \to \dfrac{1 \text{ pt}}{4} = \dfrac{1}{4} \text{ pt}
\end{array}
$$

4. 1 ft 4 in

5.
$$
\begin{array}{ll}
\quad\ 1 \text{ hr} \qquad 50 \text{ min} & \qquad\ 50 \text{ min}\\
3\overline{\smash{)}5 \text{ hr} \qquad 30 \text{ min}} & 3\overline{\smash{)}150 \text{ min}}\\
\ \underline{3}\\
\ 2 \text{ hr} = \underline{120 \text{ min}}\\
\qquad\qquad 150 \text{ min}\\
\qquad\qquad \underline{150}
\end{array}
$$

6. 1 hr 40 min

7.
$$
\begin{array}{l}
\quad\ 1 \text{ gal} \quad\ 2 \text{ qt} \qquad 1 \text{ pt}\\
3\overline{\smash{)}4 \text{ gal} \quad 3 \text{ qt} \qquad 1 \text{ pt}}\\
\ \underline{3}\\
\ 1 \text{ gal} = \underline{4 \text{ qt}}\\
\qquad\qquad 7 \text{ qt}\\
\qquad\qquad \underline{6}\\
\qquad\qquad 1 \text{ qt} = \underline{2 \text{ pt}}\\
\qquad\qquad\qquad\quad 3 \text{ pt}
\end{array}
$$

8. 1 yd 2 ft 10 in

9.
$$
\begin{array}{l}
\qquad 1 \text{ yd} \quad\ 2 \text{ ft} \qquad 4\tfrac{2}{5} \text{ in}\\
5\overline{\smash{)}8 \text{ yd} \quad 2 \text{ ft} \qquad 10 \text{ in}}\\
\ \underline{5}\\
\ 3 \text{ yd} = \underline{9 \text{ ft}}\\
\qquad\qquad 11 \text{ ft}\\
\qquad\qquad \underline{10}\\
\qquad\qquad 1 \text{ ft} = \underline{12 \text{ in}}\\
\qquad\qquad\qquad\quad 22 \text{ in}\\
\qquad\qquad\qquad\quad \underline{20}\\
\qquad\qquad\qquad\quad 2 \text{ in} \to \dfrac{2 \text{ in}}{5} = \dfrac{2}{5} \text{ in}
\end{array}
$$

10. 3 qt $1\tfrac{5}{6}$ pt

EXERCISES 709B, SET I (page 280)

1. $\dfrac{\overset{8}{\cancel{16}} \text{ ft}}{\underset{1}{\cancel{2}} \text{ ft}} = 8$

2. 8

3.
$$
\begin{array}{l}
\qquad\ \ 11\\
15¢\overline{\smash{)}175¢}\\
\quad\ \underline{15}\\
\qquad 25\\
\qquad \underline{15}\\
\qquad 10¢
\end{array}
$$
Therefore 11 postcards with 10¢ left over

4. 134 envelopes with 8¢ left over

5. $\dfrac{1 \text{ hr} \ 30 \text{ min}}{2 \text{ min}} = \dfrac{\overset{45}{\cancel{90} \text{ min}}}{\underset{1}{\cancel{2} \text{ min}}} = 45$

6. 15

REVIEW EXERCISES 710, SET I (page 281)

1. 1.5 mi = __?__ ft
   $\dfrac{1.5 \cancel{\text{ mi}}}{1}\left(\dfrac{5{,}280 \text{ ft}}{1 \cancel{\text{ mi}}}\right) = 1.5(5{,}280 \text{ ft})$
   $= 7{,}930 \text{ ft}$

2. 3.5 mi

3. 3.75 yd = __?__ in
   $\dfrac{3.75 \cancel{\text{ yd}}}{1}\left(\dfrac{36 \text{ in}}{1 \cancel{\text{ yd}}}\right) = 3.75(36 \text{ in})$
   $= 135 \text{ in}$

4. 6 yd

5. 11 gal = __?__ pt
   $\dfrac{11 \cancel{\text{ gal}}}{1}\left(\dfrac{8 \text{ pt}}{1 \cancel{\text{ gal}}}\right) = 11(8 \text{ pt}) = 88 \text{ pt}$

6. 5.5 gal

7. $2\tfrac{3}{4}$ da = __?__ hr
   $2\tfrac{3}{4} \cancel{\text{ da}} \left(\dfrac{24 \text{ hr}}{1 \cancel{\text{ da}}}\right) = \dfrac{11}{\underset{1}{\cancel{4}}}\left(\dfrac{\overset{6}{\cancel{24} \text{ hr}}}{1}\right) = 66 \text{ hr}$

8. $4\tfrac{2}{3}$ min

9. 240 oz = __?__ lb
   $240 \cancel{\text{ oz}}\left(\dfrac{1 \text{ lb}}{16 \cancel{\text{ oz}}}\right) = \dfrac{240 \cancel{\text{ oz}}}{1} = 15 \text{ lb}$

10. 7,000 lb

11. $1\tfrac{1}{2}$ cups = __?__ oz
   $1\tfrac{1}{2} \cancel{\text{ cups}}\left(\dfrac{8 \text{ oz}}{1 \cancel{\text{ cup}}}\right) = \dfrac{3}{\underset{1}{\cancel{2}}}\left(\dfrac{\overset{4}{\cancel{8} \text{ oz}}}{1}\right) = 6 \text{ oz}$

12. 3.5 cups

13. 3 hr 25 min 45 sec
    4   50   30
    $\overset{1}{}$ $\overset{1}{76}$ $\overset{}{75}$
    ───────────────────
    8 hr 16 min 15 sec

14. 6 tons 700 lb

15. $\overset{6}{\cancel{7}}$ gal $\overset{5}{\cancel{1}}$ qt 1 pt
    − 3   3   1
    ──────────────────
    3 gal 2 qt 0 pt

16. 2 lb 12 oz

17. 3 yd   2 ft   4 in
    ×   7
    ────────────────────
    21 yd 14 ft $\overset{}{28}$ in
              2   4
            $\overset{}{16}$ ft
     5   1
    ────────────────────
    26 yd   1 ft   4 in

18. 16 hr 8 min 15 sec

19.   1 hr   50 min   12 sec
    4)7 hr   20 min   48 sec
      4
    ─────
    3 hr = 180 min
           200 min
           200
    ─────────────────
             0   48 sec
                      48

20. 1 yd 2 ft 11 in

21. $\dfrac{36{,}198\ \cancel{ft}}{1}\left(\dfrac{1\ mi}{5{,}280\ \cancel{ft}}\right) = \dfrac{36{,}198}{5{,}280}$ mi $\doteq$ 6.9 mi

22. 5.5 mi (To the nearest half mile)

23. $4\tfrac{1}{2}$ mph = __?__ ft per min

    $4\tfrac{1}{2}$ mph $= \dfrac{4.5\ \cancel{mi}}{1\ \cancel{hr}}\left(\dfrac{5{,}280\ ft}{1\ \cancel{mi}}\right)\left(\dfrac{1\ \cancel{hr}}{60\ min}\right) = \dfrac{4.5(5{,}280)\ ft}{60\ min}$

    $= 396$ ft per min

24. 3,960 ft per min

25. $\dfrac{186{,}000\ mi}{1\ \cancel{sec}}\left(\dfrac{60\ \cancel{sec}}{1\ \cancel{min}}\right)\left(\dfrac{60\ \cancel{min}}{1\ \cancel{hr}}\right)\left(\dfrac{24\ \cancel{hr}}{1\ \cancel{da}}\right)\left(\dfrac{365\ \cancel{da}}{1\ \cancel{yr}}\right)\left(\dfrac{4.3\ \cancel{yr}}{1}\right)$

    $= 25{,}222{,}492{,}800{,}000$ mi $= 25{,}222{,}492.8$ million miles

    Actually the distance is more than this because a year is about 365.25 days and light travels about 186,300 mi per sec.

26. When we are talking about time, 9:60 = 10:00 on the clock.

SOLUTIONS FOR CHAPTER 7 DIAGNOSTIC TEST (page 283)

Following each problem number is the textbook section reference (in parentheses) where that kind of problem is discussed.

1. (701, 702)

   (a) $\dfrac{4\ \cancel{yd}}{1}\left(\dfrac{3\ ft}{1\ \cancel{yd}}\right) = (4 \times 3)$ ft = 12 ft

       $\dfrac{12\ \cancel{ft}}{1}\left(\dfrac{12\ in}{1\ \cancel{ft}}\right) = (12 \times 12)$ in = 144 in

   (b) $\dfrac{2\ \cancel{gal}}{1}\left(\dfrac{4\ qt}{1\ \cancel{gal}}\right) = (2 \times 4)$ qt = 8 qt

       $\dfrac{8\ \cancel{qt}}{1}\left(\dfrac{2\ pt}{1\ \cancel{qt}}\right) = (8 \times 2)$ pt = 16 pt

   (c) $\dfrac{3\ \cancel{hr}}{1}\left(\dfrac{60\ min}{1\ \cancel{hr}}\right) = (3 \times 60)$ min = 180 min

       $\dfrac{180\ \cancel{min}}{1}\left(\dfrac{60\ sec}{1\ \cancel{min}}\right) = (180 \times 60)$ sec = 10,800 sec

   (d) $\dfrac{5\ \cancel{wk}}{1}\left(\dfrac{7\ da}{1\ \cancel{wk}}\right) = (5 \times 7)$ da = 35 da

       $\dfrac{35\ \cancel{da}}{1}\left(\dfrac{24\ hr}{1\ \cancel{da}}\right) = (35 \times 24)$ hr = 840 hr

   (e) $\dfrac{10\ \cancel{lb}}{1}\left(\dfrac{16\ oz}{1\ \cancel{lb}}\right) = (10 \times 16)$ oz = 160 oz

2. (705)
   (a) 2 hr 85 min $\cancel{97}$ sec
                    1   37 ──────── 97 sec = 1 min 37 sec
                   86
            1   26 ──────── 86 min = 1 hr 26 sec
       3 hr 26 min 37 sec
       Therefore, 2 hr 85 min 97 sec = 3 hr 26 min 37 sec

   (b) 1 yr 49 wk $\cancel{67}$ da
                    9   4 ──────── 67 da = 9 wk 4 da
                   58
            1   6 ──────── 58 wk = 1 yr 6 wk
       2 yr 6 wk 4 da
       Therefore, 1 yr 49 wk 67 da = 2 yr 6 wk 4 da

   (c) 7 mi $\cancel{6{,}080}$ ft
            1   800 ──────── 6,080 ft = 1 mi 800 ft
       8 mi   800 ft
       Therefore, 7 mi 6,080 ft = 8 mi 800 ft

3. (706) 5 yd 2 ft   8 in
          3   10
          6   1   9
          1    $\overset{}{5}$ $\overset{}{27}$
          ──────────────────
          15 yd 2 ft   3 in

4. (707) $\overset{6}{\cancel{7}}$ gal   1 pt
          − 3 gal 2 qt
          ─────────────────
          3 gal 2 qt 1 pt

5. (708)   2 lb 7 oz
           ×   6
           ──────────────
           12 lb $\cancel{42}$ oz
            2   10 ──────── 42 oz = 2 lb 10 oz
           14 lb 10 oz

6. (708)

   ┌──────────────┐
   │   │ 24 ft
   └──────────────┘
     55 ft

   Area = length × width
        = 55 ft × 24 ft
        = (55 × 24) sq ft
        = 1,320 sq ft

7. (709)   1 yd   1 ft   9 in
         5)7 yd   2 ft   9 in
           5
         ─────
         2 yd = 6 ft
                8 ft
                5
              ─────
                3 ft = 36 in
                       45 in
                       45
                     ─────
                       0

8. (709) $3.15 = 315¢

    $\dfrac{315\ \cancel{¢}}{1}\left(\dfrac{1\ can}{21\ \cancel{¢}}\right) = \dfrac{315}{21}$ cans = 15 cans

9. (703)

(a) $\dfrac{3 \text{ cup}}{1}\left(\dfrac{1 \text{ pt}}{2 \text{ cup}}\right) = \dfrac{3}{2}$ pt $= 1\dfrac{1}{2}$ pt

(b) $\dfrac{6 \text{ tsp}}{1}\left(\dfrac{1 \text{ tbsp}}{3 \text{ tsp}}\right) = \dfrac{6}{3}$ tbsp $= 2$ tbsp

(c) $\dfrac{8 \text{ tbsp}}{1}\left(\dfrac{1 \text{ cup}}{16 \text{ tbsp}}\right) = \dfrac{8}{16}$ cup $= \dfrac{1}{2}$ cup

(d) $\dfrac{2 \text{ pt}}{1}\left(\dfrac{16 \text{ oz}}{1 \text{ pt}}\right) = (2 \times 16)$ oz $= 32$ oz

10. (702) $\dfrac{60 \text{ mi}}{1 \text{ hr}}\left(\dfrac{5,280 \text{ ft}}{1 \text{ mi}}\right)\left(\dfrac{1 \text{ hr}}{60 \text{ min}}\right)\left(\dfrac{1 \text{ min}}{60 \text{ sec}}\right)$

$= \dfrac{60 \times 5,280}{60 \times 60} \dfrac{\text{ft}}{\text{sec}} = 88$ ft per sec

EXERCISES 802A, SET I (page 292)

1. $1_\wedge 800. = 1,800$ m   2. 34,000 ℓ

3. $0_\wedge 249. = 249$ g   4. 5,710 m

5. $.060_\wedge 5 = 0.0605$ kl   6. 0.322 kg

7. $.275_\wedge = 0.275$ kg   8. 0.0564 kl

9. $0_\wedge 780. = 780$ m   10. 9,300 g

11. $72.350_\wedge = 72.35$ kg   12. 2.365 km

13. $\dfrac{125 \text{ ℓ}}{\text{mo}}\left(\dfrac{1 \text{ kl}}{1,000 \text{ ℓ}}\right)\left(\dfrac{12 \text{ mo}}{1 \text{ yr}}\right) = \dfrac{125 \times 12}{1,000} \dfrac{\text{kl}}{\text{yr}} = 1.5 \dfrac{\text{kl}}{\text{yr}}$

14. 180,000 m²

EXERCISES 802B, SET I (page 294)

1. $2.79_\wedge = 2.79$ m   2. 0.54 ℓ

3. $.08_\wedge 3 = 0.083$ g   4. 40.9 m

5. $2_\wedge 50. = 250$ cm   6. 72 cl

7. $3_\wedge 90.6 = 390.6$ cg   8. 84.2 cm

9. $6.32_\wedge = 6.32$ ℓ   10. 0.581 g

11. $0_\wedge 02.63 = 2.63$ cg   12. 9.2 cl

13. $1_\wedge 82. = 182$ cm   14. 143,000 cg

EXERCISES 802C, SET I (page 295)

1. $.091_\wedge = 0.091$ m   2. 5.6 ℓ

3. $.470_\wedge = 0.47$ g   4. 4.3 m

5. $2_\wedge 600. = 2,600$ mm   6. 390 ml

7. $0_\wedge 108.0 = 108$ mg   8. 82.7 mm

9. $.230_\wedge = 0.23$ ℓ   10. 9.16 g

11. $7_\wedge 040. = 7,040$ mg   12. 21,600 ml

13. $2_\wedge 000. = 2,000$ cc   14. 3,550 cc

15. $.175_\wedge = 0.175$ ℓ   16. 2.5 ℓ

17. $\dfrac{1.5 \text{ g}}{\text{da}}\left(\dfrac{7 \text{ da}}{1 \text{ wk}}\right)\left(\dfrac{1,000 \text{ mg}}{1 \text{ g}}\right) = 10,500 \dfrac{\text{mg}}{\text{wk}}$

18. $75 \dfrac{\text{ℓ}}{\text{da}}$

EXERCISES 803, SET I (page 297)

1. $\dfrac{10 \text{ in}}{1}\left(\dfrac{2.54 \text{ cm}}{1 \text{ in}}\right) = 25.4$ cm   2. 45.72 cm

3. $\dfrac{2.12 \text{ qt}}{1}\left(\dfrac{1 \text{ ℓ}}{1.06 \text{ qt}}\right) = 2$ ℓ   4. 3 ℓ

5. $\dfrac{0.55 \text{ kg}}{1}\left(\dfrac{2.20 \text{ lb}}{1 \text{ kg}}\right) = 1.21$ lb   6. 10.34 lb

7. $\dfrac{82 \text{ km}}{1}\left(\dfrac{0.621 \text{ mi}}{1 \text{ km}}\right) \doteq 50.92$ mi
(Using 1.61 km per mi: Answer $\doteq$ 50.93)

8. 86.94 (Using 0.621); 86.96 (Using 1.61)

9. $\dfrac{10 \text{ ℓ}}{1}\left(\dfrac{1.06 \text{ qt}}{1 \text{ ℓ}}\right) = 10.6$ qt   10. 7.42 qt

11. $\dfrac{31 \text{ mi}}{1}\left(\dfrac{1.61 \text{ km}}{1 \text{ mi}}\right) = 49.91$ km   12. 249.55 km

13. $\dfrac{17.6 \text{ lb}}{1}\left(\dfrac{1 \text{ kg}}{2.20 \text{ lb}}\right) = 8$ kg   14. 12 kg

15. $\dfrac{150 \text{ ha}}{1}\left(\dfrac{2.47 \text{ acres}}{1 \text{ ha}}\right) = 370.5$ acres

16. 26.32 ha

17. $\dfrac{66 \text{ in}}{1}\left(\dfrac{1 \text{ m}}{39.4 \text{ in}}\right) \doteq 1.68$ m   18. 1.88 m

19. $\dfrac{2 \text{ m}}{1}\left(\dfrac{39.4 \text{ in}}{1 \text{ m}}\right) = 78.8$ in   20. 118.2 in

21. $\dfrac{908 \text{ g}}{1}\left(\dfrac{1 \text{ lb}}{454 \text{ g}}\right) = 2$ lb   22. 3 lb

23. $\dfrac{0.75 \text{ lb}}{1}\left(\dfrac{454 \text{ g}}{1 \text{ lb}}\right) = 340.5$ g   24. 567.5 g

25. $\dfrac{1.5 \text{ yd}}{1}\left(\dfrac{36 \text{ in}}{1 \text{ yd}}\right)\left(\dfrac{2.54 \text{ cm}}{1 \text{ in}}\right) = 137.16$ cm

26. 1.09 yd

27. $\dfrac{227 \text{ g}}{1}\left(\dfrac{1 \text{ lb}}{454 \text{ g}}\right)\left(\dfrac{16 \text{ oz}}{1 \text{ lb}}\right) = 8$ oz   28. 340.5 g

29. $\dfrac{120 \text{ km}}{1}\left(\dfrac{0.621 \text{ mi}}{1 \text{ km}}\right) = 74.52$ mi   30. 18.63 mph

31. $\dfrac{85.7 \text{ kg}}{1}\left(\dfrac{2.20 \text{ lb}}{1 \text{ kg}}\right) = 188.54$ lb   32. 148.5 lb

33. $\dfrac{1,500 \text{ m}}{1}\left(\dfrac{39.4 \text{ in}}{1 \text{ m}}\right)\left(\dfrac{1 \text{ ft}}{12 \text{ in}}\right) = 4,925$ ft $\left.\right\}$ $\begin{array}{r} 5,280 \\ -\ 4,925 \\ \hline 355 \end{array}$

1 mile = 5,280 ft.
Therefore, the 1,500 m race is 355 ft shorter than the mile race.

34. The 400 m race is $6\dfrac{2}{3}$ ft ($\doteq$ 6.67 ft) shorter than the quarter mile race.

35. $\dfrac{175 \text{ mm}}{1}\left(\dfrac{1 \text{ cm}}{10 \text{ mm}}\right)\left(\dfrac{1 \text{ in}}{2.54 \text{ cm}}\right) \doteq 6.89$ in

36. $\doteq$ 1.38 in

37. $\dfrac{20\ \cancel{\ell}}{1}\left(\dfrac{1.06\ \cancel{qt}}{1\ \cancel{\ell}}\right)\left(\dfrac{1\ gal}{4\ \cancel{qt}}\right) = 5.30$ gal

38. $\doteq$ 447.22 $\dfrac{m}{s}$

<u>EXERCISES 804, SET I (page 300)</u>

1. $F = \dfrac{9}{5}\overset{4}{\underset{1}{(\cancel{20})}} + 32 = 36 + 32 = 68°F$ 2. $59°F$

3. $C = \dfrac{5}{9}(50 - 32) = \dfrac{5}{9}\overset{2}{(\cancel{18})} = 10°C$ 4. $15°C$

5. $F = \dfrac{9}{5}(8) + 32 = \dfrac{72}{5} + 32 \doteq 46°F$ 6. $\doteq 63°F$

7. $C = \dfrac{5}{9}(72 - 32) = \dfrac{5}{9}(40) = \dfrac{200}{9} \doteq 22°C$ 8. $\doteq 29°C$

9. $C = \dfrac{5}{9}(41 - 32) = \dfrac{5}{9}\overset{1}{(\cancel{9})} = 5°C$ 10. Warmer by $2°F$

<u>REVIEW EXERCISES 805, SET I (page 301)</u>

1. $2\underset{\underset{3}{\rightarrow}}{\wedge}000. = 2,000$ ml   2. $0.24\ \ell$

3. $2.000\underset{\underset{-3}{\leftarrow}}{\wedge} = 2$ kg   4. $3,860$ g

5. $\dfrac{35\ \cancel{mm}}{1}\left(\dfrac{1\ \cancel{cm}}{10\ \cancel{mm}}\right)\left(\dfrac{1\ in}{2.54\ \cancel{cm}}\right) \doteq 1.38$ in

6. 228.6 mm

7. $1\underset{\underset{2}{\rightarrow}}{\wedge}75. = 175$ cm   8. 3 m

9. $.500\underset{\underset{-3}{\leftarrow}}{\wedge} = 0.5$ g   10. 56 mg

11. $\dfrac{1,200\ \cancel{acres}}{1}\left(\dfrac{1\ ha}{2.47\ \cancel{acres}}\right) \doteq 486$ ha

12. $\doteq 1,581$ acres

13. $F = \dfrac{9}{5}(27) + 32 \doteq 81°F$ 14. $\doteq 6°C$

15. $\dfrac{24,902\ \cancel{mi}}{1}\left(\dfrac{1.61\ km}{1\ \cancel{mi}}\right) \doteq 40,092$ km

16. $6,210$ mi

17. $\left(\dfrac{80\ \cancel{km}}{hr}\right)\left(\dfrac{0.621\ mi}{1\ \cancel{km}}\right) = 49.68$ mph   18. $\doteq 112.7\ \dfrac{km}{hr}$

19. $\dfrac{120\ \cancel{lb}}{1}\left(\dfrac{1\ kg}{2.20\ \cancel{lb}}\right) \doteq 55$ kg   20. $\doteq 175$ lb

21. $\dfrac{20\ \cancel{kg}}{1}\left(\dfrac{2.20\ lb}{1\ \cancel{kg}}\right) = 44$ lb   22. $\doteq 1,649°C$

23. $\dfrac{1.75\ \cancel{\ell}}{1}\left(\dfrac{1,000\ \cancel{ml}}{1\ \cancel{\ell}}\right)\left(\dfrac{1\ cc}{1\ \cancel{ml}}\right) = 1,750$ cc

24. $\doteq 122$ in^3

25. $\dfrac{8.6\ \cancel{yr}}{1}\left(\dfrac{186,000\ mi}{1\ \cancel{sec}}\right)\left(\dfrac{60\ \cancel{sec}}{1\ \cancel{min}}\right)\left(\dfrac{60\ \cancel{min}}{1\ \cancel{hr}}\right)\left(\dfrac{24\ \cancel{hr}}{1\ \cancel{da}}\right)$
$\cdot\left(\dfrac{365\ \cancel{da}}{1\ \cancel{yr}}\right)\left(\dfrac{1.61\ km}{1\ \cancel{mi}}\right)$
    $\doteq 81,216,427$ million kilometers

26. $\doteq 1,888,754$ billion kilometers

<u>SOLUTIONS FOR CHAPTER 8 DIAGNOSTIC TEST (page 303)</u>

Following each problem number is the textbook section reference (in parentheses) where that kind of problem is discussed.

1. (802) $5\underset{\underset{2}{\rightarrow}}{\wedge}24. = 524$ cm

2. (803) $\dfrac{25\ \cancel{kg}}{1}\left(\dfrac{2.20\ lb}{1\ \cancel{kg}}\right) = 55$ lb

3. (802) 500 cc $= 500$ ml $= .500\underset{\underset{-3}{\leftarrow}}{\wedge}\ell = 0.5\ \ell$

4. (803) $\dfrac{100\ \cancel{km}}{1}\left(\dfrac{0.621\ mi}{1\ \cancel{km}}\right) = 62.1$ mi

5. (803) $\dfrac{7.42\ \cancel{qt}}{1}\left(\dfrac{1\ \ell}{1.06\ \cancel{qt}}\right) = 7\ \ell$

6. (803) $\dfrac{100\ \cancel{ha}}{1}\left(\dfrac{2.47\ acres}{1\ \cancel{ha}}\right) = 247$ acres

7. (802) $.870\underset{\underset{-3}{\leftarrow}}{\wedge} = 0.87$ kg

8. (804) $F = \dfrac{9}{5}\overset{7}{\underset{1}{(\cancel{35})}} + 32 = 63 + 32 = 95°F$

9. (802) $12.60\underset{\underset{2}{\leftarrow}}{\wedge} = 12.6$ m

10. (802) $42\underset{\underset{3}{\rightarrow}}{\wedge}500. = 42,500\ \ell$

11. (802) $1.075\underset{\underset{-3}{\leftarrow}}{\wedge} = 1.075$ g

12. (802) $1\underset{\underset{3}{\rightarrow}}{\wedge}650. = 1,650$ ml $= 1,650$ cc

13. (802) $1\underset{\underset{3}{\rightarrow}}{\wedge}300. = 1,300$ mm

14. (803) $\dfrac{5\ \cancel{lb}}{1}\left(\dfrac{454\ g}{1\ \cancel{lb}}\right) = 2,270$ g

15. (802) $5.094\underset{\underset{-3}{\leftarrow}}{\wedge} = 5.094$ km

16. (804) $C = \dfrac{5}{9}(77 - 32) = \dfrac{5}{9}\overset{5}{\underset{1}{(\cancel{45})}} = 25°C$

17. (803) $\dfrac{20\ \cancel{in}}{1}\left(\dfrac{2.54\ cm}{1\ \cancel{in}}\right) = 50.8$ cm

18. (803) $\dfrac{50\ \cancel{km}}{hr}\left(\dfrac{0.621\ mi}{1\ \cancel{km}}\right) \doteq 31$ mph

19. (803) $\dfrac{650\ \cancel{mg}}{\cancel{da}}\left(\dfrac{1\ g}{1,000\ \cancel{mg}}\right)\left(\dfrac{7\ \cancel{da}}{1\ wk}\right) = 4.55\ \dfrac{g}{wk}$

20. (803) $\dfrac{10\ \cancel{ft}}{1}\left(\dfrac{12\ \cancel{in}}{1\ \cancel{ft}}\right)\left(\dfrac{1\ m}{39.4\ \cancel{in}}\right) \doteq 3.0$ m

<u>EXERCISES 901, SET I (page 310)</u>

1. $A = 7 \times 5 = 35$ sq ft
$P = 2 \times 7 + 2 \times 5 = 14 + 10 = 24$ ft
2. 143 sq ft; 48 ft
3. $A = 4\dfrac{1}{2} \times 2\dfrac{1}{2} = \dfrac{9}{2} \times \dfrac{5}{2} = \dfrac{45}{4} = 11\dfrac{1}{4}$ sq ft

    $P = 2 \times 4\dfrac{1}{2} + 2 \times 2\dfrac{1}{2} = 9 + 5 = 14$ ft

4. 375 sq cm; 80 cm
5. $A = 75 \times 35 = 2,625$ sq cm
$P = 2 \times 75 + 2 \times 35 = 150 + 70 = 220$ cm
6. 11.5 sq in; 14.0 in
7. $A = 12 \times 12 = 144$ sq in
$A = 1 \times 1 = 1$ sq ft
$P = 2 \times 1 + 2 \times 1 = 4$ ft

8. 9 sq ft; 1 sq yd; 12 ft

9. $A = 3 \times 3 = 9$ sq yd
   $A = (3 \times 3)(3 \times 3) = 9 \times 9 = 81$ sq ft
   $P = 2 \times 3 + 2 \times 3 = 6 + 6 = 12$ yd

10. $12\frac{15}{32}$ sq ft; $15\frac{1}{4}$ ft

11.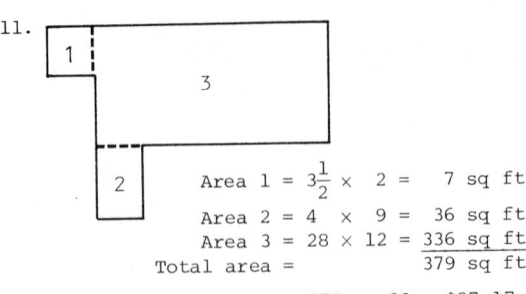

Area 1 = $3\frac{1}{2} \times 2$ =   7 sq ft
Area 2 = 4×9 = 36 sq ft
Area 3 = 28×12 = $\underline{336}$ sq ft
Total area =   379 sq ft

Cost = 379 × .23 = $87.17

12. 216 sq in

EXERCISES 902, SET I (page 313)

1. $A = \frac{1}{2} \times \overset{7}{\cancel{14}} \times 12 = 84$ sq m
   $P = 14 + 13 + 15 = 42$ m

2. (a) 756 sq in   (b) $5\frac{1}{4}$ sq ft

3. (a) $A = \frac{1}{2}(64) \times 93 = 2{,}976$ sq in

   (b) $A = \frac{1}{2} \times 5\frac{1}{3} \times 7\frac{3}{4} = \frac{1}{\cancel{2}} \times \frac{\overset{8}{\cancel{16}}}{3} \times \frac{31}{\cancel{4}} = \frac{62}{3} = 20\frac{2}{3}$ sq ft

4. (a) 20 yd 2 ft   (b) 222 sq ft

5. (a) $P = 2 \times (2\text{ ft }6\text{ in}) + 2 \times (4\text{ ft }6\text{ in}) + 3\text{ ft}$
   $= 5\text{ ft} + 9\text{ ft} + 3\text{ ft} = 17\text{ ft}$

   (b)

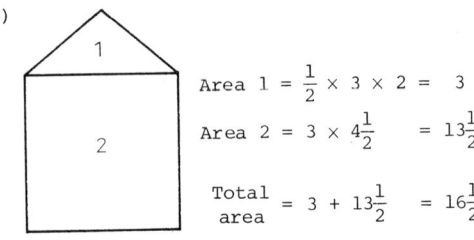

Area 1 = $\frac{1}{2} \times 3 \times 2$ =   3 sq ft
Area 2 = $3 \times 4\frac{1}{2}$ = $13\frac{1}{2}$ sq ft
Total area = $3 + 13\frac{1}{2}$ = $16\frac{1}{2}$ sq ft

6. (a) 247 ft 8 in   (b) 3,575 sq ft

7.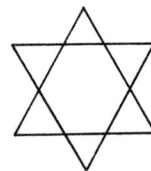

6 small triangles
$\underline{2}$ large triangles
8 triangles

EXERCISES 903A, SET I (page 316)

1. $D = 2 \times 10 = 20$ ft
   $C = \pi \times D = 3.14 \times 20 = 62.8$ ft
   $A = \pi \times R^2 = 3.14 \times 10 \times 10 = 314$ sq ft

2. $R = 4$ in
   $C \doteq 25.1$ in
   $A \doteq 50.2$ sq in

3. $D = 2 \times 3 = 6$ yd
   $C = \pi \times D = 3.14 \times 6 \doteq 18.8$ yd
   $A = \pi R^2 = 3.14 \times 3 \times 3 \doteq 28.3$ sq yd

4. $D = 5$ ft
   $C = 15.7$ ft
   $A \doteq 19.6$ sq ft

5. $R = \frac{1}{2} \times 20 = 10$ cm
   $C = \pi \times D = 3.14 \times 20 = 62.8$ cm
   $A = \pi \times R^2 = 3.14 \times 10 \times 10 = 314$ sq cm

6. $R = 4.5$ ft
   $C \doteq 28.3$ ft
   $A \doteq 63.6$ sq ft

7. $D = 2 \times 3.8 = 7.6$ in
   $C = \pi \times D = 3.14 \times 7.6 \doteq 23.9$ in
   $A = \pi R^2 = 3.14 \times 3.8 \times 3.8 \doteq 45.3$ sq in

8. $R = 3.5$ in
   $C \doteq 22.0$ in
   $A \doteq 38.5$ sq in

EXERCISES 903B, SET I (page 318)

1. $\frac{F}{f} = \left(\frac{D}{d}\right)^2 = \left(\frac{\frac{7}{8}}{\frac{1}{2}}\right)^2 = \frac{49}{64} \div \frac{1}{4} = \frac{49}{\underset{16}{\cancel{64}}} \cdot \frac{\overset{1}{\cancel{4}}}{1} = \frac{49}{16} = 3\frac{1}{16}$

   The flow in the $\frac{7}{8}$ - inch hose is $3\frac{1}{16}$ times the flow in the $\frac{1}{2}$ - inch hose.

2. The flow in the $\frac{5}{8}$ - inch hose is $1\frac{9}{16}$ times the flow in the $\frac{1}{2}$ - inch hose.

3. $\frac{F}{f} = \left(\frac{D}{d}\right)^2 = \left(\frac{2}{\frac{1}{2}}\right)^2 = 4 \div \frac{1}{4} = \frac{4}{1} \cdot \frac{4}{1} = 16$

   This means sixteen $\frac{1}{2}$ - inch lines can be run from a 2-inch main.

4. $5\frac{11}{49}$ hr

5. $\frac{F}{f} = \left(\frac{\frac{7}{8}}{\frac{1}{4}}\right)^2 = \frac{49}{64} \div \frac{1}{16} = \frac{49}{\underset{4}{\cancel{64}}} \cdot \frac{\overset{1}{\cancel{16}}}{1} = \frac{49}{4} = 12\frac{1}{4}$

   The flow in the $\frac{7}{8}$ - inch hose is $12\frac{1}{4}$ times the flow in the $\frac{1}{4}$ - inch hose.

6. The flow in the 1-inch hose is $1\frac{7}{9}$ times the flow in the $\frac{3}{4}$ - inch hose.

EXERCISES 904A, SET I (page 320)

1. $V = L \times W \times H$
   $= 22$ ft $\times 15$ ft $\times 8$ ft
   $= 2{,}640$ cu ft

2. 2,145 cu in

3. $V = L \times W \times H$
   $= 3 \times 3 \times 3$ cu ft
   $= 27$ cu ft

4. 1,728 cu in

5. (a) $V = L \times W \times H$
   $= 10 \times 9 \times 4$ cu yd
   $= 360$ cu yd

   (b) Weight = 2(360) lb
   $= 720$ lb

6. (a) 3,432 cu in   (b) $\doteq 124$ lb
   (c) $\doteq 14.86$ gal (rounded off to the nearest hundredth)

7. 0 cu ft. This is a trick question. In order to be a hole, it must be empty.

EXERCISES 904B, SET I (page 323)

1. $V = 10 \times 5 \times 3 = 150$ cu in
   Lateral area $= 2(10 \times 3) + 2(5 \times 3)$
   $= 2(30) + 2(15)$
   $= 60 + 30 = 90$ sq in
   Top area $=$ Bottom area $= 5 \times 10 = 50$ sq in
   Total surface area $=$ Lateral area $+$ Top area
   $+$ Bottom area
   $= 90 + 50 + 50 = 190$ sq in

2. $V = 5,100$ cc. Total surface area $= 1,858$ sq cm

3. $V = e^3 = 8 \times 8 \times 8 = 512$ cc
   $S = 6e^2 = 6 \times 8 \times 8 = 384$ sq cm

4. 27 cu ft/cu yd

5. (a) $V = e^3 = (12)^3 = 12 \times 12 \times 12$
   $= 1,728$ cu in/cu ft
   (b) $\dfrac{1,728}{231} \doteq 7.5$ gal

6. (a) 1,000,000 cc/cu m
   (b) 1,000 liters/cu m

7. Lateral area $= 2(11\frac{1}{2} \times 8) + 2(16 \times 8)$
   $= 2(92) + 2(128) = 184 + 256$
   $= 440$ sq ft

8. 1,172 sq in

9. $V = 30 \times 10\frac{1}{2} \times 11\frac{1}{2} = \dfrac{\overset{15}{\cancel{30}}}{1} \times \dfrac{21}{\cancel{2}} \times \dfrac{23}{2} = \dfrac{7,245}{2}$
   $= 3,622\frac{1}{2}$ cu ft

   Lateral area $= 2(30 \times 11\frac{1}{2}) + 2(10\frac{1}{2} \times 11\frac{1}{2})$
   $= 2\left(\dfrac{\overset{15}{\cancel{30}}}{1} \times \dfrac{23}{\underset{1}{\cancel{2}}}\right) + \cancel{2}\left(\dfrac{21}{\cancel{2}} \times \dfrac{23}{2}\right)$
   $= 690 + \dfrac{483}{2} = 690 + 241\frac{1}{2}$
   $= 931\frac{1}{2}$ sq ft

   Top area $=$ Bottom area $= 30 \times 10\frac{1}{2} = \dfrac{\overset{15}{\cancel{30}}}{1} \times \dfrac{21}{\underset{1}{\cancel{2}}}$
   $= 315$ sq ft
   Total surface area $=$ Lateral area $+$ Top area
   $+$ Bottom area
   $= 931\frac{1}{2} + 315 + 315$
   $= 1,561\frac{1}{2}$ sq ft

10. $V = 1,031\frac{1}{4}$ cu ft; $S = 621\frac{1}{4}$ sq ft

EXERCISES 905, SET I (page 324)

1. $V = \pi r^2 h = 3.14(5)(5)(12) = 942.0$ cu in
   Lateral area $= 2\pi r h = 2(3.14)(5)(12) = 376.8$ sq in
   Top area $=$ Bottom area $= \pi r^2 = 3.14(5)(5)$
   $= 78.5$ sq in
   Total surface area $=$ Lateral $+$ Top $+$ Bottom
   $= 376.8 + 78.5 + 78.5$
   $= 533.8$ sq in

2. $\doteq 4.6$ cu ft

3. $V = \pi r^2 h = 3.14(6)(6)(16) = 1,808.64$ cu ft
   $= 1,808.64 \times 7.48$
   $\doteq 13,528.6$ gal

4. $\doteq 68.0$ gal

EXERCISES 906, SET I (page 325)

1. $S = 4\pi r^2 = 4(3.14)(8)(8) \doteq 803.8$ sq in
   $V = \dfrac{4}{3}(3.14)(8)(8)(8) \doteq 2,143.6$ cu in

2. $S \doteq 201.0$ sq in; $V \doteq 267.9$ cu in

3. $\dfrac{V_1}{V_2} = \dfrac{2,143.6}{267.9} \doteq 8.0$. Therefore, doubling the
   diameter makes the volume
   eight times as large.
   $(2)^3 = 8$

4. $V \doteq 7,233.3$ cu in

5. $V = \dfrac{1}{\cancel{2}} \cdot \dfrac{\overset{2}{\cancel{4}}}{3}(3.14)(8)(8)(8) \doteq 1,071.8$ cu in

6. $V \doteq 244.7$ gal

7. (a) Volume of hemisphere $= \dfrac{1}{\underset{1}{\cancel{2}}} \times \dfrac{\overset{2}{\cancel{4}}}{3}(3.14)(5)(5)(5)$
   $\doteq 261.67$ cu ft
   Volume of cylinder $\doteq (3.14)(5)(5)(16)$
   $\doteq 1,256.00$ cu ft

   Total volume $\doteq 261.67 + 1,256.00$
   $\doteq 1,517.7$ cu ft
   (b) Total surface area
   $= \begin{matrix}\text{surface of} \\ \text{hemisphere}\end{matrix} + \begin{matrix}\text{lateral} \\ \text{area}\end{matrix} + \begin{matrix}\text{surface of} \\ \text{bottom}\end{matrix}$
   $= 2\pi r^2 + 2\pi r h + \pi r^2$
   $= 2(3.14)(5)(5) + 2(3.14)(5)(16) + 3.14(5)(5)$
   $= 157 + 502.4 + 78.5$
   $= 737.9$ sq ft

EXERCISES 907, SET I (page 331)

1. (a) $r = \dfrac{10}{5} = 2$; diagonal $\doteq 2(7.07) = 14.14$ in
   (b) $r = \dfrac{15}{5} = 3$; diagonal $\doteq 3(7.07) = 21.21$ in
   (c) $r = \dfrac{20}{5} = 4$; diagonal $\doteq 4(7.07) = 28.28$ in

2. (a) 2 in   (b) 6 in

3. $r = \dfrac{32}{4} = 8$; height $= 8(6) = 48$ ft

4. 32 ft

5. $r = \dfrac{5}{4}$; $V = \left(\dfrac{5}{4}\right)^3 (50) = \dfrac{125}{\underset{32}{\cancel{64}}}\left(\dfrac{\overset{25}{\cancel{50}}}{1}\right) = \dfrac{3,125}{32} = 97\frac{21}{32}$ gal

6. $23\frac{7}{16}$ tons

7. $r = \dfrac{40}{30} = \dfrac{4}{3}$; $S = \left(\dfrac{4}{3}\right)^2 (16) = \dfrac{16}{9}\left(\dfrac{16}{1}\right) = \dfrac{256}{9} = 28\frac{4}{9}$ gal

8. 640 lb

EXERCISES 908, SET I (page 334)

1. 2	2. 2	3. 4	4. 4
5. 2	6. 1	7. 3	8. 3
9. 1	10. 2	11. 3	12. 3
13. 3	14. 4	15. 4	16. 5

17. 3; accurate to tenths
18. 3; accurate to hundredths
19. 2; accurate to tens
20. 2; accurate to hundreds
21. 3; accurate to hundredths
22. 3; accurate to tenths
23. 4; accurate to hundredths
24. 4; accurate to hundredths

25. 5$\cancel{3}$4 ≐ 530 rounded off to tens   26. ≐ 390

27. 56.$\cancel{9}$7 ≐ 57.0 rounded off to one decimal place   28. ≐ 24.8

29. 2$\cancel{7}$.50 ≐ 28 rounded off to units   30. ≐ 28

31. 0.028$\cancel{5}$01 ≐ 0.029 rounded off to three decimal places   32. ≐ 0.036

33. $\overset{\overset{\text{4 sig. dig.}}{\ulcorner}}{23.75}$ + $\overset{\overset{\text{3 sig. dig}}{\ulcorner}}{66.25}$ − $\underset{\underset{\text{4 sig. dig.}}{\llcorner}}{3.60}$ ≐ $\underset{\underset{\text{3 sig. dig.}}{\llcorner}}{86.4}$

34. ≐ 91.3

35. $\overset{\overset{\text{1 sig. dig.}}{\ulcorner}}{6}$ × $\underset{\underset{\text{2 sig. dig.}}{\llcorner}}{7.8}$ × $\overset{\overset{\text{2 sig. dig.}}{\ulcorner}}{35}$ ≐ $\underset{\underset{\text{1 sig. dig.}}{\llcorner}}{2,000}$

36. ≐ 500

37. $\overset{\overset{\text{2 sig. dig.}}{\ulcorner}}{1,400}$ ÷ $\underset{\underset{\text{3 sig. dig.}}{\llcorner}}{175}$ ≐ $\overset{\overset{\text{2 sig. dig.}}{\ulcorner}}{8.0}$

38. ≐ 6.00

39. $\dfrac{\overset{\overset{\text{2 sig. dig.}}{\ulcorner}}{0.0045} \times \overset{\overset{\text{3 sig. dig.}}{\ulcorner}}{8.05}}{\underset{\underset{\text{3 sig. dig.}}{\llcorner}}{25}}$ ≐ $\underset{\underset{\text{2 sig. dig.}}{\llcorner}}{0.0014}$

40. ≐ 0.0085

41. $\overset{\overset{\text{3 sig. dig.}}{\ulcorner}}{127}$ × $\underset{\underset{\text{2 sig. dig.}}{\llcorner}}{0.43}$ × $\overset{\overset{\text{3 sig. dig.}}{\ulcorner}}{85.6}$ ≐ $\underset{\underset{\text{2 sig. dig.}}{\llcorner}}{4,700}$

42. ≐ 7,800

43. $\overset{\overset{\text{3 sig. dig.}}{\ulcorner}}{3.14}\underset{\underset{\text{3 sig. dig.}}{\llcorner}}{(12.5)}\overset{\overset{\text{1 sig. dig.}}{\ulcorner}}{8}$ ≐ $\underset{\underset{\text{1 sig. dig.}}{\llcorner}}{300}$

44. ≐ 1,900

45. $3.14\overset{\overset{\text{3 sig. dig.}}{\ulcorner}}{(9.8)}\underset{\underset{\text{2 sig. dig.}}{\llcorner}}{^2}$ ≐ 3.0 × $\overset{\overset{\text{exact number}}{\ulcorner}}{10^2}$ (2 sig. dig.)

46. ≐ 2.0 × 10²

REVIEW EXERCISES 909, SET I (page 337)

1. $P = 2(15 \text{ ft}) + 2(3 \text{ ft } 6 \text{ in}) + 2(6 \text{ ft} - 3 \text{ ft } 6 \text{ in})$
   = 30 ft + 7 ft + 5 ft = 42 ft

   $A = (15 \text{ ft})(3.5 \text{ ft}) + (3.5 \text{ ft})(2.5 \text{ ft}) = 61.25 \text{ sq ft}$

2. (a) 18 yd 2 ft   (b) 144 sq ft

3. $C = 2\pi r = 2(3.14)(2.5) = 15.7 \text{ ft}$
   $A = \pi r^2 = 3.14(2.5)(2.5) ≐ 19.6 \text{ sq ft}$

4. ≐ 1,178 cu in

5. $V = LWH = (3.75)(2)(4.5) = 33.75 ≐ 33.8 \text{ cu ft}$
   Lateral area = 2(3.75 × 4.5) + 2(2 × 4.5)
                = 33.75 + 18 = 51.75 sq ft
   Top = Bottom = 3.75 × 2 = 7.5 sq ft
   Total surface area = Lateral area + Top + Bottom
                        = 51.75 + 7.5 + 7.5
                        = 66.75 ≐ 66.8 sq ft

6. ≐ 396 cu yd

7. $\dfrac{A}{a} = \left(\dfrac{D}{d}\right)^2 = \left(\dfrac{1\frac{1}{4}}{\frac{1}{2}}\right)^2 = \left(\dfrac{\frac{5}{4}}{\frac{1}{2}}\right)^2 = \left(\dfrac{5}{4} \div \dfrac{1}{2}\right)^2 = \left(\dfrac{5}{\underset{2}{\cancel{4}}} \cdot \dfrac{\overset{1}{\cancel{2}}}{1}\right)^2$
   $= \dfrac{25}{4} = 6\frac{1}{4}$

The area of larger circle is $6\frac{1}{4}$ times the area of the smaller circle.

8. ≐ 423 gal

9. $\overset{\overset{\text{4 sig. dig.}}{\ulcorner}}{28.96}$ + 5.01 − $\underset{\underset{\text{3 sig. dig.}}{\llcorner}}{4.70}$ = 29.27 ≐ $\overset{\overset{\text{3 sig. dig.}}{\ulcorner}}{29.3}$ (3 sig. dig.)

10. ≐ 32.6

11. $\overset{\overset{\text{3 sig. dig.}}{\ulcorner}}{2.06}$ × $\underset{\underset{\text{3 sig. dig.}}{\llcorner}}{35.0}$ × $\overset{\overset{\text{1 sig. dig.}}{\ulcorner}}{0.08}$ = 5.768 ≐ 6 (1 sig. dig.)

12. ≐ 9

13. $\dfrac{\overset{\overset{\text{3 sig. dig.}}{\ulcorner}}{4.80} \times \overset{\overset{\text{3 sig. dig.}}{\ulcorner}}{0.0564}}{\underset{\underset{\text{2 sig. dig.}}{\llcorner}}{16}}$ = 0.01692 ≐ 0.017 (2 sig. dig.)

14. ≐ 0.0019

15. $\dfrac{67 \cancel{\text{ft}}}{1}\left(\dfrac{12 \cancel{\text{in}}}{1 \cancel{\text{ft}}}\right)\left(\dfrac{1 \text{ m}}{39.4 \cancel{\text{in}}}\right)$ ≐ 20.41 m

   $\dfrac{102 \cancel{\text{ft}}}{1}\left(\dfrac{12 \cancel{\text{in}}}{1 \cancel{\text{ft}}}\right)\left(\dfrac{1 \text{ m}}{39.4 \cancel{\text{in}}}\right)$ ≐ 31.07 m

   Area = LW
        = (31.07)(20.41) ≐ 634.1 ≐ 630 rounded off to tens

SOLUTIONS FOR CHAPTER 9 DIAGNOSTIC TEST (page 340)

Following each problem number is the textbook section reference (in parentheses) where that kind of problem is discussed.

1. (901) (a) 17 ft 6 in = 17.5 ft
           $A = L × W$
              = 17.5 × 12 = 210 sq ft

        (b) $P = 2L + 2W$
              = 2(17.5) + 2(12)
              = 35 + 24 = 59 ft

2. (901) (a) 1 m 27 cm = 1.27 m
           $A = s^2$
              = (1.27)² = 1.6129 sq m

        (b) $P = 4s$
              = 4(1.27) = 5.08 m

3. (902) (a) $A = \frac{1}{2}bh$
              = $\frac{1}{2}$(17)(9) = 76.5 sq in

        (b) $P = 17 + 10.3 + 15$
              = 42.3 in

4. (903) (a) $C = \pi d$
              = (3.14)(5) = 15.7 ft
        (b) $A = \pi r^2$
              = 3.14(2.5)²
              = 3.14(2.5)(2.5)
              = 19.625
              ≐ 19.6 sq ft

5. (903) Use formula $\dfrac{F}{f} = \left(\dfrac{D}{d}\right)^2$

$$\dfrac{F}{f} = \left(\dfrac{1}{\frac{7}{8}}\right)^2 = \left(1 \div \dfrac{7}{8}\right)^2 = \left(1 \cdot \dfrac{8}{7}\right)^2$$

$$= \left(\dfrac{8}{7}\right)^2 = \dfrac{64}{49} \doteq \dfrac{1.3}{1}$$

This means the flow in the 1-inch hose is

1.3 times the flow in the $\frac{7}{8}$-inch hose.

6. (904) Volume = LWH
   = (9)(5)(4)
   = 180 cu in

Area of front	= 4 × 9 =	36 sq in
Area of back	= 4 × 9 =	36 sq in
Area of right end	= 4 × 5 =	20 sq in
Area of left end	= 4 × 5 =	20 sq in
Area of top	= 5 × 9 =	45 sq in
Area of bottom	= 5 × 9 =	45 sq in
Total surface area =		202 sq in

7. (904) $V = e^3 = 10^3 = 10 \times 10 \times 10 = 1{,}000$ cc

Total surface area $= 6e^2 = 6(10^2) = 6 \times 100$
$= 600$ sq cm

8. (905) $V = \pi r^2 h = 3.14(5^2)(15) = 3.14(25)(15)$
$= 1{,}177.5$ cu in $\doteq 1{,}178$ cu in

Total surface area $= 2\pi r h + 2\pi r^2$
$= 2(3.14)(5)(15) + 2(3.14)(5^2)$
$= 471 + 157 = 628$ sq in

9. (906) Surface area $= 4\pi r^2 = 4(3.14)(6^2)$
$= 4(3.14)(36) = 452.16$
$\doteq 452$ sq in

Volume $= \dfrac{4}{3}\pi r^3 = \dfrac{4}{3}(3.14)(6^3)$

$= \dfrac{4}{3}(3.14)(216)$

$= 904.32 \doteq 904$ cu in

10. (907) (508) 5 ft 6 in = 5.5 ft

Tree	Women
$\dfrac{H \text{ height}}{18 \text{ shadow}}$	$= \dfrac{5.5 \text{ height}}{3 \text{ shadow}}$

$$\dfrac{H}{18} = \dfrac{5.5}{3}$$

$$3H = 18(5.5)$$

$$\dfrac{\cancel{3}H}{\cancel{3}} = \dfrac{\overset{6}{\cancel{18}}(5.5)}{\cancel{3}}$$

$$H = 6(5.5) = 33.0 \text{ ft}$$

11. (908) $3.14 \times 15.2 \times 7.1 = 338.8688 \doteq 340$

EXERCISES 1001, SET I (page 348)

1. $2 \cdot 10^1 + 7 \cdot 10^0$

2. $9 \cdot 10^1 + 8 \cdot 10^0$

3. $1 \cdot 10^2 + 0 \cdot 10^1 + 5 \cdot 10^0$

   (or) $1 \cdot 10^2 + 5 \cdot 10^0$

4. $5 \cdot 10^2 + 3 \cdot 10^0$

5. $8 \cdot 10^3 + 7 \cdot 10^1 + 4 \cdot 10^0$

6. $7 \cdot 10^3 + 1 \cdot 10^2 + 5 \cdot 10^0$

7. $1 \cdot 10^4 + 1 \cdot 10^3 + 1 \cdot 10^1 + 1 \cdot 10^0$

8. $1 \cdot 10^4 + 1 \cdot 10^2 + 1 \cdot 10^0$

9. $4 \cdot 10^4 + 6 \cdot 10^3 + 9 \cdot 10^2 + 3 \cdot 10^0$

10. $6 \cdot 10^4 + 5 \cdot 10^3 + 2 \cdot 10^1 + 4 \cdot 10^0$

11. $5 \cdot 10^5 + 7 \cdot 10^4 + 4 \cdot 10^2 + 6 \cdot 10^1 + 2 \cdot 10^0$

12. $8 \cdot 10^5 + 5 \cdot 10^3 + 9 \cdot 10^2 + 2 \cdot 10^0$

13. $6 \cdot 10^6 + 4 \cdot 10^4 + 1 \cdot 10^3 + 9 \cdot 10^2 + 8 \cdot 10^1$

14. $4 \cdot 10^6 + 7 \cdot 10^5 + 3 \cdot 10^3 + 6 \cdot 10^1 + 8 \cdot 10^0$

15. $1 \cdot 10^7 + 6 \cdot 10^5 + 9 \cdot 10^3 + 4 \cdot 10^1 + 8 \cdot 10^0$

16. $2 \cdot 10^7 + 1 \cdot 10^6 + 1 \cdot 10^5 + 5 \cdot 10^4 + 1 \cdot 10^2$

17. $300 + 20 = 320$ 18. 5,040

19. $600 + 3 = 603$ 20. 70,802

21. $90{,}000 + 3{,}000 + 500 = 93{,}500$ 22. 300,240

EXERCISES 1002, SET I (page 354)

1. $(111)_2 = 2^2 + 2^1 + 2^0 = 4 + 2 + 1 = (7)_{10}$

2. $(21)_{10}$

3. $(11011)_2 = 2^4 + 2^3 + 2^1 + 2^0 = 16 + 8 + 2 + 1$
$= (27)_{10}$

4. $(54)_{10}$

5. $(1010101)_2 = 2^6 + 2^4 + 2^2 + 2^0 = 64 + 16 + 4 + 1$
$= (85)_{10}$

6. $(255)_{10}$

7. First method: $12 = 8 + 4 = 2^3 + 2^2 = (1100)_2$

Second method:
```
2 | 12
  2 | 6   R0
    2 | 3   R0  ⎫  Read (1100)₂
      1   R1  ⎭
```

8. $(1111)_2$

9. First method: $(99)_{10} = (64 + 32 + 2 + 1)$
$= 2^6 + 2^5 + 2^1 + 2^0$
$= (1100011)_2$

Second method:
```
2 | 99
  2 | 49   R1  ↑
    2 | 24   R1
      2 | 12   R0   Read (1100011)₂
        2 | 6    R0
          2 | 3    R0
            1    R1
```

10. $(1101001)_2$

11.
```
2 | 489
  2 | 244   R1  ↑
    2 | 122   R0
      2 | 61    R0
        2 | 30    R1   Read (111101001)₂
          2 | 15    R0
            2 | 7     R1
              2 | 3     R1
                1     R1
```

12. $(10000110110001)_2$

13. $(1101)_2 = 1 \cdot 2^3 + 1 \cdot 2^2 + 1 \cdot 2^0$

14. $(1011)_2 = 1 \cdot 2^3 + 1 \cdot 2^1 + 1 \cdot 2^0$

15. $(10101)_2 = 1 \cdot 2^4 + 1 \cdot 2^2 + 1 \cdot 2^0$

16. $(11011)_2 = 1 \cdot 2^4 + 1 \cdot 2^3 + 1 \cdot 2^1 + 1 \cdot 2^0$

EXERCISES 1003, SET I (page 356)

1. $(104)_5 = 1 \cdot 5^2 + 0 \cdot 5^1 + 4 \cdot 5^0 = 25 + 0 + 4$
$$= (29)_{10}$$

2. $(69)_{10}$

3. $(1201)_5 = 1 \cdot 5^3 + 2 \cdot 5^2 + 0 \cdot 5^1 + 1 \cdot 5^0$
$$= 125 + 50 + 1 = (176)_{10}$$

4. $(280)_{10}$

5. $(4231)_5 = 4 \cdot 5^3 + 2 \cdot 5^2 + 3 \cdot 5^1 + 1 \cdot 5^0$
$$= 500 + 50 + 15 + 1$$
$$= (566)_{10}$$

6. $(2,102)_{10}$

7.
```
5 | 30
5 |  6  R0    Read (110)_5
   |  1  R1
```

8. $(140)_5$

9.
```
5 | 331
5 |  66  R1
5 |  13  R1    Read (2311)_5
   |   2  R3
```

10. $(4310)_5$

11.
```
5 | 7,906
5 | 1,581  R1
5 |   316  R1
5 |    63  R1    Read (223111)_5
5 |    12  R3
   |     2  R2
```

12. $(10011104)_5$

13. $(11011)_2 = (102)_5 = (27)_{10}$

$(11011)_2 = 2^4 + 2^3 + 2^1 + 2^0 = 16 + 8 + 2 + 1$
$$= (27)_{10}$$

```
5 | 27
5 |  5  R2    Read (102)_5
   |  1  R0
```

14. $(101100)_2 = (134)_5 = (44)_{10}$

15.
```
2 | 46
2 | 23  R0
2 | 11  R1    Read (101110)_2
2 |  5  R1
2 |  2  R1
   |  1  R0
```
```
5 | 46
5 |  9  R1    Read (141)_5
   |  1  R4
```

EXERCISES 1004, SET I (page 359)

1. XVII	2. XLII	3. XIX
4. XLVIII	5. LXIV	6. LIII
7. LIX	8. LXXXVII	9. CXLV
10. CCLXXXIX	11. CDXIV	12. DCCXCIX

13. MCMLXXIII 　　　　14. MMCXIX

15. $\overline{\text{IV}}$CMXCIV 　　　　16. $\overline{\text{VIII}}$CMXLVIII

17. $\overline{\text{LXXXVI}}$XLV 　　　　18. $\overline{\text{CCLV}}$DCXLVIII

19. $\overline{\text{CLVII}}$XCV 　　　　20. $\overline{\text{MMCCCLXV}}$CDXXXIV

21. CLXXIX 　　　　22. CDXCIX

23. MMMXIX 　　　　24. $\overline{\text{IX}}$DCCCLXVIII

25. $\overline{\text{XVC}}$CXLIX 　　　　26. $\overline{\text{DCXXIII}}$CDXCVII

27. 14	28. 17	29. 23
30. 19	31. 45	32. 22
33. 47	34. 69	35. 125
36. 462	37. 2,792	38. 79
39. 245	40. 474	41. 3,493
42. 6,751	43. 187,215	44. 254,319

REVIEW EXERCISES 1005, SET I (page 360)

	Base 10	Base 2	Base 5	Roman
1.	17	10001	32	XVII
2.	15	1111	30	XV
3.	96	1100000	341	XCVI
4.	69	1000101	234	LXIX
5.	123	1111011	443	CXXIII
6.	45	101101	140	XLV
7.	127	1111111	1002	CXXVII
8.	114	1110010	424	CXIV

1. Solution:
```
2 | 17
2 |  8  R1
2 |  4  R0    (10001)_2
2 |  2  R0
   |  1  R0
```
```
5 | 17
   |  3  R2    (32)_5
```

3. Solution: $(341)_5 = 3 \cdot 5^2 + 4 \cdot 5^1 + 1 \cdot 5^0$
$$= 75 + 20 + 1$$
$$= 96$$

```
2 | 96
2 | 48  R0
2 | 24  R0
2 | 12  R0    (1100000)_2
2 |  6  R0
2 |  3  R0
   |  1  R1
```

5. Solution:
```
2 | 123
2 |  61  R1
2 |  30  R1
2 |  15  R0    (1111011)_2
2 |   7  R1
2 |   3  R1
   |   1  R1
```
```
5 | 123
5 |  24  R3    (443)_5
   |   4  R4
```

7. Solution: $(1002)_5 = 1 \cdot 5^3 + 2 \cdot 5^0 = 125 + 2$
$$= 127$$

```
2 | 127
2 |  63  R1
2 |  31  R1
2 |  15  R1    (1111111)_2
2 |   7  R1
2 |   3  R1
   |   1  R1
```

9. MCMLXXIV 　　　10. Using Roman numerals: $9 = $ IX
　　　　　　　　　　　　　　　　　　　　　　　$10 = $ X

SOLUTIONS FOR CHAPTER 10 DIAGNOSTIC TEST (page 362)

Following each problem number is the textbook section reference (in parentheses) where that kind of problem is discussed.

1. (1001) (a) $475 = 4 \cdot 10^2 + 7 \cdot 10^1 + 5 \cdot 10^0$

(b) $20,306 = 2 \cdot 10^4 + 3 \cdot 10^2 + 6 \cdot 10^0$

2. (1001) (a) $500 + 30 = 530$
(b) $80,000 + 900 + 7 = 80,907$

3. (1002) $(1101)_2 = 1 \cdot 2^3 + 1 \cdot 2^2 + 1 \cdot 2^0$

4. (1002) $(10110)_2 = 2^4 + 2^2 + 2^1 = 16 + 4 + 2 = (22)_{10}$

5. (1002) $2\underline{|67}$
   $\quad\quad 2\underline{|33}$ R1
   $\quad\quad 2\underline{|16}$ R1
   $\quad\quad 2\underline{|8}$ R0 $\quad$ (1000011)$_2$
   $\quad\quad 2\underline{|4}$ R0
   $\quad\quad 2\underline{|2}$ R0
   $\quad\quad\quad 1$ R0

6. (1003) $(1204)_5 = 1 \cdot 5^3 + 2 \cdot 5^2 + 0 \cdot 5^1 + 4 \cdot 5$

7. (1003) $(342)_5 = 3 \cdot 5^2 + 4 \cdot 5^1 + 2 \cdot 5^0$
   $\quad\quad\quad\quad\quad = 75 + 20 + 2 = (97)_{10}$

8. (1003) $5\underline{|174}$
   $\quad\quad 5\underline{|34}$ R4
   $\quad\quad 5\underline{|6}$ R4 $\quad$ (1144)$_5$
   $\quad\quad\quad 1$ R1

9. (1001, 1002, 1003)

	Base 2	Base 5	Base 10
(a)	1011	21	11
(b)	110000	143	48
(c)	100111	124	39

Solutions: (a) $(1011)_2 = 2^3 + 2^1 + 2^0$
$\quad\quad\quad\quad\quad\quad\quad = 8 + 2 + 1 = (11)_{10}$

$\quad\quad 5\underline{|11}$
$\quad\quad\quad 2$ R1 $\quad$ (21)$_5$

(b) $(143)_5 = 1 \cdot 5^2 + 4 \cdot 5^1 + 3 \cdot 5^0$
$\quad\quad\quad\quad = 25 + 20 + 3 = (48)_{10}$

$\quad 2\underline{|48}$
$\quad 2\underline{|24}$ R0
$\quad 2\underline{|12}$ R0
$\quad 2\underline{|6}$ R0 $\quad$ (110000)$_2$
$\quad 2\underline{|3}$ R0
$\quad\quad 1$ R1

(c) $2\underline{|39}$
$\quad 2\underline{|19}$ R1
$\quad 2\underline{|9}$ R1
$\quad 2\underline{|4}$ R1 $\quad$ (100111)$_2$
$\quad 2\underline{|2}$ R0
$\quad\quad 1$ R0

$\quad 5\underline{|39}$
$\quad 5\underline{|7}$ R4 $\quad$ (124)$_5$
$\quad\quad 1$ R2

10. (1004) (a) XXIX $\quad\quad$ (b) CXLIV
$\quad\quad\quad$ (c) DCXCIX $\quad\quad$ (d) MCMLXXIV

11. (1004) (a) 28 $\quad\quad\quad$ (b) 149
$\quad\quad\quad$ (c) 1,675 $\quad\quad$ (d) 6,230

EXERCISES 1101, SET I (page 368)

1. Minus seventy-five or negative seventy-five.
2. Minus forty-nine or negative forty-nine.
3. -54 $\quad\quad\quad\quad\quad$ 4. -109
5. -2 because it is to the right of -4 on the number line
6. 0
7. -62 $\quad\quad\quad\quad\quad$ 8. -45°F
9. 8 > 5 $\quad\quad\quad\quad$ 10. 7 < 9
11. 0 > -3 $\quad\quad\quad\quad$ 12. -10 < 0
13. -17 < -11 $\quad\quad\quad$ 14. -10 < -4
15. -5 > -16 $\quad\quad\quad$ 16. -3 > -20

17. -1 $\quad\quad\quad\quad\quad\quad$ 18. It cannot be found.
19. -99 $\quad\quad\quad\quad\quad$ 20. -10

EXERCISES 1102, SET I (page 370)

1. 13 $\quad\quad$ 2. 17 $\quad\quad$ 3. -14 $\quad\quad$ 4. -17
5. 12 and -7 = 5. Since the signs are different, subtract the number parts and use the + sign because the larger number part is +.
6. 9 $\quad\quad$ 7. 15 $\quad\quad$ 8. 23 $\quad\quad$ 9. -8
10. -29 $\quad$ 11. -39 $\quad$ 12. -34 $\quad$ 13. 41
14. 107 $\quad$ 15. 18 $\quad$ 16. 23 $\quad$ 17. -127
18. -117 $\quad$ 19. -640 $\quad$ 20. -1,134

EXERCISES 1103, SET I (page 372)

1. $\begin{array}{r} 10 \\ +4 \\ \hline 6 \end{array}$ $\quad\quad$ 2. 7 $\quad\quad\quad$ 3. $\begin{array}{r} 8 \\ -2 \\ + \\ \hline 10 \end{array}$

4. 13 $\quad\quad\quad$ 5. $\begin{array}{r} -10 \\ -4 \\ + \\ \hline -6 \end{array}$ $\quad\quad\quad$ 6. -7

7. $\begin{array}{r} -15 \\ +11 \\ \hline -26 \end{array}$ $\quad\quad$ 8. -37 $\quad\quad\quad$ 9. $\begin{array}{r} 86 \\ +96 \\ - \\ \hline -10 \end{array}$

10. -17 $\quad\quad\quad$ 11. $\begin{array}{r} 156 \\ -97 \\ + \\ \hline 253 \end{array}$ $\quad\quad$ 12. 373

13. $\begin{array}{r} -354 \\ -286 \\ + \\ \hline -68 \end{array}$ $\quad\quad$ 14. -109 $\quad\quad$ 15. $\begin{array}{r} 780 \\ +840 \\ - \\ \hline -60 \end{array}$

16. -121 $\quad\quad\quad$ 17. $\begin{array}{r} 1,786 \\ -295 \\ + \\ \hline 2,081 \end{array}$ $\quad\quad$ 18. 4,841

19. $\begin{array}{r} -16,780 \\ +3,915 \\ \hline -20,695 \end{array}$ $\quad\quad$ 20. -56,213 $\quad\quad$ 21. $\begin{array}{r} -3,005 \\ -5,001 \\ + \\ \hline 1,996 \end{array}$

22. -4,991

23. $-5\frac{3}{4} = -5\frac{3}{4}$
$\quad 2\frac{1}{2} = +2\frac{2}{4}$
$\quad\quad\quad\quad\quad \begin{array}{r} - \\ \hline -1 \\ -\frac{5}{4} = -1\frac{1}{4} \\ \hline -8\frac{1}{4} \end{array}$ $\quad\quad\quad$ 24. $-26\frac{1}{6}$

25. $\begin{array}{r} -16.71 \\ -18.39 \\ + \\ \hline 1.68 \end{array}$ $\quad$ 26. 7.46 $\quad\quad$ 27. $3\frac{1}{5} = 2\frac{12}{10}$
$\quad\quad\quad\quad\quad\quad\quad\quad\quad\quad\quad\quad\quad \frac{7}{10} = +\frac{7}{10}$
$\quad\quad\quad\quad\quad\quad\quad\quad\quad\quad\quad\quad\quad\quad\quad \overline{2\frac{5}{10} = 2\frac{1}{2}}$

28. $3\frac{5}{16}$ $\quad\quad$ 29. $\begin{array}{r} 7.015 \\ -2.94 \\ + \\ \hline 9.955 \end{array}$ $\quad\quad$ 30. 0.375

EXERCISES 1104, SET I (page 374)

1. -32 $\quad$ 2. -45 $\quad$ 3. -63 $\quad$ 4. -54
5. 100 $\quad$ 6. 81 $\quad$ 7. -1,125 $\quad$ 8. -1,118
9. -2,625 $\quad$ 10. -200 $\quad$ 11. 24 $\quad$ 12. 90
13. -70 $\quad$ 14. -130 $\quad$ 15. -350,000
16. -1,500,000

17. $10^2(-47) = 100(-47) = -4,700$ 18. $-1,560$

19. $10^4(-2)(-5) = 10,000(-2)(-5) = (-20,000)(-5)$
$$= 100,000$$

20. -36

21. $3(-2)(2)(-5) = (-6)(-10) = 60$

22. -120

23. $\left(3\frac{1}{2}\right)\left(-\frac{8}{21}\right) = \frac{\cancel{7}^{1}}{\cancel{2}_{1}}\left(-\frac{\cancel{8}^{4}}{\cancel{21}_{3}}\right) = -\frac{4}{3} = -1\frac{1}{3}$

24. -4

25. 21.28 26. 0.854

27. $\left(-5\frac{1}{5}\right)\left(1\frac{2}{13}\right) = \frac{\cancel{-26}^{-2}}{\cancel{5}_{1}} \cdot \frac{\cancel{15}^{3}}{\cancel{13}_{1}} = -6$ 28. -36

29. -52.5 30. -6.6

EXERCISES 1105, SET I (page 375)

1. -5 2. -6 3. -4 4. -5
5. 3 6. 3 7. -3 8. -4
9. 9 10. 7 11. -15 12. -2.5
13. $2\frac{1}{2}$ 14. $2\frac{1}{4}$ 15. -8.5 16. -6
17. -3.67 18. -4.86
19. -0.0785 20. -0.985

EXERCISES 1106, SET I (page 377)

1. (a) $2(10) + 3(-5)$
   $= 20 + (-15) = 5$
   (b) $2(-10) + 3(5)$
   $= -20 + 15 = -5$
   (c) $2(-10) + 3(-5)$
   $= -20 - 15 = -35$

2. (a) 40 (b) -40 (c) 60

3. (a) $2(5) - 7(4)$
   $= 10 - 28 = -18$
   (b) $2(-8) - 7(3)$
   $= -16 - 21 = -37$
   (c) $2(-9) - 7(-8)$
   $= -18 + 56 = 38$

4. (a) -31 (b) -53 (c) -21

5. (a) $\dfrac{h(a + b)}{2} = \dfrac{6(4 + 5)}{2} = \dfrac{\cancel{6}^{3}(9)}{\cancel{2}_{1}} = 27$

   (b) $\dfrac{7(4 + 9)}{2} = \dfrac{7(13)}{2} = \dfrac{91}{2} = 45\frac{1}{2}$

   (c) $\dfrac{2\frac{1}{2}\left(3\frac{1}{2} + 9\frac{1}{2}\right)}{2} = \dfrac{\frac{5}{2}(13)}{2} = \dfrac{\frac{65}{2}}{2} = \dfrac{65}{2} \div 2 = \dfrac{65}{2} \cdot \dfrac{1}{2}$
   $$= \dfrac{65}{4} = 16\frac{1}{4}$$

6. (a) 2 (b) $\frac{1}{2}$ (c) $3\frac{1}{3}$

7. $C = \dfrac{5}{9}(F - 32) = \dfrac{5}{9}(77 - 32) = \dfrac{5}{9}(\cancel{45}^{5}) = 25$

8. 19.625 sq m

9.

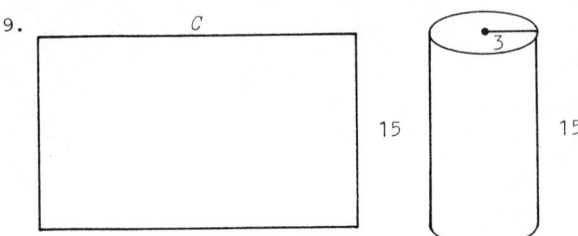

Area of top $= \pi r^2 = 3.14(3^2) = 28.26$ sq in
Area of bottom = Area of top $= 28.26$ sq in
The lateral surface is a rectangle as shown
here. The width of the rectangle is the
circumference (C). $C = 2\pi r = 2(3.14)(3)$
$$= 18.84 \text{ in}$$
Area of lateral surface $= 15(18.84)$
$$= 282.6 \text{ sq in}$$
$$\text{Area of top} = 28.26 \text{ sq in}$$
$$\text{Area of bottom} = \underline{28.26} \text{ sq in}$$

$$\text{Total area} = 339.12 \text{ sq in}$$

10. (a) 60 (b) 27 (c) 112.5

11. (a) $A = P(1 + rt)$
    $= 100\big(1 + 0.08(1.5)\big)$
    $= 100(1 + 0.12)$
    $= 100(1.12) = 112$

    (b) $A = P(1 + rt)$
    $= 500\big(1 + 0.09(2.5)\big)$
    $= 500(1 + 0.225)$
    $= 500(1.225) = 612.5$

    (c) $A = P(1 + rt)$
    $= 250\big(1 + 0.075(24)\big)$
    $= 250(1 + 1.8)$
    $= 250(2.8) = 700$ 12. 2.5

EXERCISES 1108, SET I (page 381)

1. $\begin{aligned} x + 5 &= 8 \\ -5 \quad &\underline{-5} \\ x &= 3 \end{aligned}$ Check: $x + 5 = 8$ 2. 5
   $3 + 5 = 8$
   $8 = 8$

3. $\begin{aligned} x - 3 &= 4 \\ +3 \quad &\underline{+3} \\ x &= 7 \end{aligned}$ Check: $x - 3 = 4$ 4. 9
   $7 - 3 = 4$
   $4 = 4$

5. $\begin{aligned} 3 + x &= -4 \\ -3 \quad &\underline{-3} \\ x &= -7 \end{aligned}$ Check: $3 + x = -4$ 6. -7
   $3 + (-7) = -4$
   $-4 = -4$

7. $\begin{aligned} x + 4 &= 21 \\ -4 \quad &\underline{-4} \\ x &= 17 \end{aligned}$ Check: $x + 4 = 21$ 8. 9
   $17 + 4 = 21$
   $21 = 21$

9. $\begin{aligned} x - 35 &= 7 \\ +35 \quad &\underline{+35} \\ x &= 42 \end{aligned}$ Check: $x - 35 = 7$ 10. 51
   $42 - 35 = 7$
   $7 = 7$

11. $\begin{aligned} 9 &= x + 5 \\ -5 \quad &\underline{- 5} \\ 4 &= x \end{aligned}$ Check: $9 = x + 5$ 12. 3
    $9 = 4 + 5$
    $9 = 9$

13. $\begin{aligned} 12 &= x - 11 \\ +11 \quad &\underline{+ 11} \\ 23 &= x \end{aligned}$ Check: $12 = x - 11$ 14. 29
    $12 = 23 - 11$
    $12 = 12$

15. $\begin{aligned} -17 + x &= 28 \\ +17 \quad &\underline{+17} \\ x &= 45 \end{aligned}$ Check: $-17 + x = 28$ 16. 47
    $-17 + 45 = 28$
    $28 = 28$

17. $\begin{aligned} -21 + x &= -42 \\ +21 \quad &\underline{+21} \\ x &= -21 \end{aligned}$ Check: $-21 + x = -42$
    $-21 + (-21) = -42$
    $-42 = -42$

18. -14

19. $-28 = -15 + x$ Check: $-28 = -15 + x$
     $\underline{+15 \quad\quad +15}$ $-28 = -15 + (-13)$
     $-13 = \quad\quad x$ $-28 = -28$

20. -29

EXERCISES 1109, SET I (page 384)

1. $2x = 8$ Check: $2x = 8$ 2. 5
    $\dfrac{2x}{2} = \dfrac{8}{2}$ $2(4) = 8$
                          $8 = 8$
    $x = 4$

3. $21 = 7x$ Check: $21 = 7x$ 4. 7
    $\dfrac{21}{7} = \dfrac{7x}{7}$ $21 = 7(3)$
                        $21 = 21$
    $3 = x$

5. $11x = 33$ Check: $11x = 33$ 6. 4
    $\dfrac{11x}{11} = \dfrac{33}{11}$ $11(3) = 33$
                         $33 = 33$
    $x = 3$

7. $-9x = 6$ Check: $-9x = 6$ 8. $-\dfrac{4}{5}$
    $\dfrac{-9x}{-9} = \dfrac{6}{-9}$ $-\overset{3}{\cancel{9}}\left(-\dfrac{2}{\underset{1}{\cancel{3}}}\right) = 6$
    $x = -\dfrac{2}{3}$ $6 = 6$

9. $36 = -3x$ Check: $36 = -3x$ 10. -3
    $\dfrac{36}{-3} = \dfrac{-3x}{-3}$ $36 = -3(-12)$
                        $36 = 36$
    $-12 = x$

11. $-24 = 4x$ Check: $-24 = 4x$ 12. -4
    $\dfrac{-24}{4} = \dfrac{4x}{4}$ $-24 = 4(-6)$
                        $-24 = -24$
    $-6 = x$

13. $4x + 1 = 9$ Check: $4x + 1 = 9$ 14. 2
     $\underline{\quad - 1 \quad\; -1}$ $4(2) + 1 = 9$
     $4x \quad\quad = 8$ $8 + 1 = 9$
                            $9 = 9$
     $\dfrac{4x}{4} = \dfrac{8}{4}$
     $x = 2$

15. $6x - 2 = 10$ Check: $6x - 2 = 10$ 16. 1
     $\underline{\quad + 2 \quad\; +2}$ $6(2) - 2 = 10$
     $6x \quad\quad = 12$ $12 - 2 = 10$
                            $10 = 10$
     $\dfrac{6x}{6} = \dfrac{12}{6}$
     $x = 2$

17. $2x - 15 = 11$ Check: $2x - 15 = 11$
     $\underline{\quad + 15 \quad +15}$ $2(13) - 15 = 11$
     $2x \quad\quad = 26$ $26 - 15 = 11$
                              $11 = 11$
     $\dfrac{2x}{2} = \dfrac{26}{2}$
     $x = 13$

18. 6

19. $4x + 2 = -14$ 20. -3
     $\underline{\quad - 2 \quad\; - 2}$
     $4x \quad\quad = -16$
     $\dfrac{4x}{4} = \dfrac{-16}{4}$
     $x = -4$

21. $14 = 9x - 13$ 22. 5
     $\underline{+13 \quad\quad + 13}$
     $27 = 9x$

     $\dfrac{27}{9} = \dfrac{9x}{9}$

     $3 = x$

23. $12x + 17 = 65$ 24. 2
     $\underline{\quad - 17 \quad -17}$
     $12x \quad\quad = 48$

     $\dfrac{12x}{12} = \dfrac{48}{12}$

     $x = 4$

25. $8x - 23 = 31$ 26. $10\dfrac{1}{3}$
     $\underline{\quad + 23 \quad +23}$
     $8x \quad\quad = 54$

     $\dfrac{8x}{8} = \dfrac{54}{8}$

     $x = 6\dfrac{3}{4}$

27. $-14 + 4x = 28$ 28. $10\dfrac{1}{3}$
     $\underline{+14 \quad\quad +14}$
     $4x = 42$

     $\dfrac{4x}{4} = \dfrac{42}{4}$

     $x = 10\dfrac{1}{2}$

29. $-8 = 3x - 25$ 30. $8\dfrac{1}{2}$
     $\underline{+25 \quad\quad + 25}$
     $17 = 3x$

     $\dfrac{17}{3} = \dfrac{3x}{3}$ 32. $-2\dfrac{1}{2}$

     $5\dfrac{2}{3} = x$

31. $-73 = 24x + 31$ Check: $-73 = \quad 24x \quad + 31$
     $\underline{\;-31 \quad\quad\; - 31}$ $-73 = 24\left(-4\dfrac{1}{3}\right) + 31$
     $-104 = 24x$ $\overset{8}{\quad}$
                          $-73 = \dfrac{24}{1}\left(\dfrac{-13}{\underset{1}{\cancel{3}}}\right) + 31$
     $\dfrac{-104}{24} = \dfrac{24x}{24}$
                          $-73 = -104 \quad + 31$
     $-4\dfrac{1}{3} = x$ $-73 = -73$

EXERCISES 1110, SET I (page 387)

1. $\dfrac{x}{3} = 4$ Check: $\dfrac{x}{3} = 4$

    $3\left(\dfrac{x}{3}\right) = 3(4)$ $\dfrac{12}{3} = 4$

    $x = 12$ $4 = 4$

2. 10

3. $\dfrac{x}{5} = -2$ Check: $\dfrac{x}{5} = -2$

    $5\left(\dfrac{x}{5}\right) = 5(-2)$ $\dfrac{-10}{5} = -2$

    $x = -10$ $-2 = -2$

4. -24

5. $\dfrac{x}{10} = 3.14$ Check: $\dfrac{x}{10} = 3.14$

    $10\left(\dfrac{x}{10}\right) = 10(3.14)$ $\dfrac{31.4}{10} = 3.14$

    $x = 31.4$ $3.14 = 3.14$

6. 39

7. $-4 = \frac{2x}{7}$ Check: $-4 = \frac{2x}{7}$

$7(-4) = 7\left(\frac{2x}{7}\right)$ $-4 = \frac{2(-14)}{7}$

$-28 = 2x$ $-4 = -4$

$\frac{-28}{2} = \frac{2x}{2}$

$-14 = x$

8. -2

9. $\frac{20x}{5} = 12$ Check: $\frac{20x}{5} = 12$

$\frac{20x}{5} = 12$ $\frac{20(3)}{5} = 12$

$4x = 12$ $12 = 12$

$\frac{4x}{4} = \frac{12}{4}$

$x = 3$

10. 4

11. $\frac{x}{4} + 6 = 9$ Check: $\frac{x}{4} + 6 = 9$

$\frac{-6}{} \quad \frac{-6}{}$ $\frac{12}{4} + 6 = 9$

$\frac{x}{4} = 3$ $3 + 6 = 9$

$4\left(\frac{x}{4}\right) = 4(3)$ $9 = 9$

$x = 12$

12. 25

13. $\frac{x}{10} - 5 = 13$ Check: $\frac{x}{10} - 5 = 13$

$\frac{+5}{} \quad \frac{+5}{}$ $\frac{180}{10} - 5 = 13$

$\frac{x}{10} = 18$ $18 - 5 = 13$

$10\left(\frac{x}{10}\right) = 10(18)$ $13 = 13$

$x = 180$

14. 320

15. $7 = \frac{2x}{5} + 3$ Check: $7 = \frac{2x}{5} + 3$

$\frac{-3}{} \quad \frac{-3}{}$ $7 = \frac{2(10)}{5} + 3$

$4 = \frac{2x}{5}$ $7 = 4 + 3$

$5(4) = 5\left(\frac{2x}{5}\right)$ $7 = 7$

$20 = 2x$

$\frac{20}{2} = \frac{2x}{2}$

$10 = x$

16. -4

17. $4 - \frac{7x}{5} = 11$ Check: $4 - \frac{7x}{5} = 11$

$\frac{-4}{} \quad \frac{-4}{}$ $4 - \frac{7(-5)}{5} = 11$

$-\frac{7x}{5} = 7$ $4 - 7(-1) = 11$

$5\left(-\frac{7x}{5}\right) = 5(7)$ $4 + 7 = 11$

$-7x = 35$ $11 = 11$

$\frac{-7x}{-7} = \frac{35}{-7}$

$x = -5$

18. $-40\frac{1}{2}$

19. $10^3 = 10^2 - \frac{10x}{7}$ Check: $10^3 = 10^2 - \frac{10x}{7}$

$1,000 = 100 - \frac{10x}{7}$ $1,000 = 100 - \frac{10(-630)}{7}$

$\frac{-100}{} \quad \frac{-100}{}$ $1,000 = 100 + 900$

$900 = -\frac{10x}{7}$ $1,000 = 1,000$

$7(900) = 7\left(-\frac{10x}{7}\right)$

$6,300 = -10x$

$\frac{6,300}{-10} = \frac{-10x}{-10}$

$-630 = x$

20. 810

REVIEW EXERCISES 1111, SET I (page 389)

1. (a) $\begin{array}{r} 274 \\ -\ 345 \\ \hline -71 \end{array}$ (b) $\begin{array}{r} -706 \\ -315 \\ \hline -1,021 \end{array}$

(c) $-3\frac{1}{4} = -2\frac{5}{4}$ (d) $\begin{array}{r} 7.64 \\ -9.01 \\ \hline -1.37 \end{array}$

$2\frac{1}{2} = \ \ 2\frac{2}{4}$

$\overline{ -\frac{3}{4}}$

2. (a) -208 (b) -56.3 (c) $-2\frac{9}{10}$ (d) -1.15

3. (a) $\begin{array}{r} 354 \\ +286 \\ \hline 640 \end{array}$ (b) $\begin{array}{r} -507 \\ -314 \\ \hline -193 \end{array}$

(c) $-2\frac{7}{8} = -2\frac{7}{8}$ (d) $\begin{array}{r} 2.99 \\ +3.08 \\ \hline -0.09 \end{array}$

$3\frac{3}{4} = +3\frac{6}{8}$

$\overline{ -5\frac{13}{8} = -6\frac{5}{8}}$

4. (a) 365   (b) -534 (c) $-8\frac{7}{12}$ (d) -0.48

5. (a) 1,050   (b) -462
(c) $(-2)(3)(-5) = (-6)(-5) = 30$
(d) $(4)(-6)(-25)(-3) = (-24)(-25)(-3) = (600)(-3) = -1,800$

6. (a) 1,998   (b) -420
(c) -70 (d) $-14,000$

7. (a) -8 (b) 25
(c) $\frac{3.75}{-12.5} = -\frac{375}{1,250} = -\frac{3}{10} = -0.3$

(d) $\left(-2\frac{3}{4}\right) \div 1\frac{3}{8} = -\frac{11}{4} \div \frac{11}{8} = -\frac{11}{4} \cdot \frac{8}{11} = -2$

8. (a) -6   (b) 8   (c) $-1\frac{1}{6}$ (d) $-\frac{1}{4}$

9. $\dfrac{\overset{-2}{\cancel{(4)}}\,(-5)}{\underset{1}{\cancel{-2}}} = 10$ 10. 12

11. $(-5)(10^2) = (-5)(100) = -500$ 12. -3,600

13. $-2 + \dfrac{\overset{-4}{\cancel{12}}}{\cancel{-3}} = -2 + (-4) = -2 - 4 = -6$ 14. -11

15. $I = Prt$ 16. -10

     $= (500)(0.08)\left(1\frac{1}{2}\right)$

     $= (\overset{20}{\cancel{40}})\left(\dfrac{3}{\underset{1}{\cancel{2}}}\right) = 60$

17.
$$
\begin{aligned}
x - 5 &= 7 \\
\underline{+\ 5} &\ \underline{+5} \\
x &= 12
\end{aligned}
$$
Check:
$$
\begin{aligned}
x - 5 &= 7 \\
12 - 5 &= 7 \\
7 &= 7
\end{aligned}
$$
18. 8

19.
$$
\begin{aligned}
-5 &= x + 9 \\
\underline{-9} &\quad \underline{-\ 9} \\
-14 &= x
\end{aligned}
$$
Check:
$$
\begin{aligned}
-5 &= x + 9 \\
-5 &= -14 + 9 \\
-5 &= -5
\end{aligned}
$$
20. -13

21.
$$
\begin{aligned}
5x &= 55 \\
\dfrac{\cancel{5}x}{\cancel{5}} &= \dfrac{55}{5} \\
x &= 11
\end{aligned}
$$
Check:
$$
\begin{aligned}
5x &= 55 \\
5(11) &= 55 \\
55 &= 55
\end{aligned}
$$
22. 8

23.
$$
\begin{aligned}
18 &= \dfrac{x}{7} \\
7(18) &= \cancel{7}\left(\dfrac{x}{\cancel{7}}\right) \\
126 &= x
\end{aligned}
$$
Check:
$$
\begin{aligned}
18 &= \dfrac{x}{7} \\
18 &= \dfrac{126}{7} \\
18 &= 18
\end{aligned}
$$
24. 112

25.
$$
\begin{aligned}
2x + 7 &= -15 \\
\underline{-\ 7} &\ \underline{-\ 7} \\
2x &= -22 \\
\dfrac{2x}{\cancel{2}} &= \dfrac{-22}{2} \\
x &= -11
\end{aligned}
$$
Check:
$$
\begin{aligned}
2x + 7 &= -15 \\
2(-11) + 7 &= -15 \\
-22 + 7 &= -15 \\
-15 &= -15
\end{aligned}
$$

26. -5

27.
$$
\begin{aligned}
\dfrac{5x}{4} + 5 &= 20 \\
\underline{-\ 5} &\ \underline{-5} \\
\dfrac{5x}{4} &= 15 \\
\cancel{4}\left(\dfrac{5x}{\cancel{4}}\right) &= 4(15) \\
5x &= 60 \\
\dfrac{\cancel{5}x}{\cancel{5}} &= \dfrac{60}{5} \\
x &= 12
\end{aligned}
$$
Check:
$$
\begin{aligned}
\dfrac{5x}{4} + 5 &= 20 \\
\dfrac{5(\overset{3}{\cancel{12}})}{\underset{1}{\cancel{4}}} + 5 &= 20 \\
15 + 5 &= 20 \\
20 &= 20
\end{aligned}
$$

28. 15

29.
$$
\begin{aligned}
17 &= 9 - x \\
\underline{-\ 9} &\ \underline{-9} \\
8 &= -x \\
\dfrac{8}{-1} &= \dfrac{-x}{-1} \\
-8 &= x
\end{aligned}
$$
Check:
$$
\begin{aligned}
17 &= 9 - x \\
17 &= 9 - (-8) \\
17 &= 9 + 8 \\
17 &= 17
\end{aligned}
$$

30. -4

31. $45° + 17° - 33° = 62° - 33° = 29°$

32. $45\frac{1}{2}$

Following each problem number is the textbook section reference (in parentheses) where that kind of problem is discussed.

1. (1102) (a) 4   (b) -16   (c) -5   (d) 41
           (e) -51   (f) -16   (g) 91   (h) 25

2. (1103) (a) $\begin{array}{r} 57 \\ +32 \\ \hline 89 \end{array}$ (b) $\begin{array}{r} -14 \\ +22 \\ \hline 8 \end{array}$ (c) $\begin{array}{r} -93 \\ -27 \\ \hline -120 \end{array}$ (d) $\begin{array}{r} 138 \\ +481 \\ \hline -343 \end{array}$

3. (1104) (a) -45   (b) 42   (c) -48   (d) 72
           (e) 180   (f) -280

4. (1105) (a) $\dfrac{126}{9} = 14$ (b) $\dfrac{39}{-13} = -3$

         (c) $\dfrac{-64}{-16} = 4$ (d) $\dfrac{-75}{15} = -5$

         (e) $\dfrac{84}{-12} = -7$ (f) $\dfrac{-144}{-9} = 16$

5. (1106) $3x + 5y = 3(-10) + 5(6) = -30 + 30 = 0$

6. (1106) $A = \dfrac{1}{2}bh = \dfrac{1}{\underset{1}{\cancel{2}}}(7)(\overset{9}{\cancel{18}}) = 63$

7. (1106) $A = P(1 + rt)$
           $= 600\big(1 + 0.07(1.5)\big)$
           $= 600(1 + 0.105)$
           $= 600(1.105) = 663$

8. (1106) $C = \dfrac{5}{9}(F - 32)$

           $= \dfrac{5}{9}(77 - 32)$

           $= \dfrac{5}{\underset{1}{\cancel{9}}}(\overset{5}{\cancel{45}}) = 25$

9. (a) (1108)
$$
\begin{aligned}
x - 3 &= 7 \\
\underline{+\ 3} &\ \underline{+3} \\
x &= 10
\end{aligned}
$$

   (b) (1108, 1109)
$$
\begin{aligned}
5x + 8 &= -22 \\
\underline{-\ 8} &\ \underline{-\ 8} \\
5x &= -30 \\
\dfrac{\cancel{5}x}{\cancel{5}} &= \dfrac{-30}{5} \\
x &= -6
\end{aligned}
$$

   (c) (1110)
$$
\begin{aligned}
\dfrac{x}{6} &= -2 \\
\cancel{6}\left(\dfrac{x}{\cancel{6}}\right) &= 6(-2) \\
x &= -12
\end{aligned}
$$

   (d) (1109, 1110)
$$
\begin{aligned}
3 &= \dfrac{3x}{7} + 15 \\
\underline{-15} &\quad \underline{-\ 15} \\
-12 &= \dfrac{3x}{7} \\
7(-12) &= \cancel{7}\left(\dfrac{3x}{\cancel{7}}\right) \\
-84 &= 3x \\
\dfrac{-84}{3} &= \dfrac{3x}{\cancel{3}} \\
-28 &= x
\end{aligned}
$$

10. (1108) $7 + 5x = 12$

$7 + 5\left(\frac{3}{5}\right) \stackrel{?}{=} 12$

$7 + 3 \stackrel{?}{=} 12$

$10 \neq 12$

Therefore, $\frac{3}{5}$ is not a solution of the equation.

EXERCISES 1201, SET I (page 399)

1. $10^2 \cdot 10^3 = 10^{2+3} = 10^5$ 2. 10^{10}

3. $10^8 \cdot 10^5 = 10^{8+5} = 10^{13}$ 4. 10^{17}

5. $\dfrac{10^6}{10^2} = 10^{6-2} = 10^4$ 6. 10^2

7. $\dfrac{10^2}{10^5} = 10^{2-5} = 10^{-3}$ 8. 10^{-5}

9. $10 \cdot 10^2 \cdot 10^5 = 10^{1+2+5} = 10^8$ 10. 10^{11}

11. $10^4 \cdot 10^{-2} = 10^{4+(-2)} = 10^{4-2} = 10^2$ 12. 10^3

13. $\dfrac{10^5 \cdot 10^7}{10^{-2}} = 10^{5+7-(-2)} = 10^{12+2} = 10^{14}$

14. 10^{19}

15. $\dfrac{10^{-5} \cdot 10^{-6}}{10^{-8}} = 10^{-5+(-6)-(-8)} = 10^{-5-6+8} = 10^{-3}$

16. 1

17. $\dfrac{10^8 \cdot 10^{-5}}{10^3} = 10^{8+(-5)-(3)} = 10^{8-5-3} = 10^0 = 1$

18. 10^9

19. $10^4 \cdot 10^{-6} \cdot 10^0 \cdot 10^{-1} = 10^{4+(-6)+0+(-1)}$
$= 10^{4-6-1} = 10^{-3}$

20. 10^7

EXERCISES 1202, SET I (page 401)

1. $75.4 \times 10^2 = 7\,5\,4\,0. = 7,540$

2. $3,860$

3. $88 \times 10^{-1} = 8.8 = 8.8$

4. 0.49

5. $0.075 \times 10^3 = 0\,0\,7\,5. = 75$

6. 550

7. $6,840 \times 10^{-4} = 0.6\,8\,4\,0 = 0.684$

8. 0.3785

9. $0.00086 \times 10^6 = 0\,0\,0\,0\,8\,6\,0. = 860$

10. 0.0000063

11. $15 \times 10^{12} = 1\,5\,0\,0\,0,0\,0\,0,0\,0\,0,0\,0\,0.$
$= 15,000,000,000,000$

12. $5,800,000,000,000$

EXERCISES 1203, SET I (page 403)

Common Notation	*Scientific Notation*
1. $7\,4\,8.$	7.48×10^2
2. $2\,5,0\,0\,0.$	2.5×10^4
3. $0.0\,6\,3$	6.3×10^{-2}
4. $6\,,7\,0\,0.$	6.7×10^3
5. $0.0\,0\,1\,7\,3\,2$	1.732×10^{-3}
6. $0.0\,2\,8\,1$	2.81×10^{-2}
7. $3\,,4\,7\,0,0\,0\,0.$	3.47×10^6
8. $8\,6.4\,8$	8.648×10^1
9. $0.0\,0\,0\,0\,0\,1\,9\,1$	1.91×10^{-6}
10. $5\,8\,8,0\,0\,0.$	5.88×10^5
11. $0.0\,0\,0\,0\,5\,6\,3$	5.63×10^{-5}
12. $2\,7,8\,0\,0.$	2.78×10^4
13. $0.0\,0\,0\,0\,5\,8$	5.8×10^{-5}
14. $1\,,7\,6\,1,0\,0\,0.$	1.761×10^6

15. $0.\underbrace{00000000000000000000000000}6\,5\,4\,7$
$= 6.547 \times 10^{-27}$

16. $0.0\,0\,0\,0\,0\,7\,8$	7.8×10^{-7}
17. $0.0\,0\,0\,0\,0\,0\,0\,0\,4\,7\,7$	4.77×10^{-10}
18. $5\,,7\,8\,0,0\,0\,0,0\,0\,0,0\,0\,0.$	5.78×10^{12}
19. $(50)(2,000)$	1×10^5
20. $600,000$	$(2 \times 10^2)(3 \times 10^3)$

EXERCISES 1204, SET I (page 405)

1. $400 \times 7,000 = (4 \times 10^2) \times (7 \times 10^3)$
$= 4 \times 7 \times 10^2 \times 10^3$
$= 28 \times 10^5 = 2,800,000$

2. $2,400,000$

3. $(0.05)(0.007) = (5 \times 10^{-2}) \times (7 \times 10^{-3})$
$= 5 \times 7 \times 10^{-2} \times 10^{-3}$
$= 35 \times 10^{-5} = 0.00035$

4. 0.00036

5. $50 \times 0.006 = (5 \times 10^1) \times (6 \times 10^{-3})$
$= 5 \times 6 \times 10^1 \times 10^{-3}$
$= 30 \times 10^{-2} = 0.3$

6. 0.63

7. $(20)(3,000)(0.08) = (2 \times 10^1)(3 \times 10^3)(8 \times 10^{-2})$
$= 2 \times 3 \times 8 \times 10^{1+3-2}$
$= 48 \times 10^2 = 4,800$

8. 40

9. $(0.9)(0.07)(4,000) = (9 \times 10^{-1})(7 \times 10^{-2})(4 \times 10^3)$
$= 9 \times 7 \times 4 \times 10^{-1-2+3}$
$= 252 \times 10^0 = 252$

10. 1.8

11. $\dfrac{9,000}{0.03} = \dfrac{\overset{3}{\cancel{9}} \times 10^3}{\underset{1}{\cancel{3}} \times 10^{-2}} = 3 \times 10^{3-(-2)} = 3 \times 10^5$
$= 300,000$

12. $200,000$

13. $\dfrac{0.0008}{400} = \dfrac{\overset{2}{\cancel{8}} \times 10^{-4}}{\underset{1}{\cancel{4}} \times 10^2} = 2 \times 10^{-4-2} = 2 \times 10^{-6}$
$= 0.000002$

14. 0.00002

15. $\dfrac{(80)(0.005)}{4,000} = \dfrac{(\overset{2}{\cancel{8}} \times 10^1)(5 \times 10^{-3})}{\underset{1}{\cancel{4}} \times 10^3} = 10 \times 10^{1-3-3}$
$= 10^{-4} = 0.0001$

16. 30

17. $\dfrac{(0.03)(200)(40)}{(0.008)(30)} = \dfrac{(3 \times 10^{-2})(\overset{1}{\cancel{2}} \times 10^2)(\overset{1}{\cancel{4}} \times 10^1)}{(\underset{1}{\cancel{8}} \times 10^{-3})(\underset{1}{\cancel{3}} \times 10^1)}$

$= 10^{-2+2+1-(-3)-(1)} = 10^3 = 1,000$

18. $\doteq 0.047$

19. $\dfrac{(5,000)(25)(0.08)}{(40)(0.002)} = \dfrac{(5 \times 10^3)(25)(\overset{1}{\cancel{8}} \times 10^{-2})}{(\underset{1}{\cancel{4}} \times 10^1)(\underset{1}{\cancel{2}} \times 10^{-3})}$

$= 125 \times 10^{3+(-2)-(1)-(-3)}$

$= 125 \times 10^3 = 125,000$

20. 0.01

EXERCISES 1205, SET I (page 407)

1. $(7.86)(204) \doteq (8)(200) = 8 \times 2 \times 10^2 = 16 \times 10^2$
$= 1,600$
(Actual answer = 1,603 rounded off to units.)

2. $\doteq 600$ (Actual answer = 536 rounded off to units.)

3. $0.714 \times 0.286 \doteq 0.7 \times 0.3 = (7 \times 10^{-1}) \times (3 \times 10^{-1})$
$= 21 \times 10^{-2} = 0.21$
(Actual answer = 0.204 rounded off to thousandths.)

4. $\doteq 0.32$
(Actual answer = 0.30 rounded off to hundredths.)

5. $\dfrac{75.5}{0.894} \doteq \dfrac{80}{0.9} = \dfrac{8 \times 10^1}{9 \times 10^{-1}} = \dfrac{8}{9} \times 10^{1-(-1)}$
$\doteq 0.9 \times 10^2 = 90$
(Actual answer = 84.45 rounded off to the nearest hundredth.)

6. $\doteq 0.0002$ (Actual answer = 0.00016 rounded off to five decimal places.)

7. $\dfrac{17.1}{57,000} \doteq \dfrac{20}{60,000} = \dfrac{2 \times 10^1}{6 \times 10^4} = \dfrac{1}{3} \times 10^{1-4}$
$\doteq 0.3 \times 10^{-3} = 0.0003$
(Actual answer = 0.0003 exactly.)

8. 80,000 (Actual answer = 72,407 rounded off to units.)

9. $21.9 \times 0.0071 \times 0.864 \doteq 20 \times 0.007 \times 0.9$
$= (2 \times 10^1)(7 \times 10^{-3})(9 \times 10^{-1}) = 126 \times 10^{-3} \doteq 0.1$
(Actual answer = 0.13 rounded off to hundredths.)

10. $\doteq 16,200$ (Actual answer = 16,236 rounded off to units.)

11. $328 \times 0.91 \times 0.0088 \doteq 300 \times 0.9 \times 0.009$
$= (3 \times 10^2)(9 \times 10^{-1})(9 \times 10^{-3}) = 243 \times 10^{-2}$
$= 2.43$
(Actual answer = 2.63 rounded off to hundredths.)

12. $\doteq 4.8$ (Actual answer = 4.2 rounded off to tenths.)

13. $\dfrac{45,000 \times 0.0159}{0.788} \doteq \dfrac{40,000 \times 0.02}{0.8}$
$= \dfrac{(\cancel{4} \times 10^4)(\cancel{2} \times 10^{-2})}{\underset{\cancel{}}{\cancel{8}} \times 10^{-1}} = 10^3$
$= 1,000$
(Actual answer = 908 rounded off to units.)

14. $\doteq 2.4$ (Actual answer = 2.13 rounded off to hundredths.)

15. $\dfrac{2,800 \times 0.416}{70.6} \doteq \dfrac{3,000 \times 0.4}{70}$
$= \dfrac{(3 \times 10^3)(4 \times 10^{-1})}{7 \times 10^1}$
$= \dfrac{12}{7} \times 10^{3+(-1)-(1)} \doteq 1.7 \times 10 = 17$
(Actual answer = 16.5 rounded off to tenths.)

16. $\doteq 5,000$ (Actual answer = 4,832 rounded off to units.)

17. $\dfrac{25.6 \times 2,100}{0.064 \times 550} \doteq \dfrac{30 \times 2,000}{0.06 \times 600} = \dfrac{(\overset{1}{\cancel{3}} \times 10^1)(2 \times 10^3)}{(\underset{1}{\underset{2}{\cancel{6}}} \times 10^{-2})(6 \times 10^2)}$
$= \dfrac{1}{6} \times 10^4 \doteq 0.17 \times 10^4 = 1,700$
(Actual answer = 1,527 rounded off to units.)

18. $\doteq 1,000$ (Actual answer = 780, accurate to units.)

19. $\dfrac{285 \times 1.99 \times 72,000}{0.691 \times 566 \times 34.2}$

$\doteq \dfrac{(3 \times 10^2)(2)(7 \times 10^4)}{(\underset{1}{\cancel{7}} \times 10^{-1})(\underset{3}{\cancel{6}} \times 10^2)(\underset{1}{\cancel{3}} \times 10^1)}$

$= \dfrac{1}{3} \times 10^{2+4-(-1)-(2)-1} \doteq 0.3 \times 10^4 = 3,000$

(Actual answer = 3.052 rounded off to the nearest unit.)

20. $\doteq 100$ (Actual answer = 136 rounded off to the nearest unit.)

SOLUTIONS FOR CHAPTER 12 DIAGNOSTIC TEST (page 408)

Following each problem number is the textbook section reference (in parentheses) where that kind of problem is discussed.

1. (1201) (a) $10^2 \cdot 10^4 = 10^{2+4} = 10^6$

(b) $\dfrac{10^5}{10^3} = 10^{5-3} = 10^2$

(c) $\dfrac{10^4}{10^7} = 10^{4-7} = 10^{-3}$ (or) $\dfrac{1}{10^3}$

(d) $10 \cdot 10^{-3} = 10^{1+(-3)} = 10^{-2}$ (or) $\dfrac{1}{10^2}$

(e) $\dfrac{10^4}{10^{-2}} = 10^{4-(-2)} = 10^{4+2} = 10^6$

(f) $\dfrac{10^6 \cdot 10^{-2}}{10} = 10^{6+(-2)-(1)} = 10^3$
$= 10^{4-1} = 10^3$

(g) $10^5 \cdot 10^{-3} \cdot 10^0 = 10^{5+(-3)+0} = 10^2$

2. (1202) (a) $3.46 \times 10^3 = 3\!\!\underset{\wedge}{}4\ 6\ 0. = 3,460$

(b) $0.068 \times 10^2 = 0\!\!\underset{\wedge}{}0\ 6.8 = 6.8$

(c) $742 \times 10^{-2} = 7.4\ 2 = 7.42$

(d) $54.9 \times 10^0 = 54.9 \times 1 = 54.9$

3. (1203) (a) $4\!\!\underset{\wedge}{}5\ 0\ 0. = 4.5 \times 10^3$

(b) $7\!\!\underset{\wedge}{}9.6\ 4 = 7.964 \times 10^1$

(c) $0.0\ 0\ 6\!\!\underset{\wedge}{}8\ 4 = 6.84 \times 10^{-3}$

4. (1203) (a) $4.75 \times 10^{-2} = 0.0\ 4\ 7\ 5 = 0.0475$

(b) $9.3 \times 10^6 = 9\!\!\underset{\wedge}{}3\ 0\ 0,0\ 0\ 0. = 9,300,000$

(c) $8.654 \times 10^2 = 8\!\!\underset{\wedge}{}6\ 5.4 = 865.4$

5. (1204) (a) $(300)(0.04)(50)$
$= (3 \times 10^2)(4 \times 10^{-2})(5 \times 10^1)$
$= 3 \times 4 \times 5 \times 10^{2-2+1} = 60 \times 10^1$
$= 6 \times 10^2$

(b) $\dfrac{(0.008)(1,200)}{(0.048)}$
$= \dfrac{(\overset{}{\cancel{8}} \times 10^{-3})(1.2 \times 10^3)}{\underset{1}{\underset{\cancel{}}{4.8}} \times 10^{-2}} = \dfrac{2 \times 10^{-3+3}}{10^{-2}}$
$= 2 \times 10^2$

6. (1204) (a) $186{,}300 \times 427 \doteq 200{,}000 \times 400$
$$= (2 \times 10^5)(4 \times 10^2)$$
$$= 8 \times 10^7$$

(b) $\dfrac{0.0754}{17.5} \doteq \dfrac{0.08}{20} = \dfrac{\overset{4}{\cancel{8}} \times 10^{-2}}{\underset{1}{\cancel{2}} \times 10}$
$$= 4 \times 10^{-2-1} = 4 \times 10^{-3}$$

(c) $\dfrac{(387.8)(0.05136)}{4.19} \doteq \dfrac{\overset{100}{\cancel{(400)}}(0.05)}{\underset{1}{\cancel{4}}} = 5$

(or) $100(0.05) = 10^2 \times 5 \times 10^{-2}$
$$= 5 \times 10^{2+(-2)}$$
$$= 5 \times 10^0$$

EXERCISES 1301, SET I (page 415)

1. Yes, because it is a collection of objects or things.
2. Yes
3. Yes, because they have exactly the same members. Writing a member more than once tells you no more than when you write it once—namely, that the element is a member of the set.
4. {0, 1, 2}
5. { }. Since there are no digits greater than 9.
6. {0, 1, 2}
7. {10, 11, 12, ... }. This is the way we show the set of whole numbers greater than 9. It is an infinite set.
8. {5}
9. { }, since there are no whole numbers greater than 4 and at the same time less than 5.
10. {0, 1, 2, 3}
11. 2, a, 3. Because of the way the set is written we know that these are its elements.
12. 0, 1, 2, 3, 4, 5, 6, 7, 8, 9
13. (a) $n(\{1, 1, 3, 5, 5, 5\}) = n(\{1, 3, 5\}) = 3$. This set has only three elements: 1, 3, and 5.
(b) $n(\{0\}) = 1$. This set has 1 element: 0.
(c) $n(\{a, b, g, x\}) = 4$. This set has four elements: a, b, g, and x.
(d) $n(\{0, 1, 2, 3, 4, 5, 6, 7\}) = 8$. You can count its elements and see that there are eight of them.
(e) $n(\emptyset) = 0$. The empty set has no elements.
14. $\emptyset$ has no elements; {0} has one element, namely 0. Since the sets do not have exactly the same elements, they are not equal.
15. (a) The set of digits is finite because when we count its elements the counting comes to an end. In this case the counting ends at 10.
(b) The set of whole numbers is infinite because when we attempt to count its elements the counting never comes to an end.
(c) Finite. The counting ends at 7.
(d) Finite. If we started counting the books in the ELAC library, we would eventually finish counting them.
(e) Finite. Although this is a very large set, if we had some device for counting all the fish instantaneously, the counting would end.
16. (a) True   (b) True   (c) False
(d) False   (e) True
17. (a) {7, 12} because 7 and 12 are the only elements of U that are less than 15.
(b) {23} because 23 is the only element of U that is greater than 20.
18. (a) {11}   (b) {5}

EXERCISES 1302, SET I (page 417)

1. (a) {3, 5} is a proper subset of M because each of its elements, 3 and 5, are elements of M; and M has at least one element such as 2 which is not an element of {3, 5}.
(b) {0, 1, 7} is not a subset of M because elements 0 and 7 are not elements of M.
(c) $\emptyset$ is a proper subset of M, because $\emptyset$ is a proper subset of every set except itself.
(d) {2, 4, 1, 3, 5} is an improper subset of M because each element of C is an element of M, and M has no element which is not an element of C.
2. (a) Proper   (b) Improper
(c) Proper   (d) Not a subset of P.
3. {R, G, Y}_____ All the subsets with three elements.
{R, G}, {R, Y}, {G, Y}_____ All the subsets with two elements
{R}, {G}, {Y}_____ All the subsets with one element.
{ }_____ All the subsets with no elements.
4. {□, △}, {□}, {△}, { }

EXERCISES 1303, SET I (page 420)

1. (a) {1, 5, 7} ∪ {2, 4} = {1, 5, 7, 2, 4}
{1, 5, 7} ∩ {2, 4} = { }
(b) {a, b} ∪ {x, y, z, a} = {a, b, x, y, z}
{a, b} ∩ {x, y, z, a} = {a}
(c) { } ∪ {k, 2} = {k, 2}
{ } ∩ {k, 2} = { }
(d) {river, boat} ∪ {boat, streams, down}
        = {river, boat, streams, down}
{river, boat} ∩ {boat, streams, down}
        = {boat}
2. (a) {1, 3, 5, 7, 2, 4, 6}
(b) {1, 2, 3, 4, 5, 6, 7}
(c) { }
(d) {6}
3. C and D are disjoint because they have no member in common. That is: $C \cap D = \emptyset$. All other pairs of sets have at least one member in common.
4. (a) $P = \{c, d, k\}$
(b) $Q = \{k, j, f\}$
(c) $P \cup Q = \{c, d, k, j, f\}$
(d) $P \cap Q = \{k\}$
(e) $U = \{a, e, g, h, c, d, k, j, f\}$
5. (a) $X \cap Y = \{5, 11\}$ because these are the only elements in both X and Y.
(b) $Y \cap X = \{5, 11\}$ for same reason as (a).
(c) Yes. $X \cap Y = Y \cap X$ because they have exactly the same elements.
6. (a) $K \cap L = \{4, b\}$
(b) $n(K \cap L) = 2$
(c) $L \cup M = \{m, 4, 6, b, n, 7, t\}$
(d) $n(L \cup M) = 7$

REVIEW EXERCISES 1304, SET I (page 423)

1. (a) $A \cup B = \{5, 7, 8, 2\}$
(b) $A \cup B = \{5, 7\}$
(c) $C \cap B = \{ \}$
(d) $n(U) = 8$
(e) $n(A \cup B) = 4$
(f) No, because 2 ∉ A
(g) A and C are disjoint sets because $A \cap C = \{ \}$.
(h) U is finite because $n(U) = 8$.
(i) Yes. A is a proper subset of U because all the elements of A are also elements of U, and 2 ∈ U but 2 ∉ A.

490   Answers

2. (a) $D \cap E = \{3\}$
   (b) $D \cup F = \{2, 3, 6, 7, 8\}$
   (c) $D \cap F = \{ \}$
   (d) $n(D \cup E) = 4$
   (e) $n(U)$ is not any number you know.
   (f) D and F are disjoint sets.
   (g) W is an infinite set.
   (h) E is a proper subset of W.
3. Yes, because they have exactly the same elements.
4. $\{16, 17, 18\}$
5. $\{0, 1, 2, 3\}$ because these are the only whole numbers that are less than 5 *and at the same time* less than 4.
6. Yes.
7. The set of letters in the word Mississippi = $\{M, i, s, p\}$. $n(\{M, i, s, p\}) = 4$
8. $\{\Box, \Delta\}$, $\{\Box\}$, $\{\Delta\}$, $\{ \}$
9.

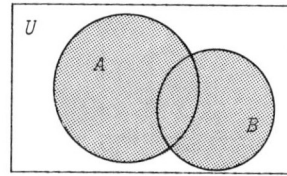

$A \cup B$

10.

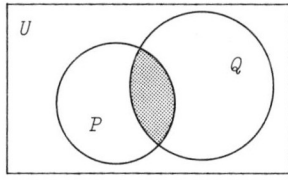

$P \cap Q$

11. $R \cap S$ because the shaded area is in both R and S.
12. $Y \cup Z$.

SOLUTIONS FOR CHAPTER 13 DIAGNOSTIC TEST (page 426)

Following each problem number is the textbook section reference (in parentheses) where that kind of problem is discussed.

1. (1301) Yes, because it is a collection of objects or things.
2. (1301) Yes, because they have exactly the same members.

3. (1301) (101) $\{5, 6, 7\}$
4. (1301) 4, because it has only four different elements.
5. (1301) (101) $\{9\}$
6. (1301) (a) Finite, because if we had some way of counting all the grains of sand on the earth today, the counting would come to an end.
   (b) Infinite, because the counting can never come to an end.
7. (1302) (a) T, because 10 is a member of set D.
   (b) F, because a is an element of set P.
   (c) F, because 11 is not a digit.
   (d) T, because $\emptyset$ has no elements.
   (e) F, because $\{ \}$ has no elements.
8. (1302) (a) Proper, because its elements 4 and 3 are elements of set R, and R has at least one element (0 or 7) which is not also an element of set A.
   (b) Not a subset of R, because 1 is an element of B and not of R.
   (c) Proper, because $\emptyset$ is a proper subset of every set except itself.
   (d) Improper, because every element of C is also an element of R, and R contains no element which is not also an element of C.
9. (1302) $\emptyset$, $\{a\}$, $\{b\}$, $\{a, b\}$
10. (1303) (a) $\{1, 3, 5, 2\}$, because these are the elements that are in A or C or in both.
    (b) $\{2\}$, because 2 is the only element in both B and C.
    (c) $\emptyset$, because there are no elements in both A and B.
    (d) $\{2, 6, 8, 1, 3\}$ because these are the elements that are in B or C or in both.
11. (1303) (a) $\{e, g, b\}$ because these are the letters in the A circle.
    (b) $\{b\}$ because b is the only letter in both the A and B circles.
    (c) $\{b, j, d, h\}$ because these are the letters in the B circle.
    (d) $\{c, e, i, g, b, j, d, h, a, f\}$ because these are all the letters in the rectangle U.
    (e) $\{e, g, b, j, d, h\}$ because these are the letters in either the A circle, the B circle, or both.
12. (1303) $S \cap T$, because the shaded area is all of the area that is in both the S and T circles.

Index